Multimodality in Mobile Computing and Mobile Devices:
Methods for Adaptable Usability

Stan Kurkovsky
Central Connecticut State University, USA

INFORMATION SCIENCE REFERENCE

Hershey · New York

Director of Editorial Content:	Kristin Klinger
Senior Managing Editor:	Jamie Snavely
Assistant Managing Editor:	Michael Brehm
Publishing Assistant:	Sean Woznicki
Typesetter:	Michael Brehm, Kurt Smith
Cover Design:	Lisa Tosheff
Printed at:	Yurchak Printing Inc.

Published in the United States of America by
   Information Science Reference (an imprint of IGI Global)
   701 E. Chocolate Avenue
   Hershey PA 17033
   Tel: 717-533-8845
   Fax: 717-533-8661
   E-mail: cust@igi-global.com
   Web site: http://www.igi-global.com/reference

Library of Congress Cataloging-in-Publication Data

Multimodality in mobile computing and mobile devices : methods for adaptable
usability / Stan Kurkovsky, editor.
      p. cm.
   Includes bibliographical references and index.
   Summary: "This book offers a variety of perspectives on multimodal user
interface design, describes a variety of novel multimodal applications and
provides several experience reports with experimental and industry-adopted
mobile multimodal applications"--Provided by publisher.
   ISBN 978-1-60566-978-6 (hardcover) -- ISBN 978-1-60566-979-3 (ebook) 1.
Mobile computing. I. Kurkovsky, Stan, 1973-
   QA76.59.M85 2010
   004.167--dc22
                              2009020551

British Cataloguing in Publication Data
A Cataloguing in Publication record for this book is available from the British Library.

All work contributed to this book is new, previously-unpublished material. The views expressed in this book are those of the authors, but not necessarily of the publisher.

Editorial Advisory Board

Table of Contents

Foreword xv

Preface xvii

Section 1
Introduction

Chapter 1
Multimodal and Multichannel Issues in Pervasive and Ubiquitous Computing 1
José Rouillard, Université de Lille 1, France

Chapter 2
Ubiquitous User Interfaces: Multimodal Adaptive Interaction for Smart Environments 24
Marco Blumendorf, Technische Universität Berlin, Germany
Grzegorz Lehmann, Technische Universität Berlin, Germany
Dirk Roscher, Technische Universität Berlin, Germany
Sahin Albayrak, Technische Universität Berlin, Germany

Section 2
Theoretical Foundations

Chapter 3
A Formal Approach to the Verification of Adaptability Properties
for Mobile Multimodal User Interfaces 53
Nadjet Kamel, University of Moncton, Canada
Sid Ahmed Selouani, University of Moncton, Canada
Habib Hamam, University of Moncton, Canada

Chapter 4
Platform Support for Multimodality on Mobile Devices.. 75
Kay Kadner, SAP AG, Germany
Martin Knechtel, SAP AG, Germany
Gerald Huebsch, TU Dresden, Germany
Thomas Springer, TU Dresden, Germany
Christoph Pohl, SAP AG, Germany

Section 3
Design Approaches

Chapter 5
Designing Mobile Multimodal Applications 106
Marco de Sá, University of Lisboa, Portugal
Carlos Duarte, University of Lisboa, Portugal
Luís Carriço, University of Lisboa, Portugal
Tiago Reis, University of Lisboa, Portugal

Chapter 6
Bodily Engagement in Multimodal Interaction: A Basis for a New Design Paradigm?..................... 137
Kai Tuuri, University of Jyväskylä, Finland
Antti Pirhonen, University of Jyväskylä, Finland
Pasi Välkkynen, VTT Technical Research Centre of Finland, Finland

Chapter 7
Two Frameworks for the Adaptive Multimodal Presentation of Information 166
Yacine Bellik, Université d'Orsay, Paris-Sud, France
Christophe Jacquet, SUPELEC, France
Cyril Rousseau, Université d'Orsay, Paris-Sud, France

Chapter 8
A Usability Framework for the Design and Evaluation of Multimodal Interaction:
Application to a Multimodal Mobile Phone 196
Jaeseung Chang, Handmade Mobile Entertainment Ltd., UK
Marie-Luce Bourguet, University of London, UK

Section 4
Applications and Field Reports

Chapter 9

Exploiting Multimodality for Intelligent Mobile Access to Pervasive Services
in Cultural Heritage Sites...217

Antonio Gentile, Università di Palermo, Italy

Antonella Santangelo, Università di Palermo, Italy

Salvatore Sorce, Università di Palermo, Italy

Agnese Augello, Università di Palermo, Italy

Giovanni Pilato, Consiglio Nazionale delle Ricerche, Italy

Alessandro Genco, Università di Palermo, Italy

Salvatore Gaglio, Consiglio Nazionale delle Ricerche, Italy

Chapter 10

Multimodal Search on Mobiles Devices: Exploring Innovative Query Modalities
for Mobile Search 242

Xin Fan, University of Sheffield, UK

Mark Sanderson, University of Sheffield, UK

Xing Xie, Microsoft Research Asia, China

Chapter 11

Simplifying the Multimodal Mobile User Experience..260

Keith Waters, Orange Labs Boston, USA

Section 5
New Directions

Chapter 12

Multimodal Cues: Exploring Pause Intervals between Haptic/Audio Cues
and Subsequent Speech Information..277

Aidan Kehoe, University College Cork, Ireland

Flaithri Neff, University College Cork, Ireland

Ian Pitt, University College Cork, Ireland

Chapter 13

Towards Multimodal Mobile GIS for the Elderly...301

Julie Doyle, University College Dublin, Ireland

Michela Bertolotto, School of Computer Science and Informatics, Ireland

David Wilson, University of North Carolina at Charlotte, USA

Chapter 14
Automatic Signature Verification on Handheld Devices 321
    Marcos Martinez-Diaz, Universidad Autonoma de Madrid, Spain
    Julian Fierrez, Universidad Autonoma de Madrid, Spain
    Javier Ortega-Garcia, Universidad Autonoma de Madrid, Spain

Compilation of References 339

About the Contributors 370

Index 380

Detailed Table of Contents

Foreword xv

Preface.. xvii

Section 1
Introduction

Chapter 1
Multimodal and Multichannel Issues in Pervasive and Ubiquitous Computing..................................... 1
José Rouillard, Université de Lille 1, France

Multimodality in mobile computing has become a very active field of research in the past few years. Soon, mobile devices will allow smooth and smart interaction with everyday life's objects, thanks to natural and multimodal interactions. In this context, this chapter introduces some concepts needed to address the topic of pervasive and ubiquitous computing. The multi-modal, multi-channel and multi-device notions are presented and are referenced by the name and partial acronym "multi-DMC". A multi-DMC referential is explained, in order to understand what kind of notions have to be sustained in such systems. Next we have three case studies that illustrate the issues faced when proposing systems able to support at the same time different modalities including voice or gesture, different devices, like PC or smartphone and different channels such as web or telephone.

Chapter 2
Ubiquitous User Interfaces: Multimodal Adaptive Interaction for Smart Environments..................... 24
Marco Blumendorf, Technische Universität Berlin, Germany
Grzegorz Lehmann, Technische Universität Berlin, Germany
Dirk Roscher, Technische Universität Berlin, Germany
Sahin Albayrak, Technische Universität Berlin, Germany

The widespread use of computing technology raises the need for interactive systems that adapt to user, device and environment. Multimodal user interfaces provide the means to support the user in various situations and to adapt the interaction to the user's needs. In this chapter we present a system utilizing design-time user interface models at runtime to provide flexible multimodal user interfaces. The server-based system allows the combination and integration of multiple devices to support multimodal interaction and the adaptation of the user interface to the used devices, the user and the environment. The utilization of the user interface models at runtime allows exploiting the design information for advanced adaptation possibilities. An implementation of the system has been successfully deployed in a smart home environment throughout the Service Centric Home project (www.sercho.de).

Section 2
Theoretical Foundations

Chapter 3
A Formal Approach to the Verification of Adaptability Properties
for Mobile Multimodal User Interfaces 53

Nadjet Kamel, University of Moncton, Canada
Sid Ahmed Selouani, University of Moncton, Canada
Habib Hamam, University of Moncton, Canada

Multimodal User Interfaces (MUIs) offer to users the possibility to interact with systems using one or more modalities. In the context of mobile systems, this will increase the flexibility of interaction and will give the choice to use the most appropriate modality. These interfaces must satisfy usability properties to guarantee that users do not reject them. Within this context, we show the benefits of using formal methods for the specification and verification of multimodal user interfaces (MUIs) for mobile systems. We focus on the usability properties and specifically on the adaptability property. We show how transition systems can be used to model the MUI and temporal logics to specify usability properties. The verification is performed by using fully automatic model-checking technique. This technique allows the verification at earlier stages of the development life cycle which decreases the high costs involved by the maintenance of such systems.

Chapter 4
Platform Support for Multimodality on Mobile Devices 75

Kay Kadner, SAP AG, Germany
Martin Knechtel, SAP AG, Germany
Gerald Huebsch, TU Dresden, Germany
Thomas Springer, TU Dresden, Germany
Christoph Pohl, SAP AG, Germany

The diversity of today's mobile technology also entails multiple interaction channels offered per device. This chapter surveys the basics of multimodal interactions in a mobility context and introduces a number of concepts for platform support. Synchronization approaches for input fusion and output fission as well as a concept for device federation as a means to leverage from heterogeneous devices are discussed with the help of an exemplary multimodal route planning application. An outlook on future trends concludes the chapter.

Section 3
Design Approaches

Chapter 5

Designing Mobile Multimodal Applications 106

Marco de Sá, University of Lisboa, Portugal
Carlos Duarte, University of Lisboa, Portugal
Luís Carriço, University of Lisboa, Portugal
Tiago Reis, University of Lisboa, Portugal

In this chapter we describe a set of techniques and tools that aim at supporting designers while creating mobile multimodal applications. We explain how the additional difficulties that designers face during this process, especially those related to multimodalities, can be tackled. In particular, we present a scenario generation and context definition framework that can be used to drive design and support evaluation within realistic settings, promoting in-situ design and richer results. In conjunction with the scenario framework, we detail a prototyping tool that was developed to support the early stage prototyping and evaluation process of mobile multimodal applications, from the first sketch-based prototypes up to the final quantitative analysis of usage results. As a case study, we describe a mobile application for accessing and reading rich digital books. The application aims at offering users, in particular blind users, means to read and annotate digital books and it was designed to be used on Pocket PCs and Smartphones, including a set of features that enhance both content and usability of traditional books.

Chapter 6

Bodily Engagement in Multimodal Interaction: A Basis for a New Design Paradigm?..................... 137

Kai Tuuri, University of Jyväskylä, Finland
Antti Pirhonen, University of Jyväskylä, Finland
Pasi Välkkynen, VTT Technical Research Centre of Finland, Finland

The creative processes of interaction design operate in terms we generally use for conceptualising human-computer interaction (HCI). Therefore the prevailing design paradigm provides a framework that essentially affects and guides the design process. We argue that the current mainstream design paradigm for multimodal user-interfaces takes human sensory-motor modalities and the related user-interface technologies as separate channels of communication between user and an application. Within such a conceptualisation, multimodality implies the use of different technical devices in interaction design. This chapter outlines an alternative design paradigm, which is based on an action-oriented perspective on human perception and meaning creation process. The proposed perspective stresses the integrated sensory-motor experience and the active embodied involvement of a subject in perception coupled as a natural part of interaction. The outlined paradigm provides a new conceptual framework for the design of multimodal user interfaces. A key motivation for this new framework is in acknowledging multi-modality as an inevitable quality of interaction and interaction design, the existence of which does not depend on, for example, the number of implemented presentation modes in an HCI application. We see that the need for such an interaction–and experience–derived perspective is amplified within the trend for computing to be moving into smaller devices of various forms which are being embedded into our everyday life. As a brief illustration of the proposed framework in practice, one case study of mobile sonic interaction design is also presented.

Chapter 7

Two Frameworks for the Adaptive Multimodal Presentation of Information 166

Yacine Bellik, Université d'Orsay, Paris-Sud, France
Christophe Jacquet, SUPELEC, France
Cyril Rousseau, Université d'Orsay, Paris-Sud, France

Our work aims at developing models and software tools that can exploit intelligently all modalities available to the system at a given moment, in order to communicate information to the user. In this chapter, we present the outcome of two research projects addressing this problem in two different areas: the first one is relative to the contextual presentation of information in a "classical" interaction situation, while the second one deals with the opportunistic presentation of information in an ambient environment. The first research work described in this chapter proposes a conceptual model for intelligent multimodal presentation of information. This model called WWHT is based on four concepts: "What", "Which", "How" and "Then". The first three concepts are about the initial presentation design while the last concept is relative to the presentation evolution. On the basis of this model, we present the ELOQUENCE software platform for the specification, the simulation and the execution of output multimodal systems. The second research work deals with the design of multimodal information systems in the framework of ambient intelligence. We propose an ubiquitous information system that is capable of providing personalized information to mobile users. Furthermore, we focus on multimodal information presentation. The proposed system architecture is based on KUP, an alternative to traditional software architecture models for human-computer interaction. The KUP model takes three logical entities into account: Knowledge, Users, and Presentation devices. It is accompanied by algorithms for choosing and instantiating dynamically interaction modalities. The model and the algorithms have been implemented within a platform called PRIAM (PResentation of Information in AMbient environment), with which we have performed experiments in pseudo-real scale. After comparing the results of both projects, we define the characteristics of an ideal multimodal output system and discuss some perspectives relative to the intelligent multimodal presentation of information.

Chapter 8

A Usability Framework for the Design and Evaluation of Multimodal Interaction:
Application to a Multimodal Mobile Phone 196

Jaeseung Chang, Handmade Mobile Entertainment Ltd., UK
Marie-Luce Bourguet, University of London, UK

Currently, a lack of reliable methodologies for the design and evaluation of usable multimodal interfaces makes developing multimodal interaction systems a big challenge. In this paper, we present a usability framework to support the design and evaluation of multimodal interaction systems. First, elementary multimodal commands are elicited using traditional usability techniques. Next, based on the CARE (Complementarity, Assignment, Redundancy, and Equivalence) properties and the FSM (Finite State Machine) formalism, the original set of elementary commands is expanded to form a comprehensive set of multimodal commands. Finally, this new set of multimodal commands is evaluated in two ways:

user-testing and error-robustness evaluation. This usability framework acts as a structured and general methodology both for the design and for the evaluation of multimodal interaction. We have implemented software tools and applied this methodology to the design of a multimodal mobile phone to illustrate the use and potential of the proposed framework.

Section 4
Applications and Field Reports

Chapter 9

Exploiting Multimodality for Intelligent Mobile Access to Pervasive Services in Cultural Heritage Sites 217

Antonio Gentile, Università di Palermo, Italy
Antonella Santangelo, Università di Palermo, Italy
Salvatore Sorce, Università di Palermo, Italy
Agnese Augello, Università di Palermo, Italy
Giovanni Pilato, Consiglio Nazionale delle Ricerche, Italy
Alessandro Genco, Università di Palermo, Italy
Salvatore Gaglio, Consiglio Nazionale delle Ricerche, Italy

In this chapter the role of multimodality in intelligent, mobile guides for cultural heritage environments is discussed. Multimodal access to information contents enables the creation of systems with an higher degree of accessibility and usability. A multimodal interaction may involve several human interaction modes, such as sight, touch and voice to navigate contents, or gestures to activate controls. We first start our discussion by presenting a timeline of cultural heritage system evolution, spanning from 2001 to 2008, which highlights design issues as intelligence and context-awareness in providing information. Then, multimodal access to contents is discussed, along with problems and corresponding solutions; an evaluation of several reviewed systems is also presented. Lastly, a case study multimodal framework termed MAGA is described, which combines intelligent conversational agents with speech recognition/synthesis technology, in a framework employing RFID based location and Wi-Fi based data exchange.

Chapter 10

Multimodal Search on Mobiles Devices: Exploring Innovative Query Modalities for Mobile Search 242

Xin Fan, University of Sheffield, UK
Mark Sanderson, University of Sheffield, UK
Xing Xie, Microsoft Research Asia, China

The increasingly popularity of powerful mobile devices, such as smart phones and PDA phones, enables users freely to perform information search on the move. However, text is still the main input modality in most of current mobile search services although some pioneering search service providers are attempting to provide voice-based mobile search solutions. In this chapter, we explore the innovative query modalities to enable the mobile devices to support richer and hybrid queries such as text, voice, image, location, and their combinations. We propose a solution to support mobile users to perform visual queries. The queries by captured pictures and text information are studied in depth. For example, the user can simply take a photo of an unfamiliar flower or surrounding buildings to find related information from Web. A set of indexing schemes are designed to achieve accurate results and fast search speed in volumes of data. Experimental results show that our prototype system achieved satisfactory performance. Also, we briefly introduce a prospective mobile search solution based on our ongoing research, which supports multimodal queries including location information, captured pictures and text information.

Chapter 11
Simplifying the Multimodal Mobile User Experience... 260
    Keith Waters, Orange Labs Boston, USA

Multimodality presents challenges within a mobile cellular network. Variable connectivity, coupled with a wide variety of handset capabilities, present significant constraints that are difficult to overcome. As a result, commercial mobile multimodal implementations have yet to reach the consumer mass market, and are considered niche services. This chapter describes multimodality with handsets in cellular mobile networks that are coupled to new opportunities in targeted Web services. Such Web services aim to simplify and speed up interactions through new user experiences. This chapter highlights some key components with respect to a few existing approaches. While the most common forms of multimodality use voice and graphics, new modes of interaction are enabled via simple access to device properties, call the Delivery Context: Client Interfaces (DCCI).

Section 5
New Directions

Chapter 12
Multimodal Cues: Exploring Pause Intervals between Haptic/Audio Cues
and Subsequent Speech Information... 277
    Aidan Kehoe, University College Cork, Ireland
    Flaithri Neff, University College Cork, Ireland
    Ian Pitt, University College Cork, Ireland

There are numerous challenges to accessing user assistance information in mobile and ubiquitous computing scenarios. For example, there may be little-or-no display real estate on which to present information visually, the user's eyes may be busy with another task (e.g., driving), it can be difficult to read text while moving, etc. Speech, together with non-speech sounds and haptic feedback can be used to make assistance information available to users in these situations. Non-speech sounds and haptic feedback can be used to cue information that is to be presented to users via speech, ensuring that the listener is prepared and that leading words are not missed. In this chapter, we report on two studies that examine user perception of the duration of a pause between a cue (which may be a variety of non-speech sounds, haptic effects or combined non-speech sound plus haptic effects) and the subsequent delivery of assistance information using speech. Based on these user studies, recommendations for use of cue pause intervals in the range of 600 ms to 800 ms are made.

Chapter 13

Towards Multimodal Mobile GIS for the Elderly.. 301

Julie Doyle, University College Dublin, Ireland
Michela Bertolotto, School of Computer Science and Informatics, Ireland
David Wilson, University of North Carolina at Charlotte, USA

Information technology can play an important role in helping the elderly to live full, healthy and independent lives. However, elders are often overlooked as a potential user group of many technologies. In particular, we are concerned with the lack of GIS applications which might be useful to the elderly population. The main underlying reasons which make it difficult to design usable applications for elders is threefold. The first concerns a lack of digital literacy within this cohort, the second involves physical and cognitive age-related impairments while the third involves a lack of knowledge on improving usability in interactive geovisualisation and spatial systems. As such, in this chapter we analyse existing literature in the fields of mobile multimodal interfaces with emphasis on GIS and the specific requirements of the elderly in relation to the use of such technologies. We also examine the potential benefits that the elderly could gain through using such technology, as well as the shortcomings that current systems have, with the aim to ensure full potential for this diverse, user group. In particular, we identify specific requirements for the design of multimodal GIS through a usage example of a system we have developed. Such a system produced very good evaluation results in terms of usability and effectiveness when tested by a different user group. However, a number of changes are necessary to ensure usability and acceptability by an elderly cohort. A discussion of these concludes the chapter.

Chapter 14

Automatic Signature Verification on Handheld Devices 321

Marcos Martinez-Diaz, Universidad Autonoma de Madrid, Spain
Julian Fierrez, Universidad Autonoma de Madrid, Spain
Javier Ortega-Garcia, Universidad Autonoma de Madrid, Spain

Automatic signature verification on handheld devices can be seen as a means to improve usability in consumer applications and a way to reduce costs in corporate environments. It can be easily integrated in touchscreen devices, for example as a part of combined handwriting and keypad-based multimodal interfaces. In the last few decades, several approaches to the problem of signature verification have been proposed. However, most research has been carried out considering signatures captured with digitizing tables, in which the quality of the captured data is much higher than in handheld devices. Signature verification on handheld devices represents a new scenario both for researchers and vendors. In this chapter, we introduce automatic signature verification as a component of multimodal interfaces; we analyze the applications and challenges of signature verification and overview available resources and research directions. A case study is also given, in which a state-of-the-art signature verification system adapted to handheld devices is presented.

Compilation of References 339

About the Contributors 370

Index 380

Foreword

Instead of holding this book in your hands, wouldn't you rather be hiking in the mountains—assuming that, while doing so, you could still be absorbing and applying the content of the book? This type of scenario has been brought within reach by recent progress in two areas: 1. mobile and wearable systems, which enable us to use computers in just about every imaginable setting; and 2. multimodal interaction, which (a) gives us alternative input and output methods, such as speech and haptic feedback, to replace the ones that are not feasible with small mobile systems and (b) allows us to choose the most suitable modalities for each setting that we find ourselves in.

Viewed in this way, the marriage between mobility and multimodality seems to have been made in heaven. But as with some real marriages, realizing the full promise of this one requires resolution of some conflicts that lie under the surface. Multimodality tends to demand more intensive and complex processing than small mobile devices can readily provide. And the users experience resource limitations of their own: in addition to having to deal with typically imperfect system processing, they have to figure out how to use the available modalities effectively in any given situation, while at the same time often interacting with their physical environment.

Realizing the full potential of mobile multimodality requires a simultaneous understanding of these and related issues on both the system and the human sides of the interaction. In the research literature available so far, we can find many research contributions, in different communities and publication venues, that address particular parts of this overall challenge. But making sense of them all is like assembling a puzzle whose small pieces are dispersed among the rooms of a house. Wouldn't it be better to have a smaller number of larger puzzle pieces, all in the same place, that can help us to see the whole picture and understand the remaining gaps?

That's where this book comes in. Most of the chapters describe ambitious approaches to significant portions of the overall challenge of mobile multimodality, in each case showing how a number of facets of the problem can be handled simultaneously; the last three chapters give the reader a chance to apply the more general concepts to specific problems of current interest. By studying and comparing these complementary perspectives on the field, just about any reader will achieve a more coherent and detailed mental model of this multifaceted topic than they had before, along with a keener awareness of what needs to be done next.

In sum, the traditional format of this book actually masks its true character as a significant part of today's progress toward a world full of smoothly used mobile multimodal systems.

Anthony Jameson
International University in Germany, Germany

Anthony Jameson *is a Principal Researcher in the Intelligent User Interfaces Department at the German Research Center for Artificial Intelligence (DFKI) and Adjunct Professor for Human-Computer Interaction at the International University in Germany. He earned degrees at Harvard, the University of Hamburg, the University of Amsterdam, and Saarland University. He began conducting and publishing research at the interface between psychology and artificial intelligence as a student in the late 1970s. Among the topics relevant to the present volume on which he has published widely are user modeling and adaptation; intelligent mobile assistants; and the usability issues raised by the design of mobile and multimodal systems.*

Preface

In the last two decades, two technological innovations produced a very significant impact on our everyday lives and habits. The first innovation, the Internet, provides new means and ways to access the vast amounts of information and an exploding number of online services available today. The Internet has revolutionized the way people communicate with each other, how they receive news, shop, and conduct other day-to-day activities. The second innovation, a mobile phone, provides users with a simple anytime and anywhere communication tool. Originally designed for interpersonal communication, today mobile phones are capable of connecting their users to a wide variety of Internet-enabled services and applications, which can vary from a simplified web browser to a GPS-enabled navigation system. Researchers and practitioners agree that a combination of these two innovations (e.g., online Internet-enabled services that can be accessed via a mobile phone) provides a revolutionary impact on the development of mobile computing systems.

However, development of mobile applications may be somewhat hindered by some features of the mobile devices, such as their screen sizes which are often too small to effectively handle the graphical content that can be found on the Internet. Mobile phones may also suffer from being too slow or having an inconvenient keyboard, making it difficult to access lengthy or media-rich information found on the web; or a relatively short battery life that may not be sufficient enough for such network traffic-intensive uses as web browsing or viewing mobile video broadcasts. So far, current research has been focusing mostly on mobile applications designed for smart phones, in which the application logic is usually placed within the mobile device. Speech, however, remains the most basic and the most efficient form of communication, and providing a form of speech-based communication remains the prevalent function of any mobile phone. Many other means of interpersonal communication and human perception, which include gestures, motion, and touch, are also finding their ways to the realm of mobile computing interfaces.

The primary functionality of any phone, no matter how basic, is to enable voice communication, which still remains the most natural and simple method of communication, ideally suited for on the go and hands-free access to information. A voice interface allows the user to speak commands and queries while receiving an audio response. Furthermore, a combination of mobile and voice technologies can lead to new venues for marketing, entertainment, news and information, and business locator services. Speech remains the most basic and the most efficient form of interpersonal communication, and facilitating voice communication remains the main function of even the simplest mobile phone. As mobile devices grow in popularity and become ever more powerful and feature-rich, they still remain constrained in terms of the screen and keyboard size, battery capacity, and processor speed. There is a wide variety of models, manufacturers, and operating systems for mobile devices, each of which may have unique input/output capabilities. This creates new challenges to the developers of mobile applications, especially if they are to embrace different interaction modalities. Current efforts to establish a standard for the multimodal interface specification still remain far from being mature and are not widely accepted by the industry. However, multimodal interface design is a rapidly evolving research area, especially in the area of mobile information services.

Applications or systems that combine multiple modalities of input and output are referred to as multimodal. For example, iPhone combines the capabilities of a traditional screen & keyboard interface, a touch interface, an accelerometer-based motion interface, and a speech interface, and all applications running on it (at least in theory) should be able to take advantage of these three modalities of input/output. The objectives of multimodal systems are two-pronged: to reach a kind of interaction that is closer to natural interpersonal human communication, and to improve the dependability of the interaction by employing complementary or redundant information. Generally, multimodal applications are more adaptable to the needs of different users in varying contexts. Multimodal applications have a stronger acceptance potential because they can generally be accessed in more than one manner (e.g. using speech and web interface) and by a broader range of users in a varying set of circumstances.

Recognizing that mobile computing is one of the most rapidly growing areas in the software market, this book explores the role of multimodality and multimodal interfaces in the area of mobile computing. Mobile computing has a very strong potential due the extremely high market penetration of mobile and smart phones, high degree of user interest in and engagement with mobile applications, and an emerging trend of integrating traditional desktop and online systems with their mobile counterparts. Multimodal interfaces play a very important role in improving the accessibility of these applications, therefore leading to their increased acceptance by the users.

This book is a collective effort of many researchers and practitioners from industry (including Orange, Microsoft, SAP, and others) and academia. It offers a variety of perspectives on multimodal user interface design, describes a variety of novel multimodal applications, and provides several experience reports with experimental and industry-adopted mobile multimodal applications.

The book opens with the Introduction, which consists of two chapters. Chapter 1, *Multimodal and Multichannel issues in pervasive and ubiquitous computing* by José Rouillard, describes the core concepts that define multi-modal, multi-channel and multi-device interactions and their role in mobile, pervasive and ubiquitous computing. This chapter also presents three case studies illustrating the issues arising in designing mobile systems that support different interaction modalities that include voice or gesture over different communication channels such as web or telephone. Chapter 2, *Ubiquitous User Interfaces: Multimodal Adaptive Interaction for Smart Environments* by Marco Blumendorf, Grzegorz Lehmann, Dirk Roscher, and Sahin Albayrak surveys the topic of multimodal user interfaces as they provide the means to support the user in various situations and to adapt the interaction to the user's needs. The authors focus on different aspects of modeling user interaction in an adaptive multimodal system and illustrate their approach with a system utilizes design-time user interface models at runtime to provide flexible multimodal user interfaces.

Theoretical foundations of multimodal interactions in the mobile environment are discussed in the second section of the book. Chapter 3, *A Formal Approach to the Verification of Adaptability Properties for Mobile Multimodal User Interfaces* by Nadjet Kamel, Sid Ahmed Selouani, and Habib Hamam discusses the benefits of using formal methods for specification and verification of multimodal user interfaces in mobile computing systems with the main emphasis on usability and adaptability. The authors present an approach that provides a formal interface verification using a fully automatic model-checking technique, which allows the verification at earlier stages of the development life cycle and decreases system maintenance costs. Chapter 4, *Platform support for multimodality on mobile devices* by Kay Kadner, Gerald Huebsch, Martin Knechtel, Thomas Springer, and Christoph Pohl surveys the basics of multimodal interactions in the context of mobility and introduces a number of concepts for platform support. This chapter also discusses different synchronization approaches for input fusion and output fission as well as a concept for device federation as a means to leverage from heterogeneous devices.

The third section of the book outlines some approaches for designing multimodal mobile applications and systems. Chapter 5, *Designing Multimodal Mobile Applications* by Marco de Sá, Carlos Duarte, Luís Carriço, and Tiago Reis describes a set of techniques and tools that aim at supporting designers while creating mobile multimodal applications. The authors present a framework for scenario generation and context definition that can be used to drive design and support evaluation within realistic settings, promoting in-situ design and richer results. This chapter also describes a prototyping tool that was developed to support the early stage prototyping and evaluation process of mobile multimodal applications, from the first sketch-based prototypes up to the final quantitative analysis of usage results. Chapter 6, *Bodily Engagement in Multimodal Interaction: A Basis for a New Design Paradigm?* by Kai Tuuri, Antti Pirhonen, and Pasi Välkkynen argues that the current mainstream design paradigm for multimodal user-interfaces takes human sensory-motor modalities and the related user-interface technologies as separate channels of communication between user and an application. This chapter outlines an alternative design paradigm, which is based on an action-oriented perspective on human perception and meaning creation process. This perspective stresses the integrated sensory-motor experience and the active embodied involvement of a subject in perception coupled as a natural part of interaction. Chapter 7, *Two Frameworks for the Adaptive Multimodal Presentation of Information* by Yacine Bellik, Christophe Jacquet, and Cyril Rousseau addresses the problem of choosing the correct communication modalities that are available to the system at the given moment. The authors consider this problem from two perspectives: as a contextual presentation of information in a "classical" interaction situation, and as an opportunistic presentation of information in an ambient environment. As a combination of the two approaches, the authors define the characteristics of an ideal multimodal output system and discuss some perspectives relative to the intelligent multimodal presentation of information. Chapter 8, *Usability Framework for the Design and Evaluation of Multimodal Interaction: Application to a Multimodal Mobile Phone* by Jaeseung Chang and Marie-Luce Bourguet, presents a usability framework to support the design and evaluation of multimodal interaction systems. This usability framework acts as a structured and general methodology both for the design and for the evaluation of multimodal interaction. The authors have implemented software tools and applied this methodology to the design of a multimodal mobile phone to illustrate the use and potential of the described framework.

The fourth section of the book describes several mobile multimodal applications and systems that have been developed and deployed. Chapter 9, *Exploiting Multimodality for Intelligent Mobile Access to Pervasive Services in Cultural Heritage Sites* by Antonio Gentile, Antonella Santangelo, Salvatore Sorce, Agnese Augello, Giovanni Pilato, Alessandro Genco, Salvatore Gaglio texamines he role of multimodality in intelligent, mobile guides for cultural heritage environments. This chapter outlines a timeline of cultural heritage system evolution, which highlights design issues as intelligence and context-awareness in providing information. The authors describe the role and advantages of using multimodal interfaces in such systems and present a case study of a multimodal framework, which combines intelligent conversational agents with speech recognition/synthesis technology within location-based framework. Chapter 10, *Multimodal Search on Mobiles Devices - Exploring Innovative Query Modalities for Mobile Search* by Xin Fan, Mark Sanderson, and Xing Xie explores innovative query modalities to enable mobile devices to support richer and hybrid queries such as text, voice, image, location, and their combinations. The authors describe a solution to support mobile users to perform visual queries, e.g. by captured pictures and textual information. Chapter 11, *Simplifying the multimodal mobile user experience* by Keith Waters, describes multimodality with handsets in cellular mobile networks that are coupled to new opportunities in targeted Web services that aim to simplify and speed up interactions through new user experiences.

The fifth section of the book presents a number of new directions in mobile multimodal interfaces that researchers and practitioners are currently exploring. Chapter 12, *Multimodal Cues: Exploring Pause Intervals between Haptic/Audio Cues and Subsequent Speech Information* by Aidan Kehoe, Flaithri Neff, and Ian Pitt focuses on addressing numerous challenges to accessing user assistance information in mobile and ubiquitous computing scenarios. Speech, together with non-speech sounds and haptic feedback, can be used to make assistance information available to users. This chapter examines the user perception of the duration of a pause between a cue that may be a variety of non-speech sounds, haptic effects or combined non-speech sound plus haptic effects, and the subsequent delivery of assistance information using speech. Chapter 13, *Towards Multimodal Mobile GIS for the Elderly* by Julie Doyle, Michela Bertolotto, and David Wilson focuses on developing technological solutions that can help the elderly to live full, healthy and independent lives. This chapter they analyze mobile multimodal interfaces with emphasis on GIS and the specific requirements of the elderly in relation to the use of assistive technologies. The authors identify specific requirements for the design of multimodal GIS through a usage example of a system that have been developed. Chapter 14, *Automatic Signature Verification on Handheld Devices* by Marcos Martinez-Diaz, Julian Fierrez, Javier Ortega-Garcia introduces automatic signature verification as a component of a multimodal interface on a mobile device with comparatively low resolution. The authors analyze applications and challenges of signature verification and review available resources and research directions. The chapter includes a case study which describes a state-of-the-art signature verification system adapted to handheld devices.

With an in-depth coverage of a variety of topics in multimodal interaction in a mobile environment, this book aims to fill in the gap in the literature catalog dedicated to mobile computing. Mobile computing receives a growing coverage in the literature due to its tremendous potential. Today, there is a growing trend of convergence of mobile computing with online services, which will lead to an increased interest in publications covering different aspects of mobile computing and ways to make mobile applications more accessible and acceptable to users. Multimodal user interfaces that can combine mutually-complementing aspects of human-computer interaction are an excellent avenue for increasing the user satisfaction with technology and leading to a higher user acceptance of a computing system.

This book will be of interest to researchers in industry and academia working in the areas of mobile computing, human-computer interaction, and interface usability, to graduate and undergraduate students, and anyone with an interest in mobile computing and human-computer interaction.

Stan Kurkovsky
Central Connecticut State University, USA

Section 1
Introduction

Chapter 1
Multimodal and Multichannel Issues in Pervasive and Ubiquitous Computing

José Rouillard
Université de Lille 1, France

ABSTRACT

Multimodality in mobile computing has become a very active field of research in the past few years. Soon, mobile devices will allow smooth and smart interaction with everyday life's objects, thanks to natural and multimodal interactions. In this context, this chapter introduces some concepts needed to address the topic of pervasive and ubiquitous computing. The multi-modal, multi-channel and multi-device notions are presented and are referenced by the name and partial acroynm "multi-DMC". A multi-DMC referential is explained, in order to understand what kind of notions have to be sustained in such systems. Next we have three case studies that illustrate the issues faced when proposing systems able to support at the same time different modalities including voice or gesture, different devices, like PC or smartphone and different channels such as web or telephone.

INTRODUCTION

For the general public, the year 1984 marks the emergence of WIMP (Windows, Icon, Menu, Pointing device) interfaces. Developed at Xerox PARC in 1973 and popularized by the Macintosh, this type of graphical user interfaces is still largely used today, on most computers. However, in recent years, numerous scientific researches focus on post-WIMP interfaces. It is no longer limited to a

single way of interacting with a computer system, but considering the different solutions to offer user interfaces as natural as possible.

With the introduction of many types of mobile devices, such as cellphones, Personal Digital Assistant (PDA), pocket PC, and the rise of their capabilities (Wifi, GPS, RFID, NFC...) designing and deploying mobile interactive software that optimize the human-computer interaction has become a fundamental challenge. Modern terminals are natively equipped with many input and output resources needed for multimodal interactions, such

DOI: 10.4018/978-1-60566-978-6.ch001

as camera, vibration, accelerometer, stylus, etc. However, the main difference between multimedia and multimodal interaction lies in the semantic interpretation and the time management.

Multimodality in mobile computing appears as an important trend, but a very few applications allow a real synergic multimodality. Yet, since the famous Bolt's ("put that there") paradigm (Bolt 1980), researchers are studying models, frameworks, infrastructure and multimodal architecture allowing relevant use of the multimodality, especially in mobile situations. Multimodality tries to combine interaction means to enhance the ability of the user interface adaptation to its context of use, without requiring costly redesign and reimplementation. Blending multiple access channels provides new possibilities of interaction to users. The multimodal interface promises to let users choose the way they would naturally interact with it. Users have the possibility to switch between interaction means or to multiple available modes of interaction in parallel.

Another field of research in which multimodality is playing an important role is in the Computer Supported Cooperative Work domain (CSCW). CSCW is commonly seen as the study of how groups of people can work together using technology in a shared time, space hardware and software relationship. "*In the context of ubiquitous and mobile computing, this situation of independent and collocated users performing unrelated tasks is however very likely to occur.*" (Kray & al. 2004). Even if there is a risk of overlapping categories, design issues are often classified into management, technical and social issues.

- Management: mainly deals with registration and later identification of users and devices as they enter and leave the workspace environment.
- Technical: issue occurs with the control of specific device features and also the technical management of services offering the possibility to introduce (discover) or remove specific components from an interaction. The design problems is then related to fusion (i.e. combining multiple input types) and fission (i.e. combining multiple output types) mechanisms, synchronization and rules management between heterogeneous devices.
- Social issues are more related to social rules and privacy matters. As we know, some devices are inherently unsuitable for supporting privacy, such as microphones, speakers and public displays.

The ubiquitous role of the computer makes each day more unsuitable for the screen-keyboard-mouse model posed on a corner of a desk. In fact, the large success and rise of the Internet networks have complemented computing communication due to the technical standards used and their adoption of languages such as HTML, WML, or VoiceXML. Yet, we observe a few incompatibilities even though promulgated standards subsist. An example of this incompatiblity can be found in computers with different operating systems that process various types of media (texts, graphics, sounds, and video). Though the information can be easily transmitted through the networks, the formats of the coded data are incompatible. As a direct result the end-user bears addional cost and time lost when trying to obtain or utilize product and service based on their particular platform. This creates the urgent need for easier access to information — whether at the office, home, or on the train, etc. This need is felt all the more with the constant new arrival of soft/hardware materials, the success of the pocket computers and mobile telephones. In fact, the current trends are leaning towards the transformation of end-user's interface for "anyone and anywhere" (Lopez & Szekely 2001).

With the multiplicity of the means of connecting to Internet, it is necessary to conceive generic interfaces and mechanisms of transformation to obtain concrete interfaces for each platform. Of course, designing for mobile is not exactly the same as designing for desktop or laptop. The main pitfall is to think that it's just a matter of screen size. That's why the W3C launched an activity in the Device Independence field.

This chapter presents some aspects of this scientific problem and is structured as follows: section two explains the background and motivation for understanding multimodality for mobile computing. Section three gives an overview about our multi-DMC referential. Section four is talking about pervasive and ubiquitous computing. Some case studies are then exposed and future trends and ideas for further work are presented in the conclusion of the chapter.

UNDERSTANDING MULTIMODALITY FOR MOBILE COMPUTING

Multimodality has been studied since the 1980's. The famous "Put-That-There" (Bolt 1980) concept was probably the first multimodal system of the field combining gesture recognition and speech input. In order to move an object in Bolt's system, the user point to it and says "Put that," then he indicates a destination and says "there." Other prototypes were built thereafter to test the combined use of direct manipulation and natural language (Cohen & al. 1989), or gestures and speech (Weimer & Ganapathy 1989). Processing techniques real-time video have been used to improve speech recognition by reading the movements of the lips and to monitor implementation of viewing panoramic images by combining eye movements and voice commands (Yang & al. 1998).

The field has reached a new dimension with technologies such as pervasive and ubiquitous computing, mobile computing, and augmented reality, which makes an extensive use of multimodality as well as providing interfaces and experiences that would not be at all possible without it. For several years, we have seen the miniaturization of electronic devices and their integration into everyday life. Mobile phones are almost all equipped with high quality camera, diverse connections to different networks (WiFi, GPRS…), "free hand" features, etc.

The trend that consists of systematically digitalizing resources and enabling access to data needed anywhere and in anytime is sometimes called, in the literature, ubiquitous. Ubiquitous computing engages many computational devices and systems simultaneously in the course of ordinary activities. There is a wide variety of terms used to describe this paradigm such as ambient intelligence, ubiquitous or pervasive computing that is contrary to the more conventional desktop one-computer, one-platform, one-person metaphor. These terms refers to the increased use of widespread processors that exchanges rapid and brief communications with each other and sensors. Thanks to their much smaller size, these sensors are being integrated into everyday objects, until they become almost invisible to users. Pervasive computing also denotes the ability to access the same service through various way of communication, depending on the needs and constraints of the user.

Ubiquitous computing considers a multitude of connected computers available at all times without interruption and geographically disseminated (building, road, street, etc.). Major problematical issues in the field of ubiquitous computing are related to space, time and scaling. Of the three, the issue of scaling is crucial to end-user in "real life" situations, which in some instances the problem is not encountered in the laboratory.

We desire natural and context-aware interfaces that facilitate a richer variety of communications capabilities between humans and machines. In this sense, it's a challenge to conceive non ad-hoc systems adapting their behaviour based on information sensed from the physical and computational environment. Identity, location, task, and other elements of the context have to be taken into account in those kinds of applications. Creating reusable representations of context is one of the remaining complex topics being addressed in Human- computing interaction (HCI).

Mobile computing is a generic term describing the ability to use computer technology while moving. This situation of mobility is stronger than the notion of portable computing, where the user often uses their laptop, for instance, in different geographical areas, but not necessarily while they are moving. Many types of mobile computers have been introduced since the 1990s, including Pocket PC, Wearable computer, PDA, Smartphone, Ultra Mobile PC (UMPC), etc.

Augmented reality (AR) is a recent field of computer research which deals with the combination of real-world and computer-generated data. In AR applications, graphics objects are blended into real footage in real time. Users interact with real and virtual environment according to numerous modalities.

The use of various ways of communications input (keyboard, mouse, speech recognition, eye tracking, gesture recognition, etc.) and in output (screen, sounds, speech synthesis, force feedback, etc.) are the so-called modalities of a multimodal system. Some of there are inherently bidirectional such as touch-screens or haptic interfaces. Multimodal interfaces offer the possibility of combining the advantages of natural (but ambiguous) inputs, such as speech, and less natural (but unambiguous) inputs by way of direct manipulation. In the domain of mobility, particularly, software and hardware are becoming more and more suitable for multimodal uses. It is now possible to have mutiple interfaces on the same traditional screen and keyboard device that include touch and speech interface capabilities. Various kinds of connections and technologies are also, sometimes natively, available on modern mobile devices, such as RFID (Radio-Frequency IDentification), NFC (Near Field Communication) or QR (Quick Response) Codes.

In situation where the user is mobile, it is very important to provide the right information at the right time in the right place. For instance, there could be a way to easily retrieve nutrition information based upon a QR Code on a burger package, paying for a can drink at a vending machine, getting bus information in real time, obtaining in-store information about a product, speeding up flight check-in and boarding etc. Other areas where this technology could be beneficial for mobile persons are may include cultural, entertainment and advertising of relevant information. Embedded camera, GPS, Wifi, Bluetooth, barcode detection, accelerometers, and other mobile devices features are primarily used individually, without possesing the ability to optimizing other applications capable of adaptating accordingly to a given context. For example, it is not easy to implement voice, stylus, or keyboard applications on a phone, where the user can choose to act and speak freely, due to the lack of standards available to achieve this goal.

We will see later in this chapter that some computer frameworks (e.g. W3C MMI) and languages have been developed especially to facilitate the development of multimodal interfaces, but when systems with a higher complexity are envisaged, researchers tried to tackle the problem thanks to component-based software engineering approaches (Bouchet & Nigay 2004). Finally, multimodality can be combined with multi-platform systems, where the user can switch between a desktop computer and a mobile phone, to offer even richer possibilities (Paternò 2004).

Mobiles devices and phone, originally created for vocal interaction between humans are nowadays being used to interact with machines. Some mobile phones devices are adapted with user interface features that are dedicated to aiding the partially sighted and blind persons. While others are used to smoothly interact with everyday smart objects such as refrigerators or interactive TV using a ubiquitous and multimodal adaptive interaction. A person using an iPhone with one finger pressure can triggers an application in charge of decoding 2D barcodes simply by aiming the camera of his phone. The barcode displayed on a poster against the wall for example, can be instantly mapped and displayed on the phone screen that shows the location of the restaurant requested. The user can then zooms in and navigate within this map. A link between the physical and the digital world is made.

But, as we explained, it is not always easy to distinguish systems in terms of modalities, channels, or features usable on any platform. Even definition of employed terms is sometimes ambiguous. For example, multimodality is often seen as the ability of a system to communicate with a user using different types of communication **channels** (Nigay & Coutaz 1993). We are presenting in the following the distinction between those words.

Multi-Device

A multi-device system allows using many devices (PC, PDA, Smartphone, Kiosk, etc.) in order to do the same type of task (Calvary & al. 2003). For example, within the framework of interface plasticity and context-aware computing (Rupnik & al. 2004), some researchers showed how to redistribute interfaces, from smartphone to PC (Ganneau & al. 2008).

This allows the system to propose to users, when it become possible, via a Bluetooth connection, for instance, a migration of his/her work environment from a mobile but restricted platform to a wired but more powerful one, with adjunction of functionalities. And of course, the migration could be done from PC to a mobile device, when the user decides to move from a place to another one.

Most of the time a confusion is made between the notion of multi-platform and multi-device systems. The term "multi-platform" is more related to a class of devices that share the same characteristic in terms of resources for the interaction: *"They range from small devices, such as interactive watches to very large flat displays. Examples of platforms are the graphical desktop, PDAs, mobile phones and vocal systems. The modalities*

available have an important role in the definition of each platform." (Paternò 2004). In our work, we are interested in multi-device systems more than in multi-platform systems, because we have to manage multiple devices, distinctively, even if there are from the same class.

Multi-Modal

Multimodal systems are characterised by human-machine interfaces that go beyond the traditional screen, keyboard, and mouse. Different kind of modalities, such as voice or gesture can be used in multimodal system, according to the user needs or constraints. Multimodality is seen as the combination of multiple input and/or output modalities in the same user interface, together with additional software components such as fusion, fission, and synchronisation engines. Since the famous Bolt's ("put that there") paradigm (Bolt 1980), researchers are studying models, frameworks, infrastructure and multimodal architecture allowing relevant use of the multimodality, especially in mobile situations. Multimodality tries to combine interaction means to enhance the ability of the user interface adaptation to its context of use, without requiring costly redesign and reimplementation.

There are different types of multimodality and in order to indicate how to combine the modalities, we often consider the four well-known CARE properties: Complementarity, Assignment, Redundancy, and Equivalence. In addition, *"The formal expression of the CARE properties relies on the notions of state, goal, modality, and temporal relationships"* (Coutaz & al. 1995). In cases where modalities can be used to achieve the same type of input, we talk about redundant modalities. In this kind of situation, it is often up to the user to decide which modality to use, according to the context. In cases where two modalities or more can be used to simultaneously or sequentially generate a message, we talk about complementary modalities. In this case, it would not have been possible to express a message by using only one of the modalities. Therefore, a fusion engine is required in such interfaces, for the combination of information coming from various and heterogeneous devices, channels and modalities. As the notion of fusion allows for combining modalities in input, the notion of fission allows for splitting the information in order to select the most appropriate way to render data to the user.

More formally, to model the expressive power of a modality m, that is, its capacity to allow an agent to reach state s' from state s in one step, we use the function Reach(s, m, s'). We can describe those properties as the following:

- **Equivalence:** Modalities of set M are equivalent for reaching s' from s, if it is necessary and sufficient to use any one of the modalities. M is assumed to contain at least two modalities.
- **Assignment:** Modality m is assigned in state s to reach s', if no other modality is used to reach s' from s. In contrast to equivalence, assignment expresses the absence of choice.
- **Redundancy:** Modalities of a set M are used redundantly to reach state s' from state s, if they have the same expressive power (they are equivalent) and if all of them are used within the same temporal window.
- **Complementarity:** Modalities of a set M must be used in a complementary way to reach state s' from state s within a temporal window, if all of them must be used to reach s' from s, i.e., none of them taken individually can cover the target state. As typical in human-computer interaction, the usability of those kinds of interfaces is proportional to the complexity needed to obtain them. We agree with Alapetite when he stated, "*Multimodal interfaces enrich human-computer interaction possibilities, and provide advantages such as an increase in robustness and flexibility. However, there are some drawbacks such as an increased*

complexity, together with more hardware and software needed" (Alapetite 2007). Some computer frameworks (e.g. W3C MMI: Multimodal Interaction Activity or OpenInterface) and languages have been proposed especially to facilitate the development of multimodal interfaces. Some examples of such languages will be given in the next paragraphs.

X+V

X + V is an XML language for developing multimodal applications. This was created to address growing demand for voice applications for mobile phones and PDAs. Technically, it is a combination of XHTML and VoiceXML (VoiceXML 2.0), with some additional facilities such as automatic synchronization between the modalities.

Unlike the VoiceXML language, X + V use both the voice and graphic, which offers new possibilities for developing interfaces for mobile devices. X + V combines XHTML and a sub-part of VoiceXML. The code of the two languages is separated, which simplifies the development and helps to create independently graphical and vocal parts. XHTML is an HTML readjusted to respect fully the XML rules. These two languages allow developers to add voice (input and output) to traditional Web pages.

Rather than inventing new tags, X+V reuse the existing tags from VoiceXML 2.0. In the context of our case studies, some limitations were identified during the automatic generation of multimodal applications development:

First, VoiceXML applications are described by context-free grammar. Then, recognized vocabulary is limited. If vocal grammar is easy to prepare for input fields of which we already know all the possible values, it is more difficult, even impossible to establish a grammar for open fields. For example, it is not possible to prepare an exhaustive grammar for the field "Name input," because all the possible responses can not be prepared for this field.

Second, it is not possible to obtain synergic multimodal applications, because, we are limited by the X+V language. Indeed, with X+V the user can choose to use a vocal or a graphical interaction, but it is not yet possible to pronounce a word and click on a object simultaneously, in order to combine the sense of those interactions. However, Alapetite has demonstrated the implementability of complementary multimodality with a simple "Put-That-There" test case using combined speech recognition and mouse that allows for user inputs such as "I want a new blue square … there" (Alapetite 2007). This prototype application runs on a multimodal browser (Opera) that supports X + V and SVG1 (SVG). Another W3C language presented in the MMI is EMMA. The subsequent section will present it briefly.

EMMA

EMMA (EMMA: Extensible MultiModal Annotation markup language) is a language that allows a semantic representation of all input information gathered within an application. In a multimodal architecture, each input, obtained by example from voice, gesture, keyboard, mouse, data glove, camera, etc., is recognised and interpreted. According to the W3C, EMMA could become a common language between the various components of a multimodal system. This would allow developers to create multimodal platforms more reliable and less expensive for the integration of each type of components. Yet, currently there are no regulations in place for the implementation of this process available.

For the last few years, the W3C have been working towards the standardizing aspect and is publishing recommendations. VoiceXML, for instance, is a vocal interaction language based on XML. It allows describing and managing vocal interactions on the Internet network. VoiceXML is a programming language, designed for human-

computer an audio dialog that contains the following features:

- synthesized speech
- digitized audio
- recognition of spoken input
- recognition of DTMF (Dual-Tone Multi-Frequency) key input
- recording of spoken input
- telephony
- mixed initiative conversations

Its major goal is to bring the advantages of web-based development and content delivery to interactive voice response applications. The next version of VoiceXML (V3.0) will be announced as multimodal compatible.

To illustrate the use of the EMMA language, that is a standard language for representing input, let's take an example, which supposes a virtual agent not visible to the user. The user speaks and listens to a disembodied voice that assists him/her to perform tasks. He/she points with a stylus or pen to the location on a map on the screen while saying "zoom in, here" to the virtual agent. The pen module records the time and point on the map to which the user points using EMMA notation.

The grammar used by the speech recognition system contains the phrase "zoom in" and "here." The speech recognition system recognizes the words in the user phrase and generates the EMMA notation such as:

```
<command>
    <action
start="1087995961111
end="1087995964444">zoom-in</ac-
tion>
    <location
start="108799595555
end="1087995968888">here</loca-
tion>
</command>
```

Many different techniques, including unification, may be used to integrate the EMMA <location> notation from the pen system with the EMMA notation for <location> from the speech recognition system. In this example, a simple time matching algorithm determines that the time the user pointed to the location matches the time that the user spoke "here" and merges the pen and speech EMMA information into a single EMMA notation:

```
<command>
      <action>zoom-in</action>
      <location>
           <point> 42, 158 </
point>
      </location>
</command>
```

The EMMA command is then transferred to the interaction manager, which determines that an enlarge command should be sent to the generation component. This language is being finalized and should evolve further. Recently (October 2008), the first returns of implementation (Implementation Report for EMMA 1.0 Specification) have been broadcasted by some companies such as Loquendo, AT & T, Nuance, Microsoft, Conversational Technologies, and in academia such as, University of Trento, in Italy or DFKI (German Research Center for Artificial Intelligence) in Germany.

Although this approach seems attractive, on paper, the W3C specifications do not describe how to carry out implementations of EMMA language. There have been developmental impedements that we have identified in our work that we have acknowledged and remain unanswered presently. These developmental impedements include mechanisms for events notification: how the web browser will be notified that it must zoom, as show in our previuos EMMA example? It is also a question of quality of time stamping during the speech recognition: Loquendo reported, for

example, that it is difficult to achieve this with precision. Finally, we do not think the architecture proposed by the W3C will support interaction for multi-modal and multi-channel and multi-device, simultaneously. We can conclude that the voice is most of the time proposed as a modality in multimodalities system.

Table 1 summarizes the main characteristics of some[2] standard languages allowing managing vocal interactions.

We can notice that the main advantage common to all those languages is the fact that there are based on XML. The main disadvantage is that they are mostly mono-device and too web (originally graphically) oriented.

Multi-Channel

For Frohlich, a channel is defined as *an interface that makes a transformation of energy* (Frohlich 1991). From a user's point of view, he distinguished voice and movement channels, and from the system's point of view he mentioned audio, visual, and haptic channels. In the HCI domain, the notion of channel is not used very often and there are very few references to multi-channel research with some exceptions such as the work of Healey and colleagues: "*Often these modalities require specialized channels to allow access modalities such as cameras, microphones, and sensors. A multimodal multi-channel system faces the challenge of accepting information from any input method and delivering information through the appropriate output methods*" (Healey & al. 2002).

In the field of direct marketing, a channel is a vector of communication between the customer and the organization. Thus, a channel could be one of the following ways of interaction: web, phone, fax, email, SMS, interactive television, etc. The term multi-channel is frequently used in the marketing world. Indeed, multi-channel marketing is considered as marketing using many different channels to reach a customer. In a multi-channel system, the database is essential

Table 1. Comparison of W3C languages for vocal interactions. © 2009 José Rouillard. Used with permission.

Languages	Type de multi-modality	Advantage	Disadvantage
VoiceXML 2.1 (current version)	Alternate (Voice, then DTMF, for instance)	- Management of a oral man-machine dialog on telephone - Exploitable traces of speech recognition - Based on XML	- Mono-Device - Too much "fill in the form approach" oriented - Limited vocal grammars
VoiceXML 3.0 (future version)	Alternate (Voice, then DTMF, for instance)	- Speaker identification - External events managent - Based on XML	- Limited vocal grammars
X+V	Alternate (keyboard/mouse then voice, for instance)	- Based on XHTML and VoiceXML	- Mono-Device - Limited vocal grammars
EMMA	Synergic (Voice and stylus simultaneously, for instance)	- Main element of the « W3C Multimodal Interaction Framework » - Based on XML	- Mono-Device - More oriented towards inputs than towards outputs
SMIL 2.1 (Synchronized Multimedia Integration Language)	Multimodality in output. Temporal and spatial description of the different componants	- Integration of multimedia contents (images, sounds, video, animations, and text flow). - Synchronization allowing the creation of multimedia presentations. - Based on XML	- Mono-Device - Output oriented. Do not manage inputs
SALT (Speech Application Language Tags)	Alternate (Voice, then DTMF, for instance)	- HTML Extension with inclusion speech and DTMF - Based on XML	- Mono-Device - Do not manage the dialog

to manage information coming from different sources. The main idea is to let the final user choose the best way to interact with the system, according to a particular context. Here, we are not talking about a choice between a kiosk and a PC (that is a device choice), or between a gesture and a speech input (that is a modality choice), but between different kind of channels, such as the web versus the phone, for example. We believe that this distinction between channel, device, and modality is essential in HCI and particularly in the field of pervasive and ubiquitous computing. Indeed, taking only into consideration a modality used on a device is not sufficient to describe precisely a rich interaction. For instance, the voice modality could be used on a desktop PC as well as on a smartphone device. On a smartphone, this voice could be used on the Internet channel for a

vocal browsing or on the telephone channel, for a man-machine dialogue. Moreover, in matter of quality of services, it becomes crucial to determine if an interaction is able to be conducted across different channels or not. A multi-channel system is able to manage a task started on one channel and terminated on another one (Chevrin & al. 2008). Obviously, the interaction will be more flexible if the user can choose herself a relevant channel according to her needs. The following section will present some frameworks for multimodality in this context.

Framework for Multimodality

In this section we will present, in a few words, W3C, and other approaches used to conceive and implement multimodal applications.

W3C Approach

In the W3C MultiModal Interaction framework (MMI), the interaction manager may invoke application specific functions and access information in a Dynamic Processing module. Finally, the interaction manager causes the result to be presented to the user via one or more output components such as audio or a display screen.

Obviously, the interaction manager of this framework is very important because it coordinates data and manages execution flow among various input and output components. It also responds to inputs from the input components, updates the interaction state and context of the application, and initiates output to one or more output components.

Larson indicates (Larson 2006) that developers use several approaches to implement interaction managers, including:

- Traditional programming languages such as C or C++;
- Speech Application Language Tags (SALT) which extends HTML by adding a handful of HTML tags to support speech recognition, speech synthesis, audio file replay, and audio capture ;
- XHTML plus Voice (often referred to simply as "X+V"), in which the VoiceXML 2.0 voice dialog control language is partitioned into modules that are embedded into HTML;
- Formal specification techniques such as state transition diagrams and Harel state charts.

Other Approaches and Related Work

Tools helping in the design and implementation of multimodal interfaces are available. We can cite, for instance, FAME (Duarte 2006), or approaches based on crucibles3 (Nigay 1994), based on rules (Bellik 1992), or those who propose more elaborate solutions, such as PetShop (Bastide & al. 1998), but that required a high level of programming in order to encode mechanisms inherent to each agent.

Designers propose toolkits allowing setting up more easily their solutions: for example, OIDE4 (Serrano & al. 2008) is a graphical tool made upon the OpenInterface kernel (OpenInterface). Sometimes, they also provide some automatic code generation systems; as is the case for the composition of components in ICARE (Bouchet & Nigay 2004)). OpenInterface primary interest is in its aims to provide an open source platform for the design and rapid development of multimodal prototyped applications as a central tool for an iterative user-centered process. The basic objects manipulated by the OpenInterface platform are called components. OpenInterface components can be composed together to create a network of components managing some advanced task. Such an inter-connection of components is called a pipeline. A pipeline must be specified in the PDCL (Pipeline Description and Configuration Language) in order to be manipulated by the OpenInterface platform (Stanciulescu 2008). Some authors mentioned the advantages and drawbacks of such tools: "*Interface builders such as the Next Interface Builder and OpenInterface are a different class of tools to aid in the design of interfaces. These tools make it very easy to construct the particular interfaces that they support, but are very poor for design exploration. Designers have to commit to particular presentation, layout and interaction techniques early in the design*" (Neches & al. 1993).

Other languages like M3L (Multimodal Markup Language) were used in dedicated project (such as SMARTKOM (Smartkom)). M3L was developed as a complete XML language that covers all data interfaces within this complex multimodal dialogue system. Various modalities such as speech, gesture, mimic, audio, graphics were supported in cars, kiosk, interactive TV and others scenarios (Wahlster 2006). Instead of using several quite different XML languages for the various data pools, the authors aimed was to integrate a coherent language specification which includes all substructures that may occur on the different pools. In order to make the specification process manageable and to provide a thematic organization, the M3L language definition has been decomposed into about 40 schema specifications.

MULTI-DMC REFERENTIAL

The Multi-DMC referential that we propose can identify a system based on three criteria: Device (D) Modal (M) and Channel (C). It has two positions (Mono or Multi) for each of the three criteria targeted (DMC). This represents 2^3 opportunities positioning. These 8 possibilities are presented on the Figure 1.

In this reference, the "device," the "modal," and "channel" respectively refer to the physical devices used by the user, the modalities used during the interaction and the logical channels employed during the man-machine information exchange. We consider for example, SMS and phone are separate channels, even though they use the same communication network.

The position (0 0 0), on the top left on Figure 1, describes a single device, single-modality and single-channel system. Such a system is running on a single device (PC, for example) with a single modality of interaction (for instance: gesture), and a single channel of communication (for example: web).

The position (1 1 1), on the bottom right of Figure 1, describes a multi-device, multi-modal and multi-channel system. This kind of system is able to run on at least two devices (PC and smartphone, for example) with at least two possible ways of interaction (voice and gesture, for example) and use at least two different communication channels (Web and the phone, for example).

The position (0 0 1), on the top right on Figure 1, represents a mono-device, mono-modal with a multi-channel system. This could be a system

Figure 1. Multi-DMC referential. © 2009 José Rouillard. Used with permission.

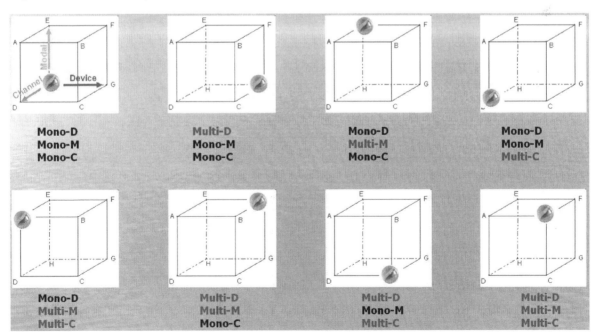

only working on a smartphone, with only audio capabilities (in input and output), but using both the web channel and the telephone channels of a smartphone. Here we have explained 3 out the possible 8 shown referential to help clarify the specifics of the system and compare systems between them. We could have added the criterion single / multi-user, etc.

As we explained, one of the main issues faced when developing multimodal and multichannel interfaces in pervasive and ubiquitous computing is the lack of standard regulation in order to create and support such usable systems. The next section will present cases studies related to pervasive and ubiquitous computing issues. We will connect each example to the multi-DMC referential previously explained, in this paper.

PERVASIVE AND UBIQUITOUS COMPUTING FOR MOBILE COMPUTING

Embedded systems are becoming ubiquitous. Most existing computers do not have traditional inputs and outputs such as a screen, a keyboard, or a mouse. Yet, they are present in our lives under numerous kinds of objects, from vehicles to domestic appliances. It is clear that in the future, computerized heterogeneous objects will be digitally linked together and will communicate with each other without humans' intervention. Hence, the multiplicity of users, proliferation of computing platforms and continuous evolution of the physical environment leads to explore new kind of usage in the domain of pervasive and ubiquitous computing.

The main concepts and key features needed for designing multimodal ubiquitous applications are quite uncomplicated to imagine in a close world. The difficulty is to prepare the integration of future devices, not yet known. For example, if it's trivial to understand that future cars will talked to each other, but also with the road and the road itself to

the city, in order to organize traffic automatically, it is not so simple to deploy such an infrastructure in a flexible and smart manner.

Basic questions are such as "How to integrate physical object of the world (screen, chair, coffee machine…) into multimodal applications, thanks to technology such as RFID, NFC, Barcodes (1D or 2D as QR Codes)?"

Making ad-hoc systems is not a satisfying method, in the long term, because it needs too many resources and will not yield long-term results that will remain a viable solution. But, from a scientific point of view, it's an interesting way of measuring requirements. For instance, the prototype presented on Figure 2 helped us to understand how to manage the switches from a modality to another. It could be done by the user that decides to interact vocally instead of using a stylus, for example. But, it could also be suggested by the system itself during the interaction, if it recognises a task or an activity. For example, while repairing a car or cooking, the user will certainly need to work with his/her two hands, so the system will propose to interact vocally with the user.

In 1991, Weiser clarifies the subject of the first orientation for pervasive computing, as he explained, *"The most profound technologies are those that disappear. They weave themselves into the fabric of everyday life until they are indistinguishable from it."* (Weiser 1991). The trend that consists of systematically digitalizing resources and enabling access to data needed anywhere and in anytime is also sometimes called, in the literature, ubiquitous. However, there is a wide variety of terms used to describe this paradigm as opposed to the more conventional desktop metaphor (one computer per person). This is known as ambient intelligence, ubiquitous and pervasive computing. This refers to the increasing use of widespread processors that exchange rapid and brief communications with each other and with sensors. Thanks to their much smaller size, these sensors will be integrated into everyday objects, until they become almost invisible to users.

Figure 2. Multimodal systems adapted to the task, (repairing a car, cooking) using voice and gesture. © 2009 José Rouillard. Used with permission.

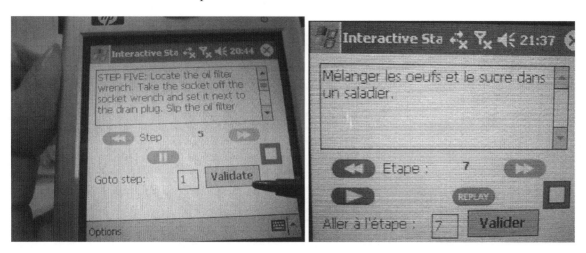

Pervasive computing refers to the ability to access the same service through various way of communication, such as a desktop computer, a PDA, a phone for using voice, phone keypad (DTMF) or SMS, depending on the needs and constraints of the user. Adam Greenfield, meanwhile, uses the word "Everyware." This neologism, formed from "everywhere" and "hard/software" is used to encompass the terms of ubiquitous computing, pervasive computing, ambient computing and tangible media. Our work is related to this field of research, and the p-LearNet project, that will be presented now, allows us to open perspectives in this domain.

In the following, we will see three case studies that illustrate our current work and researches.

Various task and scenarios are presented in those case studies in order to show the issues addressed.

Painter

The first distributed application that we developed is for drawing software, entitled "Painter", which outlines basic shapes such as rectangle, circle,

triangle, in specific colors like yellow, green, blue, red, to given coordinates. According to the multi-DMC referential presented in section 3, we have here a (1 1 0) position system. It's a multi-device, multi-modal, and mono-channel system, as we can see on Figure 3.

This kind of application has long been used in many research projects to explore multimodality.

Figure 3. The painter case study is a multi-device, multi-modal, and mono-channel system. © 2009 José Rouillard. Used with permission.

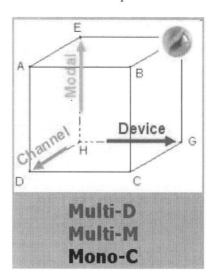

This enabled us to familiarize ourselves with the tools and test certain concepts needed later (Wizard of Oz, tracer events, delegate, etc.). It also helps us to understand the use while developing software components in an incremental development spiral cycle. Each new version is based on the previous kernel and is strengthened with new features.

Figure 4 shows a user with multiple devices and several interfaces to execute a task of drawing. He has a HTC smartphone connected in WiFi, (in his hand), a touch screen (on the left), a large display screen (on background), a speech recognition system (he wears a headset and the machine in front of him trace voice interactions programmed in VoiceXML) and a machine to supervise operations (laptop on his left).

The application that we have programmed should help drawing simple geometric shapes, using one or more ways to interact via one or more devices available. The only channel used by the paint application is the Internet. The user can choose to use a vocal interaction or a gesture interaction, with the stylus of his smartphone, for example, or a combination of available possibilities. As the messages are sent on a software bus, it is possible, for an application running on a particular device, to use a subscribe mechanism, in order to be notified if a certain event occurs. It could be used, for example, in order to give a feedback to the user, with a vibration of the smartphone that happen when a shape is drawn on the main large screen.

Another example includes pointing to 2 coordinates (x1, y1 and x2, y2) on the touch screen, then vocally demanding "draws a circle," and it will display the desired shape on the large screen. Thanks to direct manipulation on keyboard and mouse, voice and gesture capabilities, this provides a multimodal and multichannel application usable on PC or smartphone. It may be used by a single person who has a multitude of devices or by multiple users who cooperate.

Figure 4. A user that interacts with a multi-device, multi-modal, and multi-channel system. © 2009 José Rouillard. Used with permission.

Figure 5. Multimodal painter application on smartphone HTC and Wizard of Oz application. © 2009 José Rouillard. Used with permission.

Figure 5 shows, on the left, a multimodal painter application running on a smartphone under Windows Mobile 6, and on the right, a Wizard of Oz application used in the Painter project. The Wizard of Oz principle is used here to simulate the intelligence of the system in ambiguous situation. When there is not ambiguity, it's easy for the system to understand the meaning of the users. For instance, as we can see on the right of the Figure 5, the information "rectangle", "20,20 60,60" and "yellow" have been detected for the shape, the coordinates and the color slots. In this case, the system send a message on the software bus, indicating that a yellow rectangle is supposed to be painted in coordinates x=20,20 and y=60,60. But it is not always so simple. Let's say the system receives the following list of entries, more or less isolated in time, and not necessarily coming from the same agent:

1. input=shape value=circle
2. input=color value=green
3. input=shape value=rectangle
4. input=coords value=10,10 50,50

How should we interpret this? Does this mean that the user wants to draw a green circle with the coordinates 10,10 50,50, and later a rectangle whose details are not yet specified? Or maybe he wants to draw a rectangle, assuming he was deceived about the circle, or maybe he wants to draw a circle and a rectangle in the same place, probably in a different color? We can see from this example that the fusion of information is not trivial. Hence, the Wizard of Oz helps the system by providing a solution when it can't do it itself.

Controlling Remote Webcam

The second application that we are presenting here, according to the multi-DMC referential presented in section 3, is a (1 1 1) position system. It's a multi-device, multi-modal, and multi-channel system, as we can see on Figure 6. By using the infrastructure previously presented, we have created a system to validate, in another context, the use of the delegate that we have implemented. A delegate is here a software agent able to represent another agent not present on the local network.

A major question in pervasive and ubiquitous computing is how to integrate physical object of the world (screen, chair, coffee machine…) into multimodal applications, thanks to technology such as RFID, NFC, Barcodes (1D or 2D as QR Codes). This will help the users to manipulate freely virtual and real objects with commands like "identify this," "make a copy of that object, here", "move that webcam on the left," etc. We are using the notion of workflow in order to indicate to the user the tasks available at each point of the whole activity flow.

Common Knowledge (ObjectConnections) is a cross-platform business rules engine and management system that supports the capture, representation, documentation, maintenance, testing, and deployment of an organization's business rules and application logic. Common Knowledge allows the business logic to be represented in a variety of inter-operable, visual formats including: Rete rules, workflows, flowcharts, decision tables, decision trees, decision grids, state maps, scripts, and activescripts. The engine allows running, testing, and simulating the system behaviours. It can be used through many languages (such as Java, Delphi, VisualBasic, C#, DotNET) and platforms (Windows, Linux, UNIX). Figure 7 presents an example of a workflow designed graphically with this tool.

Figure 8 presents a screenshot of an application running on smartphone HTC under Windows Mobile 6, allowing to control a webcam installed in the building B6 of our university. This application coded in C #, enables a user to take control of a remote camera by the means of voice or gesture, across phone, smartphone or desktop application. It uses the notion of delegate to reach a software bus when the interaction is not situated in the same local area network. We see in our example that the user sought to move the camera up (complement = up). This interaction has been requested as a text command (see textbox in the middle of the smartphone screen). This implies that the user knows what commands to use (keyword =

Figure 6. The controlling remote webcam case study is a multi-device, multi-modal, and multi-channel system. © 2009 José Rouillard. Used with permission.

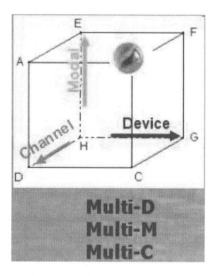

value). Similarly, the directional buttons could also be used in place of command lines sent to the delegate.

PerZoovasive

The third application that we are presenting here, according to the multi-DMC referential presented in section 3, is a (0 1 1) position system. This application called PerZoovasive, is using QR Codes and is a mono-device, multi-modal and mono-channel system, as we can see on Figure 9.

In our work, we use the notion of "Everyware" to adapt a learning environment to students. Indeed, being in a museum or zoo and receiving dense information not adapted to level or language is a problem users frequently face. This situation is not comfortable and usually the visitor's objective is not reached. To test the feasibility of our project, we use as example a complementary course done in a zoo.

A QR Code is a two-dimensional barcode introduced by the Japanese company Denso-Wave in 1994. This kind of barcode was initially used

Figure 7. Workflow designed with Studio Common Knowledge. © 2009 José Rouillard. Used with permission.

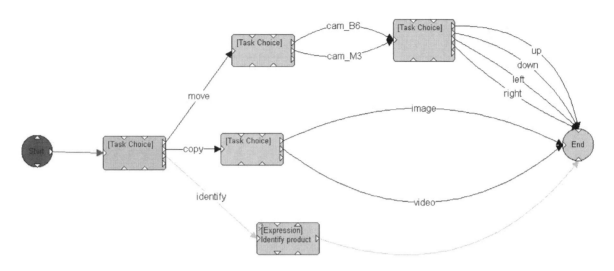

Figure 8. A smartphone capable of commanding moves of a remote webcam thanks to the IVY bus. © 2009 José Rouillard. Used with permission.

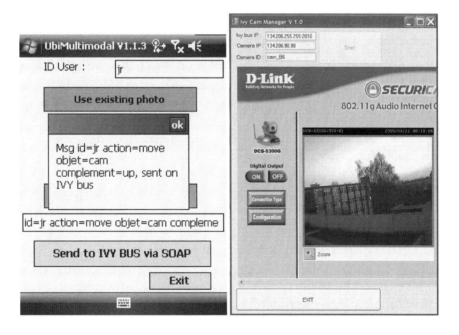

for tracking inventory in vehicle parts manufacturing and is now used in a variety of industries. QR stands for "Quick Response" as the creator intended the code to allow its contents to be decoded at high speed. In Japan, some teachers are using QR codes to distribute resources to students (Fujimura & Doi 2006).

Figure 9. PerZoovasive is a mono-device, multi-modal and mono-channel system. © 2009 José Rouillard. Used with permission.

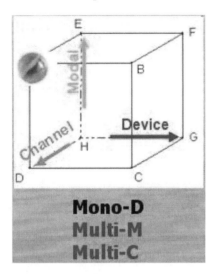

The notion of contextual QR Codes was proposed in previous recent work (Rouillard 2008). It can be defined as the following: it's the result of a fusion between a public part of information (QR Code) and a private part of information (the context) provided by the device that scanned the code.

The private part can be one or more of the following user's profile, current task, device, location, time, and environment of the interaction. The mobile device decodes the QR Code and merges it with private data obtained during the interaction. Then, the XML (Extensible Markup Language) resulting file is sent to a web service (created in our laboratory) that computes the code and returns personalized messages.

Some private information can be stored in the owner's profile of the phone (the class level for example) and some others are given directly by the user when the interaction takes place (language, class or exam, etc.).

PerZoovasive (Rouillard & Laroussi 2008) is an adaptive pervasive learning environment, based on contextual QR Codes, where information is presented to learner at the appropriate time and place, and according to a particular task. This Learning environment is called PerZoovasive, where learning activities take place in a zoo and are meant to enhance classroom activities.

Adaptivity and context awareness system are here strategies to provide support for learners in mobile pervasive situations. This work is based upon the following questions: How to adapt the information while taking into consideration the context of the student or the group? Do context-awareness applications affect information proposed to users? Which role can be played by QR Codes in a pervasive environment?

A previous study of people engaged in a location-based experience at the London zoo was reported by O'Hara and colleagues. In this experience, location-based content was collected and triggered using mobile camera phones to read 2D barcodes on signs located at particular animal enclosures around the zoo. "Each sign had an enticing caption and a data matrix code (approx. 7x7 cm) which encoded the file locations for the relevant media files." (O'Hara & al. 2007). By capturing a 2D barcode, participants extracted the file' URIs from the codes and added the corresponding preloaded content files (audio video and text) into their user's collection. The fundamental distinction between that approach and our system is that the London zoo system provides always the same content to the user, while the perZoovasive system provides information according to a particular context.

On Figure 10, we can see a teacher asking information about some turtles. She could retrieve general information about this kind of animal (species, origin, speed, food, traceability…), but also personal data about that particular turtle (name, age, birthday, parents, etc.).

In the PerZoovasive case study, the QR Code represents the number of an animal cage. For instance, 8245 is the number of the panther cage of the zoo. We can see on Figure 11 that the information provided is given according to a particular context. On the left, the French text

Figure 10. A teacher is scanning a Contextual QR Code to retrieve information about turtles. © 2009 José Rouillard. Used with permission.

Figure 11. Usage of the same QR Code for two distinct contexts. © 2009 José Rouillard. Used with permission.

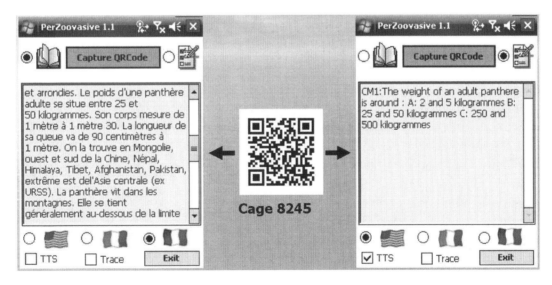

obtained is provided for a lesson task relevant for a high level of an elementary school. On the right, the English text obtained is provided for a Quiz task relevant for a middle level of an elementary school. Thus, the teacher uses a single device and multiple modalities and channels of interaction in order to complete at the zoo some notions seen in the classroom.

CONCLUSION

New emerging standards for Web services and new service-oriented design approaches are great opportunities for the re-engineering of current mobile multimodal solutions, through more flexibility and self-services possibility. Multi-channel and multimodality in the user interactions are difficult challenges from both HCI design and network interoperability viewpoint. Converting and transmitting documents across electronic networks is not sufficient. We have to deal with contents and containers simultaneously.

We are switching from "e-everything" to "u-everything" (electronic to ubiquitous). Physical objects equipped with barcodes or RFID chips, for instance, will be reachable in a numeric world. Multimodal and multi-channel adaptation will be the kernel of future ubiquitous, plastic and intelligence systems.

We have shown that Multimodal and Multichannel issues in pervasive and ubiquitous computing are mainly related to the lack of standard regulation in order to create and support such usable systems. The multi-DMC referential proposed and presented in this paper could be usable in order to understand what kind of notions have to be sustained in addressed systems. The three presented case studies illustrates the issues faced when proposing systems able to support at the same time different modalities including voice or gesture, different devices, like PC or smartphone and different channels such as web or telephone.

Our future work is oriented to the investigation of the traces/logs and original strategies to capture environment information. That will lead to modify the document on the fly, but also to better understand the behaviour of the entire system driven by multiple and sometimes contradictory policies. Technologies and solutions like IPv6, SOA (including web services approaches) are already used in order to deliver the appropriate information to the appropriate person at the appropriate place and time. But the challenge will be, for the next generation of systems, to adapt themselves smoothly and smartly (semantic aspects) to situations not yet encountered. This will be possible only with a real multimodal dialogue between the mobile systems and the users. Hence, the future ubiquitous systems will have to show their seams, instead of trying to hide them, to facilitate their usage.

ACKNOWLEDGMENT

We are grateful to ANR P-LearNet project for providing support for this research and to Object-Connections, IVY group, App-Line and Tasman, for special tools provided. The author gratefully acknowledges M. Kent Washington for his help in improving this paper.

REFERENCES

Alapetite, A. (2007). *On speech recognition during anaesthesia*. PhD thesis, Roskilde University.

Bastide, R., Palanque, P., Le, D., & Munoz, J. (1998). Integrating rendering specifications into a formalism for the design of interactive systems. In *Proceedings of DSV-IS '98*, Abington, UK. Berlin, Germany: Springer Verlag.

Bellik, Y., & Teil, D. (1992). Multimodal dialog interface. In *Proceedings of WWDU'92*, Berlin.

Bolt, R. A. (1980). Put-that-here: voice and gesture at the graphic interface. *Computer Graphics, 14,* 262–270. doi:10.1145/965105.807503

Bouchet, J., & Nigay, L. (2004). ICARE: a component-based approach for the design and development of multimodal interfaces. In *CHI '2004 Conference on Human Factors in Computing Systems,* Vienna, Austria (pp. 1325-1328).

Calvary, G., Coutaz, J., Thevenin, D., Limbourg, Q., Bouillon, L., & Vanderdonckt, J. (2003). A unifying reference framework for multi-target user interfaces. *Interacting with Computers, 15*(3), 289–308. doi:10.1016/S0953-5438(03)00010-9

Chevrin, V., & Rouillard, J. (2008). Instrumentation and measurement of multi-channel services systems. *International Journal of Internet and Enterprise Management (IJIEM), Special Issue on . Quality in Multi-Channel Services Employing Virtual Channels, 5*(4), 333–352.

Cohen, P. R., & Dalrymple, M. Moran, D. B., Pereira, F. C. & Sullivan, J. W. (1989). Synergistic use of direct manipulation and natural language. In *Proceedings of the SIGCHI conference on Human factors in computing systems* (pp. 227–233). New York: ACM Press.

Coutaz, J., Nigay, L., Salber, D., Blandford, A., May, J., & Young, M. R. (1995). Four easy pieces for assessing the usability of multimodal interaction: the CARE properties. In *INTERACT* (pp. 115-120). New York: Chapman & Hall.

Duarte, C., & Carrico, L. (2006). A conceptual framework for developing adaptive multimodal applications. In *Proceedings of IUI 2006* (pp. 132-139). New York: ACM Press.

Fujimura, N., & Doi, M. (2006). Collecting students' degree of comprehension with mobile phones. In *Proceedings of the 34th annual ACM SIGUCCS conference on User services table of contents,* Edmonton, Canada (pp. 123-127).

Ganneau, V., Calvary, G., & Demumieux, R. (2008). Learning key contexts of use in the wild for driving plastic user interfaces engineering. In *Engineering Interactive Systems 2008 (2nd Conference on Human-Centred Software Engineering (HCSE 2008) and 7th International workshop on TAsk MOdels and DIAgrams (TAMODIA 2008)),* Pisa, Italy.

Healey, J., Hosn, R., & Maes, S. H. (2002). Adaptive content for device independent multimodal browser applications. In *Lecture Notes In Computer Science; Vol. 2347, Proceedings of the Second International Conference on Adaptive Hypermedia and Adaptive Web-Based Systems* (pp. 401-405).

Kray, C., Wasinger, R., & Kortuem, G. (2004). Concepts and issues in interfaces for multiple users and multiple devices. In *Proceedings of MU3I workshop at IUI 2004* (pp. 7-11).

Larson, J. (2006). Standard languages for developing speech and multimodal applications. In *Proceedings of SpeechTEK West,* San Francisco.

Lopez, J., & Szekely, P. (2001). Automatic web page adaptation. In *Proceedings of CHI2001-Workshop "Transforming the UI for anyone anywhere",* Seattle.

Multimodal Interaction Framework. (2003). *MMI.* Retrieved February 02, 2009, from http://www.w3.org/TR/mmi-framework

Neches, R. Foley, J. Szekely, P. Sukaviriya, P., Luo, P., Kovacevic, S. & Hudson, S. (1993). Knowledgeable development environments using shared design models. In *IUI '93: Proceedings of the 1st international conference on Intelligent user interfaces.* New York: ACM Press.

Nigay, L. (1994). *Conception et modélisation logicielles des systèmes interactifs.* Thèse de l'Université Joseph Fourier, Grenoble.

Nigay, L., & Coutaz, J. (1993). A design space for multimodal systems: Concurrent processing and data fusion. In *Proceedings of ACM INTERCHI '93 Conference on Human Factors in Computing Systems, Voices and Faces* (pp. 172-178).

O'Hara, K., Kindberg, T., Glancy, M., Baptista, L., Sukumaran, B., Kahana, G., & Rowbotham, J. (2007). Collecting and sharing location-based content on mobile phones in a zoo visitor experience. [CSCW]. *Computer Supported Cooperative Work, 16*(1-2), 11–44. doi:10.1007/s10606-007-9039-2

ObjectConnections. (2008). *Common Knowledge Studio and engine provided by ObjectConnections*. Retrieved February 02, 2009, from http://www.objectconnections.com

OpenInterface European project. (n.d.). *IST Framework 6 STREP funded by the European, Commission (FP6- 35182)*. Retrieved February 02, 2009, from http://www.oiproject.org and http://www.openinterface.org

Paternò, F. (2004). Multimodality and Multi-Platform Interactive Systems. In *Building the Information Society* (pp. 421-426). Berlin, Germany: Springer.

Rouillard, J. (2008). Contextual QR codes. In *Proceedings of the Third International Multi-Conference on Computing in the Global Information Technology (ICCGI 2008)*, Athens, Greece (pp. 50-55).

Rouillard, J., & Laroussi, M. (2008). PerZoovasive: contextual pervasive QR codes as tool to provide an adaptive learning support. In *International Workshop On Context-Aware Mobile Learning - CAML'08, IEEE/ACM SIGAPP*, Paris (pp. 542-548).

Rupnik, R., Krisper, M., & Bajec, M. (2004). A new application model for mobile technologies. *International Journal of Information Technology and Management, 3*(2), 282–291. doi:10.1504/IJITM.2004.005038

SALT. (n.d.). *Speech Application Language Tags*. Retrieved February 02, 2009, from http://www.phon.ucl.ac.uk/home/mark/salt

Serrano, M., Juras, D., Ortega, M., & Nigay, L. (2008). OIDE: un outil pour la conception et le développement d'interfaces multimodales. *Quatrièmes Journées Francophones: Mobilité et Ubiquité 2008*, Saint Malo, France (pp. 91-92). New York: ACM Press.

Smartkom Project. (n.d.). *Dialog-based Human-Technology Interaction by Coordinated Analysis and Generation of Multiple Modalities*. Retrieved February 02, 2009, from http://www.smartkom.org

Stanciulescu, A. (2008). *A methodology for developing multimodal user interfaces of information system*. Ph.D. thesis, Université catholique de Louvain, Louvain-la-Neuve, Belgium.

Stanciulescu, A., Limbourg, Q., Vanderdonckt, J., Michotte, B., & Montero, F. (2005). A transformational approach for multimodal web user interfaces based on UsiXML. In *Proceedings of 7th International Conference on Multimodal Interfaces ICMI'2005* (pp. 259–266). New York: ACM Press.

SVG. Scalable Vector Graphics (n.d.). *W3C Recommendation*. Retrieved February 02, 2009, from http://www.w3.org/Graphics/SVG

Voice, X. M. L. 2.0. (2004). *W3C Recommendation*. Retrieved February 02, 2009, from http://www.w3.org/TR/voicexml20

Wahlster, W. (2006). SmartKom: Foundations of multimodal dialogue systems. *Series: Cognitive Technologies*, 644. X+V, XHTML + Voice Profile 1.2. (2004). *W3C Recommendation*. Retrieved February 02, 2009, from http://www.voicexml.org/specs/multimodal/x+v/12

Weimer, D., & Ganapathy, S. K. (1989). A synthetic visual environment with hand gesturing and voice input. *Proceedings of the SIGCHI conference on Human factors in computing systems* (pp. 235–240). New York: ACM Press.

Weiser, M. (1991). The computer for the 21st century. *Scientific American, 265*(3), 94–104.

Weiser, M. (1993). Some computer science problems in ubiquitous computing. *Communications of the ACM, 36*(7), 74–83. doi:10.1145/159544.159617

Yang, J., Stiefelhagen, R., Meier, U., & Waibel, A. (1998). Visual tracking for multimodal human computer interaction. In *Proceedings of the Conference on Human Factors in Computing Systems (CHI-98): Making the Impossible Possible* (pp. 140–147). New York: ACM Press.

ENDNOTES

[1] SVG: Scalable Vector Graphics

[2] Other languages proposed by the W3C are used by some languages mentioned in Table 1. They are, for example: SSML (Speech Synthesis Markup Language), SRGS (Speech Recognition Grammar Specification), or SISR (Semantic Interpretation for Speech Recognition).

[3] It's the metaphor of the crucible that is used here. A crucible is a cup-shaped piece of laboratory equipment used to contain chemical compounds when heated to extremely high temperatures.

[4] OIDE: OpenInterface Interaction Development Environment

Chapter 2
Ubiquitous User Interfaces:
Multimodal Adaptive Interaction for Smart Environments

Marco Blumendorf
Technische Universität Berlin, Germany

Grzegorz Lehmann
Technische Universität Berlin, Germany

Dirk Roscher
Technische Universität Berlin, Germany

Sahin Albayrak
Technische Universität Berlin, Germany

ABSTRACT

The widespread use of computing technology raises the need for interactive systems that adapt to user, device and environment. Multimodal user interfaces provide the means to support the user in various situations and to adapt the interaction to the user's needs. In this chapter we present a system utilizing design-time user interface models at runtime to provide flexible multimodal user interfaces. The server-based system allows the combination and integration of multiple devices to support multimodal interaction and the adaptation of the user interface to the used devices, the user and the environment. The utilization of the user interface models at runtime allows exploiting the design information for advanced adaptation possibilities. An implementation of the system has been successfully deployed in a smart home environment throughout the Service Centric Home project (www.sercho.de).

INTRODUCTION

Computer technology is currently changing our lives and the way we handle technology. The computer moves from a business machine dedicated to specific tasks in a well defined environment to a universal problem solver in all areas of live. Powerful mobile devices that are always online, and the ongoing paradigm shift towards ubiquitous computing concepts (Weiser, 1993) provide increasingly complex functionality and allow remote access to additional services and information. Wireless ad-hoc network technologies and the upcoming Internet of Things

DOI: 10.4018/978-1-60566-978-6.ch002

(ITU, 2005) drive the trend to local networks and smart environments. This poses challenges to applications and their user interfaces that now have to support various situations instead of the well known scenario of the user sitting in front of his desk. The widespread use of computers in all areas of life also continuously affects new groups of users. As their number grows, so does their diversity, with each user having different personal preferences, different experience levels and different capabilities.

Smart environments confront user interfaces with a variety of available (mobile) interaction resources supporting diverse modalities, and heterogeneous users with different capabilities and preferences. A user interface supporting smart environments requires a high degree of adaptability to innumerable contexts of use. Unfortunately, today's user interfaces do not sufficiently support the creation of ubiquitous systems and smart environments and a significant improvement of the communication, interaction and adaptation capabilities is required. At the same time the user must be given the power of understanding and controlling her smart environment in a flexible and comprehensible way. We therefore see the need for Ubiquitous User Interfaces (UUIs) addressing the challenges of the ubiquitous computing paradigm within the following dimensions:

- **multi-situation:** support of multiple UI layouts for different usage contexts;
- **multi-device:** support for the usage of multiple devices simultaneously (or sequentially);
- **multi-modal:** support for multiple interaction modalities according to the needs of the interaction;
- **multi-user:** support to share applications, information and interaction devices between multiple users;
- **multi-application:** support to use multiple applications per user and device simultaneously and sequentially.

Based on these five features we define Ubiquitous User Interfaces as *interfaces that are shapeable, distributable, multimodal, shareable and mergeable*. In the remainder of this chapter we focus mainly on the aspects related to multi-situation, multi-device, and multi-modal. Furthermore, we address the alteration of the user interface configuration at runtime, often denoted as adaptation. We thus address:

- The adaptation of the user interface to the used device(s), the user's needs and the environment either directly by the user or automatically by the application.
- The ability of the user interface to be distributed across multiple devices and modalities.
- The capability of the user interface to change the currently used devices by migrating the whole UI or parts of it to a different device.

Ubiquitous User Interfaces thus support the utilization of multiple modalities, devices and interaction concepts to provide robust interaction for different purposes. Furthermore they facilitate flexible interaction, mobile and stationary, within changing contexts and situations. Developing UUIs now poses the challenge to express the increasingly complex interaction concepts and to handle the adaptive distributed multimodal interaction at runtime. Model-based User Interface Development (MBUID) has been identified as a promising approach to handle such increasing complexity. Modeling technologies are utilized to formalize distinct aspects of the user interface and the interaction on different levels of abstraction, ranging from abstract tasks to final user interface elements. Currently, the underlying user interface models are mostly used to generate multiple variants of static, final user interface code, which is then executed at runtime. To also address the increasing complexity of handling user input, approaches utilizing the models at runtime recently receive increased attention. We support the idea of "keeping the models alive at runtime to make the design rationale available at runtime" (Sottet et al., 2006b) by providing a runtime system for their execution. Preserving the models and thus their knowledge at runtime allows reasoning about the designer's decisions, when she is no longer available to generate dynamic system output and reason about the meaning of multimodal user input. The natural next step is then to enable the revision of those decisions either by the user (e.g. by the means of a Meta-User Interface (Coutaz, 2006)) or intelligent adaptation components (as is the case in self-adaptive systems). Thus by altering the user interface models at runtime, the properties of Ubiquitous User Interfaces in the varying contexts of use can be assured, in other words their plasticity (Thevenin & Coutaz, 1999) can be increased.

In this chapter, we describe a runtime system for the handling of multimodal input and output. Based on the concept of executable models, we illustrate how the underlying concepts help to define models that explicitly hold the state of the interaction as well as the meaning underlying the presented user interface. This addresses the distribution across devices unknown at design time, helps solving the fusion of multimodal input and supports adaptation capabilities by adjusting the underlying models. In the reminder of the chapter, the next section describes background information in the area of Model Based User Interface Development (MBUID). Thereafter the Multi-Access Service Platform (MASP) is presented, our model-based system providing the means to create and execute Ubiquitous User Interfaces (section 3). After this overview of the MASP, its underlying (meta-) models are presented, the concept of executable models is described in detail and their meta-metamodel is introduced (section 4). Afterwards, the application of the executable models approach to user interface development is illustrated and applied to the idea of interaction modeling (section 5). The approach also facilitates

the adaptation of the user interface design to the current context of use, which is illustrated in section 6. In section 7, a case study of an application is presented to illustrate the utilization of the interaction model and its execution at runtime to create ubiquitous user interfaces. Finally, future trends are discussed (section 8) and the conclusions (section 9) eventually complete the paper.

BACKGROUND

Model-Based Software Engineering (MBSE) has recently been successfully deployed as an approach to handle the increasing complexity of current software systems. It addresses the need to define reusable building blocks that abstract from the underlying system complexity and allow the developer to focus on domain specific problems rather than to deal with implementation details. The approach is based on the idea that a description of a problem domain can be formalized as a metamodel. Different problems targeting this domain can then be solved by creating models conforming to this metamodel. The Unified Modeling Language (UML) made the idea of modeling popular by providing a common language to exchange concepts between developers. MOF and MDA provide the key concepts for the utilization of model-based software engineering to derive running systems from software models. Frameworks like the Eclipse Modeling Framework (EMF) or the Graphical Modeling Framework (GMF) provide the required infrastructure to create reusable models and metamodels. One of the basic concepts of modeling is the relation between models and their metamodels. While a model describes a specific system under study, a metamodel models all models describing systems of a specific type. The metamodel thus provides the boundaries, in which the developer of the model can act. Similarly, the meta-metamodel describes the commonalities of all metamodels.

With the advent of technologies like UML Actions or the Business Process Modeling Language (BPML) the focus of modeling shifts from static to dynamic systems. While the original static models were mainly able to present snapshot views of the systems under study and could thus only provide answers to "what is" kinds of questions, dynamic models give access to information that changes over time and are thus also able to answer "what has been" or "what if" kinds of questions (see also Breton & Bézivin (2001)). In contrast to common static models, dynamic models provide the logic that defines the dynamic behavior as part of the models, which makes them complete in the sense that they have "everything required to produce a desired functionality of a single problem domain" (Mellor, 2004). They provide the capabilities to express static elements as well as behavior and evolution of the system in one single model. Such dynamic models can be executed at runtime and thus make design decisions traceable and even allow revising them. Design knowledge is directly incorporated into the execution process. Therefore dynamic models can be described as models that provide a complete view of the system under study over time.

In contrast to the field of MBSE, which is already wide spread in industrial software development, Model-Based User Interface Development (MBUID) is currently growing as an approach to deal with the increasing complexity of user interfaces (Vanderdonckt, 2008). MBUID strives for a formalization of the domain of user interfaces by defining reusable user interface description languages (UIDL), in which different aspects of the user interface and the interaction are formalized in distinct metamodels composing the UIDL. The meta-modeling spans all levels of abstraction, ranging from tasks and domain concepts to final user interface elements, as the Cameleon Reference Framework (Calvary et al., 2002) differentiates. The (semi-) automatic generation of multiple user interfaces from one or multiple abstract models (conforming to these metamodels)

provides a maximum amount of common code, complemented by a minimal amount of specific adaptations for each device. Utilizing formal user interface models takes the design process to a computer-processable level, on which design decisions become understandable for automatic systems. Formalizing the user interface development process and the underlying models also ensures the availability of design decisions, which could be revised at any time to directly influence the generated user interfaces.

A good example of model utilization for user interface development is the UsiXML user interface description language (Limbourg et al., 2004). Composed of multiple metamodels, UsiXML provides means for the description of user interfaces with a set of models. Each model deals with a different aspect of the user interface and a mapping model declares their interdependencies and semantic relationships. Covering all levels of abstraction and accompanied by a variety of tools and transformations, UsiXML enables the multi-path development of user interfaces. The designer is thus not bound to a predefined development path and may adapt it corresponding to the requirements of the application. Through extensive utilization of MBUID concepts, UsiXML empowers the designer to apply top-down, bottom-up, middle-out or any other desired approach.

Another example of model utilization for user interface development is TERESA (Mori et al., 2004), a multi-platform UI authoring tool. Following a top-down approach, designers using TERESA start with the definition of a task model, later refined with an abstract user interface (AUI) model and finally concretized in form of a concrete user interface (CUI) model, from which the final user interface code (e.g. XHTML or Java) is generated. During the whole process TERESA maintains the mappings between models at the different levels of abstraction. This way design decisions are easier comprehensible, because the evolution of each user interface element can be traced through all models, back to the most abstract concepts in the task model.

Recently, the focus has been extended to the utilization of user interface models at runtime (Balme et al., 2004; Calvary et al., 2002; Clerckx et al., 2007; Coninx et al., 2003; Demeure et al., 2006; Sottet et al., 2006b). The main argument to do so is the possibility to exploit the information contained in the model for the adaptation of the user interface to multiple and changing contexts-of-use. Design decisions become explicitly visible instead of being concealed in generated code. Additionally, the availability of the underlying concepts allows a sophisticated interpretation of (multimodal) user input and output as the system is aware of the meaning of the interaction. Sottet et al. (2006b) propose keeping the models alive at runtime to make the design rationale available and Demeure et al. (2006) presented the Comets, which are prototypical user interface components capable of adaptations due to the application of models at runtime. Clerckx et al. (2007) extend the DynaMo-AID design process by context data evaluated at runtime, supporting UI migration and distribution. To support the linking of multiple models, model transformations, which should also be available at runtime, have been proposed (Sottet et al., 2007). Multimodal interaction based on user interface models has been realized in ICARE (Bouchet et al., 2004), Cameleon-RT (Balme et al., 2004), the Framework for Adaptive Multimodal Environments (FAME) (Duarte & Carrico, 2006) and the SmartKom system (Reithinger et al., 2003). However, none of the solutions allows identifying the common components of multiple models and the creation of links between the models, which could pave the road to interoperability between different UIDLs. There is not yet a common architecture that has been agreed on, supporting the utilization of models to address runtime issues like fusion, fission and adaptation, as well as the management of the user interface state and the more abstract definition of interaction (instead of more concrete user interfaces).

The basis for the Multi-Access Service Platform, described in this paper, is the runtime utilization of dynamic user interface models that evolve over time and describe the application, its current state and the required interaction as a whole instead of providing a static snapshot only. Runtime issues like the coordination and integration of several communication channels that operate in parallel (fusion), the partition of information sets for the generation of efficient multimodal presentations (fission), and the synchronization of the input and output to provide guidance and feedback as well as adaptation to the context of use can be supported by the utilization of user interface models.

THE MULTI-ACCESS SERVICE PLATFORM

The Multi-Access Service Platform (MASP, masp. dai-labor.de) is a model-based runtime system for the creation of Ubiquitous User Interfaces. Based on a set of dynamic user interface models that are executed at runtime, the MASP provides the means to flexibly adapt user interfaces and interaction semantics to the current needs and the context of use. As proposed by the well-accepted Cameleon Reference Framework (Calvary et al., 2002) the MASP separates four levels of abstraction: Task & Domain Model, Abstract Interaction Model and Concrete Interaction Model as well as the final user interface derived from the modeled information. Each level of abstraction is described by one or several metamodels providing the basis for the user interface models the developer can use to model UUIs. The set of reference metamodels comprises task-, domain-, service-, context- and interaction-metamodel. The developed set picks up findings from other model-based approaches (Cameleon-RT (Balme et al., 2004), DynaMo-AID (Clerckx et al., 2004), ICARE (Bouchet et al., 2004), the Framework for Adaptive Multimodal Environments (FAME) (Duarte & Carrico, 2006),

Tycoon (Martin, 1998), SmartKom (Reithinger et al., 2003)), but puts a strong focus on the issues arising during the runtime interpretation of user interface models. A more detailed view on the set of metamodels and their relations, the MASP utilizes to support multiple devices, interaction concepts and modalities (sequentially and simultaneously), is provided next by describing its architecture.

Figure 1 shows an overview of the MASP architecture comprehending the set of models, their relationships as well as components to provide advanced functionality. While the service model allows interaction with backend systems at runtime, the task and domain model describe the basic concepts underlying the user interface. Based on the defined concepts, the interaction models (Abstract Interaction-, Concrete Input and Concrete Presentation Model) define the actual communication with the user. The Abstract Interaction Model thereby defines abstract interaction elements aiming at a modality and device independent description of interaction. The Concrete Input- and Concrete Presentation Model substantiate the abstract interaction elements by specifying concrete interaction elements targeting specific modalities and devices. The interaction model has a main role to provide support for multimodal interaction. A context model provides interaction-relevant context information. The model holds information about the available interaction resources and allows their incorporation into the interaction process at runtime. Additionally it provides information about users and the environment and comprises context providers, continuously delivering context information at runtime. The model is thus continuously updated at runtime to reflect the current context of use of the interaction. Based on these models a fusion and a distribution model allow the definition of fusion and distribution rules to support multimodal

Figure 1. The MASP Runtime Architecture, comprising interaction channels, fusion, distribution and adaptation components and a comprehensive user interface model

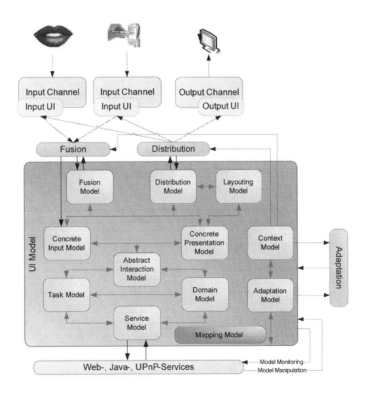

and distributed user interfaces. Finally, a layouting model allows the definition of layout constraints for different usage scenarios.

The different models are connected by mappings as described in (Blumendorf et al., 2008). The mappings are also defined via a model (mapping model) and provide the possibility to interconnect the different models and to ensure synchronization and information exchange between models. By linking the task model to service and interaction model, the execution of the task model triggers the activation of service calls and interaction elements. While service calls activate backend functions, active interaction elements are e.g. displayed on the screen and allow user interaction. They also incorporate domain model elements in their presentation and allow their manipulation through user input as defined by the mappings. The context model finally also influences the presentation of the interaction elements that are related to context information. Thus, the execution of the task model triggers a chain reaction, leading to the creation of a user interface from the defined user interface model. The structure underlying this approach also opens the possibility to add additional models or change existing models in the future.

The whole concept of the MASP is based on a client-server architecture with the MASP server managing multiple interaction resources with different interaction capabilities and modalities. The resources are connected via channels and each user interface delivered to a resource is adapted according to the capabilities of the resource and the preferences of the user. For this purpose, the MASP runtime architecture, also depicted in Figure 1, combines the models and the related mappings with a set of components. While the models reflect the state of the current interaction and thus the UI, the components provide advanced functionality. An adaptation component allows the utilization of an adaptation model to directly adapt the user interface models according to the context-of-use (including device, user

and environment information). Based on this adaptation the same application can e.g. support mobile and static devices by adapting the user interface accordingly. Multimodality is supported by a distribution component, segmenting the user interface across multiple modalities and devices if necessary, and a fusion component, matching and interpreting input from different modalities. Both components are strongly related to the underlying user interface model allowing the configuration of the components. In contrast to other multimodal approaches we do not aim at the semantic analysis of any user input but at the provisioning of a multimodal user interface guiding and restricting the possible interaction. As mentioned above the connection of the interaction resources to the MASP is done via channels. The channels identify the connection to interaction resources and provide related APIs allowing their direct incorporation into the user interface delivery and creation process. The activation of different combinations of channels also allows changing the currently supported modalities during the interaction. A channel-based delivery mechanism is used for the delivery of the created final user interfaces to the interaction devices (Blumendorf et al., 2007). The combination of multiple devices allows combining complementary devices to enhance the interaction and support multiple modalities and interaction styles. A common example for this feature is the utilization of a mobile device as a remote control for a large display. This is possible by distributing an application across the two devices. Additionally voice support can be added via a third device. Figure 1 shows this combination of mobile and static clients as an example on the very top of the image.

The main concept underlying the MASP models and architecture is the comprehensive utilization of models and their metamodels. Figure 2 shows the (meta-) models of the MASP in relation to the MOF Metadata Architecture. M1 thereby comprises the loosely coupled models while M2 provides the metamodels to which the

Figure 2. The MASP in relation to the MOF Meta-Pyramid

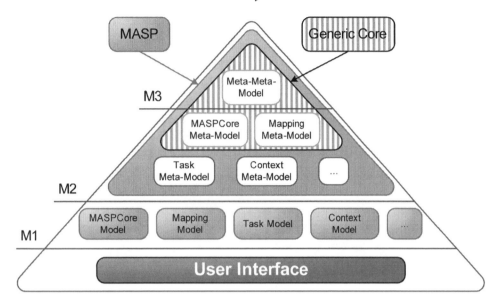

models conform. . This includes the MASP Core metamodel which provides the means to assemble and load applications as sets of models. Executable MASP Core models provide session management capabilities and bootstrapping mechanisms to start-up applications by triggering the execution of their models. Additionally it provides a basic API to access the models, making it easy to build software and management tools for the platform.

The idea underlying the network of models, the MASP is based on, is the utilization of self-consistent executable models that can be used to express the various information required to build user interfaces for smart environments and mobile applications. This allows the expression of interaction logic as well as its interpretation in the same model as illustrated in the next section.

EXECUTABLE MODELS

The executable models approach introduced in this section supports an extensive usage of models at runtime. Executable models run and have similar properties as program code. In contrast to code however, executable models provide the advantages of models like a domain-specific level of abstraction which greatly simplifies the communication with the user or customer. Besides the initial state of a system and the processing logic, dynamic models also make the model elements that change over time explicit, support the investigation of the state of the execution at any point in time and can provide well defined operations to change their structure at runtime. The idea of executable models gives the model an observable and manipulable state. Keeping design models during runtime makes design decisions traceable and even allows revising them. It incorporates design knowledge directly into the execution process. Besides the initial state of a system and the processing logic, dynamic executable models also make the model elements that change over time explicit, support the investigation of the state of the execution at any point in time and can provide well defined operations to change their structure at runtime. We can thus describe dynamic executable models as models that provide a complete view of the system under study over time.

Keeping the model(s) at runtime allows postponing design decisions to runtime and thus performing adaptations to the runtime circumstances rather than predicting all possible context situations at design time. Executable models, their meta-metamodel and the mapping metamodel allow connecting different models and concepts to build advanced user interfaces.

The Meta-Metamodel of Executable Models

Defining the shared elements of executable models (Blumendorf et al., 2008), the meta-metamodel combines the initial state of the system, the dynamic model elements that change over time and the processing logic in one model. This leads to the need to clearly distinguish definition-, situation- and execution elements. A similar classification has also been identified by Breton and Bézivin (2001). Additional construction elements allow the adaptation and alteration of the models at runtime. Figure 3 shows a conceptual sketch of the model.

Definition Elements define the static structure of the model and thus denote the constant elements that do not change over time. Definition elements are defined by the designer and represent the constants of the model, invariant over time.

Situation Elements define the current state of the model and thus identify those elements that do change over time. Situation elements are changed by the processing logic of the application when making a transition from one state to another one. Any change to a situation element can trigger an execution element.

Execution Elements define the interpretation process of the model, in other words the transitions from one state to another. In this sense execution elements are procedures or actions altering the situation elements of a model. Execution elements also provide the entry points for data exchange with entities outside of the model. Defining execution elements as part of the model allows the incorporation of semantic information and the interpretation process as part of the model itself and thus ensures consistency and an unambiguous interpretation. This approach makes an executable model complete and self-contained.

Construction Elements define manipulation processes altering the structure of the model and thus its definition elements. They do not define internal state transitions, but changes to the model itself. Construction elements obey to the meta-

Figure 3. Conceptual illustration of the meta-metamodel of executable models

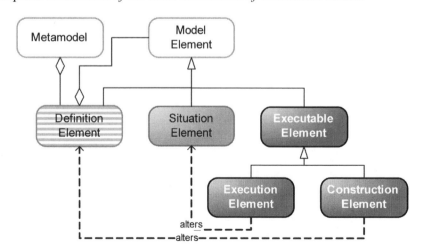

model and ensure that any performed alteration is legal with respect to this metamodel. Based on the construction elements each metamodel can be equipped with elements providing the ability to alter the definition elements of a model in a well defined manner, even at runtime.

Distinguishing these elements leads to the meta-metamodel of dynamic executable models depicted in Figure 3. The meta-metamodel provides a formal view of executable models and summarizes the common concepts the models are based on. It is positioned at M3 layer in the MOF Metadata Architecture (see also *Meta Object Facility (MOF) Specification — Version 1.4* of the Object Management Group). The clear separation of the elements provides clear boundaries for the designer, only working with the definition elements and the system architect, providing the metamodels. A definition element as the basic element finally aggregates situation-, execution- and creation elements that describe and change situation and definition elements. The difference between the execution and construction elements is essential. While the former only change the runtime state of the model without interfering with the definition elements defined by the user interface designer, the latter have the power to modify the structure of the model. Using such models in a prescriptive way (constructive rather than descriptive modeling) allows defining systems that evolve over time, reason about the past and predict future behavior. Thus, dynamic models are often used to build self-adaptive applications, as for example Rohr et al. (2006) describe. In this context, the role of the models is often that of monitoring the system.

As for complex systems (like in our case) it is very likely that the system is described by multiple models i.e. multiple levels of abstraction or different aspects of the system under study, executable models can be linked by mappings representing the relationship between elements of different models. Based on the concepts of the meta-metamodel, the mapping metamodel enables the definition of mapping logic on the metamodel level. The next section describes the mapping metamodel, which we utilize in the MASP for the expression of relations between executable models conforming to different metamodels. It shows, how the mappings allow transmitting information between models at runtime without destroying their consistency.

The Mapping Metamodel

The mapping model connects multiple executable models and allows defining relations between their elements based on the structures given by the meta-metamodel. The mappings defined in this model are the glue between the models of a multi-model architecture. The mapping metamodel is thereby located at the M2 layer of the MOF architecture. Providing an extra metamodel solely for mappings also enables to benefit from tool support and removes the problem of mappings hard-coded into the architecture, as has been already advised by Puerta and Eisenstein (1999). A mapping metamodel allows the definition of the common nature of the mappings and helps ensuring extensibility and flexibility. A mapping relates models by relating elements of the models whereas the models are not aware of their relation. An example of a mapping metamodel, consisting of a fixed set of predefined mapping types only, can also be found in UsiXML described by Limbourg (2004). Sottet et al. (2006a) have defined a mapping metamodel, which can also be used to describe transformations between model elements at runtime. However, in contrast to their approach the mapping model presented here puts a stronger focus on the specific situation at runtime and the information exchange between dynamic models. Especially interesting at runtime is the fact, that the relations can be utilized to keep models synchronized and to transport information between two or more models. The information provided by the mappings can be used to synchronize elements if the state of the source elements changes. Mellor et al. (2004) also see the main features of mappings as construction (when the target model is created from the source model) and synchronization (when data from the source model is propagated into the existing target model). Our mapping model contains mappings of the latter kind. Focusing on runtime aspects, we see a mapping as a possibility to alter an existing target model, based on changes that happen to the related source model. In contrast to the most common understanding of mappings the mappings we utilize do not transform a model into another one. Instead, they synchronize runtime data between coexisting models.

The conceptual mapping metamodel is shown in Figure 4 and combines mapping types and mappings. Mapping types are the main elements of the mapping metamodel, as they provide predefined types of mappings that can be used to define the actual mappings between elements on the M1 layer (that are elements of the meta-metamodel). A mapping type thereby consists of two definition elements as well as of well-defined links between the two. The definition elements are the source and the target of the mapping and the mapping synchronizes the runtime data defined by the links between these two elements. The links consist of a situation element, an execution element and a transformation. The situation element is the trigger

of a link. Whenever a situation element in a model changes, the link is triggered and the referenced execution logic is executed to synchronize the two definition elements of the mapping. The execution logic is thus the logical target of the link. The optional transformation associated with the link describes how the situation data, which activated the trigger, is transformed into (input) data needed by the target execution element in the other model. This transformation might be required, especially when models with distinct data types and structures are linked by mappings. To simplify the usage of the model, the metamodel supports multiple links in one mapping type, as multiple situation elements (e.g. related to the same definition element) might be relevant to trigger the execution. Supporting more than one link also allows a back linking, as some mapping types might also demand two-way links.

From the designer's point of view, the initial mapping model provides a set of available mapping types with predefined logic, defined on the metamodel level. Thus to relate two models, the user interface designer extends this initial model by creating new mappings that reference one of the available mapping types. To create such a mapping, the designer has to provide the specific source and target model elements to the mapping and define

Figure 4. Mapping metamodel

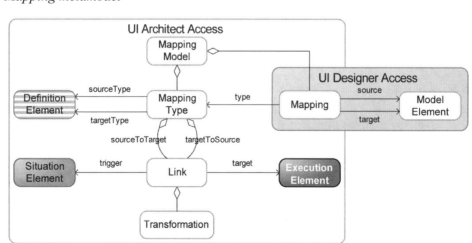

its type. This leads to a relation between the two elements and their synchronization according to the given execution logic (links).

Using our meta-metamodel we were able to define the mapping metamodel independent from the metamodels that mappings can be created between. Only the mapping models contain mapping types, which are not of generic nature, but specifically designed for the given metamodels.

INTERACTION MODELING

To illustrate the utilization of the (meta-)metamodels and the MASP runtime models, this section provides a deeper description of the MASP interaction models and their utilization. The abstract and concrete interaction metamodels presented in this section aim at the device and modality independent definition of the anticipated interaction in an abstract form and the specialization of the described interaction in a more concrete model. Figure 5 shows the basic structure of the approach (whereas the relation to other used models can be found in Figure 2). The interaction model is split into an abstract and a concrete part, which are working in conjunction with each other at runtime. While the abstract model allows the modality independent definition of the anticipated interaction, the concrete model provides a definition in terms of input styles and user interface elements. The concrete interaction separates input and output elements in order to support combinations of

multiple interaction resources. The combination of the two models describes the interaction from a system point of view, forming the mental model of the system which allows to create multiple and distributed presentations and to derive meaning from the perceived interaction.

Abstract Interaction Metamodel

The abstract interaction model (Figure 6) describes the anticipated interaction in an abstract, device and modality independent way. Its purpose is the provisioning of a common abstraction layer that all different modalities and interaction technologies adhere to. The following five AbstractInteractors provide basic interaction elements similar to UsiXML (Limbourg et al., 2004) and TERESA (Mori et al., 2004). Modality independence is thereby provided, as the elements do not refer to any modality-specific elements (buttons, voice prompts, gestures, etc.).

- *OutputOnly*, when the computer presents the user information without requiring feedback or input (e.g. images or text).
- *FreeInput*, when the user inputs (unstructured) data into the system (e.g. providing a name or a password).
- *Command*, is used when the user sends a signal to the system ordering it to perform an action (e.g. OK and Cancel buttons).
- *Choice*, provides a list to the user and denotes the possibility to choose one or

Figure 5. Basic structure of the MASP UI models, separating task, abstract and concrete level. Information exchange through mappings and interaction events are denoted by arrows.

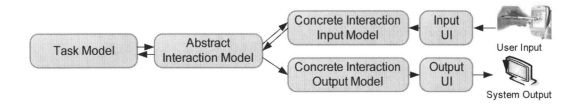

Figure 6. The Abstract Interaction Model, defining modality and device independent interaction

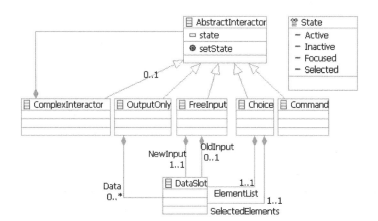

several options from this list.
- *ComplexInteractor*, allows the aggregation of multiple abstract interaction objects into a more complex type. This provides e.g. logical grouping facilities.

The additional DataSlot element provides the connection to domain data, serving as reference element for application dependent data at runtime.

From the system perspective, the interactors can be identified as basic interaction elements, covering a broad range of interaction like menus, buttons, lists of any form, free text input, text and image output, etc. Utilizing this model allows the refining of tasks in terms of interaction means. It can be used to express basic interaction semantics, independently of the used modality or device. At runtime the state (situation element) of each interactor is synchronized with the state of the related task, activating and deactivating the interactor analogous to the task. Similarly, they provide the basis for the aggregation of multiple concrete objects that convey the abstract meaning via different devices or modalities. In order to create user interfaces incorporating the formalized abstract interaction, intermediary concrete interaction models are introduced. These models

refine the interaction with respect to device and modality specifics. They allow the relation of the abstract interactions to more concrete input and output representations. In contrast to the abstract level, the concrete level separates user input and system output.

Concrete Input Metamodel

The concrete input model defines possible inputs related to the abstract interaction objects, independently from the presentation. This allows the distribution of the interactors to different devices and the free combination of multiple interaction resources to support different modalities and interaction styles. The utilized concrete interactors are designed to conceptually match the supported interaction channels and simplify the connection of new interaction resources by keeping the implementation efforts as low as possible.

Figure 7 shows the classification of *ConcreteInputInteractors*, which aim to support as many different interaction resources as possible. The concrete interactors therefore still abstract from the specific properties of the interaction resources but distinguish four input modalities: *GestureInput*, *NaturalLanguageInput*, *CharacterInput* and *PointingInput*. Additionally *Com-*

plexInput allows the combination of any of the simple input elements and adds temporal or spacial constraints to relate the aggregated elements. This allows the creation of input interactors defining a combination of multiple input elements that can be assigned to different modalities and thus complementary span multiple modalities. The main advantage of this open and device independent definition of the input capabilities for each interaction is the maximized support for various types of devices and modalities. While for system output, the application developer decides which widgets to use for which modalities, for input this decision is made by the user. This provides support for various input capabilities that can be used in conjunction (parallel or sequentially) with a minimum effort for the application developer.

Concrete Presentation Metamodel

The concrete presentation model summarizes all interactors required to present the interaction defined via the abstract interaction objects to the user. An abstract interactor can thus be mapped to input capabilities (InputInteractors) as well as to a perceivable representation (OutputInteractors). The output model addresses the different presentation needs of an interactor and allows to compose different presentation variants, allowing

the representation in specific modalities as defined by the interface developer.

Figure 8 shows the main modalities we consider in the metamodel, which are graphical and voice output. Other more limited modalities like sounds or haptic feedback (vibration) can also be considered using signal output. Each of the interactors can be associated with an *AbstractInteractor* and its functionalities. *ComplexOutput* allows the combination of any of the simple output elements and adds temporal or spacial constraints to relate the aggregated elements. This allows the creation of output interactors structuring multiple elements to a complex construct. Complex elements can also complementary span multiple modalities in which case they are distributed across multiple interaction resources. The defined interaction elements and their interconnection allow the creation of user interface descriptions that provide rendering possibilities for different modalities.

Runtime Interpretation of Interaction Models

The described interaction metamodels can be utilized by the UI developer to develop user interface models, executable at runtime. Defined mappings link the models together. Based on a task model, the abstract part of the interaction model provides

Figure 7. The Concrete Input Model, describing specific input possibilities

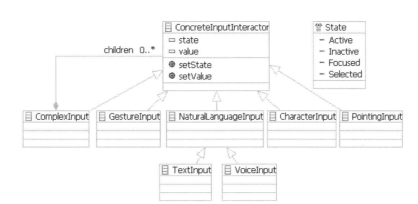

Figure 8. The Concrete Presentation Model, describing concrete output possibilities

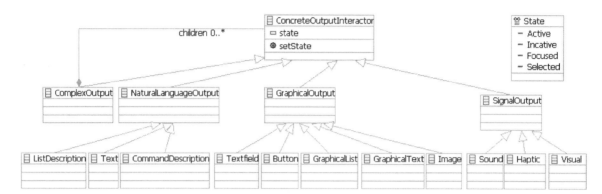

a description of the current interaction at runtime. Via the mappings between the two models, the state of the task model at runtime defines the currently active abstract interactors. Mappings between the abstract model and the concrete models additionally define which concrete elements are currently available for user interaction. A set of these elements is then selected by the distribution component, to render a (possibly distributed) user interface that is delivered to the connected devices via the channels. Input from the user is processed in the same way, by mapping concrete input elements to abstract interaction elements and further to task and domain elements. Mappings from the concrete input element can target the data slots or the state of the abstract interaction element. Data slots are in turn mapped to elements of the underlying domain model. Changing the state of an abstract interaction element can be mapped to task events (e.g. task done) leading to the recalculation of the currently possible interaction. To realize this behavior, four main states (situations) of any model element and two basic actions (execution elements) are currently defined.

Situations:

- *Inactive:* when the interactor is not activated, it cannot be interacted with and is not presented to the user.

- *Active:* when the interactor is presented to the user and perceivable/manipulable for her.

- *Focused:* an interactor reaches this state when it is focused by the user (e.g. by moving the mouse pointer over the element).

- *Selected:* a focused interactor becomes *Selected* when the user explicitly interacts with it, for example by pressing a button. The state *Selected* is only momentary, just as the user's interaction is. Therefore, after being selected the interactor instantly returns into the state *Focused*.

Execution Elements:

- *setState*, allows to set the state of the interaction element. While this is mainly used to synchronize the states of abstract and concrete interaction objects, setting the state also occurs to signal interaction. Selecting an element e.g. sets the state attribute to *"selected"*. This also holds for *"focus"* and *"unfocus"*, which is also signaled through the state.

- *setValue*, sets the value produced by the current input as each input interactor has a value situation element. Based on the produced value, different mappings to execution elements of the other models can be triggered.

To address the synchronization between the different interaction models, three types of mappings between the elements are distinguished:

- *IsStateSynchronized,* defines that two elements share a common state. This means that if one element becomes active, the other does as well. Thus activating an abstract command for example would also activate the related button and thus trigger the display of the button and the provisioning of an area to point at. The relation allows the coupling of the states of two elements either directly or with a translation function redefining the dependency of the states. The translation would thus e.g. trigger the selection of an abstract element if a concrete element is focused. The mapping can also be limited to consider certain states only. This last variant e.g. allows assigning one gesture for focusing an element and another gesture for unfocusing the element.
- *IsValueSynchronized,* addresses the need to transport additional information between the different elements and across different levels of abstraction. It allows to communicate user input to higher levels of abstraction and to configure concrete interaction objects according to information from other objects. For the communication of input each concrete input element provides a value attribute (situation element) that can be changed during the interaction. Providing a mapping based on this value allows the transportation to the abstract level. This mechanism is e.g. important to support free text input provided by the user. Receiving input from the user alters the value attribute and thus triggers a mapping to the input execution element of the

FreeformInput, resulting in an update of the related DataSlot value.

- *IsDynamicallyValueSynchronized,* addresses the creation of elements based on dynamic data. This is crucial to deal with input capabilities that depend on data unknown at design time like a list populated with dynamically generated elements where each should be directly addressable via speech at runtime. The mapping allows the creation of a relation between dynamic data and the concrete visualization or input options related to this data. The transformation method can be utilized to incorporate the dynamic data into the input construct of a given type. For output elements the mapping similarly allows the creation of output data depending on dynamic runtime information. The mapping is established between the elements of a list and a presentation object and allows the manipulation of the presentation according to the dynamic data.

Using these mappings, each abstract element is connected to one or more concrete elements to allow the modality specific definition of the anticipated interaction. The concrete elements become active whenever the state of the abstract element changes to active, which can trigger their delivery to an interaction resource. Besides the activation of concrete elements, abstract elements can also trigger the alteration of the presentation of an element. Focusing the graphical representation of a command i.e. a button, triggers a state change to *"Focused"* of the abstract interactor, which in turn changes the state of related concrete interactors to *"Focused"*, resulting in a highlighted button. Input events, triggered by user interaction work in a similar way. The user focusing or clicking a button triggers an event communicated by the channel to the concrete interaction element. This in turn triggers *"Focused"* or *"Selected"* events communicated to the abstract interaction model, causing a state change to the abstract interaction element. This is again propagated to all related elements on the task and domain level and can also be propagated back to the concrete level. This multi-level event propagation (Blumendorf et al., 2006) allows the incorporation of the information from multiple models to interpret a received input event. Additionally to the relation between the abstract and concrete level and the multi-level event propagation, there can also be direct interrelations between the concrete input and concrete presentation model on the same level. This is utilized to express modality specific input and output elements that are not required for the interaction on the abstract level. This is e.g. relevant to scroll through a list. While the abstract interaction defines to select an element from a given set of elements, a representation on a small screen might require scrolling through the list, which would then have to be added on the concrete level (e.g. in form of a scrollbar and two buttons). Expressing multimodal user interfaces using the described metamodels, requires fusion of input received from different modalities. This is widely supported by the abstract model, shared by all modalities as described in the next section.

Multimodal Fusion

Often multimodal systems aim at semantically understanding as much of a user input as possible to allow a natural dialog with the system. The MASP addresses the support of multimodal interaction with a focus on the enhancement of interaction to provide more robust and comfortable user interfaces. It does not aim at the semantic interpretation of any input, but at providing well defined multimodal interaction capabilities, constraining the interaction possibilities and guiding the user within these constrained interfaces by a system driven interaction process. Restraining the interaction is thereby similar to what humans do during conversations – limiting the area of discourse.

To realize this approach the accessibility of the interaction means in form of the interaction model at runtime is an important aspect. While it allows the derivation of multiple presentations on the one hand, it is also used to define the currently possible inputs for each interaction state. Being aware of these input possibilities and their means allows the system to categorize and classify any received input on multiple levels (according to the defined models) and thus to interpret it and derive the semantic meaning. The underlying fusion process receives an input event from any interaction resource and initially matches it to the current input user interface for this resource. This allows the immediate elimination of input not in the current area of discourse (e.g. uttering "show my balance" is ignored in a dialog that only supports "ok" or "cancel"). However, if the utterance is valid, the first processing step interprets it in relation to the concrete input model and triggers a state change of that model (e.g. uttering "ok" sets the state of the OK-button to *"Selected"*). More complex interaction, combining multiple modalities or providing ambiguous input require a more complex processing. In summary, three types of interactions can be distinguished.

- a final self-contained input
- an open input that requires additional parameters, possibly from a different modality
- a ambiguous input, leading to a list of possible interpretations

Considering these input types the fusion engine is required to combine multiple input events to resolve references and ambiguities as well as to improve robustness by handling possible recognition errors, unknown events and n-best lists of results. To do so, the fusion engine processes inputs based on the configuration, stored in the fusion model and the concrete interaction model. The result of the distribution process is also incorporated through the preprocessing by the input user interface and incorporates the relation of interaction objects and modalities/interaction resources into the process. Successful fusion produces calls to execution elements of the concrete input model. In detail the following cases can be distinguished in the fusion process.

- Final self-contained input events that can be directly mapped to an execution element of a concrete input object are directly forwarded to the related execution element.
- Open input events that require additional parameters to be processed are related to a complex input interaction object. Based on this object, the fusion engine determines the types of input that are missing to complete the complex object. Thus it stores the received open object in the fusion model and waits for additional input completing the open event. After a given timeout or the reception of another distinct open event the processing is interrupted and the object is removed from the fusion model. Events to complete an open interaction can either be another open interaction in which case the complex input object would contain another complex object, a self-contained event that can be mapped to a sub-element of the complex element or an ambiguous event that matches the open slot of the complex object. Open input events e.g. receiving a voice command with deictic references that is resolved by a pointing event. Instead of waiting for additional events, the fusion component can also directly query a channel for the latest event, which can e.g. be used to receive the latest pointer position when resolving a deictic reference. In the same way the interaction history can be queried, to resolve references to earlier utterances if necessary.
- Ambiguous events that are produced whenever an input user interface allows to find two internal representations for a single input event. In this case both interpretations are provided to the fusion engine, which has to resolve the ambiguity based on context information, recent interactions from the interaction history or additional events it received like a waiting open event.

In summary the described fusion process allows the processing of multimodal events, predetermined by a designer. It can be compared with a frame-based slot-filling process but can be extended via more complex algorithms. Besides the creation of multimodal output and the processing of multimodal input, the adaptation of the user interface and thus the adaptation of the interaction capabilities to the context-of-use is an important aspect of the Multi-Access Service Platform.

USER INTERFACE ADAPTATION

Similar to the possibilities to model interaction using executable models, it also provides the possibility to model adaptation means by providing an adaptation model. The adaptation model defines manipulations to the provided user interface models and is thus able to alter the design of the user interface. Creation elements of the meta-metamodel provide the means to adapt any kind of model (obeying to an executable metamodel). The adaptation model therefore defines context situations as triggers and references the elements to adapt as well as the adaptation rules in form of construction elements. The basis for an adaptation to the current context-of-use is provided by the context model, sensing context information and making it accessible for the application. Additionally the developer has to specify relevant context information or context situations that are then matched against the available context information at runtime. Based on this context information an adaptation is selected and executed as defined by the adaptation model. Model alteration thereby takes place via construction elements defined for each of the different types of models. The utilization of the adaptation model, the construction elements and the context information is described in the following. The adaptation metamodel is depicted in Figure 9.

As Figure 9 shows, the adaptation model (represented by the AdaptationModel element) is composed of ContextSituations and Adaptations. Similar to Sottet et al. (2007) we have defined the adaptation metamodel according to the Event-Condition-Action principle, and thus the ContextSituations act as triggers for the Adaptations. An Adaptation consists of multiple steps (represented by AdaptationStep elements) applied to model elements referenced in form of XPath-like queries (modeled as the rootQueries). The elements, of which an AdaptationStep is composed, are described in detail in the following.

- The target query represents the left hand side of the adaptation step by identifying the nodes the adaptation has to be applied to. The query allows referencing any element or set of elements in a model.
- The executableElement provides the right hand side of the adaptation step by referencing execution or construction elements of the metamodel, to which the adapted model conforms. When the adaptation step is executed the specified executable element is invoked upon the model element referenced by the target query. The utilization of executable elements for the modification of models ensures a metamodel conformant adaptation.
- The executable elements referenced by adaptation steps can be parameterized with different variables. A Variable can be a Query to some model elements, a PrimitiveVariable holding fixed values (for example a predefined integer) or a Result of the execution of an AdaptationStep. During execution the variables are evaluated and

their values are passed to the executable elements.
- If the executable element of an adaptation step returns a value, it can be stored in the *result* variable and reused in the following adaptation steps.

In summary an adaptation of a model (or multiple models) is defined by the node(s) to apply the transformations to and a description of the alteration of these nodes. The alteration of the nodes is defined in form of construction elements that are provided by the metamodels itself and are specific for each given metamodel.

Utilizing executable elements defined by the metamodel of the target model ensures compliance to the metamodel, as any modification applied is well defined and by definition conforms to the metamodel. It is thus the responsibility of the developer of the metamodel to ensure that no operations performed by an execution or construction element leads to an invalid model. This becomes even more important from the runtime perspective, where models are adapted while being executed. The utilization of creation elements guarantees that the model remains consistent and executable after being adapted. In combination with our executable models the execution of the adaptations at runtime also allows building adaptations that take the situation elements of the model into account. This allows to flexibly react to the current state of the application to improve the quality of the adaptation, especially regarding the user experience.

The defined adaptation model additionally allows the creation of reversible adaptations by the means of undoSteps. This allows applying adaptations based on a recognized situation and applying counter adaptations if a situation is not active any more. As a result, it is possible to define adaptations, which ensure that the original state of the application can be reached again.

Figure 9. Adaptation metamodel

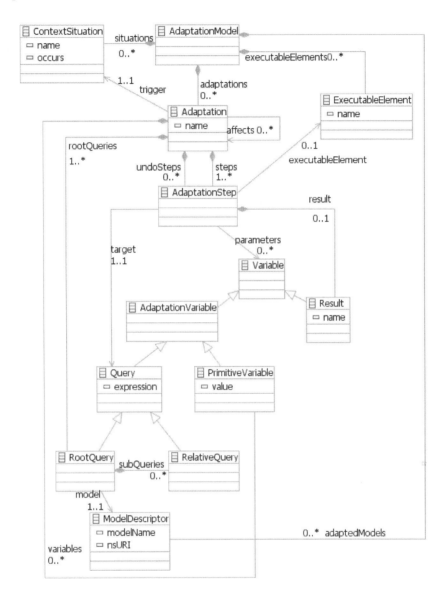

One remaining issue are the interdependencies between adaptations. It is possible that an adaptation *a* produces model elements, which match with the queries of a second adaptation *b*. If *b* has already been performed (because its triggering situation occurred) before *a*, it should be reapplied on the new elements produced by *a*. We have thus introduced the *affects* relation between *Adaptation* elements, denoting adaptations, which have to be reapplied after the application of an adaptation. We could thus express that *a affects b*, meaning that after adaptation *a* has been performed, the context situation of *b* should be checked. If it is valid, *b* must be reapplied.

In the next section we illustrate, how we utilized the described concepts to implement a multimodal cooking assistant service.

APPLICATIONS

Utilizing the MASP and its executable models, several case studies have been conducted to evaluate the feasibility of the approach. In the following a multimodal cooking assistant (CA) is described, serving as an example to illustrate the approach and the capabilities of the MASP.

The cooking assistant has been developed as a MASP application to demonstrate Ubiquitous User Interfaces in smart environments. The CA was therefore installed and tested in a four room apartment, a smart home testbed at the DAI-Labor of the Technische Universität Berlin. Using a broad range of both mobile and stationary interaction devices, the CA aims at supporting the user while preparing a meal. The interaction with the CA has been split into four main interaction steps. First the user selects a recipe, either recommended by the CA or the result of a recipe database search. Afterwards, the ingredients required for cooking are listed. Based on their availability at home, a shopping list is created. The list can be distributed to mobile devices to have it still available while outside. Thus, while obtaining the missing ingredients in the supermarket, purchased ingredients can be checked off the list step by step. Finally, back at home, the CA guides the cooking process with step by step instructions. It also supports the user by controlling and programming the available appliances in the kitchen.

Like for any other application developed with the current set of reference models, the central model of the CA is the task model, defining the underlying workflow of the application. Based on the task model, all needed objects have been modeled in the domain model. For application tasks, service calls to the backend (e.g. to retrieve the list of recipes or to control kitchen appliances) have been defined in the service model and the relationships between tasks and calls have been defined via mappings held in the mapping model. Thus as soon as an application task becomes active, the corresponding mapping is triggered, which in turn leads to the execution of the related service call. Furthermore, the domain objects serving as input and output for the service calls are related to these via mappings. In a similar way, for each interaction task, interactors have been defined in the interaction model (on the abstract and the concrete level). Interactors in the interaction model thus become activated via mappings as soon as an interaction task becomes enabled in the task model. The activation of an interactor triggers the delivery of its representation to a (sub-) set of the available interaction resources, as computed by the distribution component. The information about the available interaction resources is gathered by the MASP at runtime and stored in the context model. Interaction resources can either be automatically discovered by the MASP (via UPnP) or can be registered manually by users. Typically, interaction resources are not registered individually, but as groups composing interaction devices, e.g. a smartphone or a PC. The MASP establishes one interaction channel to each interaction resource providing the basis to utilize the interaction resource for interaction. On the interaction device, the channels are handled by a client that connects to the MASP. This is either a web browser, a SIP client (for voice, PSTN or GSM can be realized via a SIP Gateway) or pre-installed software. The channels to the interaction resources are then combined on the server to support multimodal interaction (e.g. via a mobile web browser and GSM telephony). Taking advantage of the capabilities of mobile devices, voice support could also be realized via a channel utilizing voice recognition and synthesis software running on the mobile device.

The CA supports multimodal interaction in any interaction step. Based on the modality independent task model and the related abstract definition of the interaction in the interaction model, the modality dependent parts are defined through concrete interaction elements in the interaction model. At runtime, the set of active tasks determines the set of active abstract interactors and thus the description of the currently possible interaction.

Via the related set of concrete interactors for the different modalities and further information from the context model (like the available interaction resources and their capabilities), the distribution component is able to calculate, which interactors are sent, to which interaction resources. Thereby, the distribution component tries to include a broad set of interaction possibilities to give the user the freedom of interacting the way she wants. The layouting model provides constraints and a layout configuration for the selected set of interaction resources. Input from the user through any interaction resource is sent through the interaction channels to the MASP on the server. Based on the targeted concrete interactor and its relationship to the abstract interactor, the effect of the input is calculated. If the event propagation mechanisms lead to the termination of a task, the new set of active tasks is calculated and the presentation on the interaction resources as well as the possible input is updated through the described mapping chain. At this moment, the interaction model of the CA supports haptic and voice input accompanied by graphic and speech output. Thus the application can be utilized with any interaction device or combination of devices supporting one of the input and one of the output modalities. The server-side synchronization allows supporting any combination of interaction resources (mobile or fixed) and utilizing them in any dynamic configuration to best match the current context of use. For example, the interaction with the CA could be started on a mobile device to search for a recipe using only the screen and haptic input through a browser. With the capability of the MASP to dynamically register and utilize interaction resources at any time during the interaction, voice support could be added by establishing further communication channels from the mobile device to the MASP.

After the user has choosen a recipe and walked into the kitchen to look for the necessary ingredients, she can migrate the CA from the mobile screen to a stationary display in the kitchen. Furthermore, the CA can utilize the sound system available in the kitchen when communicating with the user via voice. While the application can migrate completely from mobile to stationary devices and back again at any time, the system also supports the distribution of part of the user interface to different devices. In the second interaction step of the CA the shopping list of the CA can be distributed to a mobile device. This way the user may continue interacting with the application, while shopping. Figure 10 shows the distribution of the list in the right image. While the list itself migrates to the mobile device, buttons to get the list back onto the screen and to continue with the cooking process remain on the main screen as they are not needed on the mobile device. It can be seen that the layouting configuration allows the adaptation of the remaining buttons for maximal screen space usage. The distribution is in this case activated by the user via a button of the application. The triggered service call of the CA application then utilizes an API to communicate with the MASP distribution component. Through this API the inclusion of an available mobile device into the interaction can be triggered. Thereon, the context model is checked for registered mobile devices of the current user and, if applicable, the distribution model is modified to include the available mobile device into the interaction. This results in removing the corresponding user interface elements from the screen and sending them to the mobile device via the established interaction channel. As the distribution is changed, the layouting configuration is also recalculated and updated for both interaction resources. Besides moving the user interface elements to the device, they can also be copied to the device, allowing the simultaneous interaction with the touch screen and the mobile device. In the case of the shopping list, one person can check the items bought, while someone at home can still add new ingredients to the list. The presentation on both devices is updated as the status of the application is always kept on the server side – in form of stateful, executable models (mainly the task model and the domain model).

The adaptation of the CA application is supported in various ways. The screen size and the interaction capabilities of the used interaction resources are incorporated to calculate the layout information (Feuerstack et al., 2008). Additionally, information about the interaction device (e.g. whether it is mobile or stationary), the environment (e.g. distinction between public or private context), the user and the content of the application is considered to calculate the distribution of interaction elements among the available interaction resources. The incorporation of such additional context information can be demonstrated with the example of the shopping list. The MASP is capable of adapting the presentation of the shopping list in relation to the distance between the user and the screen. Figure 11 shows three different versions of the list, dynamically generated depending on the current context of use and the state of the application. It must be noted that the presented adaptation not only considers the distance to the user, but also prioritizes user interface elements holding the most important information at a given moment. While for the presented application this adaptation is used on a fixed device in the kitchen, where an Ubisense localization system provides positioning information, it can also be used for interaction on a mobile device, assuming the

system can detect the distance to the user (e.g. through face detection).

The same adaptation mechanisms can also be used to adapt to user preferences (e.g. bad eyes) or interaction context (e.g. rotation of the device). Technically, the adaptation model defines different sizes for the presentation elements in this example. The size of the main output element, the list, is increased depending on the users distance, so that the user can read the required ingredients even if she is further away from the display. Such adaptations can be defined on an abstract level as the information in the models are available at runtime and can thus be utilized transparently at design time.

FUTURE TRENDS

The model-based runtime system and the underlying notion of executable models, we described in this chapter, form the basis to handle the increased complexity of user interface design. While software engineering continuously develops methodologies, tools and best practices, user interface design is still a highly creative process and yet not suitably backed up by engineering practices. Human information processing and

Figure 10. Distribution of the shopping list to a mobile device

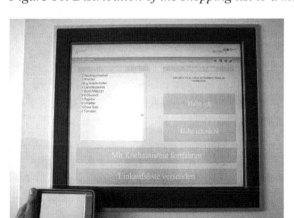
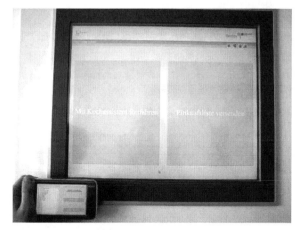

Figure 11. Adaptation of the shopping list, according to the distance of the user to the screen (ltr: close, medium, far)

cognitive processes have to be better understood to fully take them into account in the big picture of user interface design. User interfaces, currently highly artistic and sometimes fun to use, need a shift towards the ability to cope with changing situations and diverse users (while still keeping its appeal).

Smart environments raise the need of highly adaptive user interfaces, supporting multiple situations, devices, modalities, users and application. With current state of the art technologies, the complexity of such user interfaces is almost impossible to handle. Existing languages, tools and methodologies are not expressive enough and usually target specialized contexts instead of the broad adaptive applications. Further research is needed in engineering factors like expressive and standardized languages, runtime adaptation, self-aware and self-adaptive systems, multimodal input processing, the understanding of the meaning of interaction and the handling of the required world knowledge as well as human factors as the usage and combination of (multiple) modalities, natural and robust interaction or the collaborative usage of applications and resources.

From the perspective of mobile computing, future trends include the seamless integration of mobile and personal devices with their surrounding environment. Thus mobile devices and especially their applications are required to make use of varying available (wireless) connected resources and to expose information and services to their human users. Additionally, to more dynamic service combination and utilization, new interaction concepts are needed that on the one hand make these increasingly complex and distributed systems accessible and usable for human users and on the other hand better exploit the broad spectrum of human communication capabilities. While current interaction resources like keyboard, mouse and even touch screen are well adapted to the needs of interactive systems, they do not reflect human communication means like speech, sounds, gestures or facial and body expressions.

Moving the computer from a business machine dedicated to specific tasks in a well defined environment to a universal problem solver in all areas of live continues to pose great challenges to technicians and scientists. Making computers more powerful does not automatically makes them smarter. Making machines ubiquitously available does not automatically makes them usable and enjoyable. The main goal for future developments has to be to prevent systems from being a hassle to users and to build systems providing supportive and enjoyable interaction for users.

CONCLUSION

This chapter provided an overview the Multi-Access Service Platform and its utilization to create future Ubiquitous User Interfaces. The underlying concepts of executable models allow the creation of metamodels to define user interface characteristics as well as their evolution over time. Combining definition-, situation-, execution- and creation elements in the meta-metamodel provides the means to make all relevant information explicitly accessible and changeable and also helps separating the parts of the models relevant for the UI designer. In combination with the mapping model, this approach allows an easy integration of multiple models at runtime to build complex systems. The loose coupling of models also provides a very flexible structure that can easily be extended and adapted to different needs. This also addresses the problem that there are currently no standard or widely accepted UI models. Combined with development and debugging tools this approach allows to inspect and analyze the behavior of the interactive system on various levels of detail. Based on the developed metamodels the MASP runtime system executes UI models to support multimodal interaction, fusion, fission and adaptation. Functionalities are configured and influenced by the models, but realized as components working externally on the defined model. User interface modeling also uses task, domain, service and interaction models and mappings between these models at runtime to interpret the modeled information and derive a user interface. The approach has been successfully deployed to create mobile and stationary user interfaces for smart home environments in the Service Centric Home project (www.sercho.de), but can also be utilized in other scenarios.

While the approach is usable already, some work will be conducted in the future to define a set of multimodal widgets that reduce the efforts necessary to build user interfaces based on the different levels of abstraction of the interaction models. Additionally the manageability and configurability of the provided features by the developer and also the control by the user will be improved. Further evaluations of the performance of the EMF- and Java-based implementation to op-

timize the system are planned, although its current implementation shows that the systems perform very well. While the current combination of the different models already seems suitable, they will be further refined to address additional use cases. Especially the current interaction model gives room for extensions and enhancements. Finally, the possibility to build self-aware systems using executable models is a fascinating feature that needs further evaluation. Utilizing the models at runtime does not solve all problems of model-based user interface development, but it gives possibilities to overcome the technical challenges in addressing these problems.

REFERENCES

Balme, L., Demeure, A., Barralon, N., Coutaz, J., & Calvary, G. (2004). Cameleon-rt: A software architecture reference model for distributed, migratable, and plastic user interfaces. In P. Markopoulos, B. Eggen, E. Aarts, J. L. Crowley (Eds.), *Proceedings of the European Symposium on Ambient Intelligence 2004* (pp. 291–302). Berlin, Germany: Springer.

Blumendorf, M., Feuerstack, S., & Albayrak, S. (2006). Event-based synchronization of model-based multimodal user interfaces. In A. Pleuss, J. V. den Bergh, H. Hussmann, S. Sauer, & A. Boedcher, (Eds.), *MDDAUI '06 - Model Driven Development of Advanced User Interfaces 2006, Proceedings of the 9th International Conference on Model-Driven Engineering Languages and Systems: Workshop on Model Driven Development of Advanced User Interfaces*, CEUR Workshop Proceedings.

Blumendorf, M., Feuerstack, S., & Albayrak, S. (2007). Multimodal user interaction in smart environments: Delivering distributed user interfaces. In M. Mühlhäuser, A. Ferscha, & E. Aitenbichler, (Eds.), *Constructing Ambient Intelligence: AmI 2007 Workshops Darmstadt* (pp. 113-120). Berlin, Germany: Springer.

Blumendorf, M., Lehmann, G., Feuerstack, S., & Albayrak, S. (2008). Executable models for human-computer interaction. In T. C. N. Graham & P. Palanque, (Eds.), *Interactive Systems - Design, Specification, and Verification* (pp. 238-251). Berlin, Germany: Springer.

Bouchet, J., Nigay, L., & Ganille, T. (2004). Icare software components for rapidly developing multimodal interfaces. In R. Sharma, T. Darrell, M. P. Harper, G. Lazzari & M. Turk (Eds.), *Proceedings of the 6th international conference on Multimodal interfaces 2004* (pp. 251–258). New York: ACM Press.

Breton, E., & Bézivin, J. (2001). Towards an understanding of model executability. In N. Guarino, B. Smith & C. Welty (Eds.) *Proceedings of the international conference on Formal Ontology in Information Systems 2001* (pp. 70–80). New York: ACM Press.

Calvary, G., Coutaz, J., Thevenin, D., Limbourg, Q., Souchon, N., Bouillon, L., et al. (2002). Plasticity of user interfaces: A revised reference framework. In C. Pribeanu & J. Vanderdonckt (Eds.), *Proceedings of the First International Workshop on Task Models and Diagrams for User Interface Design 2002* (pp. 127–134). Bucharest, Rumania: INFOREC Publishing House Bucharest.

Clerckx, T., Luyten, K., & Coninx, K. (2004). DynaMo-AID: A design process and a runtime architecture for dynamic model-based user interface development. In R. Bastide, P. Palanque & J. Roth (Eds.), *Engineering Human Computer Interaction and Interactive Systems: Joint Working Conferences EHCI-DSVIS 2004* (pp. 77–95). Berlin, Germany: Springer.

Clerckx, T., Vandervelpen, C., & Coninx, K. (2007). *Task-based design and runtime support for multimodal user interface distribution.* Paper presented at Engineering Interactive Systems 2007 (EHCI-HCSE-DSVIS'07), Salamanca, Spain.

Coninx, K., Luyten, K., Vandervelpen, C., den Bergh, J. V., & Creemers, B. (2003). Dygimes: Dynamically generating interfaces for mobile computing devices and embedded systems. In L. Chittaro (Ed.), *Human-Computer Interaction with Mobile Devices and Services 2003* (pp. 256–270). Berlin, Germany: Springer.

Coutaz, J. (2006). Meta-User Interfaces for Ambient Spaces. In K. Coninx, K. Luyten & K.A. Schneider (Eds.), *Task Models and Diagrams for Users Interface Design 2006* (pp. 1-15). Berlin, Germany: Springer.

Demeure, A., Calvary, G., Coutaz, J., & Vanderdonckt, J. (2006). The comets inspector: Towards run time plasticity control based on a sematic network. In K. Coninx, K. Luyten & K.A. Schneider (Eds.), *Task Models and Diagrams for Users Interface Design 2006* (pp. 324-338). Berlin, Germany: Springer.

Demeure, A., Sottet, J.-S., Calvary, G., Coutaz, J., Ganneau, V., & Vanderdonkt, J. (2008). The 4c reference model for distributed user interfaces. In D. Greenwood, M. Grottke, H. Lutfiyya & M. Popescu (Eds.), *The Fourth International Conference on Autonomic and Autonomous Systems 2008* (pp. 61-69). Gosier, Guadeloupe: IEEE Computer Society Press.

Duarte, C., & Carriço, L. (2006). A conceptual framework for developing adaptive multimodal applications. In E. Edmonds, D. Riecken, C. L. Paris & C. L. Sidner (Eds.), *Proceedings of the 11th international conference on Intelligent user interfaces 2006* (pp. 132–139). New York: ACM Press.

Feuerstack, S., Blumendorf, M., Schwartze, V., & Albayrak, S. (2008). Model-based layout generation. In P. Bottoni & S. Levialdi (Ed.), *Proceedings of the working conference on Advanced visual interfaces* (pp. 217-224). New York: ACM Press.

International Telecommunication Union. (2005). *ITU Internet Reports 2005: The Internet of Things.* Geneva, Switzerland: ITU.

Klug, T., & Kangasharju, J. (2005). Executable task models. In A. Dix & A. Dittmar (Eds.), *Proceedings of the 4th international workshop on Task models and diagrams* (pp. 119–122). New York: ACM Press.

Limbourg, Q., Vanderdonckt, J., Michotte, B., Bouillon, L., & López-Jaquero, V. (2004). Usixml: A language supporting multi-path development of user interfaces. In R. Bastide, P. A. Palanque & J. Roth (Eds.), *Engineering Human Computer Interaction and Interactive Systems* (pp. 200–220). Berlin, Germany: Springer.

Martin, J.-C. (1998). Tycoon: Theoretical framework and software tools for multimodal interfaces. In J. Lee (Ed.), *Intelligence and Multimodality in Multimedia interfaces.* Menlo Park, CA: AAAI Press.

Mellor, S. J. (2004). *Agile MDA.* Retrieved January 21, 2008, from http://www.omg.org/mda/mda_files/Agile_MDA.pdf

Mellor, S. J., Scott, K., Uhl, A., & Weise, D. (2004). *MDA Distilled: Principles of Model-Driven Architecture.* Boston: Addison-Wesley.

Mori, G., Paternò, F., & Santoro, C. (2004). Design and development of multidevice user interfaces through multiple logical descriptions. [Piscataway, NJ: IEEE Press.]. *IEEE Transactions on Software Engineering*, 507–520. doi:10.1109/TSE.2004.40

Puerta, A. R., & Eisenstein, J. (1999). Towards a general computational framework for model-based interface development systems. In M. Maybury, P. Szekely & C. G. Thomas (Eds.), *Proceedings of the 4th international conference on Intelligent user interfaces 1999* (pp. 171–178). Los Angeles: ACM Press.

Reithinger, N., Alexandersson, J., Becker, T., Blocher, A., Engel, R., Löckelt, M., et al. (2003). Smartkom: adaptive and flexible multimodal access to multiple applications. In S. Oviatt, T. Darrell, M. Maybury & W. Wahlster, *Proceedings of the 5th international conference on Multimodal interfaces 2003* (pp. 101–108). New York: ACM Press.

Rohr, M., Boskovic, M., Giesecke, S., & Hasselbring, W. (2006). *Model-driven development of self-managing software systems.* Paper presented at the "Models@run.time" Workshop at the 9th International Conference on Model Driven Engineering Languages and Systems 2006, Genoa, Italy.

Sottet, J.-S., Calvary, G., & Favre, J.-M. (2006a). Mapping model: A first step to ensure usability for sustaining user interface plasticity. In A. Pleuss, J. V. den Bergh, H. Hussmann, S. Sauer, & A. Boedcher, (Eds.), *MDDAUI '06 - Model Driven Development of Advanced User Interfaces 2006, Proceedings of the 9th International Conference on Model-Driven Engineering Languages and Systems: Workshop on Model Driven Development of Advanced User Interfaces* (pp 51–54), CEUR Workshop Proceedings.

Sottet, J.-S., Calvary, G., & Favre, J.-M. (2006b). *Models at runtime for sustaining user interface plasticity.* Retrieved January 21, 2008, from http://www.comp.lancs.ac.uk/~bencomo/MRT06/

Sottet, J.-S., Ganneau, V., Calvary, G., Coutaz, J., Demeure, A., Favre, J.-M., & Demumieux, R. (2007). Model-driven adaptation for plastic user interfaces. In C. Baranauskas, P. Palanque, J. Abascal & S. D. J. Barbosa (Eds.), *Human-Computer Interaction – INTERACT 2007* (pp 397–410). Berlin, Germany: Springer.

Thevenin, D., & Coutaz, J. (1999). Plasticity of user interfaces: Framework and research agenda. In C. Baranauskas, P. Palanque, J. Abascal & S. D. J. Barbosa (Eds.), *Human-Computer Interaction – INTERACT 2007* (pp 397–410). Berlin, Germany: Springer.

Vanderdonckt, J. (2008). *Model-driven engineering of user interfaces: Promises, successes, failures, and challenges.* Paper presented at 5th Romanian Conference on Human-Computer Interaction, Bucuresti, Romania.

Weiser, M. (1993). Some computer science issues in ubiquitous computing. *Communications of the ACM, 36*(7), 75–84. doi:10.1145/159544.159617

Section 2
Theoretical Foundations

Chapter 3
A Formal Approach to the Verification of Adaptability Properties for Mobile Multimodal User Interfaces

Nadjet Kamel
University of Moncton, Canada

Sid Ahmed Selouani
University of Moncton, Canada

Habib Hamam
University of Moncton, Canada

ABSTRACT

Multimodal User Interfaces (MUIs) offer to users the possibility to interact with systems using one or more modalities. In the context of mobile systems, this will increase the flexibility of interaction and will give the choice to use the most appropriate modality. These interfaces must satisfy usability properties to guarantee that users do not reject them. Within this context, we show the benefits of using formal methods for the specification and verification of multimodal user interfaces (MUIs) for mobile systems. We focus on the usability properties and specifically on the adaptability property. We show how transition systems can be used to model the MUI and temporal logics to specify usability properties. The verification is performed by using fully automatic model-checking technique. This technique allows the verification at earlier stages of the development life cycle which decreases the high costs involved by the maintenance of such systems.

INTRODUCTION

The development of new interactive devices and technologies such as speech recognition, visual recognition and gesture recognition enriches User Interfaces (UIs) interactions and allows user to interact with systems in a more natural way. Multimodal User Interfaces (MUIs) are characterized by several possibilities, defined as modalities, to interact with the systems. The users can choose

DOI: 10.4018/978-1-60566-978-6.ch003

the most appropriate modality for a given situation which allows them interacting with systems anytime and anywhere. They can also use more than one modality to accomplish a given task. For example, they can use speech and gesture to move a graphical object from an initial to a target position. This was studied by R. Bold in (Bold, 1980) through the command "Put That There" using speech and gesture modalities. The gesture is used to indicate the initial and the target positions of the object to be moved.

With the emerging technologies such as ambient, intelligent and ubiquitous or pervasive computing, various input and output modalities are necessary to enable efficient interaction with these systems. The latter are mainly characterized by the multiplicity of users, variety of use contexts and user mobility. At present, MUIs are developed for many of such systems using mobile devices like PDAs and mobile phones. The degrees of freedom, offered to users to interact with these systems, enhance their flexibility and adaptability to different kinds of users, such as people with special needs, as well as to dynamic environments and ambient systems. The choice given to use one or more modalities improves their accessibility to various usage contexts, and enhances the performance stability, the robustness, as well as the efficiency of communication. For example, the user may interact with the system by using *speech* or, if the environment is too noisy, he/she can use *press keys*.

The development of such systems has been studied through different aspects. The usability of User Interfaces is a one of these aspects. It is a measure of the efficiency and satisfaction with which users can achieve their goals through these UIs. It is captured through a set of properties called usability properties. Many usability properties were defined for WIMP (Window, Icon, Mouse, and Pointing device) interfaces and MUIs, but the characteristics of mobile systems require the consideration of new properties that must be satisfied to increase their usability. For instance, in

a mobile environment users can switch between modalities according to their efficiency in a given environment. In an ambient system, users having different profiles must be able to interact with the system. The user profile and environment variations involve changes that must be considered for the interaction. This forces the UI developers to take into account new properties such as *adaptability*. The adaptability property states that the UI remains usable even if some changes have occurred within the interaction environment, like the disability of an interaction device or the user has special needs. This property is very useful for mobile systems since it increases the users' satisfaction.

Many empirical approaches have been used to evaluate these systems. They are based on experimentations which mean that the system must be implemented before it is evaluated. To make this evaluation, observations are collected while participants are using the systems to accomplish some predefined scenarios. These observations are analyzed to deduce whether the desired properties are satisfied. The main disadvantage of such approaches is their imprecision and that they are not exhaustive. For example, we cannot state how many participants are sufficient to state that the required properties are satisfied. Besides, the used scenarios do not cover all possible situations. So, while the results are satisfactory nothing guarantees that the system is usable in every situation which is an issue for the security of these systems. As a consequent, new approaches based on formal methods are proposed as alternatives and in some circumstances as complementary to these empirical ones. Formal methods are based on mathematical models that allow modeling and specifying systems without ambiguities. Formal methods allow designers to conceive models by using analyzing techniques at an abstract level before the implementation. Properties can be verified at earlier stages which decreases the cost of the software development. The advantage of these methods is that they can be implemented in tools for automatic or semi-automatic analysis. The analysis are exhaustive which guarantees that the properties are satisfied for every situation. They can be used at earlier stage of development which helps fixing error before implementing the system. They were successfully used for many systems such as in concurrent programs (Chetali, 1998), embedded (Große & al., 2006) and in real-time systems (Ruf & Kropf, 2003). These techniques have been used for Human Computer Interfaces but a little work is done for Multimodal interfaces.

The aim of this chapter is to show how formal models can be used to verify usability properties for MUIs, at the design step, before implementing these systems. We present different formal approaches to specify and verify usability properties for MUIs. We focus on the usability properties related on the modality use. We propose to model and verify some usability properties shared by fixed and mobile systems. The properties, we intend to formalize and verify, concern MUIs for mobile systems as well as for fixed ones. We focus also on adaptability property which characterizes more mobile systems. We consider the case where one modality becomes disabled while the user is interacting with the system to perform a given task. This makes the system more adaptable to situations of interactive device failure. The objective, in this case, is to answer, at the design step, the following question: If a given property is satisfied by the interface, is this property still satisfied by the interface when a modality becomes unavailable? The study shows how this property can be checked automatically at the design step, allowing the designer to fix errors at the earlier stages of the software development life cycle.

This chapter is organized as follows. The next section gives a background on MUIs, their usability properties as well as their evaluation. Section 3 presents research works made to specify and verify user interfaces as well MUIs using formal methods. Our approach is presented in Section 4 where formal models to describe MUIs behaviors and usability properties are presented. An illus-tration is given in this section by presenting a case study. Finally, Section 5 and Section 6, give respectively future trends and conclusion.

BACKGROUND

This section presents an overview of the works related to the development of multimodal user interfaces and particularly the evaluation of their usability properties. Earlier studies deal with the definition of the multimodality concepts such channel, modality etc... (Niagay & Coutaz, 1996). Then, other studies have been proposed to cover the development process of these systems such as modalities fusion strategies, multimodal dialogue strategies and software architectures. Modalities fusion deals with the composition of data and events produced by several modalities to perform a same command. Indeed, these events and data must be fused to process the semantics of the command. An example of such command is the famous 'Put That There' command performed using speech and gesture modalities. In (Nigay & Coutaz, 1995), the authors proposed a structure called 'melting pots' to deal with modalities fusion. In this work, simple input events are fused through microtemporal, macrotemporal or contextual fusion procedures to provide 'melting pots'. The fusion has been also studied in (Tue Vo & Waibel, 1997). In this work, the authors use a semantic model and a multimodal grammar structure that embeds this semantic model in a context-free modeling language. They use a set of grammar-based Java tools to facilitate the construction of input processing in multimodal applications. The tool set includes a connectionist-network-based parsing engine capable of segmenting and labeling multimodal input streams to ease the multimodal interpretation task. In (Portillo & al., 2006), a new hybrid strategy for modalities fusion was proposed. This strategy merges two fusion approaches at two levels: the multimodal grammar level and the dialogue level. It requires the inclusion of multimodal grammar entries, temporal constraints and additional information at dialogue level.

The multimodal dialogue is still one challenging issue addressed in many research works. It deals with the strategies that manage and control the dialogue between the user and the system. In (Harchani & al., 2007), the authors propose to generate cooperative responses in an appropriate multimodal form, highlighting the intertwined relation of content and presentation. They identify a key component, the dialogic strategy component, as a mediator between the natural dialogue management and the multimodal presentation.

Usability is one fundamental aspect of most interactive applications. First, this issue was studied for interactive systems, and later for MUIs. Several usability properties have been identified for traditional interfaces of interactive systems such as *consistency*, *observability*, and *preemptiveness*. Nevertheless, these usability properties do not cover the specific aspects of new interaction technologies such as MUIs. To address such aspects, a set of properties that characterize MUIs was defined. Since the development of MUIs for mobile systems is growing, new usability properties related to these new environments are also studied. They mainly deal with the adaptability property. As this chapter deals with the evaluation of usability properties, more details related to these concepts are presented in what follows. First, we present some usability properties defined for MUIs that we propose to model formally later. These properties characterize the usability of multimodal interfaces and particularly those for mobile systems. Then, we present some research works dealing with the evaluation of user interfaces. We present both the works using empirical approaches and those using formal approaches. We conclude by a discussion that compares these two approaches by giving their advantages as well as their disadvantage.

Usability of Multimodal User Interfaces

CARE Properties

Complementarity, Assignment, Redundancy and Equivalence (CARE) are four ways to use modalities to interact with a multimodal system. In (Coutaz & al., 1995), the authors show how CARE of modalities can affect the usability of a MUI. These properties are defined according to a set of modalities and two states of the considered MUI. These states identify the initial and the final state of a user interactive task T. We assume that *InitState* and *FinalState* identify, respectively, the initial state and the final state of the task T. We illustrate the CARE definitions on the task that consists on filling a text field by the city name *"New York"*.

- *Complementarity.* Modalities are complementary for the task T if they may be used to reach a state *FinalState*, starting from a state *InitState*. Moreover, a single modality shall not allow doing this. For example, the user clicks on the text field using the mouse, and then he/she pronounces the words *'New York'*. In this case, the modalities *direct manipulation* and *Speech* are complementary for the task T. The complementarity combines modalities to increase the efficiency of the interaction. For example, the use of *speech* and *gesture* to accomplish the task 'Put That There' allows for moving an object from one location to another in an efficient way. Indeed the gesture modality is more efficient to point the locations in a precise way, while speech is more efficient to express a command in some words (three words in this example).
- *Assignation.* A modality m is assigned to the task T if it is the only modality that allows user to reach *FinalState* from *InitState*

and no other modality allows it. For example, if the user can only fill the text field by pronouncing the sentence *'Fill New York'*, then the modality *Speech* is assigned to the task *T*. The assignation expresses the absence of the choice. To accomplish the task *T*, the user must use the modality *m* assigned to this task. This increases the system preemptiveness. Indeed, the system controls the choice of the modality to be used.

- *Equivalence.* Modalities are equivalent for a task *T*, if each modality allows the user to reach the state *FinalState* starting from the state *InitState*. For example, the user can either click on the text field, and then on the item *'New York'* on the list of cities, or he can pronounce the sentence *'Fill New York'*. In this example, the modalities *Speech* and *Direct manipulation* are equivalent for the task *T*. The equivalence property expresses the availability of choice between several modalities. This will increase the flexibility and the robustness of the system. Indeed, if one modality becomes not available for some reasons, the user can continue to interact with the system using another equivalent modality.

- *Redundancy.* Modalities are redundant for the task *T*, if they are equivalent and are used in parallel to reach the state *FinalState* starting from the state *InitState*. The redundancy increases the perceivability when it is used for the outputs. To guaranty that alert information will be perceived by the user, it is more interesting to use more than one modality to output it. For example, using the voice to announce the alert, and using a lighting message to display it, will arise the chance for the user to perceive it.

Adaptability Property

Many studies were achieved to deal with the adaptability of user interfaces according to a multitude of characteristics that may determine an adaptive interface's success or failure. These characteristics may be the user profiles, the context of use, or the platform on which the application is executed, etc... In addition, MUIs for mobile systems introduce new challenges for their adaptability. These systems are characterized by the diversity of context of use and interactive devices. The changes involved by the environment and the context of use have effects on the user system interactions. For example, interacting with the system using the speech is not recommended in a noisy environment. The user interfaces that detect these changes and adapt the presentation and the interaction according to these changes will be more usable.

Adaptation to the users profile takes into consideration the various skills and physical abilities of users. This allows each kind of user to interact differently by incommoding the content of the interface, the style of the presented information, and the interaction modalities to the user profile. Many studies have been achieved to adapt to users' needs and preferences. For instance, in (Stent, 2002 & De Bra, 1999) the authors focus on how to customize content based on user models.

Since the same application may be executed on several platforms (PC, PDA, Mobile Phone, etc...), the adaptability of the interfaces to this diversity of platforms was also studied. This is a part of the plasticity concept which characterizes user interfaces that are capable to adapt to physical or environmental changes by remaining usable (Thevenin & Coutaz, 1999). For example, the interface adapts the presentation to the size of screen that is not necessary the same for all the devices. Many studied has been achieved to deal with this issue (Findlater & McGrenere, 2008; Bridle & McCreath, 2006).

Other works propose to integrate all these aspects, to be considered in the development process of adaptive interfaces, in a framework. An example of such works is the study presented in (Duarte & Carriço, 2006). The authors propose a model-based architecture for adaptive multimodal applications (FAME). The framework proposes to guide the development of adaptive multimodal application by providing a conceptual basis that relates the different aspects of an adaptive multimodal system, and a set of guidelines for conducting the development process. The applicability of this framework was demonstrated through the development of an adaptive Digital Talking Book player. This application supports interaction through voice, keyboard and mouse inputs, and audio and visual outputs. The player adapts to devices' properties, reproduction environment, and users' behaviors and physical characteristics.

Evaluation of Usability Properties for Multimodal User Interfaces

Empirical Approaches

The evaluation of multimodal user interfaces usability is important to guarantee the satisfaction of users. If users are not satisfied, they will reject the system even it is robust and efficient. The evaluation of multimodal interactive systems on mobile systems remains a difficult task. Many works deal with the usability evaluation of these systems. They are based on experimentation and tests (Been-Lirn Duh & al., 2006; Serrano & al, 2006; Pascoe & al., 2000; Beringer 2002). To use these methods, first the system must be implemented, and then users are asked to perform some tasks interacting with the system. After that, user comments and evaluator observations are collected and used to evaluate these systems. We report here on some examples on the usability evaluation of some multimodal systems:

In (Alexander & al., 2004), the authors deal with the evaluation of a Virtual personal assistant (VPA). The VPA is a multilingual multimodal and multi-device application that provides management facilities through a conversation interfaces. The main goal of the evaluation was to evaluate the usability of VPA. Representative users were asked to complete a set of typical e-mail tasks and measures were taken of effectiveness, efficiency and satisfaction. In (Branco & al., 2006) the same system was evaluated in a laboratory environment, with 12 native Portuguese speakers interacting with VPA. A set of tasks was defined to be accomplished by these participants. Two evaluators observed the participants using the system to assist them and to note the issue with it. To collect participant comments, a set of questions was used for each task. The participant comments are then analyzed to evaluate the system.

In (Petridis & al., 2006), the authors evaluate the usability of a multimodal interface by comparing it with two input devices: the Magellan SpaceMouse, and a 'black box'. The evaluation compares a tangible interface in the form of a cultural artifact replica with the Magellan SpaceMouse as well as with a plain black box. The black box contains the same electronics as the multimodal interface but without the tactile feedback offered by the 'Kromstaf' replica. To conduct the evaluation, 54 participants were recruited. They were separated into three groups of 18. Each group was charged to test one type of interface. Task measures were used to conduct the evaluation. In addition, user satisfaction questionnaires were defined to collect user comments. These questionnaires show that there is a difference between the female and the male interfaces satisfaction.

In (Serranno & al., 2006), the authors address the problem of evaluating multimodal interfaces on mobile phone using the platform ACICARE. This platform results from the combination of two complementary tools: ICARE and ACIDU. ICARE is a component-based platform for rapidly developing multimodal interfaces (Bouchet & al., 2005). It includes an editor that enables the system designer to graphically manipulate and assemble JavaBeans components in order to create a multimodal interaction for a given task. The code for the multimodal interaction is automatically generated from these graphical specifications. ACIDU is a tool implemented on mobile phones (Demumieux & al., 2005). It allows collecting objective data and used functionalities (e.g. camera and calendar), durations of use, and navigation (e.g. the opened windows). ACIDU has been implemented as an embedded application. ICARE was connected to ACIDU to enable the rapid development of multimodal interfaces as well as the automatic capture of multimodal usage for in-field evaluations.

Formal Method Approaches

Several formal methods and techniques were used to verify and validate usability properties for Human Computer Interfaces. In general, they are performed thought the following steps:

1. Modeling the user interface using a formal model;
2. Specification of the properties, to be checked, using a formal model;
3. Verification of the properties specified in step 2 on the system modeled in step 1.

To perform the third step (3), a procedure is followed according to the used technique. There are three main techniques: model-checking technique, proof technique and test technique. The model-checking technique consists of modeling the system using a finite state model and expressing the property using a temporal logic, and then using a procedure which explores all the system states to check whether the specified property is satisfied. This technique permits to completely analyze the space of reachable states of a specified system. The proof technique uses a theorem prover, at the syntactic level, to check if the system satisfies the specified constraints. Finally, the test technique use scenarios of use

that are executed by the system. According to the results of these executions, conclusions are made about the satisfaction of the properties by the modeled system.

In the next section, we give an overview on research works that use these formal techniques to specify and verify usability properties of Human Computer Interfaces. Earlier works were carried out for user interfaces and then for multimodal user interfaces.

Formal Methods for UI

Earlier research works achieved in this area was done for WIMP (Windows, Icons, Menus and Pointing devices) user interfaces. Both the formal model-checking technique and the formal proof technique were used to specify and verify this kind of interfaces.

The model-checking technique based on temporal logics was used in (Abowd, 1995). In this work, usability properties related to Human-Machine dialogue such as *Deadlock freedom* and *task completeness* were expressed in CTL (Emerson & Clarke, 1982) temporal logic and checked using the Symbolic Model Verifier (SMV) model-checker. The same model-checking technique was used in (Ausbourg, 1998), (Campos & Harrisson, 2001), (Loer & Harrisson, 2002) and (Campos & al., 2004) to verify usability properties of user interfaces. In (Paterno & al., 1994), the authors used ACTL (Action Computation Tree Logic) temporal logic to specify the same usability properties in generic formulae. The Lite (Van-Eijk, 1991) model-checker was used to verify their satisfaction. Petri Nets were used in (Palanque & al., 1995) and (Kolski & al., 2002) to design interactive systems. In (Markopoulos, 1995), both the system and the properties are specified as processes expressed in Lotos process algebra (Bolognesi & Brinksma, 1987). A property is expressed as a process describing an undesired behavior. In this case, the verification of the property consists of verifying that the behavior of the system process is different from the behavior of the property process. The CADP (Garavel & al., 2007) model-checker was used to implement this approach.

Among the formal proof techniques that were used for interactive systems, we cite the VDM and Z methods that have been used for defining atomic structures like interactors (Duke & Harrisson, 1993), and the B method (Abriel, 1996) that has been used in (Ait-Ameur & al., 1998).

Formal Methods for MUI

Unfortunately too little works using formal models were accomplished for multimodal systems. This is mainly due to the fact that these systems are not mature yet. Both model-checking technique, test technique and proof technique were used to verify usability properties for MUIs.

The model-checking technique was used in (Palanque & Schyn, 2003). In this work, the authors use the Interactive Cooperative Objects (ICO) based on Petri nets to deal with the fusion of modalities. ICO is a formal description technique mainly dedicated to dialogue modeling. This work shows how ICO can be used to fuse basic interactive actions performed using two different modalities: speech and gesture, to produce more complex ones. This formal technique uses the oriented object concepts (dynamic instantiation, encapsulation, etc...) to describe the static part of the system and the Petri net to describe its dynamic part. The dynamic part corresponds to the behavior of the system. ICO was extended to capture the specificities of multimodal interactions. A set of mechanisms such as the communication by production and the consumption of the events were added to the ICO formalism. These mechanisms allow for synchronizing a set of transitions on an event, and allow sending an event when a transition is performed. This approach was used in (Navarre & al., 2006). In this work two approaches for the development of multimodal interactive systems are integrated. The first approach is based on ICoM (a data-flow model dedicated to low-level input modeling) and its environment ICon which allows for editing and simulating ICoM models. The second approach is based on ICO and its environment PetShop which allows for editing, simulating and verifying ICO models.

The formal test technique was used in (Jourde & al., 2006; Bouchet & al., 2007) to validate the CARE (Complementarity, Assignment, Redundancy and Equivalence) usability properties. In these works, the Lutess tool is used to verify if a multimodal interactive system developed using the ICARE framework (Bouchet & al., 2005) satisfies the CARE properties specified in Lustre. Lutess is a testing environment handling specifications written in Lustre language. It requires the software under test, the environment description and the test oracle. The test oracles consist of properties that must be checked. The CARE properties were expressed as Lustre expressions and included in an automatic oracle. Lutess automatically builds a generator that generates and randomly selects an input vector and sends it to the software under test. The software reacts with an output vector which is sent to the generator. The generator produces a new input for a new cycle. The oracle observes the inputs and the outputs of the software under test, and determines whether the software properties are violated. The software under test and the environment are written in Lustre.

The formal proof method was used in (Ait-Ameur & al., 2006). In this work, the B method (Abrial, 1996), based on the refinement technique, is used to verify the CARE usability properties. The MUI interactions are encoded in the event B language and the properties are described as invariant in the first order logic. In this approach, the development is done through a downward step. First, an abstraction of the system is modeled and checked. Then, at each step, the system is refined by adding more conceptual details.

Discussion

The evaluation approaches based on empirical methods such experimentation and test is generally costly. Indeed, it requires many participants who run the system to conduct the tests. Recruiting these participants is not an easy task since, for many cases, these participants must be remunerated. In addition, these people must be taught to how use these systems and they should have different educational and cultural backgrounds, age and technology experiences. It is not always obvious to find a sufficient number of participants satisfying the required conditions. All these tasks make the evaluation difficult to accomplish. In the case when the system does not satisfy the required property, the modification can yield high cost of development since the updates may require rebuilding the whole system from the design step. This is due to the use of these techniques late in the development process. In addition, it not possible to determine how many tests and how many participants are needed to conduct these tests. Thus, the outcomes of these results are not sure. Indeed, we cannot determine how many participants are required to test the system to have a conclusive feedback about the usability of the system. The conducted tests are not exhaustive because the scenarios defined to be run by the participants do not cover all possible operations of the system. This may lead to a serious problem of security.

These reasons motivated researchers, including ourselves, to study the use of model-based methods and particularly formal methods in order to prove their usefulness to conduct automatic verification of multimodal systems. The research works presented above show the applicability of these techniques in the Human Computer Interfaces domain. The advantage of proof approaches such as the B method is that the properties satisfied at the abstract level are preserved in the refined level. This approach has the disadvantage to be partially automatic because it needs an expert to perform some mathematical proofs not decidable automatically. This is not an easy task for non experts like user interfaces designers. Contrary to the proof techniques, the model-checking technique is fully automatic and can be implemented in tools to help

designers analyze their specifications. As outlined above, research works for MUIs are still at the beginning and a lot must be done. Even though the works based on ICO enable modeling multimodal interactions, and performing some analysis allowed by the formalism of Petri Nets, such as reachability and deadlock free, they do not deal with the usability properties such as CARE properties or those related to systems mobility. The only work that deal with the verification of the CARE properties is (Jourde & al., 2006; Bouchet & al., 2007), but the used approach is not exhaustive since it is based on the verification of scenarios. Despite the use of formal notations, this approach shares the same inconvenience with the empirical approach. This is because it is used at the end of the development cycle. This requires that the system must be implemented before it can be tested. For these reasons we propose modeling and verifying CARE and adaptability properties.

THE PROPOSED APPROACH

Our approach is based on model-checking technique. We define a formal model for the multimodal interaction. The semantics of this model is based on transition systems. We propose to express the usability properties as generic formulas in the CTL temporal logic (Emerson & Clarke, 1982). The verification of these properties can be done using the SMV model-checker that implements the CTL temporal model-checking technique. We propose to express the following properties: *equivalence*, *assignment*, *complementarity* and the *adaptatbility* properties. We could not express the *redundancy* in the CTL temporal logic. It needs a logic that expresses the parallelism.

To verify these properties, the following steps are followed:

1.   The multimodal interactions are described using the syntax of the formal model, and then the corresponding transition system is

generated. This transition system describes the dynamic behavior of the multimodal user interface.

2. The required properties are expressed using the CTL generic formulas.
3. The transition system obtained in step 1 is encoded in the input language of SMV, and then the SMV tool is used to check whether the property expressed in step 2 is satisfied by the transition system under analysis.

To illustrate our approach, we use a part of a mobile phone multimodal interface. We assume that the user can interact with the interface by using *Speech* or by clicking on the *press Keys* of the mobile phone to scroll and select menu items. A set of hierarchical menu items are specified allowing the user to access a set of tasks. A schematic view of a part of this interface is shown in Figure 1.

Formal Model for Multimodal Interaction

In this section, we present the formal model for the multimodal interaction systems.

Syntax

The syntax is defined through a grammar with two rules. Each rule may have a left-hand side defining a non terminal symbol (S and E). The symbol S defines user tasks, and the symbol E represents multimodal events which are elementary user tasks. Each rule has a right hand-side, describing the composition possibilities of user tasks, multimodal events and interactive actions.

S::= S[]S | S>>S | S ||| S | S || S | Disable(S,mi) | E

E::= a;E | a ||| E | a || E | δ

The right-hand side is a production as in the traditional BNF grammar. It is composed of one

Figure 1. Some elements of the hierarchical menu

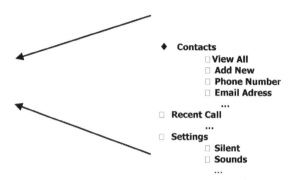

or more alternatives separated by the symbol "|". Each alternative contains non-terminal elements (S and E), compositional operators, and terminal elements which are interactive actions (a). The rule S generates the multimodal interactions model at a higher level by composing multimodal interactive events (E). The rule E generates the multimodal interactive events using interactive actions a of the set A. A set of operators are used to compose multimodal interactive actions and events: ";", ">>", "[]", "|||" and "||". The symbols ";" and ">>" stand for sequence operators. The symbols "[]", "|||" and "||" stand, respectively, for choice, interleaving and parallel operators. The symbol $δ$ stands for a term that does not do anything. It expresses the end of a process. The operator *Disable(S,mi)* allows for disabling the modality *mi* in the system *S*.

The grammar rules allow for encoding constraints of the modalities use. The grammar allows for the generation of a set of formal sentences expressed through the terminals (interactive actions, compositional operators and $δ$). Each sentence describes the syntactical expression of a possible MUI model.

For example, the expression *(a_1;a_2; $δ$) [] (a_3; $δ$)* models the choice between two tasks. The first task is performed by the action a_1 followed by the action a_2. The second task is performed by the action a_3.

Semantics

The semantics corresponding to each syntactical expression, generated by the above grammar, is defined through a transition system. A transition system is basically a graph whose nodes represent the reachable states of the system and whose edges represent state transitions. The initial state of the transition system of a given term is identified by the term itself. Let P and Q be two terms of the previous grammar, and a, a_1 and a_2 be interactive actions of the set A. Then, the transition $P \xrightarrow{a} Q$ expresses that the term P performs the interactive action a, and then behaves like the term Q. The notation $P\neg \xrightarrow{a}$ expresses that the process P cannot perform the action a. In other words, there is no transition, labeled by a, starting from P. Using this notation, the operational semantics is formally expressed by transition rules expressing the behavior of each operator of the previous grammar. According to Plotkin (Plotkin, 1981), each rule of the form $\dfrac{Pr\,emises}{Conclusion}$ expresses that when the premises hold, then the conclusion holds.

- *Stop.* δ does not perform any transition;
- *Prefix operator* ;. The term $a;P$ performs the action a and then behaves like the term P; $a;P \xrightarrow{a} P$

  Example: $a_1;a_2;a_3;\delta \xrightarrow{a_1} a_2;a_3;\delta$

- *Sequence operator* $>>$. If the term P performs an action a, and then behaves like the term P', then the term $P>>Q$ performs the same action and then behaves like the term $P'>>Q$. This is defined by the following rule: $\dfrac{P \xrightarrow{a} P' \text{ and } P' \neq \delta}{P >> Q \xrightarrow{a} P' >> Q}$

  Example:

  $(a_1;a_2;\delta) >> (a_3;\delta) \xrightarrow{a_1} (a_2;\delta) >> (a_3;\delta)$

  If the term P performs an action a, and finishes, then the term $P>>Q$ performs the same action and then behaves like the term Q. This is defined by the following rule: $\dfrac{P \xrightarrow{a} P' \text{ and } P' = \delta}{P >> Q \xrightarrow{a} Q}$

  Example: $(a_2;\delta) >> (a_3;\delta) \xrightarrow{a_2} (a_3;\delta)$

- *Choice operator* []. The first rule asserts that if the term P performs an action a, and then behaves like the term P', then the term $P[]Q$ performs the same action and then behaves like the term P' (the second rule is defined in the reverse way).

  $\dfrac{P \xrightarrow{a} P'}{P[]Q \xrightarrow{a} P'} \quad \dfrac{Q \xrightarrow{a} Q'}{P[]Q \xrightarrow{a} Q'}$

  Example: $(a_1;a_2;\delta)[](a_3;a_4;\delta) \xrightarrow{a_1} (a_2;\delta)$

  $(a_1;a_2;\delta)[](a_3;a_4;\delta) \xrightarrow{a_3} (a_4;\delta)$

- Interleaving operator |||. It allows interleaving (asynchronous) two transition systems (left and right). The first rule, asserts that if the left system transits from the state identified by the term P to the state identified by the term P', by performing an action a_1, then the composed state $P|||Q$ transits to $P'|||Q$ by performing the same action.

  $\dfrac{P \xrightarrow{a} P'}{P ||| Q \xrightarrow{a} P' ||| Q}$

  Example:

  $(a_1;a_2;\delta) ||| (a_3;a_4;\delta) \xrightarrow{a_1} (a_2;\delta) ||| (a_3;a_4;\delta)$

  In the same way, the second rule expresses the behavior of the composed system resulting from the behavior of the right term (Q).

  $\dfrac{Q \xrightarrow{a} Q'}{P ||| Q \xrightarrow{a} P ||| Q'}$

  Example:

  $(a_1;a_2;\delta) ||| (a_3;a_4;\delta) \xrightarrow{a_3} (a_1;a_2;\delta) ||| (a_4;\delta)$

- Parallel operator ||. It allows running two transition systems (left and right) in parallel (synchronous). The two first rules express the interleaving between actions. $\dfrac{P \xrightarrow{a} P'}{P ||| Q \xrightarrow{a} P' ||| Q}$

$$\frac{Q \xrightarrow{\;a\;} Q'}{P \,\|\, Q \xrightarrow{\;a\;} P \,\|\, Q'}$$ The third rule expresses that if the term P performs an action a_1 and behaves like the term P', and the term Q performs an action a_2 and behaves like the term Q', then the term $P\|Q$ performs the two actions in one step and behaves like $P'\|Q'$. We assume that the system cannot perform, at the same time, more than one action produced by the same modality. For this reason, we use the function *mod* that assigns, to each interactive action, the modality that it produces it.

$$\frac{P \xrightarrow{\;a_1\;} P' \;\text{ et }\; Q \xrightarrow{\;a_2\;} Q' \quad \mathrm{mod}(a_1) \neq \mathrm{mod}(a_2)}{P \,\|\, Q \xrightarrow{\;(a_1,a_2)\;} P' \,\|\, Q'}$$

Example:
$$(a_1;a_2;\delta) \,\|\, (a_3;a_4;\delta) \xrightarrow{\;(a_1;a_3)\;} (a_2;\delta) \,\|\, (a_4;\delta)$$

The transition system corresponding to any term of the grammar is obtained by applying the previous rules inductively.

- *Disabling operator. Disable* is an operator that disables a modality in a system. *Disable(P,mi)* is the term resulting from disabling the modality *mi* in the system P. The semantics of this operator is given by the following rules: The first rule states that if a term P performs an action a produced by the modality *mi* and behaves like P', then the term *Disable(P,mi)* cannot perform the action a. $$\frac{P \xrightarrow{\;a\;} P' \;\text{ and }\; \mathrm{mod}(a) = mi}{Disable(P,mi) \neg \xrightarrow{\;\;a\;\;}}$$

The second rule states that if a term P performs an action a produced by any modality other than the modality *mi* and behaves like P', then the term *Disable(P',mi)* performs the same action and behaves like the term *Disable(P',mi)*.

$$\frac{P \xrightarrow{\;a\;} P' \;\text{ and }\; \mathrm{mod}(a) \neq mi}{Disable(P,mi) \xrightarrow{\;a\;} Disable(P',mi)}$$

Example

We illustrate the formal model *MobileInteraction* of the MUI of the mobile phone presented previously. First, we determine the set of interactive actions A, and their composition using the operators of the formal model. The interactive actions are produced by the modalities *Speech* and *Press Keys*.

$A = \{$"up", "down", "select",..., C_{up}, C_{down}, Ok, C_{right}, C_{left}, ...$\}$

Where *"up"*, *"down"* and *"select"* are the words pronounced by the user and recognized by the system as actions that, respectively, move up, move down in the menu, and select an item. C_{up}, C_{down}, *Ok*, C_{right} and C_{left} are actions of clicking, respectively, on the keys: *up*, *down*, *Ok*, *right* and *left*. These actions, respectively, allow the user to move up, move down, select, open menu and close menu.

These interactive actions can be composed using the operators according to the grammar rules of the formal model. First, the interactive actions are composed to define user tasks, and then the user tasks are composed to define the user interface. To illustrate our approach we focus on the tasks T_1, T_2 and T_3. These allow user to, respectively, view the contact list, select the ring volume and view recent call. We assume that initially, the cursor is on the *Menu* item. The tasks T_1, T_2 and T_3 are defined by composing the interactive actions as follows:

T1 = (C_{down} ; C_{right} ; C_{down} ; Ok) [] ("Contacts"; "view list"; "Ok")

T2 = (C_{down} ; C_{down} ; C_{down} ; C_{right} ; C_{down} ; C_{down}; Ok)

[] ("Settings"; "Sounds"; "Ringer Volume"; "Ok")

Figure 2. Transition system of MobilInteraction

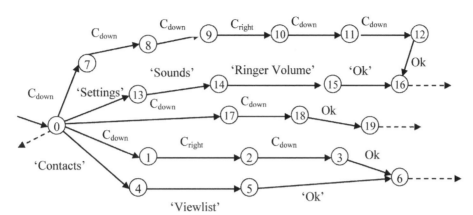

$T3 = C_{down} ; C_{down} ; Ok$

To perform the task T_1 or T_2, the user can use *Speech* or *Keys* on the press keys. The task T_3 can be performed only by clicking on the press keys. The expression of the interaction model is given as follows:

MobilInteraction = $T_1[]T_2[]T_3$

It expresses that the user has the choice to perform the task T_1 or T_2 or T_3.

The transition systems of the tasks T_1, T_2, T_3 are generated using the semantic rules of the operator ";". These transition systems are composed using the rule of the choice operator []. The resulting transition system describes the behavior *MobileInteraction*. It is given in Figure 2. The state *0* is the initial state of both the tasks T_1, T_2, T_3 and *MobileInteraction*. The states *1, 2, 3, 4, 5* and *6* are states of the task T_1. The states *7, 8, 9, 10, 11, 12, 13, 14, 15* and *16* are states of the task T_2. The states *17, 18* and *19* are states of the task T_3. The dashed arrows from the states *0, 6, 19* and *16* indicate that other tasks can be performed in alternative to the previous tasks as well as after that they have been accomplished.

Expression of Usability Properties in CTL Temporal Logic

EQUIVALENCE

Let $E = \{m_1, m_2, ... m_n\}$ be a set of modalities. *Equiv_E* denotes the equivalence of the modalities of the set E for the task T. It is expressed by the following CTL temporal logic formula:

(AG (((state= Initstate) ∧ (modality = m_1)) ⇒ E((modality = m_1) ∪ (state=Finalstate)))   (1)

∧

(AG (((state= Initstate) ∧(modality = m_2)) ⇒ E((modality = m_2) ∪ (state=Finalstate)))   (2)

∧

...

∧

(AG (((state=Initstate) ∧ (modality = m_n)) ⇒ E((modality = m_n) ∪ (state= Finalstate))) (n)

This formula states that all states (*G*) of all paths (*A*) of the MUI transition system, if any of these states is identified as the initial state of the task *T* (*state= Initstate*), then there exists at least one path (*E*), starting form this state, whose states are reached by the modality m_i ($m_i \in E$), until ($\cup$) the final state of the task (*state= Finalstate*) is reached. This is expressed by the parts *(1)*, *(2)*,..., and *(n)* of the formula.

Complementarity

Let *Cm*= *{m_1, m_2,..., m_n}* be a set of modalities. *Compl$_{Cm}$* denotes the complementarity of the set C_m and it is expressed in the CTL temporal logic as follows:

EG ((state =Initstate) $\Rightarrow$ E(((modality=m1) $\vee$ (modality = m2)

$\vee$ ...$\vee$ (modality = mn)) $\cup$ (state=Finalstate)))

(1)

$\wedge$

¬EG ((state=Initstate) $\Rightarrow$E ((modality =m1) $\cup$ (state =Finalstate)))   (2)

$\wedge$

¬EG ((state=Initstate) $\Rightarrow$E((modality=m2) $\cup$ (state=Finalstate)))   (3)

$\wedge$

...

$\wedge$

¬ EG ((state=Initstate) $\Rightarrow$E((modality=mn) $\cup$ (state=Finalstate))) (n+1)

This formula asserts that there exists at least one path in the MUI transition system, such that

if any of all its states is identified as the initial state of the task *T* (*state = Initstate*), then all the next states are reached using one of the modalities m_1, m_2, ..., m_n, until the final state is reached (*state=Finalstate*). This is expressed by the part *(1)* of the formula. It asserts, also, that there is no path starting from the state *Initstate*, and finishing at the state *Finalstate*, such that all the transitions are performed using only the modality m_i ($1 \leq i \leq n$). This is expressed by the parts *(2)*, *(3)*,..., and *(n+1)* of the formula.

Assignment

Assig$_{mi}$ denotes the assignement of the modality m_i to the task *T*. It is expressed by the following CTL temporal logic formula:

Assig$_{mi}$ = AG ((((state= Initstate) $\wedge$ (modality=m_i)) $\Rightarrow$ A((modality=m_i) $\cup$ (state=Finalstate)))

This formula asserts that all states (*G*) of all paths (*A*), starting from the state *Initstate* (*state=Initstate*), are reached using the modality m_i (*modality=m_i*), until ($\cup$) the state *Finalstate* is reached.

Adaptability

We consider the adaptability according to a given task *T*. If the user was interacting with the system to accomplish a given task *T* and the environment of the system changes, the user must be able to continue interacting with this system to perform the task *T*. For instance, if the user can perform a given task using the *Speech* modality, and the environment of the interaction changes and becomes very noisy, then the user must be able to use another modality to accomplish this task.

We define the adaptability property by using the *reachability* property. The *reachability* property defines the possibility to perform the *T*, It states that: starting from *Initstate*, identified as the initial state of the task T, the user can reach *Finalstate*, identified as the final state of the task T. This property is used to verify if the task T is feasible. We consider that if the *reachability* property is satisfied even if a modality is disabled, then the *adaptability* property is satisfied. The *reachability* property can be expressed for the task T through the following generic CTL temporal logic formula:

$$AG((state=Initstate) \Rightarrow EF(state=Finalstate))$$

Informally, this formula states that in all states (G) of all paths (A), if a state is identified as the initial state of the task (*state=Initstate*), then there exists (E) at least one path starting from this state and reaching, in the future (F), a state identified as the final state of the task (*state= Finalstate*).

The adaptability property is satisfied by the system if for each modality m_i, the *reachability* property is satisfied by the system *Disable(S, m_i)*. This means that the task T is feasible even the modality m_i is disabled in the system S.

Verification

The aim of this section is to show how the verification of the usability properties described above is processed. Once the MUI is modeled using the formal model, the corresponding transition system is generated automatically using the semantic rules. Then, the usability properties are expressed using the generic formulas given below. The formula of a given property for a given task is obtained by replacing the values m_i and m_j, in the generic formula, by the concrete modalities, and the values *Initstate* and *Finalstate* by the initial and the final state of the task under consideration. The verification is done by using a model-checker. A model-checker is a tool that implements the model-checking technique. In our case, we use the SMV model-checker (Mc Millan, 1992), but another model-checker could be used. The SMV model-checker allows for checking finite state

transition systems against specifications in the CTL temporal logic (Clarke, 1986). The transition system is encoded in the input language of SMV, and the properties are expressed as CTL temporal logic formulae.

Example

The experimentation deals with the CARE properties as well as the adaptability property. The aim is to check whether the properties are satisfied for the tasks T_1, T_2 and T_3 of the user interface of the mobile phone described previously.

Verification of the CARE Properties

The CARE properties are checked on the transition system of the user interface (Figure 2.). First, this transition system is encoded into the input language of SVM tool, and then the properties are expressed using the generic formulas defined in the previous section.

- The *equivalence* of the modalities *Speech* and *Keys* for the task T_2 is expressed by the following CTL formula:
  (AG (((state= 0) $\wedge$ (modality = Speech)) $\Rightarrow$ E((modality = Speech) $\cup$ (state=16)))
  $\wedge$
  (AG (((state= 0) $\wedge$(modality = Keys)) $\Rightarrow$ E((modality = Keys) $\cup$ (state=16)))

- The *assignment* of the modality T_3 is expressed as follows:
  AG (((state= 0) $\wedge$ (modality=Keys)) $\Rightarrow$ A((modality=Keys) $\cup$ (state=19)))

- The *complementarity* of the modalities *Speech* and *Keys* for the task T_2 is expressed by the following formula:
  EG ((state =0) $\Rightarrow$ E(((modality=Speech) $\vee$ (modality = Keys)) $\cup$ (state=16)))
  $\wedge$
  $\neg$EG ((state=0) $\Rightarrow$E ((modality =Speech) $\cup$ (state =16)))
  $\wedge$

Figure 3. Transition system of Disable(MobilInteraction, Speech)

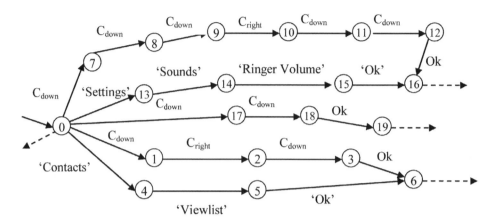

$\neg$EG ((state=0) $\Rightarrow$E((modality=Keys) $\cup$ (state=16)))

The results reported by the SMV tool indicate that the *equivalence* and the *assignation* expressions are satisfied, but the *complementarity* expression is not satisfied.

Verification of the Adaptability Property

The aim is to verify if is always possible to perform the task T_2 even if one of the modalities *Speech* or *Keys* becomes disabled. To do this, we check if the task T_2 is feasible in the two following situations:

- The user is in a noisy environment and the modality *Speech* is disabled. In this case, the user can only use the press keys to interact with the system. *Disable(ModelInteraction, Speech)* describes this new system. The corresponding transition system is illustrated by Figure 3. It is obtained by applying the semantic rule of the operator *Disable* by replacing P and m_i, respectively, by *ModelInteraction* and *Speech*.

- The user cannot click on press keys. He can only use *Speech* to interact with the system. In this case all the interactions produced by the *Keys* are disabled which provides a

new system. The behavior of this system is described by the transition system of *Disable(ModelInteraction, Keys)*. The corresponding transition system is illustrated by Figure 4. It is obtained by applying the semantic rule of the operator *Disable* by replacing P and m_i, respectively, by *ModelInteraction* and *Keys*.

Each of the previous transition systems is encoded in the input language of SMV. The feasibility of the task T_3 is expressed using the following reachability CTL formula.

AG((state=0) $\Rightarrow$ EF(state=19))

The SMV tools reports that this property is satisfied only by the system D*isable(ModelInteraction, Speech)*. The transition system of *Disable(ModelInteraction, Keys)* does not satisfy the property because there is no path starting from the state 0 and reaching state 19. This will let us conclude that the system *ModelInteraction* is not adaptable to the changes of interaction environment for the task T_3.

Figure 4. Transition system of Disable(MobilInteraction, Keys)

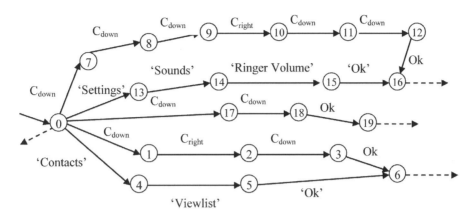

FUTURE TRENDS

The study of multimodal user interfaces and their evaluation in a mobile environment is an emerging research area. The diversity of mobile platforms (PDA, mobile phone, etc…), user profiles including their physical abilities and backgrounds, and environments of use have raised new research areas. To face this diversity, we believe that a model-based development approach using a refinement process will be a fast growing research area. For instance, to develop multimodal interfaces for many platforms, considering the implementation and the design as two separated steps is the most appropriate method. The interface must be designed at a high level of abstraction where only common concepts to all platforms are captured. Then, a deployment of the interface on a given platform can be made by translating these concepts on the considered platform. To do this, new abstract languages and models are emerging (Paterno & al., 2008).

As it was outlined in this chapter, many studies have been achieved to propose methods for evaluating user interfaces for mobile systems. Some methods are done in laboratories and are called laboratory tests. Others are done in real context of use and they are called field tests. A comparison between the two approaches can be found in (Duh & al., 2006). Some trends tend to fuse both of them. For instance, in (Singh & al., 2006), the authors propose to reproduce the realistic environment in the laboratory. They capture videos at the site of the intended deployment of a location-based service. This video is used to simulate a realistic environment in the laboratory. The systems are then tested in this simulated environment. We believe that both of the fields test, laboratory tests and the hybrid methods are empirical and their results cannot be exhaustive. For these reasons, model-based and formal methods will be established in near future to help designers evaluating multimodal systems (Amant & al., 2007). These approaches are characterized by the complexity of the notations. Indeed, formal models such as temporal logics and transition systems may be difficult to be mastered by user interface designers. This will discourage them to adopt such approaches. To face this problem, we believe that these methods must be supported by tools that implement them in an interactive way. In this case, the complexity of these methods will be hidden. This trend was used for WIMP-based interfaces in (Paterno & al., 2003; Mari & al., 2003) and we believe that it must be extended to the field of mobile multimodal interfaces.

CONCLUSION

In this chapter we have presented and discussed the benefits of using formal methods to model and verify usability properties of MUIs. We have explained how transition systems can be generated automatically from a syntactical expression modeling the MUI. This transition system is generated using the defined semantic rules of the various compositional operators. We have shown how CARE usability properties as well as the adaptability, to situations where a modality becomes disabled, are expressed using temporal logic formulas. These formulas are checked automatically using a model-checking technique implemented in the SMV tool. Our experimentation using a case study has shown how we can verify these properties at an abstract level. If the properties are not satisfied, the designer can change the MUI model before the implementation stage. In this chapter we have studied the CARE and the adaptability properties. These properties characterize mobile systems as well as fixed systems. The adaptability property studied in our approach is more appropriate to mobile systems because it deals with environment changes. Other properties characterizing mobile systems have not been studied by our approach and will be considered in future work.

Even though formal methods have the benefits to be rigorous to model the concepts and they make an exhaustive verification, we are aware that their adoption by MUI designers is not easy. Indeed, these methods need expertise and require much effort which discourages designers to use them. To face this problem, we are currently continuing the effort to develop an interface that allows designers to manipulate graphical objects corresponding to the syntactical components of the formal model. These objects represent tasks, interactive actions and compositional operators. The designer composes these objects interactively. The grammar rules will be implemented to control the composition operation. Once the MUI is designed at the syntactical level, the corresponding transition system is generated automatically. This interface hides the formal notations of these formal models. In this way, the designer does not deal with the formal notations, but he only composes interactive actions and tasks at a semantic level. Then, the generated transition system can be encoded in a model-checker tool input language.

REFERENCES

Abowd, G. D., Wang, H. M., & Monk, A. F. (1995). A formal technique for automated dialogue development. *The first symposium of designing interactive systems - DIS'95*, (219-226). New York: ACM Press.

Abrial, J. R. (Ed.). (1996). *The B book: Assigning programs to meanings*. New York: Cambridge University Press.

Abrial, J.-R. (1996). Extending B without changing it for developing distributed systems. *First conference on the B-Method*, 169-190.

Ait-Ameur, Y., & Ait-Sadoune, I., Baron, M. (2006). Étude et comparaison de scénarios de développements formels d'interfaces multimodales fondés sur la preuve et le raffinement. *MOSIM 2006 - 6ème Conférence Francophone de Modélisation et Simulation. Modélisation, Optimisation et Simulation des Systèmes: Défis et Opportunités*.

Ait-Ameur, Y., Girard, P., & Jambon, F. (1998). A uniform approach for the specification and design of interactive systems: the B method. In P. Markopoulos & P. Johnson (Eds.), *Eurographics Workshop on Design, Specification, and Verification of Interactive Systems (DSV-IS'98)*, 333-352.

Alexader, T. & Dixon, E., (2004). Usability Evaluation report for VPA 2. FASIL deliverable D.3.3.3.

Amant, R., Horton, T., & Ritter, F. (2007). Model-based evaluation of expert cell phone menu interaction. *ACM Transactions on Computer-Human Interaction, 14*(1), 347–371. doi:10.1145/1229855.1229856

Been-Lirn, D. H., & Tan, C. B. G., & Hsueh-hua Chen, V. (2006). Usability evaluation for mobile device: A comparison of laboratory and field tests. *8th Conference on Human-Computer Interaction with Mobile Device and Services, Mobile HCI'06,* 181-186. New York: ACM.

Beringer, N., Kartal, U., Louka, K., Schiel, F., & Turk, U. (2002). A procedure for multimodal interactive system evaluation. *LREC Workshop on Multimodal Resources and Multimodal System Evaluation,* 77-80.

Bold, R. (1980). Put-that-there: voice and gesture at the graphics interface. *Computer Graphics, 14*(3), 262–270. doi:10.1145/965105.807503

Bolognesi, T., & Brinksma, E. (1987). Introduction to the ISO specification language LOTOS. *Computer Networks and ISDN Systems, 14*(1), 25–59. doi:10.1016/0169-7552(87)90085-7

Bouchet, J., Madani, L., Nigay, L., Oriat, C., & Parissis, I. (2007). Formal testing of multimodal interactive systems. *DSV-IS2007, the XIII International Workshop on Design, Specification and Verification of interactive systems,* Lecture Notes in Computer Science. Berlin, Germany: Springer-Verlag.

Bouchet, J., Nigay, L., & Ganille, T. (2005). The ICARE Component-based approach for multimodal input interaction: application to real-time military aircraft cockpits. *The 11th International Conference on Human-Computer Interaction.*

Branco, G., Almeida, L., Beires, N., & Gomes, R. (2006). Evaluation of a multimodal virtual personal assistant. Proceedings from *20th International Symposium on Human Factors in Telecommunication.*

Campos, J., Harrison, M. D., & Loer, K. (2004). Verifying user interface behaviour with model checking. *VVEIS,* (pp. 87-96).

Campos, J., & Harrisson, M. D. (2001). Model checking interactor specifications. *Automated Software Engineering, 8,* 275–310. doi:10.1023/A:1011265604021

Chetali, B. (1998). Formal verification of concurrent programs using the Larch prover. *IEEE Transactions on Software Engineering, 24*(1), 46–62. doi:10.1109/32.663997

Clarke, E. M., Emerson, E. A., & Sistla, A. P. (1986). Automatic verification of finite-state concurrent systems using temporal logic specifications. *ACM Transactions on Programming Languages and Systems, 2*(8), 244–263. doi:10.1145/5397.5399

Coutaz, J., Nigay, L., Salber, D., Blandford, A., May, J., & Young, R. (1995). Four easy pieces for asserting the usability of multimodal interaction: the CARE properties. *Human Computer Interaction - INTERACT'95* (pp.115-120).

d'Ausbourg, B. (1998). Using model checking for the automatic validation of user interface systems. *DSV-IS* (pp. 242-260).

De Bra, P., Brusilovsky, P., & Houben, G. J. (1999). Adaptive Hypermedia: From Systems to Framework. [CSUR]. *ACM Computing Surveys, 31,* 1–6. doi:10.1145/345966.345996

Demumieux, R., & Losquin, P. (2005). Gathering customers's real usage on mobile phones. *Proceedings of MobileHCI'05.* New York: ACM.

du Bousquet, L., Ouabdesselam, F., Richier, J.-L., & Zuanon, N. (1999). Lutess: a specification driven testing environment for synchronous software. *International Conference of Software Engineering* (pp. 267-276). New York: ACM.

Duarte, C., & Carrico, L. (2006). A conceptual framework for developing adaptive multimodal applications. *IUI '06: Proceedings of the 11th international conference on intelligent user interfaces* (pp. 132-139), ACM.

Duh, H. B., Tan, G. C. B., & Chen, V. H. (2006). Usability evaluation for mobile devices: A comparison of laboratory and field tests. [New York: ACM.]. *MobileHCI*, *06*, 181–185. doi:10.1145/1152215.1152254

Duke, D., & Harrison, M. D. (1993). Abstract interaction objects. In *Eurographics conference and computer graphics forum, 12* (3), 25-36.

Duke, D., & Harrison, M. D. (1995). Event model of human-system interaction. *IEEE Software Engineering Journal*, *12*(1), 3–10.

Emerson, E. A., & Clarke, E. M. (1982). Using branching time temporal logic to synthesize synchronization skeleton. *Science of Computer Programming*.

Garavel, H., Lang, F., Mateescu, R., & Serwe, W. (2007). CADP: A toolbox for the construction and analysis of distributed processes. In *19th International Conference on Computer Aided Verification CAV 07* (pp. 158-163).

Große, D., Kuhne, U., & Drechsler, R. (2006). HW/SW co-verification of embedded systems using bounded model checking. In *The 16th ACM Great Lakes symposium on VLSI*, (pp 43-48). New York: ACM.

Harchani, M., Niagy, L., & Panaget, F. (2007). A platform for output dialogic strategies in natural multimodal dialogue systems. In *IUI'07*, (pp. 206-215). New York: ACM.

Jourde, F., Nigay, L., & Parissis, I. (2006). Test formel de systèmes interactifs multimodaux: couplage ICARE – Lutess. *ICSSEA 2006, 19ème journées Internationales génie logiciel & Ingènierie de Systèmes et leurs Applications, Globalisation des services et des systèmes*.

Loer, K., & Harrison, M. D. (2002). Towards usable and relevant model checking techniques for the analysis of dependable interactive systems. *ASE*, (pp. 223-226).

MacColl & Carrington, D. (1998). Testing MATIS: a case study on specification based testing of interactive systems. *FAHCI*.

Markopoulos, P. (1995). On the expression of interaction properties within an interactor model. In *DSV-IS'95: Design, Specification, Verification of Interactive Systems* (pp. 294-311). Berlin, Germany: Springer-Verlag.

McMillan, K. (1992). *The SMV system*. Pittsburgh, PA: Carnegie Mellon University.

Mori, G., Paterno, F., & Santoro, C. (2003). Tool support for designing nomadic applications. In *IUI '03: Proceedings of the 8th international conference on Intelligent user interfaces* (pp.141—148). New York: ACM.

Moussa, F., Riahi, M., Kolski, C., & Moalla, M. (2002). Interpreted Petri Nets used for human-machine dialogue specification. *Integrated Computer-Aided Eng.*, *9*(1), 87–98.

Navarre, D., Palanque, P., Dragicevic, P., & Bastide, R. (2006). An approach integrating two complementary model-based environments for the construction of multimodal interactive applications. *Interacting with Computers*, *18*(5), 910–941. doi:10.1016/j.intcom.2006.03.002

Nigay, L., & Coutaz, J. (1995). A generic platform for addressing the multimodal challenge. *International Conference on Human-Computer Interaction*, (pp.98-105) ACM.

Nigay, L., & Coutaz, J. (1996). Espaces conceptuels pour l'interaction multimédia et multimodale. *TSI, spéciale multimédia et collecticiel*, 15(9), 1195-1225.

Palanque, P., Bastide, R., & Sengès, V. (1995). Validating interactive system design through the verification of formal task and system models. In L.J. Bass & C. Unger (Eds.), *IFIP TC2/WG2.7 Working Conference on Engineering for Human-Computer Interaction (EHCI'95)* (pp. 189-212). New York: Chapman & Hall.

Palanque, P., & Schyn, A. (2003). A model-based for engineering multimodal interactive systems. *9th IFIP TC13 International Conference on Human Computer Interaction (Interact'2003)*.

Pascoe, J., Ryan, N., & Morse, D. (2000). Using while moving: human-computer interaction issues in fieldwork environments. *Annual Conference Meeting Transactions on Computer-Human Interaction*, (pp. 417-437). New York: ACM.

Patermo, F., Santoro, C., Mäntyjärvi, J., Mori, G., & Sansone, S. (2008). Autoring pervasive multimodal user interfaces. *International Journal Web Engineering and Technology*, 4(2), 235–261. doi:10.1504/IJWET.2008.018099

Paterno, F., & Mezzanotte, M. (1994). Analysing MATIS by interactors and actl. *Amodeus Esprit Basic Research Project 7040, System Modelling/WP36*.

Paterno, F., & Santoro, C. (2003). Support for reasoning about interactive systems through human-computer interaction designers' representations. *The Computer Journal*, 46(4), 340–357. doi:10.1093/comjnl/46.4.340

Petridis, P., Mania, K., Pletinckx, D., & White, M. (2006). Usability evaluation of the EPOCH multimodal user interface: designing 3D tangible interactions. In *VRST '06: Proceedings of the ACM symposium on Virtual reality software and technology*, (pp. 116-122). New York: ACM Press.

Plotkin, G., (1981). A structural approach to operational semantics. Technical report, Departement of Computer Science, University of Arhus DAIMI FN 19.

Portillo, P. M., Garcia, G. P., & Carredano, G. A. (2006). Multimodal fusion: A new hybrid strategy for dialogue systems. In *ICMI'2006*, (pp. 357-363). New York: ACM.

Ruf, J., & Kropf, T. (2003). Symbolic verification and analysis of discrete timed systems. *Formal Methods in System Design*, 23(1), 67–108. doi:10.1023/A:1024437214071

Serrano, M., Nigay, L., Demumieux, R., Descos, J., & Losquin, P. (2006). Multimodal interaction on mobile phones: Development an evaluation using ACICARE. In *8th Conference on Human-Computer Interaction with Mobile Device and Services, Mobile HCI'06* (pp. 129-136). New York: ACM.

Singh, P., Ha, H. N., Kuang, Z., Oliver, P., Kray, C., Blythe, P., & James, P. (2006). Immersive video as a rapid prototyping and evaluation tool for mobile and ambient applications. *Proceedings of the 8th conference on Human-computer interaction with mobile devices and services*. New York: ACM.

Spivey, J. M. (1988). Understanding Z: A specification language and its formal semantics. *Cambridge Tracts in Theoretical Computer Science*. New York: Cambridge University Press.

Stent, A., Walker, M., Whittaker, S., & Maloor, P. (2002). User-tailored generation for spoken dialogue: An experiment. In *ICSLP '02* (pp. 1281-84).

Thevenin, D., & Coutaz, J. (1999). Plasticity of user interfaces: Framework and research agenda. In A. Sasse & C. Johnson (Eds.), *IFIP INTER-ACT'99: Human-Computer Interaction, 1999 (Mobile systems)*, (pp. 110-117). Amsterdam: IOS Press.

Van-Eijk, P. (1991). The lotosphere integrated environment. In *4th international conference on formal description technique (FORTE91)*, (pp.473-476).

Chapter 4
Platform Support for Multimodality on Mobile Devices

Kay Kadner
SAP AG, Germany

Martin Knechtel
SAP AG, Germany

Gerald Huebsch
TU Dresden, Germany

Thomas Springer
TU Dresden, Germany

Christoph Pohl
SAP AG, Germany

ABSTRACT

The diversity of today's mobile technology also entails multiple interaction channels offered per device. This chapter surveys the basics of multimodal interactions in a mobility context and introduces a number of concepts for platform support. Synchronization approaches for input fusion and output fission, as well as a concept for device federation as a means to leverage from heterogeneous devices, are discussed with the help of an exemplary multimodal route planning application. An outlook on future trends concludes the chapter.

INTRODUCTION

Mobile devices have become an essential part of our daily life. They help us to fulfill various tasks like communicating with others, gathering data while working mobile, accessing information and content on the Internet or a company's intranet, or scheduling business or private activities. One of the most important aspects of mobile devices in general is their vast heterogeneity; mobile devices are available in very different shapes and form factors belonging to a wide range of device

DOI: 10.4018/978-1-60566-978-6.ch004

classes like mobile phones, smartphones, PDAs, subnotebooks, or notebooks. This has far reaching implications on the interaction between the user and the mobile device. Each device is built of different technological components, providing different input and output capabilities (e.g., keyboard, numeric keyboard, pen, display, or speech) and communication interfaces (like WiFi, Bluetooth, etc.). Moreover, users may be in different situations while using mobile devices (e.g., in a noisy urban environment, environments with bad lighting conditions, at work or in-car with no hands free to use a keyboard). Thus, the way a user interacts with an application on mobile devices strongly depends on the user's situation and preferences, the modalities and capabilities of the used device, and the type of application or task that should be performed.

Exploiting the available input and output capabilities of devices requires a set of functionalities that an application has to provide. These functionalities comprise the access and management of available input and output devices, the recognition, interpretation, and integration of different input streams, and the invocation of application functions due to particular user interactions. Furthermore, output has to be handled in terms of generation of output as well as styling and rendering of output according to the requirements of different output devices. Most of this functionality can be regarded as essential and thus, has to be considered in almost every multimodal application.

Instead of implementing this functionality from scratch for each new application, a generic multimodal platform can enable widespread reuse of common functionality, more efficient development in terms of time and cost, more stable and reliable implementations due to the reuse of tested code as well as standardization. Aspects such a platform would have to cope with include:

Heterogeneous input and output devices: Abstracting from the underlying, mostly heterogeneous hardware to allow reusable and cost-efficient implementation of components for multimodal interaction;

Arbitrary/Flexible combination of input and output modalities: Allowing the user to interact with multiple devices in parallel to achieve different levels of multimodality like parallel and/or complementary user input and output;

Seperation of application logic from user interactions: A clear separation of concerns by the provision of appropriate abstractions for the association of multimodal interactions with the internal logic and data model of an application supports the development of applications independent from particular considerations about interaction modalities.

Provision of general functionality for high reusability: Offering a standardized execution environment, that provides general functions for managing and controlling multimodal interaction components.

Support for distribution of components: Distributing components within a network of computing devices to exploit resources and capabilities of available devices to improve user interactions and to meet the components' requirements on hardware resources;

A brief overview about current standards and standardization activities in the area of multimodality is given in Section 2. The Multimodal Interaction Working Group (MMI) of the World Wide Web Consortium (W3C) has created a number of official documents covering basic aspects of multimodal platforms. These documents are introduced as a foundation for the understanding of platform related issues of multimodal systems. Moreover, related work dealing with platform related issues is discussed.

In Section 3, our multimodal platform and its concepts are described in detail. We introduce the Multimodal Route Planning application as an example to illustrate the features of our platform. All introduced concepts contribute to the fulfillment of the requirements for multimodal platforms listed above. In particular, we introduce an architecture with abstractions for a clear separation of application logic and data at the one hand and the handling of multimodal interaction based on a heterogeneous set of input and output devices on the other hand. Major aspects considered by the introduced abstraction layers are the management of input and output modalities which might reside on a set of interconnected devices, the fusion of parallel input streams based on the adoption of Petri-nets a description model as well as the fission of output streams and their simultaneous output in heterogeneous network and device infrastructures.

For the design and implementation of our platform we adopt a component-oriented approach, allowing for an easier separation of functional business logic from the multiple non-functional aspects of a multimodal system. In our solution, components are loosely coupled and thus easily distributable in a ubiquitous environment. Different implementations and deployment locations of the same functional service interface can be seamlessly selected according to non-functional criteria. As an extension of component distribution, the concept of *federated devices* based on OSGi (Open Services Gateway initiative) was implemented as a component platform. Each device can manage its components. This includes installing, uninstalling, migrating, and replacing components on-the-fly depending on the current availability of devices and modalities. The overall composition context is maintained with the help of service-oriented directories. The chapter closes with a look at future trends in Section 4 and a summary in Section 5.

BACKGROUND

The Multimodal Interaction Activity (Multimodal Interaction Working Group, 2002a) driven by the Multimodal Interaction Working Group of the World Wide Web Consortium strives for extending the Web to allow users to use modes of interaction that fit to their current goals, environment, and requirements. This especially includes accessibility support for impaired users as they often can only use a limited set of modalities and have additional requirements. As a result of that activity a set of documents has been created covering different aspects of multimodality. The documents comprise use cases and requirements for multimodal interactions, markup languages as well as documents dealing with the architecture of multimodal systems. Major results of the Multimodal Interaction Activity relevant to the platform aspects are the Multimodal Interaction Framework and the Multimodal Architecture. These specifications describe basic principles for multimodal systems and will be described as a foundation for the understanding of the concepts for multimodal platforms introduced in the main part in this chapter. Moreover, related work is discussed in the second part of this section.

Multimodal Interaction Framework

The Multimodal Interaction Framework (MMI-Framework) (Multimodal Interaction Working Group, 2002b) defines a conceptual framework for multimodal systems, identifying the necessary components, their interactions with each other, and functional as well as non-functional aspects. Although it is only a W3C Note, it represents a framework, which is implemented in one or the other way by every multimodal application and thus, can serve as a basis for understanding and implementing multimodal systems.

In particular, the MMI-Framework specifies a set of functions that are required in a multimodal system. It is organized in categories for input processing, output processing, and system functions (see Figure 1). Input from the user must first be *recognized* to make it accessible within a computing system. Because multiple input modalities might be used in parallel, multiple recognition processes have to be performed. Each recognition instance is specific to the input device and

Figure 1. Multimodal Interaction Framework interpreted as role model

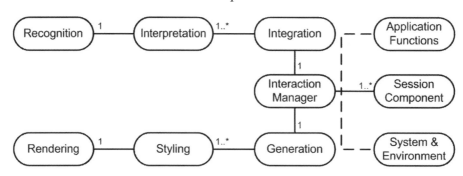

modality used for interaction. This includes the access of heterogeneous input devices as one of the tasks the application has to fulfill. Following the recognition, a semantic *interpretation* of the input data has to be performed. For example, an application could support keyboard and speech input in parallel. The speech has to be recognized from an audio stream captured by a microphone and interpreted according to a given grammar. In parallel the input stream from the keyboard has to be recognized in terms of system internal key codes and interpreted according to data input expected by the application.

The interpreted data from both input streams can than be *integrated*, possibly with input from further modalities. As a result of the integration precise instructions for the *Interaction Manager* are identified, which connects user interface interactions to *Application Functions*.

The following example explains the categories (roles in a role model) of constituents of the MMI-Framework and their purpose. Consider a mobile web browser presenting a web page of a book catalogue, which provides visual and vocal output of its content and mouse, keyboard and voice for input interactions. If the user says "open", it is *recognized* as raw audio by the microphone and its driver software. It has to be *interpreted*, which means that the raw audio is transformed to a command. The same applies to mouse and keyboard input. The user hovers his mouse cursor over a hyperlink and the coordinates are transmitted to

being *integrated* with other input. The integration component merges the acoustic "open" command and the cursor position to a composite command that opens a particular hyperlink in the browser. This process is called **input fusion**. The *interaction manager* executes this command, retrieves the response from the *application functions* and forwards it to the *generation*. The response is transformed according to the used modalities (visual and voice). This process is also referred to as **output fission** of modalities. It creates one document for the visual representation that shows the book's summary and purchase details and one representation for the audio output that reads a sample paragraph of the book. The visual document is *styled* with small fonts for being readable on the mobile device and a fast speaking female voice is used for styling the audio document according to the user's preferences. Finally, these documents are processed by the browser's parts for visual and voice output, which means that they are *rendered* to the user.

The MMI-Framework suggests a variety of W3C standards for information exchange between some of the framework components like EMMA, XHTML, SVG, SSML, and CSS. However, the MMI-Framework does not give a hint of which user interface description language can be used between *application functions*, *interaction manager*, *generation* and subsequently *styling*. They refer to it as an *internal representation*, which leaves room for interpretation and proprietary solutions.

An interesting property of the MMI-Framework was discovered in the project SNOW (Burmeister, Pohl, Bublitz & Hugues, 2006): Especially when considering very modular implementations with each necessary role of the MMI-Framework being assigned to a software component, it is useful to think of the MMI-Framework as a *role model* (Pereira, Hartmann & Kadner, 2007). Each component fulfills a particular role and has connections of variable multiplicity to other roles. The result is depicted in Figure 1. However, the MMI-Framework uses the term *components* to refer to these roles, even though they would not necessarily be realized by separate software components. During the implementation of the SNOW prototype, indirect communication between roles that are directly connected in the MMI-Framework was observed. As example: The styling role is directly connected to the generation role. In the SNOW prototype, the component for styling of voice output was located on a separate server and received its input from the browser on the mobile device. However, this browser is responsible for rendering visual output and not generation as suggested by the MMI-Framework. This leads to inconsistent representations of the system architecture in relation to the MMI-Framework and might cause confusions about component roles, their connection and interaction. Therefore, using the slightly more abstract view of a role model to maintain the distinction between the MMI-Framework roles and the software components that realize those roles is more suitable. Therefore, the MMI-Framework is easier to understand and obey when developing a multimodal system.

Multimodal Architecture

The Multimodal Architecture (MMA) (Multimodal Interaction Working Group, 2006) is an architecture specification implementing the MMI-Framework. The MMI Working Group describes interfaces and interactions between roles of the MMI-Framework on a high level because the MMA should be general and flexible enough to allow interoperability between components from different vendors. Therefore, MMA is referred to as a *framework* by the MMI working group, which should obviously be interpreted in a technical sense.

Basic goals of the MMA are *Encapsulation* (components are black boxes), *Distribution* (support distributed implementations), *Extensibility* (add new modality components at runtime), *Recursiveness* (package components of the architecture to be included in higher level architecture) and *Modularity* (Model-View-Controller paradigm). These goals are not surprising since they are more or less relevant in any software project envisioning a component-based implementation without restrictions on distribution.

However, because of the MMI Working Group's origin as a W3C Working Group, most of the content is considered to be realized using document-oriented descriptions of the user interface and control flow. Nevertheless, this decision is left to the actual realization. The most comprehensively specified aspect of MMA (besides the architecture itself) is the eventing and life cycle of modality components. The MMA defines that modality components must handle asynchronous events, which are delivered reliably and in the order in which they were sent. Registration and general management aspects of modality components at the runtime framework are specified by the life cycle events.

To draw a conclusion, the MMA is a more specific framework for application developers than the MMI-Framework but it still has a number of open issues. It is open how the runtime behaviour of such a system regarding synchronization of multiple user interfaces has to be treated, how components are dispersed in a distributed implementation and which problems can occur in a distributed and multi-device environment. These aspects, among others, are discussed in this chapter.

SMIL for Synchronizing Distributed User Interfaces

SMIL (Synchronized Multimedia Working Group, 2008) is a language recommended by the MMI-framework for coordinating the presentation of multiple heterogeneous media steams. It provides special facilities to temporally align output of information. For instance, a video could be specified to start 10 seconds after its textual abstract was shown. Thus, SMIL can be leveraged for synchronized rendering of multiple output modalities. However, SMIL was designed for synchronizing output presented on one (local) device, which poses a severe limitation for potentially distributed multimodal systems. If one renderer is located on a different device, at least the network latency must be taken into account. We will discuss ways to handle such synchronization issues in Section 3.

Related Work

Multimodal interaction – especially with mobile devices – is an actively researched area. Many approaches, concepts, and prototypes exist for various aspects of multimodal interaction systems. Main aspects include multimodal input handling in particular application scenarios, distribution of multimodal systems, and handling of multiple devices. We will discuss these approaches in more detail in the following paragraphs.

A variety of systems support multimodal interaction based on special use cases. *QuickSet* (Cohen et al., 1997), for instance, was intended to support military training by allowing a multimodal way of issuing commands to control troops from a headquarter. *Embassi* (Elting et al., 2003) aims to enable people with limited computer knowledge to interact with their home environment like consumer electronic as convenient as possible. An extension by arbitrary applications or new modalities causes major implementation effort. *Nightingale* (West et al., 2004) is a ubiq-uitous infrastructure with support for distributed applications. Goal of Nightingale is to decouple multimodal applications from particular input devices. Although this is a more generic approach, it provides only a static platform with predistributed components and requires to implement detailed aspects of the multimodal system that could be realized by platform components (like a grammar for voice recognition). All aforementioned systems are more or less aligned to basic concepts of the MMI-Framework.

The *Cameleon Runtime Architecture* (Balme et al., 2004) is a reference model for the technical infrastructure of distributed, plastic and migratable user interfaces. The architecture defines a platform layer and an interactive systems layer. These layers are connected by the DMP (Distribution, Migration, Plasticity) middleware. The reference model focuses on a technical infrastructure for the distribution of user interfaces over multiple devices and the migration of the user interface between devices while preserving their continuous usage by adapting them to external constraints. The Cameleon middleware is therefore quite heavyweight and does not focus on the peculiarities of the distribution and adaptation of multimodal user interfaces for mobile devices.

Another set of systems enables accessing one application through multiple devices, which is also an important aspect of our architecture. This kind of application access is often called "Co-Browsing" because it allows users to cooperatively browse through content. Examples are the *Multimodal Browsing Framework* (Coles, Deliot, Melamed & Lansard, 2003) and the *Composite Device Computing Environment* (Pham, Schneider & Goose, 2000). The first example is realized as a set of independent browsers, which are hidden from the application by a dedicated proxy. Their coordination (e.g., of user input) is performed directly among each other, which assumes that every browser is reachable from every other browser. An integration of parallel input is not possible since the system has no central component for that and decentralized integration could lead to different results and eventually inconsistent representation. The latter example integrates devices in the user's vicinity with his device. The user can choose which device has to show which information. The distribution is coordinated by a central gateway that represents a bottleneck of this approach because it is mandatory and cannot be moved within the infrastructure. Finally, the Pebbles system allows developing extensions for applications that allow the user to remotely control a desktop application from his mobile phone (Nichols & Myers, 2006). A special extension has to be developed for each application to be controlled. A flexible configuration allowing control of arbitrary applications is not possible.

One approach of creating device federations is the Mundo system (Aitenbichler, Kangasharju & Mühlhäuser, 2007). Mundo is a ubiquitous infrastructure, which integrates the hardware resources of different devices. By that, it allows the CPU of a Laptop to be used by a federated PDA for complex tasks, for instance. However, the system does not consider multimodality and the specific requirements like special treatment of user input and output, synchronization and flexible component deployment.

All of the presented works contain a variety of interesting viewpoints and tackle important aspects but do not qualify as a platform for multimodal applications.

PLATFORM CONCEPTS

This section describes the various concepts developed for a multimodal interaction platform, which is suitable for use with mobile devices. The platform supports the development of special purpose applications tailored for specific needs as well as general purpose applications, which can be used to access a variety of target applications (e.g., the application logic in form of a Web application, which has a separated user interface that can be easily exchanged by a multimodal front end). Most of the concepts are agnostic of the actual type of application. To understand the platform concepts better, the *Multimodal Route Planner* as a typical mobile application is introduced first. Different platform concepts are then explained with the help of this example.

Multimodal Route Planner

The Multimodal Route Planner is a typical example of a mobile application running on a PDA. Its main functions comprise calculating driving directions (routes) between points-of-interest (POI) and the distance between two POI, as well as retrieving detailed information about POI. A POI can be any place of importance, such as a city or a historic site. Figure 2 shows an annotated screenshot of the Multimodal Route Planner.

In our scenario, the user wants to drive from Leipzig to Dresden to visit the Church of Our Lady. In her car, she places her PDA in a cradle and initiates a device federation between an in-car PC and her PDA. The in-car PC features a voice output system and a voice recognition system. The device federation allows the Multimodal Route Planner application on the PDA to use voice input and output capabilities of the in-car PC. A Wireless LAN provides the physical connection between the in-car PC and the PDA.

Before the journey starts, the user initiates the calculation of the driving directions from Leipzig to Dresden using the input modalities provided by the PDA and the in-car PC in a synergistic way. In parallel to the utterance "Calculate a route from here to here", she points to the POI symbols that represent the cities of Leipzig and Dresden on the map displayed on the PDA's touch screen display using a stylus (Figure 3).

During the ride, the in-car PC provides directions via the car's audio system and processes voice commands targeted at the Multimodal Route Planner application running on the PDA. The PDA displays the directions on an electronic map.

Figure 2. Multimodal route planner application

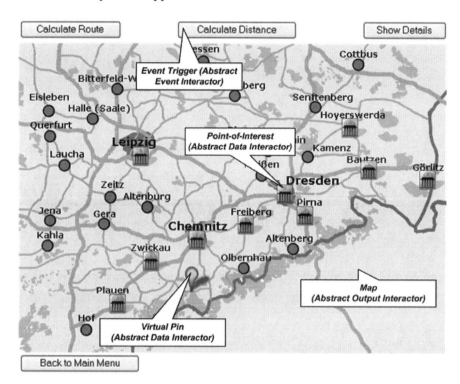

Figure 3. Example of a multimodal user interaction initiating a route calculation

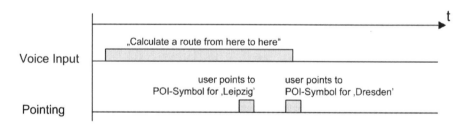

To enter locations that are not displayed as Points-of-Interest, the user has the possibility to use the virtual pin. The virtual pin can be moved on the map by dragging it with the stylus. Alternatively, it can be moved by voice commands like "move north", "move south", etc. To move the virtual pin, the user is free to choose one of the available input modalities.

Arriving in Dresden, she leaves her car at a place for park and ride to continue her travel unsing public transport. The device federation between the in-car PC and her PDA is terminated. Now the PDA supports her with visual and vocal information about walking directions and suggests bus and tram connections.

Multiple challenges that require platform support arise from our scenario. A device federation is a distributed system in which input and output components are distributed over multiple mobile devices, in our example an in-car PC and a PDA. These components provide input and output modalities for target applications and perform distributed input interpretation or output rendering. They are executed as parallel distributed processes that communicate via a wireless network like Wireless LAN, GSM, or UMTS. In our scenario, the components are the foundation of the distributed multimodal user interface of the Multimodal Route Planner application. Furthermore, the availability of input and output components varies depending on the context of use. The Multimodal Route Planner application must adapt to dynamic changes of the context of use by utilizing the voice input and output components on the in-car PC when they become available in the device federation.

Our Multimodal Route Planner application benefits from platform support at multiple points. Input and output components appear dynamically and must be integrated in the target application when they become available. Furthermore, the user interface must be *extended* to these new components. Equally, the user interface must be modified to account for components that become unavailable. Ideally, such changes in the availability of modalities should affect the user and the application as little as possible. Parts of the multimodal user interface may sometimes need to be *migrated* from one component to another. In our scenario, the voice output is migrated from the in-car PC to the PDA when the user leaves the car without disrupting the user interaction or the execution of the Multimodal Route Planner application. Neither the application logic nor its developer can foresee all possible configurations of available modalities and component distributions. Therefore, the dynamic extension and migration of the multimodal user interface goes hand-in-hand with the need for a generic abstraction concept that makes target applications independent of the available modalities. Furthermore, mechanisms for input fusion that account for the limited resources of mobile devices must be provided by the platform to support inputs like the route calculation example.

Another key challenge that requires platform support is the synchronization of multiple output streams, which should be presented on a set of federated devices. A federation is created, if two or more devices are used together to achieve a common goal (Kadner, 2008). This means that they are connected on a semantic level, for instance, because they partially render a user interface belonging to the same dialog and the same application. These devices are considered to be independent from each other, which means that they can be run and used independently. For instance, a PDA serving voice output and a laptop serving visual output of the same application are called *federated devices* (Braun & Mühlhäuser, 2005). Heterogeneous networks technologies with different data rates and latencies and devices with different capabilities in terms of processing time required to render a particular output stream prevent multimodal output from being synchronous. Therefore, dedicated synchronization techniques are required in distributed multimodal systems in general and device federations in particular to compensate these effects.

In the following sections, a concrete solution is presented that builds upon a clear separation of concerns, abstraction concepts for multimodal user interfaces, and a modular architecture to meet these challenges. First, our abstraction concept and the distributed processing of input and output are introduced. Afterwards, focus is put on the lightweight fusion of multimodal input and the synchronization of multimodal output is addressed in detail. Finally, the migration of components between different devices is discussed.

Platform Architecture and Abstraction Concepts

In this section, the architecture of our multimodal platform and the abstract concepts are presented. Figure 4 shows the architecture of our multimodal platform. The application logic and the data model of a target application form the core of the platform architecture. The core is wrapped by an *Abstract User Interface* layer, which isolates the application logic and the data model from the input and output processing in different modalities and the synchronization mechanisms. The synchronization mechanisms that are necessary for the distributed processing of multimodal input and output on federated devices are encapsulated in the synchronization layer. The outer layer of our architecture is the *Concrete User Interface* layer. It consists of input components and output components that are distributed over federated devices. Each input component implements one input modality. It consists of an input recognizer and a semantic interpretation layer for interpreting unimodal user input. Each output component implements one output modality. It consists of an output renderer and a styling layer. Input components and output components can be plugged in and

removed from our multimodal platform without disrupting the operation of the target application. The fusion of multimodal input is implemented in the fusion layer of our architecture. The rest of this section discusses the layers of our architecture in detail.

Concerns are separated by decomposing an application into three parts, closely resembling the Model-View-Controller pattern (Fowler, 2003). They are represented by the multimodal user interface, the application logic, and the data model. The application logic provides for instance the route calculation in the Multimodal Route Planner example. The data model contains all variables that can be presented and manipulated via the multimodal user interface. In the Multimodal Route Planner example, the position of the virtual pin on the map is described by a *(longitude, latitude)* pair. The value pairs are stored in a variable of type GeoCoordinate in the data model.

Figure 4. Overall architecture of a multimodal system with the view layer split into an abstract user interface and multiple concrete user interface realizations. The abstract user interface layer isolates the application logic from modality-specific input and output processing.

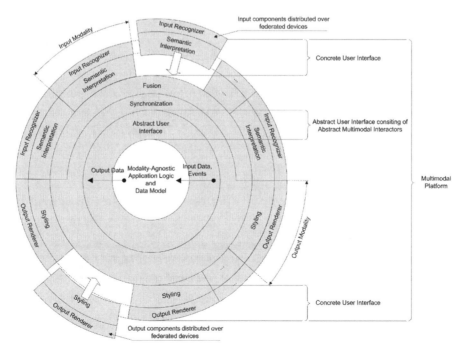

Moving the pin on the map changes the value of the variable in the data model.

In the example above, neither the input modality used to manipulate the pin's position nor the way the GeoCoordinate is presented to the user is of interest for the data model or the application logic. In our approach, the MMI-Framework concepts (see Section 2) are followed and the application logic as well as the data model of a multimodal application remains *modality-agnostic*, i.e., isolated from the peculiarities of multimodal input processing and output generation. This isolation has a multitude of advantages: The development of the application logic and the data model can focus on the aspects relevant to the domain of the application. It does not need to consider the modalities that will be available to the user. Flexibility is increased since the application logic and data model can be employed with arbitrary combinations of modalities. Consequently, the modalities that will be available to the user can be chosen freely by a user interface designer or even based on their availability at runtime.

To implement isolation in our platform, an Abstract User Interface is introduced as an abstraction layer between the modalities on the one side and the application logic and data model on the other side (see Figure 4). The Abstract User Interface is abstract in terms of its representation in a concrete modality and in terms of the functions of its constituents. The constituents of the Abstract User Interface are *Abstract Multimodal Interactors*. Abstract Multimodal Interactors are classified according to their function into three interactor types: *Abstract Data Interactors* for entering and manipulating data, *Abstract Event Interactors* for triggering events, and *Abstract Output Interactors*.

Each Abstract Data Interactor is bound to a variable v in the data model. The function of the Abstract Data Interactor is to make the value of v perceivable for the user and to provide the user with means to manipulate the value of v. The function of Abstract Event Interactors is to trigger events,

which are passed to event handlers provided by the application logic. Abstract Output Interactors are "output only". They are bound to a variable in the data model and make its value perceivable to the user. In contrast to Abstract Data Interactors, they do not allow the manipulation of the variable.

Since the Abstract User Interface abstracts from concrete representations, the user can not interact directly with it. To enable user interaction, the Abstract User Interface must be associated with a concrete user interface. The concrete user interface layer in Figure 5 spans multiple input and output modalities. These modalities are implemented by input and output components that are distributed across federated devices. In our approach, the concrete user interface is dynamically generated from an Abstract User Interface at runtime. For each Abstract Multimodal Interactor in the Abstract User Interface, multiple *Modality-specific Representations* are generated by a transformation and associated with their Abstract Multimodal Interactor (see Figure 5). Examples for Modality-specific Representations of Abstract Multimodal Interactors are shown in Figure 2 with the interactor type in parentheses.

The Modality-specific Representations act as proxies for their Abstract Multimodal Interactor in the different modalities. The state of the Modality-specific Representations of their Abstract Multimodal Interactor must be continuously synchronized in reaction to events occurring on Abstract Multimodal Interactors and Modality-specific Representations.

State changes in response to events can occur on Abstract Multimodal Interactors and on Modality-specific Representations. In both cases, state changes are received by the Abstract Multimodal Interactor and propagated to its Modality-specific Representations based on a bidirectional state synchronization protocol, which is informally presented in the following.

Events occurring on Abstract Multimodal Interactors are activation and deactivation events as well as changes of the variable value bound

Figure 5. An Abstract Data Interactor and its Modality-specific Representations

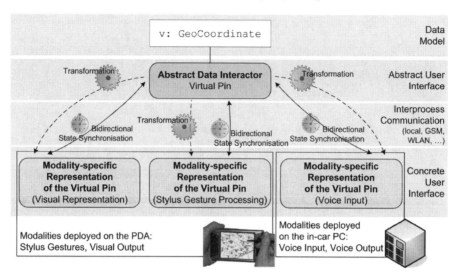

to Abstract Data Interactors and Abstract Output Interactors. Just like graying out an interactor in classical GUIs, the activation state of an Abstract Multimodal Interactor centrally enables or disables user interaction with its Modality-specific Representations. The synchronization protocol therefore activates or deactivates the Modality-specific Representations when their associated Abstract Multimodal Interactor is activated or deactivated.

Events occurring on a Modality-specific Representation are input events. They are generated by the Modality-specific Representation after the user completes an input and the Modality-specific Interpretation of the input (see below) is performed. The interpretation result is wrapped in an input event which is sent to the Abstract Multimodal Interactor. In the case of Abstract Data Interactors, input events change a variable in the data model. This change must be reflected by all Modality-specific Representations of the Abstract Multimodal Interactor. Therefore, the value of variables bound to Abstract Multimodal Interactors is constantly monitored. Upon a change, the affected Abstract Multimodal Interactors are notified and propagate the value change

to all Modality-specific Representations using the bidirectional state synchronization protocol.

In the architecture (see Figure 4), the synchronization layer between the Abstract User Interface layer and the concrete user interface layer implements the bidirectional state synchronization. At this point, the advantages of the central user interface state held by the Abstract User Interface layer are obvious. Due to the fact that the Abstract User Interface holds the activation state of all Abstract Multimodal Interactors as well as the data model state, the state synchronization protocol can be utilized to handle the addition and removal of new modalities to a device federation at runtime and transparently. Whenever a new modality becomes available, the Modality-specific Representations for this new modality can easily be synchronized with the state of the Abstract User Interface and the data model. Together with the modular architecture of the concrete user interface layer, our multimodal platform is open for different renderer and recognizer implementations that can be plugged-in, removed, or replaced depending on their availability at runtime.

As pointed out earlier in this chapter, the synchronization mechanisms presented above must also take the heterogeneous parameters of the network connections between federated devices and the processing power of these devices into account in order to prevent multimodal output from being asynchronous. The timing of multimodal output must therefore be considered in the implementation of the state synchronization protocol. An approach that accounts for these network and device parameters is presented in the following. Afterwards, our approach for the modality-specific processing of input and output as well as the fusion of multimodal input is presented.

Fission and Output Synchronization

The PDA and the in-car PC of the Multimodal Route Planner example are used simultaneously to provide voice directions via the car audio system and a map fragment on the PDA display. Both output streams need to be synchronized. Even for simple interactions like text input, a synchronized feedback is needed. In our scenario, the user can type in a search field on the PDA to find a gas station along the route, and the feedback for the typed text is provided on the PDA as well as via the car sound system. As response to user input, the application logic typically generates a user interface description document. As described above, this document should abstract from possible modalities, thus being modality independent. Several approaches of how to create one user interface and map it to multiple devices have been investigated in the literature in addition to the one described above, (please refer to Göbel, Hartmann, Kadner & Pohl, 2006, OASIS User Interface Markup Language (UIML) Technical Committee, 2008, or Mori, Paternó & Santoro, 2004). However, this document must always be transformed by some component (the *Generation* role in MMI-Framework, see Section 2) into modality specific representations according to the modalities that are currently employed by the user. Such transformations can reach from simple document splitting to complex transforma-

tions of the document's semantics. For instance, composite (different input is composed to higher level input) vs. simultaneous (different input is treated independently) input may influence how output is presented.

Resulting modality specific user interface descriptions are sent from the *Generation* role on their way to the user through a chain of *Styling* and *Rendering* roles. Assuming a high degree of modularization, all roles of the MMI-Framework may be realized by physically separate components and there might be multiple output and input modalities, as depicted in Figure 1.

In distributed environments, one might face heterogeneous network technologies and devices, which all have different processing capabilities regarding to the time required for processing a particular amount of data. For one modality, an in-car PC with the application server on the same device might be used, while modalities might be provided by a mobile phone being connected via a wireless network (e.g., UMTS, WiFi). The heterogeneity may cause the distributed user interface to be updated asynchronously, as depicted in Figure 6. Asynchronous updates confuse the user, since the individual partial updates are the response of one input interaction (like a mouse click) and belong to one dialog step. Therefore, a concept to align user interface updates is needed to ensure a synchronized user experience. As an example, the concept of forward synchronization of distributed user interfaces (Kadner & Pohl, 2008) is elaborated in the following paragraphs.

The Forward Synchronization Approach

The basic idea of forward synchronization is to determine each modality's *modality time* T_{mod}, which defines the required delay for a particular amount of data representing a user interface description to be presented to the user. The largest modality time represents the reference time T_{ref}, because it is the earliest point in time for synchronous updates.

Figure 6. Asynchronous update of distributed user interfaces

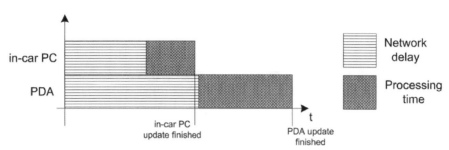

Based on this information, the faster modalities are slowed down so that their update is done simultaneously with the slowest modality.

Response Time of User Interfaces

Latency is an important parameter for usability. The user interface reacts after a specific amount of time to user input. This *response time* should be below thresholds described by usability engineering in (Nielsen, 1994, Chap. 5). Up to $0.1s$ the feedback seems instantaneously, up to $1.0s$ the user retains his train of thought, but the delay is perceived. With increasing response time, the causality between action and feedback gets lost increasingly and above $10s$ some progress indicator like an hourglass or a progress bar is adequate.

Nah (Nah, 2004) empirically measured the tolerable waiting time (TWT) users are willing to accept when interacting with Web pages. Since there is a broad range of thresholds stated in the literature ranging from $1s$ to $41s$, Nah carried out a user study with 70 students. The study revealed that simpler progress indicators are already sufficient to increase the TWT, a sophisticated progress bar is not essential. He concludes that generally a waiting time of $2s$ should not be exceeded. Rich Internet Applications (RIA) try to reduce the response time for Web applications by moving parts of the application logic to the local client.

These results for single user interfaces motivated us to consider response time also for distributed user interfaces. In the next section, different update types for distributed user interfaces are discussed.

Update Types for Distributed User Interfaces

In the following, without loss of generality our description is limited to replicated user interfaces, which means that the distributed parts of a user interface show the same content. This fission strategy is used because it is easy to implement. However, this is only one way of distributing the application's user interface to multiple devices. The synchronization concept is independent of its actual content, so it can be used for other fission strategies as well (e.g., like a split user interface, where the dialog is split into specific parts according to the available modalities).

Update types for distributed user interfaces can be distinguished in three dimensions which are synchronization granularity, simultaneity, and distribution strategy. Figure 7 shows these three dimensions in a cuboid with one sub cube for each combination. The grey shaded sub cubes represent combinations that turned out to be most appropriate. The reason to decide for them is given below in the experimental results.

Synchronization granularity levels are adopted from the W3C Multimodal Interaction

Requirements (Multimodal Interaction Working Group, 2003). Some of the specified levels are *event, field, form, page* and *session*. Keystrokes to write text in an input field need synchronization on event level, whereas the presentation of a new window needs synchronization on page level.

Simultaneity User interface feedback to user input can be presented on the devices of the distributed user interface synchronously or at different points in time.

Distribution strategy can be tentative, so that one device can present tentative feedback until a synchronized feedback can be delivered. With controlled updates, only the synchronized feedback is delivered.

In the following, the focus lies on text input and dialog changes because these input methods are considered to be the most relevant interactions. As explained below, for text input fields, tentative asynchronous update on event level is appropriate since keyboard input requires immediate feedback. For dialog changes, synchronous controlled update on page level is appropriate since the user accepts a longer response time for dialog changes but needs a consistent view over all devices of the distributed user interface.

Distribution Strategies

As already mentioned, there are two different distribution strategies for distributing events. In the following, they are explained in detail.

Asynchronous Tentative Updates

The intention of tentative update is to present an intermediate feedback until a synchronized result is distributed to all devices. In Figure 8, tentative update is compared to controlled update. While controlled updates require an input to be reported to the Abstract Multimodal Interactor, which distributes it equally to all Modality-specific Representations, tentative update allows an intermediate presentation until the event is distributed.

Synchronous Controlled Updates

The presented synchronization mechanism can be used to distribute the events to the modality-specific representations which are placed on the devices in a federation (step 3 for controlled updates, step 4 for tentative updates in Figure 8) so that their effect is rendered at the same point in time. Each modality has a specific *modality time* T_{mod}, which is needed to receive and display an update. The *reference time* period T_{ref} is the shortest time period after which all devices could update synchronously. It is calculated as the maximum over all modality times of the devices in the federation. The network delay consists of the transmission between Generation and Styling $T_{networkdelay_{GS}}$ as well as Styling and Rendering $T_{networkdelay_{SR}}$. Same applies to the transmission time. An overlay l_{ov} adds to the dialog update message length to take message encoding, e.g. for Web service calls, into account. The expansion factor e_s lengthens the dialog update message since styling information is added at the styling component.

A linear relationship is assumed between l_{dialog} and T_{mod}. This allows to predict T_{mod_n} for l_{dialog_n} based on $T_{mod_{n-1}}$ and $l_{dialog_{n-1}}$. Each device waits the *correction time* period T_{corr} before updating the user interface. Since T_{ref} is an estimation value

Figure 7. Update types in three dimensions (granularity, simultaneity, distribution strategy)

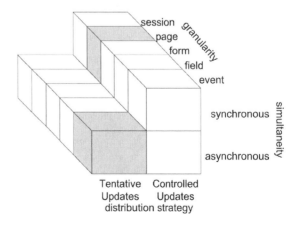

Box. 1

$$T_{mod} = T_{networkdelay} + T_{transmission} + T_{processing}$$

$$= T_{networkdelay_{GS}} + T_{transmission_{GS}} + T_{processing_S} + T_{networkdelay_{SR}} + T_{transmission_{SR}} + T_{processing_R}$$

$$= T_{networkdelay_{GS}} + \frac{l_{dialog} + l_{ov}}{v_{transmission_{GS}}} + \frac{l_{dialog}}{v_{processing_S}} + T_{networkdelay_{SR}} + \frac{e_S \times l_{dialog} + l_{ov}}{v_{transmission_{SR}}} + \frac{e_S \times l_{dialog}}{v_{processing_R}}$$

$$T_{ref} = \max_i \left\{ T_{mod_i} \right\}$$

$$T_{corr} = \max \left\{ 0, T_{ref} - T_{mod} \right\}$$

based on previous measurements which can be lower than T_{mod} measured for the current update, the correction time can be negative. Since a negative correction time is of no practical use, a delay of $0ms$ is assumed in this case. In the ideal case, when the estimation for T_{ref} was appropriate, all devices are updated at the same point in time. In summary, the following equations are used. (See Box 1)

As it was validated with our prototype for output synchronization, the ideal case where all devices perform the update at the same point in time is almost reached by our proposal. In the next section, the prototype for tentative updates on text input and controlled update on dialog changes is presented.

Output Synchronization Prototype

The Multimodal Route Planner comprises a device federation of PDA and in-car PC to display a map fragment on the PDA and to provide voice direction via the car sound system. A distributed browser was implemented to display user input and application feedback on different devices. Our prototype did not use voice output but only text output. Since the browser engine on one of the devices could also be replaced by a voice engine, which "renders" the application feedback, our results are applicable to the scenario.

A distributed Web browser application was implemented to validate our synchronization concept. In order to investigate heterogeneous devices and network connections, a Desktop PC and a PDA platform was used. A test dialog was constructed, containing text input fields and a submit button to load the next dialog. The dialog description is stored in XHTML format. The Java-based XHTML rendering engine Cobra 0.96 from the Lobo Project was reused and the graphical components were ported to AWT by replacing all Swing components. This was necessary in order to run the prototype on a Windows Mobile Platform in the Java Micro Edition virtual machine "IBM WebSphere Everyplace Micro Environment 6.1.1 CDC/PP1.1", often referred to as "J9". This java environment does not support graphical components from the Swing library.

User input is propagated to all devices of the distributed user interface. Events for text input and clicks on the "submit" button to retrieve a new dialog are intercepted and distributed n:m from all to all devices. Both, tentative and controlled update were tested for text input and compared against each other. In a second experiment the time difference between dialog updates was compared between controlled synchronous update and best effort (=controlled asynchronous) update where each device displays the new dialog as soon as possible.

Figure 8. Distribution strategies, controlled update (left) and tentative update (right)

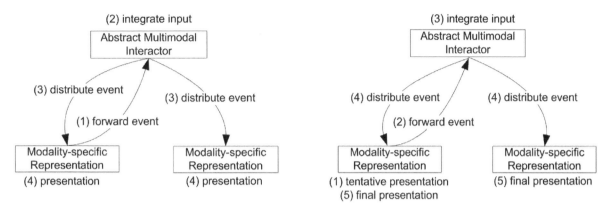

The first experiment showed that tentative update is better for text input than controlled update. Tentative updates for text input allow immediate feedback to text input, so that the text appears in the text box while typing but the text can change until the controller distributes the final version. The advantage for the user is that he can immediately see what he has typed independently of the synchronization going on in the background. For simultaneous input on multiple devices, the abstract interactor would have to merge inputs, which was neglected for this experiment.

To confirm our hypothesis of an increased usability with tentative update, 13 test subjects were asked for the highest synchronization delay they thought was acceptable to perceive smooth feedback. For this purpose, the synchronization delay added by the abstract interactor before the distribution of the keyboard events was adjustable arbitrarily by the test subjects. They were asked to determine the accepted delay for both distribution strategies. With tentative update, the subjects reported a three times higher accepted delay ($955ms$ in average) than with controlled update ($322ms$ in average). However, the practical use of tentative updates depends on the use case. There might be applications where text input must not change from a tentative representation to a final presentation but remain as it appeared

first, so that the user does not have to check the contents twice.

Controlled updates were tested in the following setting. Two instances of the replicated user interface were running on one PC. One instance waited an additional artificial delay before rendering the dialog and therefore prolonged $T_{processing}$. This did not influence network delay and transmission time, which are already rather low due to the local host communication. The artificial delay is uniformly distributed in the interval $720\text{-}970ms$ with interval length $250ms$ by the pseudo-random function Math.random() of Java. This interval reflects observed delays of a PDA with Bluetooth network connection to a PC and is therefore a realistic model. The delay was measured in a Bluetooth Personal Area Network consisting of a PC with Widcomm 1.3.2.7 Bluetooth stack and a PDA with Broadcom 1.7.1 Bluetooth stack. As stated before, for the prototype constellation both instances of the distributed browser run on the same PC to ensure precise measurements. 50 dialog updates were performed, one every $2s$.

The second experiment showed that controlled synchronous update is better for dialog changes than best effort update. As depicted in Figure 9 the controlled update synchronization reduces the time interval between the dialog updates to $\approx 80ms$. For the best effort case the interval of $720\text{-}970ms$ is reflected in the measured values. The remain-

Figure 9. Interval (in ms) between first and last dialog update, for dialog updates without synchronization and with controlled update synchronization

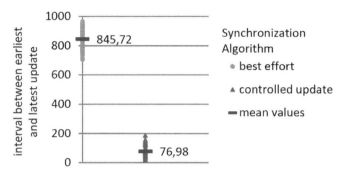

output synch.	best effort	controlled update
average	845.72	76.98
std. dev.	71.15	35.53

ing inaccuracy even for controlled updates can be explained by the stochastic artificial delay. The modality time T_{mod} measured for a dialog update is the estimated value for the next dialog update, which can deviate from the actual value within the range 0-250ms. Furthermore, Java is no real time programming language and for example the indeterministically running garbage collection delays program execution. To conclude our results, our proposal for controlled updates performs with good results in a practical setting.

Modality-Specific Processing of Input and Output

The interpretation of the output produced by recognizers as well as modality-specific styling of output is crucial for multimodal systems (see Section 2). In the following, our notion of modality-specific input interpretation and output styling and the mechanism that our solution bases on is briefly explained.

In our multimodal platform, modality-specific input interpretation is seen as the process of determining the application-specific semantics of user input. Application-specific semantics is understood as the effects that user input has on the state of the application. In the Multimodal Route Planner example, the semantics of the voice input "move west" is the change of the pin's longitude relative to its current position. The change of the application state is the modification of the variable value *GeoCoordinate.longitude* bound to the Abstract Data Interactor "Virtual Pin". More formally, the input interpretation for the Modality-specific Representations of an Abstract Data Interactor in a modality M can be expressed as a function $f_{sem_D,M}$. Given the output of the recognizer and the value $v_{current}$ assigned to the variable bound to the Abstract Data Interactor, it determines the new variable value v_{new}. Once v_{new}. has been determined by M, the state synchronization between the Modality-specific Representation that received the input and the Abstract Data Interactor is utilized to send v_{new}.to the Abstract User Interface from where it is written to the data model (see Figure 10). A similar interpretation approach is used for Abstract Event Interactors. A function $f_{sem_E,M}$ is bound to each Modality-specific Representation of an Abstract Event Interactor. It maps user input to an event in the Abstract User Interface. This event triggers a functional call in the application logic. Concrete realizations of both functions must be provided by the developer of a target application.

Two forms of output styling are distinguished in our multimodal platform. The first form is the transformation of Abstract User Interface descriptions into modality-specific user interfaces. The second form is the transformation of output data into perceivable artifacts. Output data is meant to

Figure 10. Interpretation and output styling components

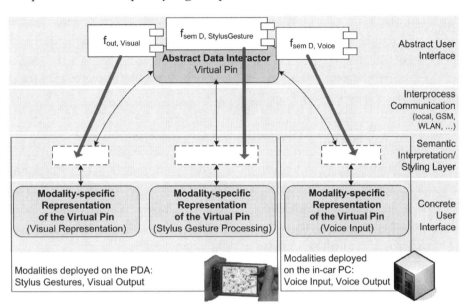

be the value of the variable bound to an Abstract Data Interactor or an Abstract Output Interactor that is to be presented to the user. The perceivable artifact is a suitable representation of the variable's value in an output modality. Assuming that, for example, the result of the distance calculation in the Multimodal Route Planner is put into an integer variable v_{dist} and that this variable is bound an Abstract Output Interactor *dist_output*, the visual presentation of *dist_output*, may be a simple text field. Before this text field is filled with the value of v_{dist}, it is desirable to append the unit (m, km, …) to the numerical value. Besides appending the unit, a spoken output of v_{dist} requires the transformation of the integer value into a numeral before it can be sent to the speech synthesizer. The speech representation of v_{dist} could be "four miles", its visual representation a text field showing "4 mi". Our approach of transforming variable values into perceivable artifacts is similar to the input interpretation approach. Formally, the output styling of variable values in a modality M can be expressed as a function $f_{out,M}$ that maps a variable value to a modality-specific perceivable artifact. Concrete realizations of $f_{out,M}$ are modality- and

application-specific. They must be provided by the developer of a target application.

The input interpretation components and the output styling components that implement the functions $f_{sem_D,M}$, $f_{sem_E,M}$ and $f_{out,M}$ are bound to Abstract Multimodal Interactors. When an Abstract Multimodal Interactor and its Modality-specific Representations are instantiated, the input interpretation and output styling components are distributed to the input and output components distributed over the federated devices and placed in the semantic interpretation layer or the styling layer of "their" modality M. The layers are configured by inserting the input interpretation and output styling components into the synchronization link between the Abstract Multimodal Interactor and its Modality-specific Representations. The processing step performed by these components is transparent to the Abstract Multimodal Interactor as well as to the Modality-specific Representations. Consequently, the implementations of Modality-specific Representations and Abstract Multimodal Interactors are generic and can be reused in different applications simply by providing different input interpretation/output styling components.

Fusion of Multimodal Input

A multimodal input consists of two or more inputs via different input modalities that need to be combined in order to determine their semantics. Multimodal input can be performed by the alternating use (only one modality at a time) or the synergistic use (multiple inputs are performed in parallel) of modalities. Two major problems arise here. First, unimodal inputs that need to be combined have to be identified and distinguished from, for example, simple voice commands. Second, their semantics must be determined. In the following, a solution for these problems that builds upon the basic concepts of Abstract Multimodal Interactors and Modality-specific Representations approach will be presented.

In our multimodal platform, the semantics of a multimodal input is a function f_{app} of the target application. Its call parameters are the constituents of the combined unimodal inputs. This notion is similar to the approach presented in (Vo & Waibel, 1997). An example of f_{app} in the Multimodal Route Planner is the route calculation function. Its call parameters are the POIs which the user identifies by pointing (see Figure 3).

The identification of inputs that need to be combined is usually based on a multimodal grammar. Multiple approaches such as Typed Feature Structures (Johnston et al., 1997), multi-band finite state transducers (Johnston & Bangalore, 2005), Mutual Information Networks (Vo & Waibel, 1997), and extensions of classic ATN (temporal ATN) (Latoschik, 2002) are proposed on the literature. None of these solutions is applicable to our approach and is lightweight enough to meet the resource constraint of mobile devices. Typed Feature Structures require processing power, since the computing time can grow exponentially with the number of inputs to be fused (Johnston, 1998) and are presumably difficult to handle for

developers without a background in computer linguistics. Mutual Information Networks need to be trained and are also computationally intensive since they use an algorithm similar to Viterbi to determine the interpretation hypothesis with the highest score. Finite State Transducers are lightweight enough for mobile devices, but the approach presented in (Johnston & Bangalore, 2005) presumes the availability of a predefined set of input modalities. A similar situation exists with temporal ATN. However, in our approach this assumption does not hold since our use cases require support for a set of input modalities that changes over time.

Our proposition is to use an extension of the *Timed Petri-Nets* described in (Baumgarten, 1996) as a multimodal grammar for multimodal user interfaces that build on our abstraction concepts. The advantage of Timed Petri-Nets over existing approaches is their simplicity, the possibility to link them to Abstract Multimodal Interactors, and their low requirements in terms of processing power.

Briefly, the following three types of transition provided by Timed Petri-Nets are used: *timer transitions*, *timer-triggered transitions*, and *as-soon-as-possible transitions*. Timer transitions are associated with a timer. This timer is initialized and started when the timer transitions fires. A timer-triggered transition is also associated with a timer. It fires when it is enabled and its associated timer reaches a predefined value. An as-soon-as-possible transition fires as soon as it is enabled.

To utilize Timed Petri-Nets as a multimodal grammar, they are extended at two points. As a first extension an *input-triggered timer transition* is introduced. Input-triggered timer transitions link Abstract Multimodal Interactors to Timed Petri-Nets. Input-triggered timer transitions are basically timer transitions that are associated with at least one Abstract Multimodal Interactor. An enabled input-triggered timer transition fires upon an input event on one of its associated Abstract Multimodal Interactors. Each constituent of the multimodal input, which is sent from the Modality-specific Representations to the Abstract Multimodal Interactor together with the input event, is stored for later use.

The second extension is a mapping function f_{semMM} and its association with an as-soon-as-possible transition. The function processes the stored constituents of the multimodal input by mapping them to the call parameters of f_{app} and invokes the function afterwards. The mapping function is invoked when the as-soon-as-possible transition fires. The developer of a Timed Petri-Net must integrate the as-soon-as-possible transition so that it is enabled when the user completes the multimodal input.

Timer triggered transitions serve as a timeout mechanism that is needed for modeling multimodal input. Two inputs may only be interpreted as constituents of one multimodal input if the interval between them is, for example, shorter than five seconds. Alternatively, the user may interrupt a multimodal input at any time.

Figure 11 shows how an extended Timed Petri-Net is used to model the multimodal input for initiating a route calculation in the Multimodal Route Planner. The transitions 1, 2 and 3 are input-triggered timer transitions, associated with the Abstract Data Interactors "POI" on the map and the Abstract Event Interactor "Calculate Route". Transition 7 is an as-soon-as-possible transition which is enabled after the user selected the second POI and completed the interaction of the Abstract Event Interactor. The transitions 4, 5 and 6 are timer-triggered transitions.

The Timed Petri-Nets models the multimodal input for initiating the route calculation (see Figure 3) as follows. The user must select two POI. After selecting the first POI, he has four seconds to select the second POI. After selecting the second POI, the interaction event from the Abstract Multimodal Interactor "Calculate Route" must occur within two seconds. If the interaction event on the "Calculate Route" Abstract Multimodal Interactor

occurs before the second POI is selected, the user is expected to complete the POI selection at most five seconds later.

Support for Federated Devices

Device federations are one way to overcome the hardware-related limitations of current devices. There is no device supporting each and every modality. However, it is reasonable to use the different modalities of devices the user already owns. This allows the user to interact with several devices (and several modalities) for accessing exactly one target application. For instance, a PDA for voice interaction can be used together with a Laptop for visual output and haptic input (keyboard, mouse).

Using federated devices has some further requirements on the multimodal interaction platform, especially if mobile devices with limited processing power and storage capacity are considered. Those devices are often connected to the device federation via wireless networks like Bluetooth or WiFi. Although they might not be connected to an ad-hoc network in the technical sense, their connection may be transient and not stable over time. The user for instance might move to a spot where the network's signal strength is low causing low transfer rate and high latency.

Typical web applications are unavailable from time to time due to a variety of reasons like missing network connection or the desired host being offline. However, the user interface (the Browser) is always available because it is running on the user's machine only. By employing a distributed user interface based on federated devices, the user interface moves a step towards the network, i.e., that its availability depends on the used network. Therefore, special mechanisms have to be considered to make the distributed user interface tolerant to some of the possible faults. However, it is very hard or almost impossible to achieve the same dependability of a distributed

Figure 11. Timed Petri-Net with input triggered transitions as a multimodal grammar for the Multimodal Route Planner route calculation example

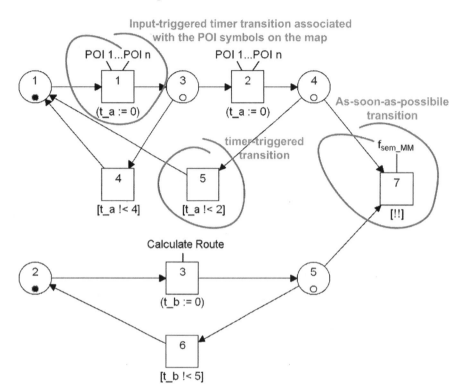

application compared to a stand-alone desktop application.

In order to increase the dependability of the distributed user interface, three mechanisms that are implemented by the platform were specified: Deployment, Migration and Replacement. The mechanisms simplify how components are handled in a distributed system by making them tangible, movable and flexibly deployable. They are described in the following. The general idea of using federated devices for multimodal user interaction has been investigated in (Kadner, 2008).

Deployment

Since most mobile devices have limited storage capabilities, the multimodal interaction platform must have a central database or similar access point, where all available components can be

downloaded from. The end user can browse this database and select components for interaction, which are deployed instantly to the device selected by the user. The device might be any device available for access by the multimodal interaction platform. This is required in cases where the desired installation device does not offer any input capabilities so that remote installation is crucial.

Further components are installed automatically. If a component needs a particular feature, it sends a request to a service directory, which manages all installed components and inherently their services and can be compared to a service registry (e.g., a UDDI directory). The service directory initiates the installation procedure of the new component. Part of the installation procedure is the determination of a suitable device. This is influenced by the user's preferences, i.e., if the user prefers a distributed or centralized component distribution. The system also matches the component profile against the device profiles. The component profile describes the required hardware resources among others whereas the device profile contains a specification of the device's hardware and software environment. Once a device is found to be suitable, the component is installed on that device and after it has registered its services, the reference is returned to the component that started the request for a feature in the beginning.

The features a component can request are a semantic description of a component's functionality. In the domain of multimodal interaction, this might refer to the roles defined by the MMI-Framework, e.g., an HTML browser component (which is renderer and recognizer at the same time) needs an integration component for performing the integration of multiple input sources. Thus, the browser requests such a feature via the above mentioned process.

Migration

The migration of components allows them to continue their operation on another host. This is can be used as a rescue action if the old host is about to

get unavailable or if the user simply needs certain information on another host in the network. Various kinds of migration are known in the literature. The most important types are strong and weak migration (Fuggetta, Picco & Vigna, 1998). Strong migration allows a component to migrate at any given time, because it is migrated with its binary data, current stack, heap, pointers, threads, and everything else necessary for executing the code. Weak migration limits the component's mobility to certain points in time, where only the binary data and some initialization information will be migrated. In other words, the component enters a check point, the initialization data is determined, and is used on the target machine to initialize the component accordingly. It is obvious that strong migration provides a higher flexibility, which is paid by a tremendous implementation effort. Weak migration is less flexible but easy to implement.

Components can be in different states, when a migration is initiated. No special treatment is necessary, if the component is not doing anything. If there are currently open requests that the component processes, it can either finish them and migrate afterwards or cancel & redirect them and migrate immediately depending on the situation.

Figure 12. Catch-up migration, a mechanism to migrate data stream receiving components (solid line- data messages, dashed line- management messages

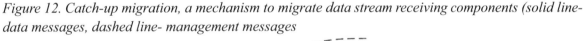
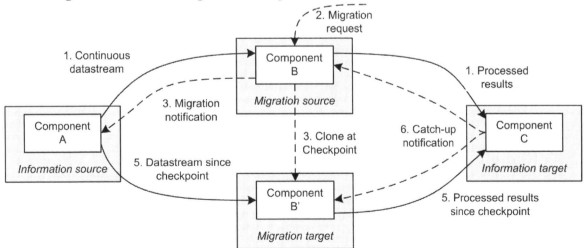

The last option is interruptive because it might imply delays on the user interface. However, the first option implies that new requests are redirected until the old ones are finished. But there has to be a component in the meantime that processes those requests, so that two components have to exist, which has further implications on their data consistency.

Finally, a component can be in the situation of receiving and processing a data stream. To cope with that, a mechanism called *catch-up migration* was developed. The catch-up migration works as follows (see Figure 12).

The system consists of three platforms, the information source, the information target, and a platform in between that does the processing of the data stream (here the migration source). The results of this processing are sent to the information target once they are ready. When component B receives the migration request, it determines a check point as soon as possible. At this checkpoint, the state of component B is extracted and moved to the migration target platform along with the component's binary data where a clone (B') of component B is created. In the meantime, component B notifies its data stream source (component A) about the upcoming migration and the checkpoint. Component A continues the stream to B normally and additionally streams its data to component B' since the specified checkpoint. The system now has two components B doing the same work, although B' is expected to be behind the current state of B. B' tries to catch up to B by receiving the data stream in a faster and maybe compressed way. Both B and B' send their results to component C, which is aware that it receives double information and ignores the delayed results from B'. C measures the time difference between the results from B and B' and once this difference is small enough, it notifies both of the successful alignment. Consequently, B stops itself and thus makes B' the original B, which finishes the migration process.

It is assumed that the data stream's structure supports the definition of checkpoints and that the new component (the migration result) can receive and process the data stream faster than the original one. Otherwise, it will never catch up to the original state. The use case behind this mechanism is that of a separate voice recognition (component A), voice interpretation (component B) and input integration (component C). The voice recognition streams audio data to the interpretation, which sends its results to an integration component. During the interpretation, B will have some internal and partially processed data, which must be saved for migration otherwise the user interaction might be interrupted.

However, the catch-up migration might not always be used, e.g, because the environment does not allow the required time for migration. A rigid migration with possible interruptions of user interaction or continuous checkpointing can be used instead.

Replacement

Several versions of components may exist in a system, which offer different implementations of the same functionality but with different Quality-of-Service properties. For instance, a voice browser might have an integrated voice interpretation with restricted capabilities due to its scope as mobile application. If the user wants a more sophisticated voice interpretation, he can replace the voice browser with interpretation by two separate components for the voice user interface and voice interpretation respectively. Besides that one-to-many replacement, the other way around and one-to-one replacements are also possible. Many-to-many can be broken down to a set of the former types.

The inner state of components must not be lost due to replacing the components. Therefore, the components have to support this by allowing the extraction and insertion of their inner state (a.k.a. session data). In the multimodal interaction domain and in relation to the MMI-Framework, the old components have to specify their role in the MMI-Framework along with their inner state. Based on that, the replacement coordinator creates the respective initialization data for the new components. This allows for an easy transition of session data from the old to the new components of a replacement process. The concrete format of the session data still needs to be specified, especially if multiple parties contribute components to the multimodal interaction platform.

Prototype

The described concepts for federated devices and output synchronization were implemented in a prototypical environment described in (Kadner, 2008). The prototype is based on OSGi and consists of separate components for various roles of the MMI-Framework like *interpretation*, *integration* and *interaction manager*. Others were merged into one component, e.g., the HTML browser does visual *rendering* and input *recognition* as well as partial *interpretation*. All components were realized as OSGi bundles, which means that they can be easily started and stopped during runtime. They communicate over web services and use service directories for finding other services, devices and users. The implementation was tested on a regular Laptop and on a mobile phone and several performance investigations have been done. As this prototype is intended to be a proof-of-concept instead of a productive platform, a proprietary implementation for transporting the recognized input instead of EMMA was used, which has no publicly available implementation. Although these concepts were implemented in a separate prototype, they could be easily integrated in the Multimodal Route Planner.

FUTURE TRENDS

Multimodality with its full feature set is by far not state of the art in current applications. Speech in combination with visual input and output is supported by technologies like X+V browsers, which integrate XHTML and VoiceXML (VoiceXML Forum, 2004). An arbitrary integration of other modalities is currently still a topic for research.

Driven by recent technology developments, the availability of new input modalities in mainstream mobile devices is increasing. For instance, accelerometers and gyroscopes for measuring orientation and acceleration are used in smart phones, PDAs, and gaming consoles to recognize hand and body movements, thus allowing users to interact via gestures. Moreover, an increasing number of mobile devices are equipped with digital cameras, which can also be used to recognize user gestures like head or hand movements, facial expressions, or movements of the devices, based on image processing, e.g., (Li, Taskiran & Danielsen, 2007, Winkler, Rangaswamy, Tedjokusumo & Zhou, 2007).

Touch screens are an alternative way to capture gestures from users. Devices equipped with touch screens are readily available and allow new, appealing forms of interaction. Among other technologies, resistive and capacitive touch screens are widely used. Examples include the Apple iPhone and Microsoft surface screens.

Current trends for 3D user interface introduce also additional complexity for handling visual input and output. Even for mobile devices rendering libraries are available and should be supported by a generic platform (Rodrigues, Barbosa & Mendonca, 2006). An example is the Mobile 3D Graphics API for the Java Micro Edition specified as JSR 184 (Sun Microsystems, 2003).

All these trends add to the challenges for generic platform support. A platform should integrate all these forms of devices and modalities in an application transparent way. The concepts introduced in this chapter, especially the abstractions and the component-based platform for device federations are ready to dynamically integrate new devices and modalities. The trade-off between generic modality support and the exploitation of modality-specific features remains a challenge. Moreover, a platform should minimize the effort of application developers for creating new applications. Thus, the introduced concepts should be extended with the new modalities.

A further important trend is the integration of sensors not only in mobile devices but also buildings or even clothes and the human body. Such sensors can be used by a multimodal platform to sense the environment of the user, derive his current situation, and adapt the interactions according to that situation. For instance, in a noisy environment speech interaction might be impossible and gestures or visual input and output have to be used. Major challenges for platform support are the integration and management of sensor devices and the abstraction of the sensed information to derive the users' situation. Concepts and technologies from the area of context-awareness can be adopted but have to be integrated with multimodal platforms.

Also driven by technological improvements is the fact that mobile devices are equipped with more resources in terms of processing power and memory, which allow more complex processing on the mobile device. This is exploited, for instance, in speech recognition to perform feature extraction partially on the mobile device (Zaykovskiy, 2006, European Telecommunications Standards Institute (ETSI), 2003). As a result the amount of data transfered between mobile device and infrastructure via a wireless link can be significantly reduced.

A different effect of more powerful mobile devices is the advent of new programming platforms for these devices. On the one hand, heterogeneity increases with the availability of platforms like Android, Openmoko and LiMo (Google, 2008, Openmoko, 2008, LiMo Foundation, 2008). On the other hand, piggybacking on their momentum

could drive the market adoption of platform concepts for multimodal applications as presented in this chapter. However, the challenge for a generic, unified execution environment for mobile multimodal applications remains large.

CONCLUSION

This chapter showed that the creation and execution of multimodal applications is a complex task: multiple heterogeneous input and output devices have to be handled, fusion and fission of input and output streams have to be synchronized, and appropriate application functions have to be invoked in reaction to particular user interactions. A platform that generically supports this functionality can dramatically increase the efficiency and thus reduce the costs of developing multimodal applications.

The abstraction concepts introduced in Section 3 enable a clear separation of concerns in terms of application logic, data model, and modality handling. They are orthogonal to the component and role model of the MMI framework discussed in Section 2. In combination, these concepts allow application logic to be decoupled from used modalities and enable a seamless and transparent integration of multiple modality-specific components. This modularization makes a platform open for different renderer and recognizer implementations, which may be introduced and removed dynamically in mobile settings. It allows distributing components with subsets of functionality between federations of mobile and static devices. Depending on current connectivity and resource availability, components can be placed and even rearranged dynamically.

Moreover, the introduced abstractions also support the development of concrete multimodal applications. Using generic transformations to produce modality-specific representations of the Abstract User Interface, applications can be built independently from concrete modalities, changes in availability of modalities, and even the concrete user interface technology. This addresses especially the heterogeneity of mobile computing infrastructures in terms of device classes, their characteristics, and application runtime environments.

Modularization is also applied to modality specific components. As discussed in Section 3, application developers have to provide concrete realizations of modality-specific input interpretation and output styling. The platform provides the framework that decides when and how these realizations are integrated. Hence, a dynamic introduction and removal of such components is easily achievable.

Input and output modalities can be used alternatively or synergistically. This requires input streams to be "fused" to extract the application functionality and parameters. The synchronization aspect is especially challenging on resource constrained mobile devices. In Section 3, an expressive and powerful fusion approach based on extended Timed Petri-Nets has been presented, which is yet simple enough to be implemented on mobile devices. On the output side, synchronization requires special solutions with respect to distributed processing and rendering of multiple output streams. Distribution strategies depending on the type of interaction were proposed. Tentative updates are proposed for interactions requiring immediate feedback and controlled updates are proposed for interactions requiring synchronized state among federated devices. An approach that monitors transfer and processing times for styling and rendering for achieving synchronicity by adapting delays of output streams was introduced.

In Section 3, an integrated infrastructure for federated devices was presented. This is of particular interest for multimodal applications to exploit a broader range of available modalities on distributed, potentially mobile devices in the user's vicinity. However, such an infrastructure has to solve additional challenges in terms of dynamics, reliability, and heterogeneity. These issues have been addressed by a component-based platform, which especially supports dynamic placement and migration of components in a multimodal system.

All of the presented concepts were validated by respective prototypes. These were not exclusively based on the use case of a Multimodal Route Planner. The prototype investigating the concepts related to components in federated devices was focused on the specific issues of that domain. However, integrating those is possible and just a matter of implementation effort. Besides that, the implementations always targeted at Java on PDAs because this combination offers a comprehensive and well documented API on small devices. Other mobile platforms are possible, but might cause restricted features of the multimodal platform.

To sum up, this chapter discussed major challenges in multimodal systems in the context of mobility and proposed general abstractions and concepts for a generic platform support as a solution. Our approaches have been adopted and evaluated in the context of large projects and proven to be feasible and adoptable in a broad range of application scenarios. Even though particular implementations for our validation were discussed, the concepts are reusable enough for implementing multimodal systems in general.

REFERENCES

Aitenbichler, E., Kangasharju, J., & Mühlhäuser, M. (2007). MundoCore: A Light-weight Infrastructure for Pervasive Computing. *Pervasive and Mobile Computing.*

Baumgarten, B. (1996). *Petri-Netze - Grundlagen und Anwendungen.* Berlin, Germany: Spektrum Verlag.

Braun, E., & Mühlhäuser, M. (2005). Interacting with Federated Devices. In A. Ferscha, R. Mayrhofer, T. Strang, C. Linnhoff-Popien, A. Dey, A. Butz, & A. Schmidt (Eds.), *Advances in pervasive computing, adjunct proceedings of the third international conference on pervasive computing* (pp. 153–160). Österreich, Austria: Austrian Computer Society.

Burmeister, R., Pohl, C., Bublitz, S., & Hugues, P. (2006). SNOW - a multimodal approach for mobile maintenance applications. In *Proceedings of the 15th IEEE International Workshops on Enabling Technologies: Infrastructures for Collaborative Enterprises (WETICE 2006)* (pp. 131–136). Manchester, United Kingdom: IEEE Computer Society.

Cohen, P. R., Johnston, M., McGee, D., Oviatt, S., Pittman, J., Smith, I., et al. (1997). Quickset: multimodal interaction for simulation set-up and control. In *Proceedings of the fifth conference on applied natural language processing* (pp. 20–24). San Francisco, CA: Morgan Kaufmann Publishers Inc.

Coles, A., Deliot, E., Melamed, T., & Lansard, K. (2003). A framework for coordinated multimodal browsing with multiple clients. In *WWW '03: Proceedings of the 12th International Conference on World Wide Web* (pp. 718–726). New York: ACM Press.

Elting, C., Rapp, S., & Strube, G. M. M. (2003). Architecture and implementation of multimodal plug and play. In *Proceedings of the 5th International Conference on Multimodal Interfaces* (pp. 93-100). New York: ACM Press.

European Telecommunications Standards Institute (ETSI). (2003). Distributed speech recognition; front-end feature extraction algorithm; compression algorithms. *ES 201 108.*

Fowler, M. (2003). *Patterns of enterprise application architecture.* Addison-Wesley: Reading, MA.

Fuggetta, A., Picco, G. P., & Vigna, G. (1998). Understanding code mobility. *IEEE Transactions on Software Engineering, 24*(5). doi:10.1109/32.685258

Göbel, S., Hartmann, F., Kadner, K., & Pohl, C. (2006). A Device-independent multimodal mark-up language. In *INFORMATIK 2006: Informatik für Menschen, Band 2* (pp. 170–177). Dresden, Germany.

Google. (2008). *Android.* Retrieved on November 12, 2008, from http://www.android.com/

Hübsch, G. (2008). *Systemunterstützung für Interaktorbasierte Multimodale Benutzungsschnittstellen.* Doctoral dissertation, Dresden University of Technology, Germany.

Johnston, M. (1998). Unification-based multimodal parsing. In *COLINGACL* (pp. 624–630).

Johnston, M., & Bangalore, S. (2005). Finite-state multimodal integration and understanding. *Natural Language Engineering, 11*(2), 159–187. doi:10.1017/S1351324904003572

Johnston, M., Cohen, P. R., McGee, D., Oviatt, S. L., Pittman, J. A., & Smith, I. (1997). Unification-based multimodal integration. In P. R. Cohen & W. Wahlster (Eds.), *Proceedings of the thirty-fifth annual meeting of the Association for Computational Linguistics and eighth conference of the European chapter of the Association for Computational Linguistics* (pp. 281–288). Somerset, NJ: Association for Computational Linguistics.

Kadner, K. (2008). *Erweiterung einer Komponentenplattform zur Unterstützung multimodaler Anwendungen mit föderierten Endgeräten.* TUDpress Verlag der Wissenschaft.

Kadner, K., & Pohl, C. (2008). *Synchronization of distributed user interfaces.* European Patent no. 1892925, European Patent Bulletin 08/42, granted as of 2008-10-15.

Latoschik, M. (2002). Designing transition networks for multimodal VR-interactions using a markup language. In *Proceedings of the Fourth IEEE International Conference on Multimodal Interfaces* (pp. 411-416).

Li, R., Taskiran, C., & Danielsen, M. (2007). Head pose tracking and gesture detection using block motion vectors on mobile devices. In *Proceedings of the 4th international conference on mobile technology, applications, and systems and the 1st international symposium on computer human interaction in mobile technology* (pp. 572–575). New York: ACM.

LiMo Foundation. (2008). *LiMo Foundation Platform.* Retrieved on November 12, 2208, from http://www.limofoundation.org/

Mori, G., Paternó, F., & Santoro, C. (2004). Design and Development of Multidevice User Interfaces through Multiple Logical Descriptions. *IEEE Transactions on Software Engineering, 30*(8), 507–520. doi:10.1109/TSE.2004.40

Multimodal Interaction Working Group. (2002a). *Multimodal Interaction Activity.* Retrieved on November 7, 2008, from http://www.w3.org/2002/mmi/

Multimodal Interaction Working Group. (2002b). *Multimodal Interaction Framework.* Retrieved on November 7, 2008, from http://www.w3.org/TR/2003/NOTE-mmi-framework- 20030506/

Multimodal Interaction Working Group. (2003). *Multimodal Interaction Requirements.* Retrieved on November 7, 2008, from http://www.w3.org/TR/2003/NOTE-mmi-reqs-20030108/

Multimodal Interaction Working Group. (2008). *Multimodal Architecture and Interfaces.* Retrieved on November 7, 2008, from http://www.w3.org/TR/2008/WD-mmi-arch-20081016/

Nah, F. F.-H. (2004). A study on tolerable waiting time: how long are web users willing to wait? *Behaviour & Information Technology, 23*(3), 153–163. doi:10.1080/01449290410001669914

Nichols, J., & Myers, B. A. (2006). Controlling home and office appliances with smart phones. *IEEE Pervasive Computing / IEEE Computer Society [and] IEEE Communications Society, 5*(3), 60–67. doi:10.1109/MPRV.2006.48

Nielsen, J. (1994). *Usability engineering (excerpt from chapter 5).* San Francisco: Morgan Kaufmann. Retrieved September 7, 2008, from http://www.useit.com/papers/ responsetime.html

Openmoko. (2008). *Openmoko development portal.* Retrieved November 28, 2008, from http://www.openmoko.org/

Pereira, A. C., Hartmann, F., & Kadner, K. (2007). A distributed staged architecture for multimodal applications. In F. Oquendo (Ed.), *European Conference on Software Architecture (ECSA) 2007* (pp. 195–206). Berlin, Germany: Springer Verlag.

Pham, T.-L., Schneider, G., & Goose, S. (2000). A situated computing framework for mobile and ubiquitous multimedia access using small screen and composite devices. In *Multimedia '00: Proceedings of the eighth acm international conference on multimedia* (pp. 323–331). New York: ACM Press.

Rodrigues, M. A. F., Barbosa, R. G., & Mendonca, N. C. (2006). Interactive mobile 3D graphics for on-the-go visualization and walkthroughs. In *Proceedings of the 2006 ACM symposium on applied computing* (pp. 1002–1007). New York: ACM.

Sun Microsystems. (2003). *JSR-184: Mobile 3D graphics API for J2ME*. Retrieved November 13, 2008, from http://www.jcp.org/en/jsr/detail?id=184

Synchronized Multimedia Working Group. (2008). *Synchronized Multimedia Integration Language (SMIL 3.0)*. Retrieved November 12, 2008, from http://www.w3.org/TR/2008/PR-SMIL3-20081006/

User Interface Markup Language, O. A. S. I. S. (UIML) Technical Committee. (2008). *User Interface Markup Language (UIML)*. Retrieved from http://www.oasis-open.org/committees/uiml/

Vo, M. T., & Waibel, A. (1997). Modeling and interpreting multimodal inputs: A semantic integration approach. *Technical Report CMU-CS-97-192*. Pittsburgh: Carnegie Mellon University.

Voice, X. M. L. Forum. (2004). *XHTML + voice profile 1.2*. Retrieved on November 12, 2008, from http://www.voicexml.org/specs/multimodal/x+v/12/

West, D., Apted, T., & Quigley, A. (2004). A context inference and multi-modal approach to mobile information access. In *Artificial intelligence in mobile systems* (pp. 28–35). Nottingham, England.

Winkler, S., Rangaswamy, K., Tedjokusumo, J., & Zhou, Z. (2007). Intuitive application-specific user interfaces for mobile devices. In *Proceedings of the 4th International Conference on Mobile Technology, Applications, and Systems and the 1st International Symposium on Computer Human Interaction in Mobile Technology* (pp. 576–582). New York: ACM.

Zaykovskiy, D. (2006). Survey of the speech recognition techniques for mobile devices. In *Proceedings of the 11th International Conference on Speech and Computer (SPECOM)* (pp. 88–93). St. Petersburg, Russia

Section 3
Design Approaches

Chapter 5
Designing Mobile Multimodal Applications

Marco de Sá
University of Lisboa, Portugal

Carlos Duarte
University of Lisboa, Portugal

Luís Carriço
University of Lisboa, Portugal

Tiago Reis
University of Lisboa, Portugal

ABSTRACT

In this chapter we describe a set of techniques and tools that aim at supporting designers while creating mobile multimodal applications. We explain how the additional difficulties that designers face during this process, especially those related to multimodalities, can be tackled. In particular, we present a scenario generation and context definition framework that can be used to drive design and support evaluation within realistic settings, promoting in-situ design and richer results. In conjunction with the scenario framework, we detail a prototyping tool that was developed to support the early stage prototyping and evaluation process of mobile multimodal applications, from the first sketch-based prototypes up to the final quantitative analysis of usage results. As a case study, we describe a mobile application for accessing and reading rich digital books. The application aims at offering users, in particular blind users, means to read and annotate digital books and it was designed to be used on Pocket PCs and Smartphones, including a set of features that enhance both content and usability of traditional books.

INTRODUCTION

The advances of mobile devices and technology have permitted their inclusion on our daily activities, supporting and propelling productivity and communication on several levels. Moreover, the ever-growing diversity and potential of these devices extends their usage through multiple settings, accentuating their ubiquitous and pervasive nature. One of the major contributions to this is undoubtedly the multiplicity of interaction possibilities.

DOI: 10.4018/978-1-60566-978-6.ch005

These open opportunities for the introduction of multimodal applications that can offer new levels of usability and adequateness to the multiple scenarios in which mobile devices play key roles. In fact, within ubiquitous activities, multimodal interfaces are paramount. They provide alternatives to usual interaction streams that can cope with the challenges that emerge in different occasions. Users can adapt interaction to the context of use by selecting the interaction modalities that are more adequate to the current situation [0]. On another dimension, multimodalities also enlarge the target audience of mobile applications. They provide the access impaired users require to common tools and information. Moreover, the mobility and multimodal dimensions pave the way to the design of new applications that aid them greatly while completing their daily tasks.

As usual, these advantages do not come without drawbacks. In fact, multimodal applications and interfaces, used in mobile contexts by an extended audience, are susceptible to (1) more environmental variables, such as noise; (2) extra device characteristics and diversity, for instance the availability of a camera or microphone; and (3) a myriad of user abilities or impairments, such as normally sighted or visually impaired users. Accordingly, besides the existing challenges that hinder the design process of mobile applications (e.g., device's size, lack of keyboard/mouse, screen resolution, mobile use, frequently as an accessory to parallel activities), multimodal applications require specific attention to the surrounding context of use and how it affects usability. Moreover, allying this with the mobility and infinite usage contexts that mobile devices provide, the design process becomes extremely demanding and challenging.

In order to mitigate these difficulties and to cope with the added challenges that designing mobile multimodal applications (MMAs) brings, the design process needs to be carefully addressed. First and foremost, the process has to be taken out of the lab, into the real world. Only there designers and users can grasp the details and problems that affect interaction and the usability of MMAs. Then, to capture these difficulties at the early design stages, prototyping and evaluation must be adjusted to the complexity of the envisaged contexts and modes of interaction. Evaluation sessions of mobile applications generally require testers and designers to follow users during tests on the real world, leading to privacy issues and biased results, also proving to be difficult to conduct. Finally, the selection of proper locations and scenarios for evaluation must be considered.

This chapter introduces a set of studies that address the added difficulties of designing multimodal applications and user interfaces for mobile devices. The first contribution that resulted from these studies is a scenario generation and context selection framework that aims at supporting the design of MMAs and their evaluation in-situ. It offers designers a set of guidelines that focus details that are paramount during the design and evaluation of mobile and, in particular, multimodal applications. The framework delineates variables that can affect the usability of MMAs and offers a systematic approach to the selection and generation of scenarios that cover a wide set of details and transitions between contexts and settings which are commonly overlooked.

The second contribution that emerged during these studies is MobPro, a prototyping and evaluation tool which enables designers and users (with no programming experience) to quickly create mixed-fidelity prototypes of multimodal applications. Furthermore, it supports the ability to evaluate them in-situ, while using actual mobile devices. The tool comprises a prototype designer, a mobile runtime environment and an evaluation/analysis tool.

The prototype designer allows users to create sketch-based low-fidelity prototypes or component-based high-fidelity prototypes for mobile devices. Overall, the tool conveys a set of pre-existent categories of prototypes, including multimodal ones, supporting audio and video components and interaction. Alternatively, it contains features which support the use of hand drawn sketches, which can be augmented with behaviour, images, sound and video, in order to quickly test design concepts and simulate multimodal interfaces at very early design stages. The mobile runtime environment is the designer tool's counterpart and allows users to interact with the created prototypes on actual mobile devices, also passively gathering usage data (e.g., through logs) or actively, by prompting questionnaires that allow users to provide their personal input through different modalities while using the prototypes on-the-go.

Complementing the two former tools, we present an evaluation tool which is able to recreate every event that took place while the user was interacting with the prototypes. The tool is composed by a log player which is able to display a video-like representation of the interaction that took place, also offering search and filter mechanisms which allow testers and designers to target specific usability concerns, reviewing behaviour patterns even without following users around. The tools offer support to the design of MMAs from the initial low-fidelity prototypes up to software component-based prototypes, promoting user participation and propelling in-situ, real world, evaluation. Globally, the developed framework integrates several techniques that have been tested for mobile devices with the multimedia features that enable designers to apply these techniques, with mixed-fidelity prototypes, at very early stages. The framework is grounded on prototyping and evaluation techniques that have shown great results for such a demanding design process and serves as a validation for these same methodological advances [0,0], offering the ability to overcome difficulties introduced both by mobility and multidimensionality of its interaction features, allowing designers to engage on an effective design process without increasing costs and from a very early design stage.

As a case study, the chapter also details the design process of a mobile multimodal digital book player in which the methodology, its underlying techniques and the prototyping tool were used by other designers, validating its application to the design of MMAs. The book player was directed to a large user universe but aimed at providing sufficient features to maintain its usability for visually impaired users as well. This goal was achieved through extensive use of multiple complementary and cooperative modalities, including touch and speech, in addition to the traditional modalities available in mobile devices. This resulted in an application which is universally accessible, capable of performing adequately for users with diverse characteristics, and also in contexts which would, otherwise, impose restrictions on such an application's usability. The scenarios and prototypes that led to the final application were developed with the aid of the aforementioned frameworks.

In this chapter we present the stated contributions and address the design process of the book player, stressing how the framework was used and validated while defining scenarios that drove design and facilitated evaluation. We elicit the details that proved to be essential while selecting the various variables that composed each scenario and how they were reflected in the design solutions that emerged from this process. The usage of MobPro is also detailed, explaining the advantages that emerged during participatory design and evaluation sessions that took place with both normally sighted and visually-impaired users. The various tests that took place over the course of the design, the design concepts that were defined and how they performed in various contexts and scenarios are presented, focusing, in particular, the design guidelines that were extracted from this experience. Finally, we discuss the contributions that resulted from this study, how they apply to the design of mobile multimodal applications and present future work.

MOBILE MULTIMODALITY, DESIGN CHALLENGES AND MAIN GOALS

Multimodal interfaces provide the adaptability that is needed to naturally accommodate the continuously changing conditions of mobile use settings [0,0]. Systems involving speech, pen or touch input, and graphical or speech output, are suitable for mobile tasks, and, when combined, users can shift among these modalities as environmental conditions change [0,0,0]. For example, the user of an in-vehicle application may frequently be unable to use manual or gaze input and graphical output, although speech is relatively more available for input and output. A multimodal interface permits users to switch between modalities as needed during the changing usage conditions.

Another major reason for developing multimodal interfaces is to improve the performance stability and robustness of recognition-based systems [0]. From a usability standpoint, multimodal systems offer a flexible interface in which people can exercise intelligence about how to use input modes effectively so that errors are avoided.

One particularly advantageous feature of multimodal interfaces is their superior error handling, both in terms of error avoidance and graceful recovery from errors [0]. There are user-centred and system-centred reasons why multimodal systems facilitate error recovery. First, in a multimodal interface users may select the input mode that is less error prone for particular lexical content, which tends to lead to error avoidance [0]. For example, users may prefer faster speech input, but will switch to pen input to communicate a foreign surname. Secondly, by allowing users to combine modalities in their input commands, the information carried by each modality is simplified when interacting multimodally, which can substantially reduce the complexity of the work requested to recognizers and thereby reduce recognition errors [0]. For example, while in a unimodal system a user may say "select the house near the lakeshore", in a multimodal system the user might point to the house and utter "select this house". Thirdly, users have a strong tendency to switch modes after system recognition errors, which facilitates error recovery [0].

In addition to these user-centred reasons for better error avoidance and resolution, there also are system-centred reasons for superior error handling. A well designed multimodal architecture with two semantically rich input modes can support mutual disambiguation of input signals. Mutual disambiguation involves recovery from unimodal recognition errors within a multimodal architecture, because semantic information from each input mode supplies partial disambiguation of the other mode, thereby leading to more stable and robust overall system performance [0].

Additionally, a large body of data documents that multimodal interfaces satisfy higher levels of user preference when interacting with simulated or real computer systems. Users have a strong preference to interact multimodally, rather than unimodally, across a wide variety of different application domains, although this preference is most pronounced in spatial domains [0,0].

These advantages of multimodal interaction are counterbalanced by the design challenges these pose. Diverse challenges need to be addressed during the development of multimodal applications, and particularly mobile multimodal applications, given the changeable interaction contexts of these settings. As stated above people have a strong preference to interact multimodally rather than unimodally. However, this is no guarantee that they will issue every command to a system multimodally, given the particular type of multimodal interface available. Therefore, the first nontrivial question that arises during input processing is whether a user is communicating unimodally or multimodally.

Predicting whether a user will express a command multimodally depends on the type of action performed. Additionally, in mobile scenarios, other constraints need to be taken into consideration: is the user moving, are her hands occupied, is she performing another task, is her visual attention focused elsewhere, are some examples of constraints that need to be considered. The knowledge of the type of action and of the necessary constraints should be used to assist the system in deciding what type of communication the user is establishing. If we offer designers a tool that is able to show them in what situations users chose to interact unimodally or multimodally, they will be able to design a better end-product, by imparting that knowledge onto the multimodal fusion component.

When users interact multimodally, there actually can be large individual differences in integration patterns [0]. Besides their natural integration patterns, user behaviour is influenced by the surrounding contextual environment. If a multimodal system can detect and adapt to a user's dominant integration pattern it could lead to considerably improved recognition rates. If a designer is aware of how the surrounding interaction context influences the user's integration patterns, once again a more informed and efficient multimodal fusion component can be developed. Accordingly, the ability to identify and frame the concerns that affect the selection of modalities and overall usability within mobile contexts is paramount. Nevertheless, given the complexity of pervasive and ubiquitous activities that can be carried on with the use of mobile devices, in particular with multimodal applications, this process is demanding and requires support in order to be properly achieved.

Other concerns must be considered for the presentation generation. A multimodal system should be able to flexibly generate various presentations for one and the same information content in order to meet the individual requirements of users, the resource limitations of the computing system, and specific physical and social characteristics of the interaction environment. Besides the presentation generation, even the presentation content should be selected according to the user's information needs, and the selection of which media to use should be made based on the context defined by user, device and environment. In a multimodal system, the fission component is the one that chooses the output to be produced on each of the output channels, and then coordinates the output across the channels. For a designer it's fundamental to have knowledge of what medium is the best to transmit each piece of information, and what is the most effective manner to coordinate the different medium used. A tool that allows the designer to evaluate different options of medium selection and combination in real-world settings, in a realistic manner should prove a powerful ally.

The methodology described in the next sections endows designers with the information required to better develop the different components of a mobile multimodal application.

Firstly, it introduces a context and scenario generation framework which supports designers while identifying key concerns for the design process by offering a method to focus these concerns while establishing design goals and scenarios that drive and frame the design process.

Secondly, it integrates a set of techniques for prototyping and multimodal simulation that clearly overcome the limitations of traditional low-fidelity prototypes even at very early design stages. Moreover, it completes the prototyping techniques with adequate means to gather information and interaction data with and without the user's intervention, supporting an in-situ, pervasive, evaluation process. These techniques are accommodated into the methodology and materialized by a comprehensive prototyping and evaluation framework that targets these concerns, paying particular attention to the challenges that emerge from the combination of multimodalities and mobility.

The following sections describe each of these contributions in detail and how the methodology and framework were validated through a design case study.

DEFINING SCENARIOS AND SELECTING CONTEXTS

Designing interactive applications is a demanding process that requires a deep understanding of users, their behaviour and the conditions in which they will be interacting with the envisioned solutions. Accordingly, and in order to support the design process, several techniques and methods have been introduced through the years targeting various issues and concerns. In particular, those that demonstrated solid results place significant focus on users and their needs[0,0], relying on different techniques to characterize them and the working settings, creating solutions that fit the usage context, providing an easy and fulfilling user experience.

However, mobile devices possess unique characteristics that equip them with ubiquity and portability features which introduce a new set of challenges into the design process [0,0,0,0,0]. Opposed to traditional software, the settings in which users interact with a mobile application are highly dynamic and can mutate very quickly. Moreover, as it has been studied [0,0], users often conduct one specific task that spans through various settings and different contexts. Still, usability has to be maintained throughout the entire usage experience and, in particular, depending on the specific requirements of demanding and austere settings, it even has to be increased. This is particularly important when multiple modalities are involved and at the same time, a basis motivation for their use as they can overcome some of the contextual difficulties by introducing alternative or complementary interaction channels.

Given the ubiquity of mobile systems, the selection and generation of scenarios and personas during the design process, is an implicitly enormous task [0], including an endless variety of combinations and presenting itself as a difficult and demanding chore for designers [0]. However, as strongly suggested by recent literature that support in-situ real world design of mobile applications [0,0,0,0,00], taking into account the actual settings in which interaction will most likely take place, context selection while defining concepts or designing solutions is paramount. Moreover, given privacy issues, with the impossibility of visiting and observing all the possible usage settings and contexts, the generation of scenarios is many times mandatory and indispensable during early design stages [0]. Although it does not provide the reality feel that in-situ design introduces, it enhances the design process by raising designers' attention to particular details that are generally of utter importance when it comes to mobile interaction and, most importantly, when using multiple modalities in real world settings [0].

However, because of the designers' "outsider" view and the infinite usage possibilities, and despite the amount of generated scenarios, important details might be discarded, retracting from the design process. Furthermore, the pervasiveness that mobile systems provide generally extends the scope of activities through various contexts, settings and scenarios. This problem does not usually exist in fixed environments and systems and is even more relevant in the context of multimodal applications.

Accordingly, in order to effectively generate scenarios and manage their complexity during the design process as well as during the evaluation, we conceived a conceptual framework which takes into consideration determinant factors that highlight possible problems and point relevant information to be used on later design stages. Although not exhaustive, the framework and examples introduce specific extensions to common design concerns that proved to be particularly relevant to the mobile experience and its design, paying special attention to those elements of the environment and context that are more prone to affect usability while taking advantage of available multimodal features. Overall, it aims at aiding designers throughout the selection and generation of scenarios by eliciting the details that must be focused, creating richer and comprehensive descriptions of real or possible usage settings.

These scenarios will provide the initial grounds and crucial concerns while engaging on the design process, by focusing the details that will and can affect the usability of the final application and by alerting designers to issues that would only be detected at later stages of design.

To achieve so, the framework defines three main modular concepts that can be arranged, composing what we regard as complete scenarios. This systematic approach allows for the composition and arrangement of different atomic parts that, either autonomously or in concert with each other, define realistic usage contexts. Schematics and models can be defined utilising these concepts as shown in figures 5 and 6. The three main concepts can be defined as follows:

- **Contextual Scenarios:** are scenarios composed by a set of variables (e.g., location; persona; device). Globally, a contextual scenario is defined by a single setting, a user/persona, and a set of variables that characterise each intervenient during a usage period.
- **Scenario Transitions:** are transitions between contextual scenarios (e.g., user moves from his/her bedroom to the kitchen and starts using the finger instead of the stylus). Transitions are necessary and of utter importance in order to simulate or enact the movement that is attainable while interacting with mobile devices through different settings, especially when achieving a single task (e.g., writing an SMS while moving around school).
- **Scenario variables:** are specific details that compose each contextual scenario. These can be categorised into five main aspects, namely:
  - **Locations and Settings:** Here, designers should pay particular care to the lighting, noise, and weather conditions since all can affect the way users interact with the mobile devices, in particular the accuracy of voice commands or the ability to view screen content or listen to audio channels. Moreover, the social environment in which the user is (e.g., surrounded by friends or unknown people, thus the user does not want to issue voice commands) might also affect how the device is used.
  - **Movement and Posture:** Users that are standing have different accuracy levels towards the touch screen and interactive components than users that are walking or sitting and these variables should be carefully assessed while defining the UI and the layout/size of the various elements that compose it.
  - **Workloads, Distractions and Activities:** Cognitive distractions (e.g., conversation), physical distractions (e.g., obstacles, use of another device), simultaneous activities, critical activities, should also be considered while designing a MMA. Here, an appropriate use of multiple modalities can play a crucial role, attenuating the required attention or use of limbs by providing alternative interaction channels; they can also be used depending on the criticality and needed accuracy, in certain cases, as a redundant and confirmatory mode.
  - **Devices and Usages:** The characteristics of a device are paramount while defining a solution, especially if it is a multimodal based one. Accordingly, special attention has to be raised to understand what input and output modalities the device is able to support, including camera and built-in speaker, microphone, virtual or physical keyboards, keypads, volume controls, etc; Additionally, the way the user is holding and interacting with

the device – single-handed; dual-handed, left hand, right hand, finger, stylus – is also extremely important when arranging the visual user interface of the application. For instance, as commented in one of the following sections in this chapter, when users interact with fingers, buttons and interactive components generally have to be larger than when users interact with the stylus of a PDA.

○ **Users and Personas:** Different users have different capabilities. For instance, characteristics like age, or gender, are determinant while using audio modalities. Moreover, the user's size, movement impairments, visual impairments, speech impairments, heterogeneity, abilities and disabilities are also paramount and can motivate the appearance or disappearance of alternative or redundant interaction modalities and can shape the layout of the UI.

For each of the above dimensions, variables can be instantiated with different values according to the settings that better suit the domain and scenarios that are being defined. More details and guidelines, as well as examples for specific variables that can be instantiated and how they affect usage and interaction within ubiquitous settings can be found in [0]. Globally, the framework is sufficiently broad and loose to allow for the creation and definition of a wide set of scenarios and contexts but it also is systematic and detailed enough to alert designers and focus concerns to which mobile and multimodal interaction is particularly sensitive. In summary, the ability to define rich and comprehensive contexts and scenarios, when accompanied by updated prototyping and evaluation techniques, can provide very useful results at very early design stages and with very few costs. This conclusion emphasized the need

to adjust existing prototyping techniques or tools so that they allow designers to conduct the design and evaluation process on real-world scenarios.

MOBPRO: MOBILE MIXED-FIDELITY SOFTWARE PROTOTYPES

Conducting prototyping sessions and developing early stage mock-ups during the design of a particular software application is clearly recognized as a highly effective way to test design ideas and share visions [0,0,0,0,0,0,0], especially with final users. However, given the unique characteristics of mobile devices, special care has to be taken in order to avoid misleading users [0,0]. The use of typical prototyping techniques during the design of mobile applications can reflect this fact and often starts to mislead users and hindering the overall design and evaluation process [0,0]. With this in mind, extensive work has been trying to overcome these problems by introducing new methods and techniques that facilitate the creation of realistic prototypes that enhance in-situ evaluation [0,0,0].

Nevertheless, despite the positive results that these experiences have provided, some effort is required in order to follow users, apply some of the procedures and, more importantly, simulate multimodal events. Traditional paper-based low-fidelity prototypes have strong limitations when it comes to test features such as voice commands, gesture recognition or even haptic feedback, which are particularly relevant within the context of mobile multimodal applications.

To solve some of these issues, new techniques and orientations, particularly for low-fidelity prototypes, have been introduced [0]. These suggest the need for more detailed and carefully built prototypes that offer a more resembling picture of final solutions and their characteristics [0] and a more realistic user experience. On these aspects, prototyping tools can play a paramount role, allowing designers to maintain their sketching and writing practices while creating prototypes that can actually run giving users a more tangible and realistic feel of the future application.

DENIM [0] and SILK [0] are two prototyping tools that give designers the ability to quickly create sketch-based prototypes and interact with them on the computer, also including the possibility of replacing drawn components with actual programmatic components. More recently, systems such as SketchWizard [0] or SUEDE [0] have also emerged, supporting new modalities and interaction modes such as pen-based input on the former and speech user-interfaces on the latter. Still, despite the useful functionalities and features provided by these systems, and the fact that they include sketching and quick prototyping mechanisms, the integration with the evaluation stages is rarely addressed. Moreover, the evolution from early based sketches to more advanced prototypes is only present on SILK and DENIM which lack crucial components (e.g., sound) and active behaviour or are deeply focused on specific domains. Furthermore, none addresses the specific needs of mobile devices or provides aids to designers while creating their prototypes. Nevertheless, the automatic support for Wizard-of-Oz prototypes and the ability to animate hand drawn sketches has shown very positive results.

Naturally, problems that result from mobility and multimodality are felt again when evaluating the developed prototypes. Although some recent studies reflect an increasing amount of attention towards contextual evaluation, out of the lab, its relative inexistence contrasts with the importance and benefits it presents to mobile devices [0,0] and multimodal applications. Existing examples usually point guidelines on how to emulate real world settings within labs [0,15] or provide solutions [0] that are useful as a complement but, even if obtaining positive results, do not address specific usability problems, do not provide quantitative data and focus mainly on user satisfaction. Furthermore, they show little regarding user interaction towards the applications.

Some recent approaches have also addressed this stage of design, focusing methods to gather usage data remotely through active (e.g., ESM, Diary Studies) and passive modes (e.g., Logging). For instance, with close goals to MobPro regarding evaluation, the Momento [0], and the MyExperience [0] systems provide support for remote data gathering. The first relies on text messaging and media messaging to distribute data. It gathers usage information and prompts questionnaires as required, sending them to a server where an experimenter manages the received data through a desktop GUI. On the second, user activities on Mobile Phones are logged and stored on the device. These are then synchronized depending on connection availability. The logging mechanism detects several events and active evaluation techniques can be triggered according to contextual settings. However, once again, these solutions provide restrictions that hinder the evaluation process and are not integrated with the prototyping stage. The ability to carry the evaluation process while using different techniques in concert and the usage of actual mobile devices for evaluation purposes, collecting and analyzing usage data in configurable ways, at very early design stages is never offered. Additionally, these tools offer little flexibility towards the fidelity of the used prototypes and generally require functioning applications, which fall out of the goals and scope of the design methodology and tools, especially at early stages, which are being discussed in this chapter.

To cope with these early design stage difficulties, added effort and inevitable limitations of the low-fidelity prototyping techniques, and to overcome the limitations of existing tools, which pertain both to prototyping and consequent evaluation, we developed a software mixed-fidelity prototyping framework called MobPro. The framework's features cover the prototyping and evaluation stages, supporting an iterative and participatory design that facilitates the transition between them and builds on the results of

techniques that enhanced these procedures when carried on mobile settings with mobile devices.

Its umbrella goal is to support the early design stages of multimodal applications for mobile devices, overcoming the difficulties that are found while using the lower-cost paper-based techniques. Like some existing frameworks [0,0,0] it provides designers with tools to quickly create prototypes and evaluate them, focusing specifically on mobile and handheld devices and multimodal applications. However, additionally, it supports in-situ and participatory design and enables designers to use both passive and active evaluation methods. The framework allows the construction of low, mid and high-fidelity prototypes, as well as mixed-fidelity prototypes (e.g., combining sketches with multimedia features and visual components) and extends its automatic Wizard-of-Oz usage through their evaluation, also providing means to analyze the gathered data. It is noteworthy that MobPro's main concern is the user interface, prototype's usability and the overall user experience. The available features target these concerns and offer designers and developers the ability to create prototypes that can be tested and used for simulations rather than functional applications that can be provided to end-users. Nevertheless, the included features and techniques allow for the construction of prototypes that are functional to a certain degree and can be interacted with by end-users without any intervention from the designer or developer.

Concept and Features

The framework takes into account existing work and the available literature and combines different techniques in order to provide extensive support to the design process through different stages. The main concepts behind its functionalities are:

- **Prototyping with mixed-fidelities (thus analyzing different usability dimensions).** Hand drawn sketches or interactive visual pre-programmed components can compose different prototypes with varied levels of visual refinement, depth of functionality and richness of interactivity [0], comparing different design alternatives, evaluating button sizes, screen arrangements, element placement, interaction types, navigation schemes, audio icons, and interaction modalities, among others.

- **Multimedia content.** With the inclusion of multiple media components and elements (e.g., video and audio clips and recording), users are able to create prototypes that can combine different visual and interactive features and concepts as well as non-visual prototypes that can be used and tested on the devices. The possibility to combine different media content using simple rules in accordance to predefined scenarios affords the possibility to simulate, evaluate and decide on multimodal presentation mechanisms to include in the final applications.

- **Avoid the cargo cult syndrome [0].** By using actual devices, problems regarding the device's characteristics (e.g., size, weight, screen resolution, shape) emulation are solved, allowing their utilization on realistic settings. This provides users a much more tangible and realistic usage experience. Moreover, it provides a seamless integration of features exploring the use of audio and video, speech recognition and gestures (and other modalities which might be available in the target devices), increasing the designers understanding of real world usage of the different modalities, thus contributing to a better definition of multimodal integration patterns, which are shaped not only in accordance to the device characteristics but also by the user [30].

- **Automatically support the Wizard-of-Oz technique.** By adding behaviour to the digitalized sketches or by using visual

components, users can navigate through the prototype without having to explicitly replace the screens by hand or without the presence of a designer to do so.

- **Gather data through passive and active techniques.** On the former, every action that the user takes is automatically logged with customized granularities. On the latter, the use of ESM [0] and diary studies [0], integrated within the tool, provides another source of data and usability information. Integrated questionnaires can be popped during or immediately after using the prototype, or even automatically during the day according to specific settings (e.g., if the user is unable to achieve a specific goal or is continuously failing to press a small button).

- Finally, the framework includes a set of analysis features, from which the most innovative one is a log player which re-enacts (through a video-like mode) all the users' activities with accurate timing and interaction details, attenuating the need for direct observation.

Architecture and Prototyping Steps

The framework is divided into several tools, each aiding designers and users through a particular activity and including features and functionalities that support the various design techniques that it includes.

Prototype Builder

The first tool is the prototype builder. It supports the design and creation of the interactive prototypes that can be later used in a real mobile device. Each prototype is composed by a set of screens or cards, closely emulating the traditional low-fidelity prototyping approach. Screens/cards can be composed by several interactive elements, or, alternatively, by low-fidelity sketches that can be augmented with behaviour. They allow designers to create different and interactive prototypes that function as a final end-user application would.

As shown in Figure 1, users can create screens individually, organizing them sequentially or spatially and customizing them according to their needs. Screens can be designed by easily dragging

Figure 1. Prototype building tool

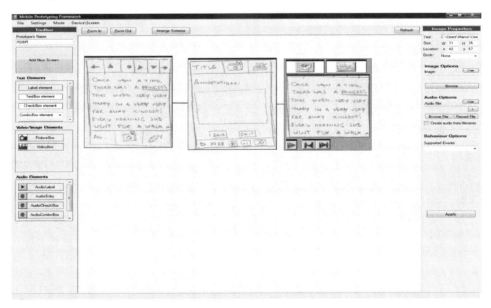

and dropping the selected interactive components, using hand drawn sketches, pictures or images for multiple screens, also arranging the "prototype's wireframe" or storyboard.

Interactive Elements

Interactive output elements/components (e.g., combo-boxes, labels, images, audio and video files) can be used. Each component assumes its usual behaviour (e.g., text label shows a short text; an audio clips plays the corresponding sound) unless otherwise indicated by the designer.

Input elements/components (e.g., text data entries, sound recorders, video recorders, buttons) are also available. Once again, these assume their usual behaviour and support the collection of data through the various modalities that they refer to, or the navigation between different parts (e.g., screens) of the prototype.

Composed elements can also be created. In general, these combine text and audio modalities, video and text, images and audio or any other set that the user wishes to use. The use and combination of these components enables users to create mid to high-fidelity or even mixed prototypes. By arranging them within a screen and adding new actions to their usual behaviour, users are able to define complex animations that function as a final application would and simulate multimodal output presentations through the concerted use of different multimedia elements. For each component, different configurations are also available (e.g., multiple-choices through radio buttons, check-boxes or combo-boxes).

For low-fidelity prototypes, sketches (hand-drawn and scanned or digital drawings) can be easily imported and their behaviour adjusted (Figure 1). For example, transparent clickable areas can be placed on top of sketches and images and used as buttons. This allows designers to animate and augment the low-fidelity sketches with actual behaviour and some degree of functionality.

Finally, as it supports multimedia elements, the framework also allows the creation of prototypes that do not use any visual components. The same building mechanisms and architecture (based on pages and elements) is used but navigation mechanisms are adjusted during interaction so that elements can be navigated sequentially without the need to change screens and without any visual aids. As elements are navigated, audio descriptions of each are automatically played. On these settings, while building the prototype, each element that contains a textual description or textual content is automatically recreated in an audio counterpart, making use of text-to-speech libraries (see Fig. 2).

Each prototype is specified in XML and stored within a file that contains its specification which can be transferred and updated even without using any specific tool. The tool's modularity allows different components to run on different devices and systems.

Behaviour and Actions to Augment the Prototypes

In order to support an automatic Wizard-of-Oz evaluation and usage of the prototypes (e.g., replacing the human that acts as the system by a mechanism that supports navigation through the prototype's user interface and cards), the framework provides an integrated rule engine that facilitates the definition of rules, composed by a combination of conditions, triggers and correspondent actions that bring the prototype to life. The prototype's behaviour can be defined within three levels: component/element, screen and global behaviour. On the first users can define the behaviour when using an individual component (e.g., a button press displays a warning). The second defines the behaviour for the entire screen (e.g., the user missed two of the screen's components and these are highlighted) and the third for the entire prototype (e.g., a questionnaire is popped once the user reaches the fifth screen).

Figure 2. a) Wizard to create a multiple-choice (e.g., radio-buttons, combo-box) audio element. b) Configuration options that allow designers to create prototypes with particular features and targeting specific devices.

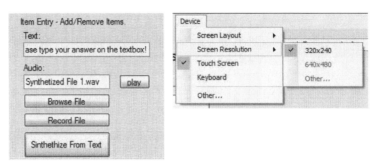

Available triggers focus time, interaction and content:

- **Time-based triggers:** can be configured to prompt warnings or change elements according to the time the user is taking to review/complete it. These can also be used to simulate multimodal output generation, by creating timelines with different multimedia components.
- **Interaction triggers:** analyze the user's interaction with the device by counting clicks, detecting where on the screen the user has interacted or which input modalities have been used in specific scenarios. These can also be used to simulate multimodal integration patterns. For such an end, just the modalities employed, and not the content expressed, are used in the simulation, since most prototypes will lack the ability to interpret the contents of recognition based modalities. Through the framework's logging mechanism the designer will have access to the content provided in each modality and will be able to determine and interpret the integration patterns used by different individuals in different contextual situations.
- **Content based triggers:** activate actions depending on the content of the items. For

example, if the user chooses a correct value or a specific option from a list or a value within a defined threshold, a certain action can be triggered.

In concert with the triggers and conditions that are defined for each prototype, two different types of actions can be selected. The first one prompts a message (e.g., audio, video, or text) that is composed by the designer while the second jumps from the current element or screen to another element/screen within the prototype (e.g., the user taps a button or element and a sound is played or the following screen is shown – thus supporting navigation within the various sketches/screens). This supports the simulation of a multimodal application, by combining, for instance, user selections on a touch screen, with an audio capture of the user's speech. Although not being able to interpret the speech, a carefully designed prototype, grounded on well defined scenarios, can expect the user to input a given command in every particular scenario, and thus proceed in accordance to what is expected. Even though this approach cannot be expected to be correct in every testing run, it provides designers with audio recordings of unconstrained vocabulary in real world usage, thus affording them the possibility to make informed design decisions when deciding on the application's vocabulary support, for example.

Simulating and Prototyping Adaptation

The behaviour mechanisms that are included in the building tool can also be used to simulate and prototype the application/prototype's adaptation. Although there is no user model and adaptations are generally hard coded and associated to specific events and trigger pre-defined actions, the overall mechanism permits designers to emulate adaptation aspects regarding both recognition based modalities and visual components. For example, if the designer defines a rule that is triggered when the user fails to hit a button three times in a row, it can prompt a warning requesting the user to issue a voice command instead.

Using the Prototypes

The counterpart of the previous tool is the runtime environment that is responsible for recreating the prototypes on the targeted devices. Basically, this application reads the XML-based specification and reconstructs the prototype on the target device. Currently we have a runtime environment for Windows Mobile (Figure 7), Palm OS and SymbianOS. A Windows version has also been created to allow testing on TabletPCs.

Real-World Evaluation

In order to evaluate the usability of the prototypes, data has to be gathered while users interact with them on natural usage scenarios. To achieve this we included means to gather qualitative and quantitative data through two different approaches, namely passive and active.

Passive Techniques

Integrated within the runtime environment there is a logging engine which is responsible for the passive data gathering (i.e., without the user's explicit intervention). The logging engine stores every event that is triggered by the user's interaction with the prototype and device or by the time constraints associated to each element/screen. Events range from each tap on the screen, each button press or even each character that was typed by the user. Each event is saved with a timestamp, allowing its reproduction for the re-enactment of the usage behaviour. Other details such as the type of interaction, location of the screen tap, etc., are also stored for every event.

However, taking into consideration mobile devices' limited memory and battery, the granularity of the logged events can be easily configured both during usage and during the prototype's construction. Limiting the amount of logged events reduces the size of logs and the processing that logging requires.

Moreover, the adjustment of the logging granularity also serves analysis and evaluation purposes. For instance, if the evaluator is particularly interested in understanding how the user navigates between the existing screens that compose the prototype, but has no interest in collecting data regarding the locations of taps on the screen, the latter event can be ignored, creating logs that are focused to particular events. If no adjustment is made, by default all events are automatically logged.

The definition of the log granularity is also important when taking into consideration the several modalities that are available within a particular prototype. Here, the selection of specific events, which pertain to specific modalities (e.g., play, pause of an audio track or video, voice commands) is also paramount in order to facilitate analysis and evaluation of usage logs (e.g., if the application is to be used by a visually impaired user, the designer might select to log only the audio modality and the user's voice commands and – accidental – taps on the screen or keypad presses can be discarded).

Overall, the logging engine and mechanism support a configurable data gathering tool that can be adjusted to the evaluation purposes and goals.

Active Techniques

To support active data gathering, users that interact with the prototypes assume an active role and are responsible for providing usage or context information. This type of data gathering has been widely used on mobile devices, supporting techniques such as the Experience Sampling Method [0] or Diary Studies [0]. The main medium to gather information with these techniques is questionnaires. At particular times or events, users are required to fill-in a questionnaire, responding to questions that pertain to the action that they are performing, their location or any other relevant detail. This provides a qualitative opinion and data regarding the usability of the system that is being used. However, users are usually required to carry paper questionnaires along with them while performing their activity, which often leads to users not remembering to complete the questionnaires or hinders the activity at hands [0]. Here, the digital support plays a key role since questionnaires can be prompted automatically when necessary [0].

MobPro supports active data gathering by offering means for designers to include, within their prototypes, questionnaires that can be completed by users while interacting with the prototypes (figure 3). Using the same mechanism and interactive elements that are used to build the high-fidelity prototypes, questionnaires can be easily configured to focus the details and goals of the evaluation.

The various input and output modalities that prototypes can include extend the scope and traditional resources used for evaluation and allow users to gather data in diverse formats. For instance, when a questionnaire is popped, and in order to allow the user to continue with his/her activity while responding to the questions, these can be answered using voice or text, depending on the user's activity. This increases the flexibility and ease of responding to usability questionnaires and provides richer data. Moreover, it also allows designers to have an idea of the environment in which the user is interacting with the prototype (e.g., quiet/noisy, alone or accompanied by other users). For instance, if the user completes the questionnaire by recording his/her answers with an audio recording element, the surrounding noises can or might also be recorded. Furthermore, if needed, users are also able to film or take pictures (if the device includes a camera) of the environ-

Figure 3. Active data gathering questionnaires. Questionnaires can be composed with the same elements that are used to create the software/high-fidelity prototypes. Once completed, they can be browsed directly on the mobile device.

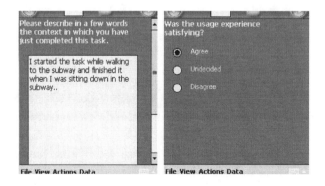

ment in which the activity is taking place or where the usability issue was detected.

Taking advantage of the behaviour engine and respective conditions and actions, the framework also provides means for users to define specific conditions or settings in which these questionnaires can or should be presented to users. This technique, if well used, provides support for intelligent active data gathering since usability questionnaires can be prompted according to time, location or behaviour triggers. For instance, if the user misses a specific screen location or button within a screen several times, or is taking more than 1 minute to respond to a question, a questionnaire can be automatically popped up. Here users can be requested to explain the reason behind the low accuracy or describe the setting in which they were working (e.g., while jogging, seating on the sofa at home). Also, if targeting navigational issues or the user interface's structure, questionnaires can be configured to appear when certain cards/screens are reached.

This contextual information is extremely valuable, providing information on the user's location and the environment and can be used in a posterior analysis relating it with the data that is gathered passively through the logging engine.

Analysis Tools

The framework includes two different approaches for the analysis of data. The first provides designers with quantitative lists of interaction details and browsing features over stored results while the second provides a video-like reviewing mechanism that re-enacts every interaction detail and event that took place while the user interacted with the prototype.

Browsing Results

All the data that is gathered by end-users, through ESM, diary studies, annotations and questionnaires can be directly reviewed and browsed on the runtime environment. As seen in figure 3, results of completed questionnaires are stored on the mobile device and can be loaded together with the questionnaires. On this facet the runtime environment allows no input from the user and only enables navigation and browsing within the results.

Interaction logs that are automatically stored can also be consulted directly on the mobile device. All the results and logs are stored in XML files and can be reviewed and searched in any text editor.

Log Player

In order to evaluate the users' behaviour and the logged interaction on real scenarios, we intended to replace direct observation, as far as possible, with a similar mechanism. Accordingly, we developed a log player that resembles a "movie player" which re-enacts every action that took place while the user was interacting with the prototype. Every event (e.g., tapping the screen, pressing buttons, playing sounds, writing text) that was raised when using the mobile device and that is stored on the log is recreated sequentially and according to the time stamps that the log contains.

To facilitate analysis, the adjustment of the speed in which events are (re)played is also possible (e.g., fast-forward; double speed). Other visualization options for the usage logs are available (e.g., event lists, selection tables). As mentioned, events can be played sequentially according to the time-stamps that were recorded or they can be aggregated and searched by type (e.g., heat maps that show all the taps in one screen or browsing every "next screen" event).

Although these logs are limited to the direct interaction that the user has with the device, they still present enough detail to indicate whether a button needs to be enlarged or if the screen arrangement should be changed or even what type of element or modality is preferable in certain situations [0].

For non-visual prototypes, logs are also available and although they do not present a visual recreation of events, they can be analysed through the same mechanism. However, on this dimension, only the audio commands, inputted data and other events that were logged will be replayed.

MULTIMODAL RICH DIGITAL BOOK PLAYER: RESULTING APPLICATION

In this section, we present the validation process for the stated contributions and illustrate how using the scenario framework and MobPro impacted the development of a multimodal mobile rich Digital Talking Book (DTB) player. We start by providing a short introduction to the DTB field, which is followed by a brief description of the prior desktop version. After this, we detail the most significant aspects and features of the mobile version.

Digital Talking Books

Digital recordings of book narrations synchronized with their textual counterpart allow for the development of DTBs, supporting advanced navigation and searching capabilities, with the potential to improve the book reading experience for visually impaired users. By introducing the possibility to present, using different output media, the different elements comprising a book (text, tables, and images) we reach the notion of Rich Digital Book [0]. These books, in addition to presenting visually or audibly the book's textual content, also present the other elements, and offer support for creating and reading annotations.

Current DTB players do not explore all the possibilities that the DTB format offers. The more advanced players are executed on PC platforms, and require visual interaction for all but the most basic operations, behaving like screen readers, and defeating the purpose to serve blind users [0].

The DTB format, possessing similarities with HTML, has, nevertheless, some advantages from an application building perspective. The most important one is the complete separation of document structure from presentation. Presentation is completely handled by the player, and absent from the digital book document. This enables the design of a player that, by adequately taking advantage of the possibilities offered by multimodal inputs and outputs, increases the usability and accessibility of such an application, while supporting different media content presentation. Navigation wise, the user should be able to move freely inside the book, and access its content at a fine level of detail. The table of contents should also be navigable. One major difference between a DTB player and a HTML browser is the support offered for annotating content. Mechanisms to prevent the reader becoming lost inside the book, and to raise awareness to the presence of annotations and other elements, like images, are also needed.

The Desktop Multimodal Rich Book Player

By combining the possibilities offered by multimodal interaction and interface adaptability we have developed the Rich Book Player, an adaptive multimodal Digital Talking Book player [0] for desktop PCs. This player can present book content visually and audibly, in an independent or synchronized fashion. The audio presentation can be based on previously recorded narrations or on synthesized speech. The player also supports user annotations, and the presentation of accompanying media, like other sounds and images. In addition to keyboard and mouse inputs, speech recognition is also supported. Due to the adaptive nature of the player, the use of each modality can be enabled or disabled during the reading experience.

Figure 4 shows the visual interface of the Rich Book Player. All the main presentation components are visible in the figure: the book's main content, the table of contents, the figures' panel and the annotations' panel. Their arrangement (size and position) can be changed by the reader, or as a result of the player's adaptation. Highlights are used in the main content to indicate the presence of annotated text and of text referencing images. The table of contents, figures and the annotations panels can be shown or hidden. This decision can be taken by the user and by the system, with the system behaviour adapting to the user behaviour through its adaptation mechanisms. Whenever there is a figure or an annotation to present and the corresponding panel is hidden, the system may choose to present it immediately or may choose to warn the user to its presence. The warnings are done in both visual and audio modalities.

All the visual interaction components have a corresponding audio interaction element, meaning all commands can be given using either the visual elements or speech commands, and also all the book's elements can be presented visually or aurally.

The Mobile Version of the Multimodal Rich Book Player

The primary concern of a DTB player is to increase visually impaired people's accessibility to literary content. By endowing our desktop DTB player with multimodal capability, besides meeting that target, we also increased the accessibility and usability of that application for other users, with other impairments, and even non-impaired, by offering multiple interaction modalities. The final step towards achieving a universally accessible DTB player was porting it to a mobile platform, in order to have a DTB player that could be used anywhere, anytime, and ideally by anyone.

Accordingly, the mobile version of the Rich Book Player was developed with three main goals in mind: 1) Allow for an anytime, anywhere entertaining and pleasant reading experience; 2) Retain as most as possible of the features available in the desktop version; 3) Support a similar look and feel and foster coherence between both applications.

Figure 4. The Rich Book Player's interface. The centre window presents the book's main content. On the top left is the table of contents. On the bottom left is the annotations panel. On the right is the figures panel.

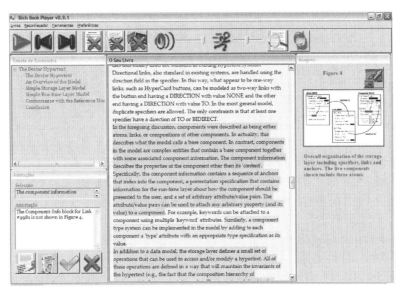

Figure 5. One of the scenarios used during the design of the rich digital book player

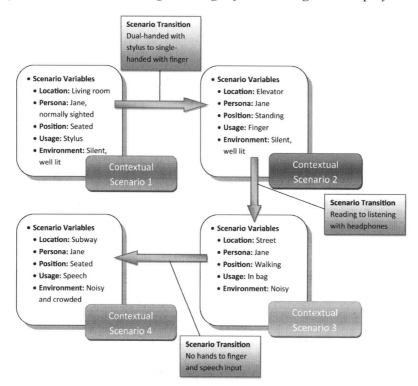

To achieve these goals, architectural and interaction changes had to be made with regard to the desktop version. The major distinctions of the desktop and mobile platforms are given by mobile platform's limited screen size and processing power, the distinct means of interaction available, and the influence of the dynamic surrounding environment.

Those differences encouraged us to begin the development process of the mobile version as a new project, grounded on the requirements of achieving a consistent look and feel between both applications, and also of having interchangeable books.

The design team that was responsible for developing the mobile version of the rich book player followed the design process detailed in this chapter. During the initial stages of the design of the mobile multimodal rich digital book reader, the framework played an essential role, highlighting details within scenarios that would affect the us-

ability of the application while in different settings. Moreover, it provided, at very early stages, the awareness on the target users that was necessary to properly design the tool in order to suit both visually impaired and non-visually impaired users with a universally accessible interface. Throughout this process numerous scenarios were defined and used to drive design. Figure 5 and figure 6 show two different scenarios that emerged during this stage and that were especially useful to the definition of the tool's features as well as to its usability concerns.

In figure 5, Jane, a persona for a non-visually impaired, young adult, goes through three scenario transitions. The first contextual scenario sees Jane sitting in her living room, selecting the book to read during her morning trip to work. She relies on visual output and the stylus for input during this operation. Contextual scenario 2 shows Jane in the elevator while starting the book playback. Having already stored the stylus, she uses one finger to

Figure 6. Complete scenario composed by three contextual-scenarios

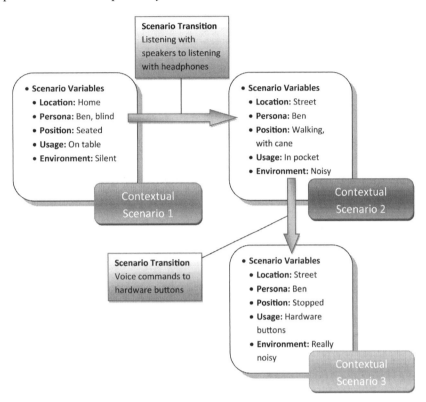

initiate the playback. She visually skims the book's current position to make sure she remembers the plot from when she last stopped reading. In the third contextual scenario, Jane reached the street and walks to the subway entrance. Still using her fingers she starts the book playback, turns off the screen, and listens to the narration with her headphones. While on the subway on the way to work, she listens to an excerpt that strikes her as important, and she wishes to annotate the book to remind her to check it later. She takes the mobile book player from her bag. Operating it with just one finger, she turns on the screen, selects new annotation, and dictates the annotation to the player. Had she managed to find a seat during the trip, she might have used the stylus to enter the annotation through the device's virtual keyboard, but as she is standing, and doesn't mind the people nearby listening to her annotation, she decided

using speech would be faster and less prone to errors than writing with the stylus.

In figure 6, Ben, a persona for a blind man, is listening to a book at home. Due to the silent environment he's in (contextual scenario 1), the book player stands atop a table, playing back the book's narration, and Ben can use voice commands to control the application. When Ben goes out (contextual scenario 2), he plugs the headphones to the player's jack, and places it in his pocket. In this fashion, Ben can continue to listen to the book while walking. Before leaving he takes precautions to lower the playback volume in order to be able to still listen to surrounding sounds in the street. When Ben reaches a particularly crowded and noisy street (contextual scenario 3) he decides it is time to stop the book playback. With such a noise level, Ben is aware that voice commands are not effective. He takes the book player from his pocket, and using the device's hardware but-

Figure 7. Mixed-fidelity prototypes employed in the early stages of design of the multimodal mobile rich book player

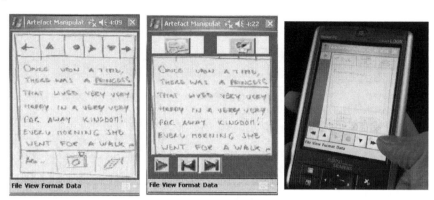

tons he issues commands to stop playback and terminate the application.

These scenarios, together with the above presented goals, guided designers on some early design decisions regarding the use of multiple modalities on the mobile DTB player. The first decision made concerns the use of multiple modalities to create the output presentation. In order to reach all target audiences, the book's content has to be presented both visually and audibly. Thus, it was decided to use those modalities in an equivalent fashion. Nevertheless, a multimodal mechanism was required to guarantee both representations of the content would not contradict themselves. This was required whenever the user changes the presentation settings of one modality, which impacts the other. For instance, a change in reading speed requires a different page turning rhythm, which is also impacted by changing font size or screen orientation, rending useless a simple synchronization mechanism. The second decision concerned the integration of input modalities. It can be assumed every device makes available a touch screen and a set of hardware buttons. These modalities were used equivalently, with the guarantee that every command was made available through the hardware buttons, because of accessibility concerns. On some devices, speech recognition is also available. This allowed the

exploration of complementary relations between modalities, mainly for annotation creation and management. For instance, the user can select a page, or a text excerpt inside a page, and issue a voice-based command to create an annotation.

These scenarios were later used as actual contexts and were the basis for the creation of scripts that were used during the evaluation of the prototypes that were created.

In the following paragraphs we detail some of the more relevant aspects of the application, giving higher importance to the interaction based ones.

Visual Presentation of Components

Figure 4 presented the main components of the Rich Book Player: main content, table of contents, annotations and images windows. On the desktop version it is possible to display all the components simultaneously and users can find the arrangement that best suits them. This approach was deemed unsuccessful on the early design stages due to the much smaller screen size of the mobile devices. The low-fidelity prototypes presented above in figure 7 validated an approach where each of the main contents was presented to the users in a different tab, and the screen was portioned in four different interaction areas.

Figure 8 presents the main content view of the mobile version of the Rich Book Player. The four main areas of interaction can be seen in the figure. On the top, three tabs allow for the selection of the current view. The left tab opens the content view (figure 8, left). The tab header is used to display the current chapter number, which, in this way, is quickly available to the user. The middle tab displays the annotations view. This is the view used to read previously entered annotations and write new ones. The right tab is the images view (figure 8, right). In this tab users can see the images that are part of the book, together with their title and caption. All these contents, book text, annotations and images are displayed in the biggest of the interaction areas. Bellow this area is a toolbar which displays the commands to control book playback, navigation, and other features. The final interaction area is the menu bar located at the bottom of the screen. Besides being used to present the menu, the menu bar is also used to display command buttons whenever necessary.

Of the four main components of the desktop version, three of them were already mentioned and although they cannot be displayed simultaneously as in the desktop version, they all can be displayed on their own view. The component not yet mentioned is the table of contents. This component was downgraded to a menu entry and retained just one of its functions. In the desktop version, the table of contents was used to display the current chapter being read, by highlighting its entry, and as a navigation mechanism, allowing users to jump to a particular chapter by selecting its entry. In the mobile version, only the navigation function was retained. The current chapter feedback is now provided as the header of the main content tab.

As seen in figure 7 and figure 8, the navigation bars were replaced and exchanged on the screen. This resulted from interaction problems that emerged when using the device with only one hand, which was quite frequent (as defined on the first scenario – figure 5). Since users had difficulties reaching the upper part of the screen when using just one hand (particularly in prototypes with longer screens), the navigation bar, used more often than the tab bar was moved to the lower part of the screen, where users could interact with it by using their thumbs while holding the device vertically.

The annotation creation process can be triggered by two different approaches, resorting to different combinations of modalities, which will be detailed in the next section. Independently of the approach followed, the process continues with

Figure 8. The mobile Rich Book Player. Main content view on the left and images view on the right.

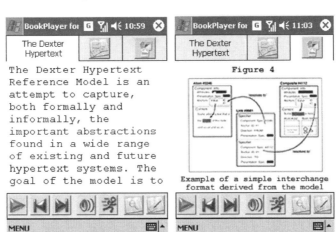

the user in the annotations view in create annotation mode (figure 9, left).

In the annotations view, the user is still able to see the selected text while entering the annotation. When the text box is selected, the virtual keyboard is displayed, and the confirm and cancel buttons are reallocated to the menu bar (figure 9, right). The confirm and cancel buttons were also adjusted and evolved since the initial prototypes. Initial ones were text-based but were smaller and difficult to read in well lit conditions where contrast was insufficient to provide a clear image, especially when placed on the lower gray-bar.

Figure 10 presents the annotations view in annotation display mode. When the user changes to the annotations tab, the annotations menu (figure 10, left) displays all the existing annotations in a tree view. By selecting one of the annotations the user is taken to the annotations detail view (figure 10, right). In this view the user can read the current annotation, edit the annotation (which means going to annotation creation mode), delete the annotation, and navigate to the text that has been annotated. The navigation buttons in the toolbar, which in the content view navigate to the next and previous pages, in this view navigate to the next and previous annotations.

The Use of Multiple Modalities

The main feature distinguishing a Digital Book Player from an e-book player is the possibility to present the book's content using speech, either recorded or synthesized. The desktop version of the Rich Book Player supports both modes of speech presentation. The mobile version currently supports the presentation of recorded speech.

Speech presentation opens up interaction possibilities that are not available with a visual only interface. With speech, users can change tabs and view images or read annotations while listening to the narration, thus avoiding the forced pause in reading if speech had not been available. Speech also allows users to access the book content without having to look at the device, allowing usage scenarios that, up until now, were available only with portable music devices, but without the limitations of those. For example, performing a search in such a device is extremely cumbersome. In comparison, with the mobile Rich Book Player,

Figure 9. Annotations view during an annotation creation process. On the right, buttons are reallocated to the menu bar when using the virtual keyboard.

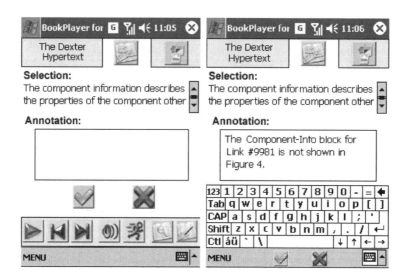

search is extremely simple due to the presence of a digital version of the book's text.

With the benefits of incorporating speech into the interface new challenges are uncovered. With speech comes the need for synchronization mechanisms. The application needs these to be able to know when and what images and annotations to present. Designers were able to port the synchronization mechanism of the desktop version to the mobile version without losing synchronization granularity, meaning the mobile version also supports word synchronization. However, due to several features the application is expected to support a simple synchronization between displayed and narrated text is not enough, and a mechanism to guarantee coherence between both streams had to be designed. The two main reasons why this multimodal fission mechanisms is required are discussed in the following paragraphs.

One of the main presentation and interaction differences between the desktop and mobile versions of the Rich Book Player is the introduction of the page concept in the mobile version. The main motivation behind this decision was given by the difficulties observed when users interacted with

prototypes of the mobile rich book player. It was noticed that, with moderate to large books, even small scroll bar movements might give origin to large displacements of the text being displayed, which would quickly turn into a usability problem. In accordance it was decided to avoid the use of scroll bars to read the book, and the page concept was introduced in the mobile version.

Additionally, to support changing font and font sizes and changing screen orientation from portrait to landscape and vice-versa, while reading the book, the pagination algorithm had to be executed in real-time [0]. The need to introduce this feature came about while using an actual device on the bus. As the user was viewing real images that were associated to a small text, some of the used images were better viewed in a landscape mode (which was unavailable on that prototype but suggested the need to add it to the final applications). Moreover, as he was comfortably seated, he provided annotations stating that if he was able to use the player in landscape mode (Figure 11), it would show more text and it would be easier to read.

Figure 10. Annotations view in annotation creation mode. Annotations menu on the left, and details of an annotation on the right.

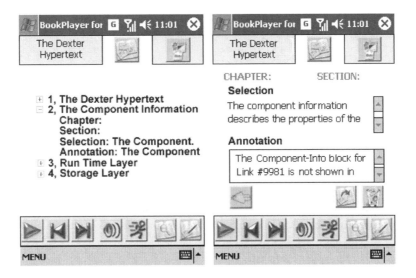

Images and annotations require a notification system to alert the user to their presence. Users just listening to the narration need to be alerted through sound signals, while users reading the text need to be alerted through visual signals. Both mechanisms coexist also in the mobile version of the Rich Book Player.

Here, mixed-fidelity prototypes were also used to evaluate different alternatives of audio based awareness mechanisms. Speech recordings, auditory icons [0] and earcons [0] were compared. These tests led to the adoption of auditory icons for audio awareness mechanisms in the mobile DTB player. Following these results, auditory icons are played when the narration reaches a page with annotations or associated images. Visually, when the user reaches such a page, the annotations or images tabs flash to indicate the presence of an annotation or image.

During the design stages, MobPro played a crucial part in this process. In previous attempts to simulate interaction with the player and while using the typical Wizard-of-Oz approach, users were constantly confusing the warnings and auditory icons within the tool with the text and content of the book. The main reason behind such confusion was related to the evaluator's (the wizard) tone while recreating the various sounds and the fact that all the audio was produced by the same voice.

As MobPro supports the design and development of prototypes with different multimedia elements, it was used to create both low-fidelity and mid-fidelity prototypes that could simulate some of the multimodal features that were necessary for the final application. Moreover, with the ability to create rules that can define and restrict navigation through the several screens and user interfaces that are being tested, the Wizard-of-Oz techniques is automatically supplied by the underlying structure of the prototype. Besides providing a realistic usage experience to end-users, by offering them the possibility to test the prototypes (even sketch-based ones) on actual

Figure 11. The Rich Book Player layout in landscape mode

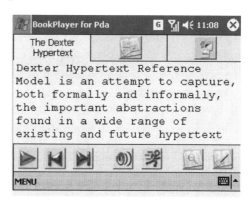

devices, it also enriches the testing sessions and gathered results by allowing designers to augment the sketches and visual elements with simulations of multimodal components.

Accordingly, when using MobPro, different and realistic sounds (e.g., generated by the designers or downloaded from the web) were used and professionally dictated excerpts from audio books were included, providing users with a much more realistic experience and far better results and acceptance during the evaluation stages. Moreover, it was possible to detect whether the sounds were audible or not within loud locations, testing the limits of the used device and the adequacy of particular sounds to particular settings. For instance, on the first scenario (where Jane uses the player in the subway) most of the auditory icons weren't noticeable and had to be replaced by louder and more acute sounds. This information was noticeable when using the analysis mechanisms that MobPro includes. In particular, during log revision, situations where audio warnings were systematically ignored clearly emphasized the need to replace and improve the used auditory icons. Naturally, this information was confirmed by the annotations and registers that users provided together with their automatically collected usage results.

The multimodal presentation mechanism is, thus, responsible for the real-time pagination algorithm and for managing the awareness raising mechanisms, ensuring the real-time generation of the presentation using both visual and audio modalities. A multimodal fission mechanism allows the application to turn to the next page when the narration reaches the end of the current one. It is also used to visually highlight the word currently being spoken, in order to make the narration easier to follow when the user chooses to both read and listen to the book, and to fire the events required to make the user aware to the presence of images or annotations. This mechanism is capable of guaranteeing synchronized presentation of both contents by changing in real time the parameters guiding the presentation in each modality in order to accommodate changes made by the user, like setting a new narration speed, or choosing a different font size.

Input Alternatives

The features described so far have dealt primarily with the output side of the application, based on visual and audio output. The input side of the application also takes advantage of the multiple modalities available: On-screen buttons operated with the stylus or fingers for input commands, speech, images and video for annotations.

One of the book player features, annotation creation, allowed to explore different relationships between modalities. It should be stressed that annotation creation in the mobile device was one of the more troublesome features to design. Two approaches were made available using mixed fidelity prototypes. In the first approach, the user first expressed the desire to create an annotation by issuing the assigned command (through any of the available modalities) and then selected the text to be annotated. In the second approach, the user first selected the text, and then issued the command to annotate it. This is, probably, the clearest example of different multimodal integration pat-

terns seen during the development of the mobile DTB player. The generated logs allowed designers to detect two different patterns of usage. The first pattern showed that when users employed stylus based interaction the first approach was adopted. The second pattern was perceived when voice commands or the devices' physical buttons were favoured. In this situation the second approach was adopted. Thus, through the use of the Mob-Pro framework, it was possible to identify two multimodal integration patters, and design the multimodal fusion component accordingly.

Another evolution was prompted by the different scenarios that, from the early stages of design, allowed designers to see that input commands based on on-screen buttons would be insufficient for all the proposed interaction objectives. Visually impaired users would be unable to know where to press. The same can be said of non-visually impaired users in scenarios where their visual attention must be focused elsewhere. To overcome this limitation, the possibility to use speech recognition was introduced for devices where it is available. Since this is not a standard feature of current mobile devices, another input alternative had to be devised.

It can be safely assumed that all mobile devices possess a minimal set of physical buttons: a joystick or four directional buttons, with the accompanying selection button, and two more selection buttons. Some recent devices, like the iPhone, do not meet these requirements, but we can safely expect that a visually impaired person would not own one of these.

To enable a non-visual operation of the mobile rich book player all the application's commands were mapped to these seven buttons. Since there are more than seven commands in the Rich Book Player interface, it was necessary to map the interface to different states, where each button has a different meaning. To fully map all the operations different state diagrams were defined. These state diagrams were defined with the help of mixed-fidelity prototypes, which proved invaluable for this design decision. The MobPro framework allowed to quickly develop different prototypes with several mapping alternatives, which were used to understand what mappings the users felt more comfortable with.

Figure 12 presents an example of the key mappings when the interface is in the normal playback condition. Keys 1, 2 and 7 are selection buttons. Keys 3 to 6 are the directional buttons. As can be seen in the figure, during playback the directional keys are used to navigate the content. Up and down keys (keys 3 and 4) are used to go back or advance one chapter. Left and right keys (keys 5 and 6) are used to advance or go back one page. The user can pause the playback by pressing key 7. The same key will resume playback when paused. To access the main menu, the user can press the right selection key (key 2). This takes the user to another set of key binding states. During playback, whenever the user wishes to create an annotation, he or she should press the left selection key (key 1). The same key is used to listen to an annotation, whenever the playback reaches a point when there is one available to listen. Annotation creation and annotation listening are also states with different key bindings.

Evaluations performed with the mixed-fidelity prototypes allowed the optimization of the key bindings, and ensure consistency, not only between the different states, but also between the visual and non-visual modes of operation. An example of the kind of problems that these prototypes allowed to detect early in the design stage was an inconsistency between the key mappings in the normal playback state and the table of contents state. In the initial design, in the normal playback state the up directional key took the user to next chapter, and the down to the previous chapter. In the table of contents, inspired by the visual presentation of the table of contents on screen, the up key took the user one chapter back and the down key to the next. User's commented on this inconsistency, and allowed its correction early on.

CONCLUSION AND FUTURE WORK

Designing mobile applications that take advantage of the multimodal and multimedia capabilities of

Figure 12. Key mappings and state changes for normal playback conditions

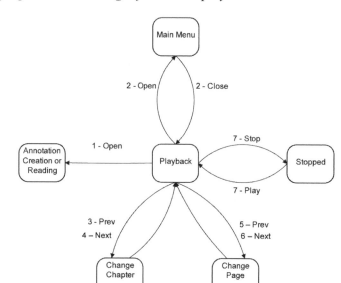

newer mobile devices is a demanding task. Currently, most design methodologies are ill-equipped to deal with the added challenges that ubiquity and the peculiar characteristics of mobile devices bring to the design process, let along deal with the dimensions that can affect the use of multimodalities when used in mobile scenarios. This lack of adequate techniques is felt particularly at the initial stages of design, affecting especially the early prototyping and evaluation procedures.

This chapter presented a set of contributions that emerged during the design of various MMAs and that can be applied on similar design processes, facilitating them and providing better final results. Firstly, we detailed a scenario generation and context selection framework that can play a key role during the bootstrap of design projects, highlighting details that are crucial during design while affecting interaction and usability during the utilisation of the envisioned solution. Moreover, the framework has also proved to be highly beneficial when used on posterior stages, especially during the evaluation of the created prototypes, raising attention to details that derive from the ubiquitous usage settings and illustrating how these influence the usage of multimodalities.

Motivated by the abovementioned framework and the evidence that taking the design out of the lab, into the real world, brings added value to the design process, we developed a prototyping tool that incorporates updated prototyping and evaluation techniques that overcome most of the challenges that designing in-situ for mobile devices can entail. Globally, the tool offers means to quickly create realistic prototypes, including multimedia elements and augmenting them with behaviour, thus enabling users to interact freely with them using actual devices and simulating multimodal operation.

Both contributions were used by designers on the design of a mobile multimodal rich digital book player and provided solid evidence of their impact on its design, facilitating the detection of usability issues and suggesting the adjustment of the user interface, especially regarding the use of multiple modalities, since the initial stages of the process. Throughout this process, designers' feedback was extremely positive and validated its contribution.

Building on these positive results, further endeavours are currently being undertaken aiming at completing both the scenario framework and the prototyping/evaluation tool. On the former, new dimensions and variables are being studied in order to extend its scope to specific domains (e.g., health care, education, cooperative work). On the latter, efforts are being directed towards the inclusion of new modalities and features ranging from gesture recognition to the automatic suggestion of design alternatives and the use of embedded design guidelines. Finally, a mobile version of the prototyping building tool is also underway, aiming especially at in-situ on-the-fly adjustments to the prototypes that are being tested and at the guided support of participatory design.

REFERENCES

Ballagas, R., Memon, F., Reiners, R., & Borchers, J. (2006). iStuff Mobile: prototyping interactions for mobile phones in interactive spaces. In *PERMID 2006, Workshop at Pervasive 2006*, Dublin, Ireland.

Barnard, L., Yi, J., Jacko, J., & Sears, A. (2007). Capturing the effects of context on human performance in mobile computing systems. *Personal and Ubiquitous Computing, 11*(2), 81–96. doi:10.1007/s00779-006-0063-x

Beyer, H., & Holtzblatt, K. (1998). *Contextual design: A customer centered approach to systems design*. San Francisco, CA: Academic Press.

Brewster, S., Wright, P., & Edwards, A. (1995). Experimentally derived guidelines for the creation of earcons. In *Proceedings of the British Computer Society Human-Computer Interaction Group Annual Conference* (pp. 155-159). Huddersfield, UK: Cambridge University Press.

Carriço, L., Duarte, C., Lopes, R., Rodrigues, M., & Guimarães, N. (2005). Building rich user interfaces for digital talking books. In Jacob, R.; Limbourg, Q. & Vanderdonckt, J., (Eds.), *Computer-Aided Design of User Interfaces IV* (pp. 335-348). Berlin, Germany: Springer-Verlag

Carter, S., Mankoff, J., & Heer, J. (2007). Momento: Support for situated ubicomp experimentation. In *Proceedings of the SIGCHI Conference on Human Factors in Computing Systems* (pp. 125-134). New York: ACM Press.

Consolvo, S., & Walker, M. (2003). Using the experience sampling method to evaluate ubicomp applications. *IEEE Pervasive Computing / IEEE Computer Society [and] IEEE Communications Society*, 2(2), 24–31. doi:10.1109/MPRV.2003.1203750

Davis, R. C., Saponas, T. S., Shilman, M., & Landay, J. A. (2007). SketchWizard: Wizard of Oz prototyping of pen-based user interfaces. In *Proceedings of the 20th Annual ACM Symposium on User interface Software and Technology* (pp. 119-128). New York: ACM Press.

Duarte, C., & Carriço, L. (2005). Users and Usage Driven Adaptation of Digital Talking Books. In *Proceedings of the 11th International Conference on Human-Computer Interaction*. Las Vegas, NV: Lawrence Erlbaum Associates, Inc.

Duarte, C., & Carriço, L. (2006). A conceptual framework for developing adaptive multimodal applications. In *Proceedings of the 11th ACM International Conference on Intelligent User Interfaces* (pp. 132-139). New York: ACM Press.

Duarte, C., Carriço, L., & Morgado, F. (2007). Playback of rich digital books on mobile devices. In *Proceedings of the 12th International Conference on Human-Computer Interaction (HCI International 2007)*. Berlin, Germany: Springer.

Duh, H. B., Tan, G. C., & Chen, V. H. (2006). Usability evaluation for mobile device: A comparison of laboratory and field tests. In *Proceedings of the 8th Conference on Human-Computer interaction with Mobile Devices and Services* (pp. 181-186). New York: ACM Press.

Froehlich, J., Chen, M. Y., Consolvo, S., Harrison, B., & Landay, J. A. (2007). MyExperience: A system for in-situ tracing and capturing of user feedback on mobile phones. In *Proceedings of the 5th international Conference on Mobile Systems, Applications and Services* (pp. 57-70). New York: ACM Press.

Gaver, W. (1997). Auditory interfaces. In Helander, M., Landauer, T. & Prabhu, P., (Ed.) *Handbook of Human-Computer Interaction, 2nd edition* (pp. 1003-1041). Amsterdam: Elsevier.

Hagen, P., Robertson, T., Kan, M., & Sadler, K. (2005). Emerging research methods for understanding mobile technology use. In *Proceedings of the 17th Australia Conference on Computer-Human interaction: Citizens online: Considerations For Today and the Future* (pp. 1-10). New York: ACM Press.

Holmquist, L. (2005). Prototyping: generating ideas or cargo cult designs? *Interaction, 12*(2), 48–54. doi:10.1145/1052438.1052465

Holzman, T. G. (1999). Computer-human interface solutions for emergency medical care. *Interaction, 6*, 13–24. doi:10.1145/301153.301160

Hulkko, S., Mattelmäki, T., Virtanen, K., & Keinonen, T. (2004). Mobile Probes. In *Proceedings of the Third Nordic Conference on Human-Computer interaction* (pp. 43-51). New York: ACM Press.

Kjeldskov, J., & Graham, C. (2003). A review of mobile HCI research methods. In *Human-Computer Interaction with Mobile Devices and Services* (pp. 317-335). Berlin, Germany: Springer.

Kjeldskov, J., & Stage, J. (2004). New techniques for usability evaluation of mobile systems. *International Journal of Human-Computer Studies, 60*(5-6), 599–620. doi:10.1016/j.ijhcs.2003.11.001

Klemmer, S. R., Sinha, A. K., Chen, J., Landay, J. A., Aboobaker, N., & Wang, A. (2000). SUEDE: A Wizard of Oz prototyping tool for speech user interfaces. In *Proceedings of the 13th Annual ACM Symposium on User interface Software and Technology* (pp. 1-10). New York: ACM Press.

Landay, J. (1996) SILK: Sketching interfaces like krazy. In *Conference Companion on Human Factors in Computing Systems: Common Ground* (pp.398-399). New York: ACM Press.

Li, Y., Hong, J., & Landay, J. (2007). Design challenges and principles for Wizard of Oz testing of location-enhanced applications. *IEEE Pervasive Computing / IEEE Computer Society [and] IEEE Communications Society, 6*(2), 70–75. doi:10.1109/MPRV.2007.28

Lin, J., Newman, M. W., Hong, J. I., & Landay, J. A. (2000). Denim: Finding a tighter fit between tools and practice for Web site design. In *Proceedings of the SIGCHI Conference on Human Factors in Computing Systems* (pp. 510-517). New York: ACM Press.

Mayhew, D. J. (1999). *The usability engineering lifecycle*. San Francisco, CA: Morgan Kaufmann.

McCurdy, M., Connors, C., Pyrzak, G., Kanefsky, B., & Vera, A. (2006). Breaking the fidelity barrier: An examination of our current characterization of prototypes and an example of a mixed-fidelity success. In *Proceedings of the SIGCHI Conference on Human Factors in Computing Systems* (pp. 1233-1242). New York: ACM Press.

Nielsen, C. M., Overgaard, M., Pedersen, M. B., Stage, J., & Stenild, S. (2006). It's worth the hassle! The added value of evaluating the usability of mobile systems in the field. In *Proceedings of the 4th Nordic Conference on Human-Computer interaction: Changing Roles* (pp. 272-280). New York: ACM Press.

Oviatt, S. (1997). Multimodal interactive maps: Designing for human performance. *Human-Computer Interaction, 12*, 93–129. doi:10.1207/s15327051hci1201&2_4

Oviatt, S. (1999). Mutual disambiguation of recognition errors in a multimodal architecture. In *Proceedings of the SIGCHI conference on Human factors in computing systems* (pp. 576–583). New York: ACM Press.

Oviatt, S. (2000). Multimodal signal processing in naturalistic noisy environments. In Yuan, B., Huang, T., & Tang, X. (Eds.) *Proceedings of the International Conference on Spoken Language Processing* (pp. 696–699). Beijing, China: Chinese Friendship Publishers.

Oviatt, S. (2000). Taming recognition errors with a multimodal interface. *Communications of the ACM, 43*, 45–51.

Oviatt, S. (2003). Multimodal interfaces. In Jacko, J., & Sears, A. (Ed), *Human-Computer interaction Handbook: Fundamentals, Evolving Technologies and Emerging Applications* (pp. 286-304). Hillsdale, NJ: L. Erlbaum Associates.

Oviatt, S., Cohen, P., Wu, L., Duncan, L., Suhm, B., & Bers, J. (2000). Designing the user interface for multimodal speech and pen-based gesture applications: State-of-the-art systems and future research directions. *Human-Computer Interaction, 15*, 263–322. doi:10.1207/S15327051HCI1504_1

Oviatt, S., DeAngeli, A., & Kuhn, K. (1997). Integration and synchronization of input modes during multimodal human-computer interaction. In *Proceedings of the SIGCHI Conference on Human Factors in Computing Systems* (pp. 415-422). New York: ACM Press.

Oviatt, S., & Kuhn, K. (1998). Referential features and linguistic indirection in multimodal language. In *Proceedings of the International Conference on Spoken Language Processing* (pp. 2339–2342). Sydney, Australia: ASSTA, Inc.

Oviatt, S., & VanGent, R. (1996). Error resolution during multimodal human computer interaction. In *Proceedings of the Fourth International Conference on Spoken Language* (pp. 204–207).

Reichl, P., Froehlich, P., Baillie, L., Schatz, R., & Dantcheva, A. (2007). The LiLiPUT prototype: a wearable lab environment for user tests of mobile telecommunication applications. In *CHI '07 Extended Abstracts on Human Factors in Computing Systems* (pp. 1833-1838). New York: ACM Press.

Reis, T., Sá, M., & Carriço, L. (2008). Multimodal interaction: Real context studies on mobile digital artefacts. In *Proceedings of the 3rd international Workshop on Haptic and Audio interaction Design* (pp. 60-69). Jyväskylä, Finland: Springer-Verlag.

Sá, M., & Carriço, L. (2006) Low-fi prototyping for mobile devices. In *CHI '06 Extended Abstracts on Human Factors in Computing Systems* (pp. 694-699). New York: ACM Press.

Sá, M., & Carriço, L. (2008). Lessons from early stages design of mobile applications. In *Proceedings of the 10th international Conference on Human Computer interaction with Mobile Devices and Services* (pp. 127-136). New York: ACM Press.

Sá, M., Carriço, L., & Duarte, C. (2008). Mobile interaction design: Techniques for early stage In-Situ design. In Asai, K. (Ed.), *Human-Computer Interaction, New Developments* (pp. 191-216). Vienna, Austria: In-teh, I-Tech Education and Publishing KG.

Sá, M., Carriço, L., Duarte, L., & Reis, T. (2008). A mixed-fidelity prototyping tool for mobile devices. In *Proceedings of the Working Conference on Advanced Visual interfaces* (pp. 225-232). New York: ACM Press.

Sohn, T., Li, K. A., Griswold, W. G., & Hollan, J. D. (2008). A diary study of mobile information needs. In *Proceeding of the Twenty-Sixth Annual SIGCHI Conference on Human Factors in Computing Systems* (pp. 433-442). New York: ACM Press.

Svanaes, D., & Seland, G. (2004). Putting the users center stage: role playing and low-fi prototyping enable end users to design mobile systems. In *Proceedings of the SIGCHI Conference on Human Factors in Computing Systems* (pp. 479-486). New York: ACM Press.

Chapter 6
Bodily Engagement in Multimodal Interaction:
A Basis for a New Design Paradigm?

Kai Tuuri
University of Jyväskylä, Finland

Antti Pirhonen
University of Jyväskylä, Finland

Pasi Välkkynen
VTT Technical Research Centre of Finland, Finland

ABSTRACT

The creative processes of interaction design operate in terms we generally use for conceptualising human-computer interaction (HCI). Therefore the prevailing design paradigm provides a framework that essentially affects and guides the design process. We argue that the current mainstream design paradigm for multimodal user-interfaces takes human sensory-motor modalities and the related user-interface technologies as separate channels of communication between user and an application. Within such a conceptualisation, multimodality implies the use of different technical devices in interaction design. This chapter outlines an alternative design paradigm, which is based on an action-oriented perspective on human perception and meaning creation process. The proposed perspective stresses the integrated sensory-motor experience and the active embodied involvement of a subject in perception coupled as a natural part of interaction. The outlined paradigm provides a new conceptual framework for the design of multimodal user interfaces. A key motivation for this new framework is in acknowledging multimodality as an inevitable quality of interaction and interaction design, the existence of which does not depend on, for example, the number of implemented presentation modes in an HCI application. We see that the need for such an interaction- and experience-derived perspective is amplified within the trend for computing to be moving into smaller devices of various forms which are being embedded into our everyday life. As a brief illustration of the proposed framework in practice, one case study of sonic interaction design is presented.

DOI: 10.4018/978-1-60566-978-6.ch006

INTRODUCTION

In early days of human-computer interaction (HCI), the paradigm was mainly seen as a means to "synchronise" the human being and a computer (Card et. al., 1983). While the number of computer users rose rapidly and computers were suddenly in the hands of "the man in the street", there was an evident need to make computers easier to use than when used by experts in computing. Psychologists were challenged to model the human mind and behaviour for the needs of user-interface design. It was thought that if we knew how the human mind works, user interfaces (UIs) could be designed to be compatible with it. To understand the human mind, computer metaphor was used. Correspondingly, multimodality has often meant that in interaction with a computer, several senses ("input devices") and several motor systems ("output devices") are utilised. This kind of *cognitivist* conceptualisation of the human being as a "smart device" with separate systems for input, central (symbolic) processing and motor activity has indeed been appealing from the perspective of HCI practices. However, contemporary trends of cognitive science have drifted away from such computer-based input-output model towards the idea of mind as emergent system which is structurally coupled with the environment as the result of the history of the system itself (Varela et. al, 1991). We thus argue that as a conceptual framework for HCI the traditional cognitivist approach is limited, as it conflicts with the contemporary view of the human mind and also with the common sense knowledge of the way we interact with our everyday environment (see Varela et. al., 1991; Noë, 2004; Lakoff & Johnson, 1999; Clark, 1997; Searle, 2004). One of the shortcomings of the traditional input-output scheme is that it implies that the capacity for perception could be disassociated from the capacities of thought and action (Noë, 2004).

For a long time, the development of UIs had been strongly focused on textual and graphical forms of presentation and interaction in terms of the traditional desktop setting. However, as computing becomes increasingly embedded into various everyday devices and activities, a clear need has been recognised to learn about the interaction between a user and a technical device when there is no keyboard, large display or mouse available. Therefore, the need to widen the scope of human-computer interaction design to exploit multiple modalities of interaction is generally acknowledged.

Within the mainstream paradigm of HCI design, conceptions of multimodality tend to make clear distinctions between interaction modalities (see, e.g. Bernsen, 1995). There is the fundamental division between perceiving (gaining feedback presentation from the system) and acting/doing (providing input to the system). These, in turn, have been split into several modality categories. Of course traditional distinctions of modalities have proved their usefulness as conceptual tools and have thus served many practical needs, as they make the analysis and development of HCI applications straightforward. However, too analytic and distinctive emphasis on interaction modalities may promote (or reflect) design practices where interaction between a user and an application is conceptualised in terms of technical instrumentation representing different input and output modalities. Such an approach also potentially encourages conceptualising modalities as channels of information transmission (Shannon & Weaver, 1949). Channel-orientation is also related to the ideal that information in interaction could be handled independently from its form and could thus be interchangeably allocated and coded into any technically available "channels". We see that, in its application to practical design, the traditional paradigm for multimodality may hinder the design potential of truly multimodal interaction. We argue that in the design of HCI, it is not necessarily appropriate to handle interaction modalities in isolation, apart from each other. For instance, the use of haptics and audio in interaction, though referring to different perceptual systems, benefits from these modalities being considered together (Cañadas-Quesada & Reyes-Lecuona, 2006; Bresciani et. al., 2005; Lederman et. al., 2002). However, even recent HCI-studies of cross-modal interaction, although concerning the integration of modalities, still seem to possess the information-centric ideal of interchangeable channels (see e.g. Hoggan & Brewster, 2007).

In this study, we are looking for an alternative paradigm for multimodality. Although modalities relating to perceptual awareness and motor activity differ in their low-level qualities, they have interconnections and share properties, which make them highly suitable for study and handling within a shared conceptual framework. This chapter outlines a propositional basis for such a framework, gaining ingredients from the literature of embodied cognition, cross-modal integration, ecological perception and phenomenology. In the *embodied* approach (Varela et. al., 1991), the human mind is inseparable from the sensory-motor experiencing of the physical world and cognition is best described in terms of embodied interaction with the world. The resulting framework is bound up with a concept of physical embodiment, which has been utilised within several scientific disciplines to reveal the role of bodily experience as the core of meaning-creation. In the course of this chapter, we will present arguments to support the following basic claims as cornerstones of the proposed paradigm:

1)   *Interaction is always multimodal in nature*

The embodied approach to human cognition implies understanding and meaning as being based on our interactions with the world. Understanding is thus seen as arising inherently from an experiential background of constant encounters and interaction with the world by using our bodies. Although our sensory modalities depend on different perceptual systems, human awareness is not channel-oriented. Instead, it is oriented to actions in the environment: objects and agents involved in actions and our own action possibilities (see e.g. Gibson, 1979; Varela et. al., 1991; Clark, 1997; Gallese & Lakoff, 2005). In *action-oriented* ontology, multimodality and multimodal experience appear as inseparable characteristic of interaction. This view is supported by recent research in neuroscience; in perception and thinking, the neural linkages of motor control and perception as well as the integration of sensory modalities appear to be extensive (Gallese & Lakoff, 2005).

2)   *Design arises from mental images and results in mental images*

We argue that the design of UI elements for human-computer interaction involves the communication of *action-relevant mental imagery*. This imagery, being bonded to embodied experiences, is essentially multimodal in nature. Hence, we argue that the starting point of UI-element design should not be rigidly any specific "channel" – presentation or input modality – but the embodied nature of interaction itself, and the subjective exploration of the "imagery" of its action-related meanings. Of course, "channels" do exist in terms of a technical medium. But they should be primarily responsible for supporting the construction of contextually coherent (action-relevant) mental imagery. This kind of construction of imagery, as a gestalt process, could be called *amodal* completion. But unlike the traditional cognitivist view (Fodor, 1975), we propose that amodality is not symbolic in nature but inseparably bound up with our sensory-motor system. Therefore is makes more sense to call these constructed mental images multimodal, whether they are mental images of the designer or mental images of the user.

3)   *Action-relevant mental imagery can be communicated by manifesting action-relevant attributions in the contextual appearance of UI*

As already implied above, we propose that the appearance of elements of a UI can convey attributions of a certain type of mental imagery. These attributions are action-relevant because they propose the occurrence or afforded potential of some activity, and are also indexed to contextual activity. Because of the situated nature of UI appearances, the interpretation of action-relevant attributes is likely to be highly context dependent. Unlike traditional linguistic approaches, our approach to the semantic content of perception is embodied, i.e., it stresses a) the sensory-motoric experiential background/skills of a perceiver b) the action-oriented bias of perception, c) the perceiver's own activity as an integral part of the perception and d) the situated context of perception. Semiotics of a linguistic tradition tend to consider the relation between appearance and meaning as arbitrary or symbolic (Saussure, 1983). However, the peircean school of semiotics has also acknowledged non-symbolic (iconic, indexical) meaning relations (Peirce, 1998), which clearly are related to the action-oriented perspective because both views imply the existence of "the world" (its appearances and laws) as a familiar semantic reference.

4) *Acknowledging the bodily nature of interaction as a basis in design inevitably results in support for multimodal interaction*

If designers take into account our naturally multimodal and action-oriented bias to perceive the world in terms of situationally embodied meanings, it would ultimately provide at least a fraction of the "easiness" of everyday interactions to the design of human-computer interaction. However, such meanings are often invisible to us, and camouflaged as common sense. The challenge for our conceptual analysis is to explicate them.

The claims above define a paradigm for *multimodal interaction design*. The outlined paradigm provides a new conceptual framework for the design of multimodal UIs, which is based on a sound theoretical foundation. Within the framework, this chapter also explores how to utilise the concept of multimodal mental imagery in UI design. We thus aim to provide conceptual tools to understand the relations between meaningful subjective action-related experiences and concrete physical properties of a UI. In addition to theoretical discussion we expose this suggested paradigm to a real-world design case. The brief case study of sound design for mobile application is meant to illustrate how the new framework is realised in design practices and in the resulting design.

CONCEPTUALISING MULTIMODALITY

The term multimodal and the related, more technical and presentation-oriented concept multimedia have been conceptualised in numerous ways. This section provides a comprehensive summary of how multimodality has been handled in the literature. In addition to these previous accounts, we analyse the concept of multimodality in terms of some recent research on embodied cognition and discuss how multimodality could be conceptualised within that framework. As will be discussed, much of the previous conceptualisations is still relevant in the framework of embodied cognition, but some aspects deserve a critical look.

Perspectives on Multimodality

Multimodality as a Technical Opportunity

In the development of information and communication technology (ICT) products, a typical driver is the emergence of new technical opportunities. Visual displays and the related display processors, for instance, have rapidly developed. However, the development of visual display technology is mainly due to the investments in the research and development of that technology, not the needs of multimodal interaction. While the amount of data presented in contemporary displays is several hundred times greater than its 20 year old ancestor, the user of a personal computer has pretty much the same typewriter-derived means to control an application as her parents had in the 70's. Thus it can be seen that the development of technology for multimodal interaction has not been ruled by the needs to enhance human-computer interaction, but the merely commercial assumptions about what consumers want to buy.

Once the technology is there, whatever motivated its development, we have to find uses for it. Much of what is marketed as multimedia are products resulting from this kind of approach. Especially in the early stages of multimedia, the producers were under pressure to show their technical sophistication by supporting all available means of interaction – which, as discussed above, mainly meant ever fancier screen layouts.

As soon as a critical mass had been reached in sales, multimedia products can be argued to have become part of our everyday life. The next step was to elaborate the multimedia technology. An essential part of the elaboration was to legitimate the technology in terms of human-computer interaction. Advantages were sought from multimodal interaction. The models of human cognition, on which the multimodality conception was based, were very simple. A typical example is an idea of a free cognitive resource; for instance when information was presented via a visual display, other sensory systems were thought of as free resources. When this claim was empirically found unsustainable, human ability to process information from multiple sources and in multiple modalities became a central issue. An important source of information was attention studies. In them, human ability to process information had been under intensive research since the 1950's, when the rapidly growing air traffic made the cognitive capacity of air-traffic controllers the bottle-neck of fluent flight organisation. These studies resulted in models which either modelled the structure of those mechanisms which define attention (Broadbent, 1958; Deutsch & Deutsch, 1963), or models which analysed human capacity (Wickens, 1984).

In mobile applications, the technical challenges for multimodal interaction differ from what they used to be in the static context. However, we argue, even in mobile computing it is the new technical opportunities which are the driving force for developing mobile multimedia. A good example is the current addition of accelerometers to various mobile devices. The popularity of the Wii gaming console with its innovative control methods might have something to do with recently grown interest in accelerometer-based gestural control. Now that we have similar technology included in our mobile phones, interaction designers of mobile applications have been challenged to utilise it. In the near future, we will see whether the application of accelerometer technology in mobile phones turns out to be just another technology driven craze or a useful opportunity resulting in novel ways of interacting with mobile devices.

Multimodality Provides Options

The use of multimedia or designing multimodal applications is often thought of as a selection of means of interaction. For instance, it is easy to find texts which give an impression that a given piece of information can be presented in various forms; e.g., text, speech, picture or video (e.g. Waterworth & Chignell, 1997). The underlying idea is that there is the content and there is the form that is independent of it. However, this notion has been found untenable in various disciplines. In the context of information presentation in UIs, it has been found that paralleling sound and an image, for instance, is extremely complicated. When trying to trace meaning creation on the basis of non-speech sound by asking the participants in an experiment to pair sounds and images, it was found that conclusions could be made only when the images were simple symbols indicating a clearly identifiable piece of information, such as physical direction (Pirhonen, 2007; Pirhonen & Palomäki, 2008). In other words, the idea that the designer is free to choose in which modality to present certain information is a gross over-simplification and lacks support from the studies concerning semantics. Worn phrases like "the medium is the message" (McLuhan, 1966) or that "a picture is worth a thousand words", still hold in the multimodal context. Referring to the sub-heading, i.e., multimodality provides options, it should be understood as that technology provides modality-related options but that these options are qualitatively different from each other. The process of choosing an interaction modality is not independent of other design efforts.

Multimodality Provides Redundancy

In mathematical information theory (Shannon & Weaver, 1949), redundancy was introduced as something to get rid of. Redundancy unnecessarily uses the resources of a communication channel, thus lowering the efficiency of an information system. However, in the context of information systems, the concept of redundancy has also more positive connotations; redundancy can be seen as a way of increasing system stability by provid-ing backup. This idea, which originates from the mathematical theory of communication, has been applied to human-computer interaction in the era of multimedia. It has been argued that if information is delivered in multiple formats, the message is more reliably received. A classic example is users with disabilities; if information is provided both in an audio and visual format, for example, the same application can be used by users with vision impairment as well as by those with impairment in hearing (Edwards, 1992).

As discussed in the previous section, the in-separability of form and content of information inevitably questions the endeavour to present "the same" information in multiple formats. However, as the research on cross-modal design of UI feed-back indicates (e.g., Hoggan & Brewster, 2007), such cross-sensorical information can – to some extent – be defined and thus be interchangeably attributed to multiple formats. Even though these kinds of redundant combinations undoubtedly are beneficial in many applications, they should not be seen as a straightforward, universal solution for multimodal interfaces. But we admit that, in the mobile or ubiquitous context, redundant information presentation can provide valuable flexibility. While the actual context of use is hard or impossible to anticipate, it is important that there are options for interacting with the ap-plication. In one situation, visual presentation is the best form, in some other situation, audio or haptics works best.

Natural Interaction is Multimodal

All the approaches discussed above are techni-cally oriented in that they analyse modalities in terms of available technology. However, when conceptualising human-computer interaction from a technical perspective, the essential difference between interaction with the real world and vir-tual objects has to be noted. When constructing a virtual object we as designers split our mental image of the whole object into its constituents. For instance, when creating a virtual dog we consider separately its appearance, sound and how the user could control it in the application. The division into constituents (in this case image, sound, and control elements) is based on technical facilities. Techni-cally, all constituents are separate entities. Only when linked and synchronised with each other, is the illusion of a virtual dog able to emerge.

In contrast to a virtual dog which is analytically constructed from separate parts, a real world dog is one single physical object. It causes many kinds of perceptions; we can see, hear, and smell it. We can also communicate with it. In other words, we are in multimodal interaction with the dog, even if it doesn't have separate devices to cause stimulus in different sensory modalities or provide control.

This common sense notion that interaction with the real world is always multimodal by nature, has often been used as a rationale for multimodal UIs in various application areas (see, e.g. Oviatt & Cohen, 2000).

However, the suggested naturalness through multimodality cannot be achieved by burying the application under a heap of visuals, sounds and vibrations. The naturalness can only be achieved by designing objects which are not in the first place "sounds" or "images" or anything else which primarily refers to a certain technology. Multimodality should not be an end in itself. Rather, it should be treated as an inevitable way of interaction. This is important to understand in all human-computer interaction design, because multimodality is a basic quality of our way of interacting with our environment, whether "real" or "virtual".

An oversimplification, which may result from striving towards natural-like virtual objects, is the mechanical imitation of their real-world counterparts. When creating a multimodal, virtual dog, the best strategy is not necessarily to go with a camera and sound recorder to a real dog – unless you are sure about what kind of recordings you will *exactly* need for your purpose. As will be discussed in the next section, filmmakers have learned long ago that the (action-relevant) expressive qualities of sound are much more important than the "natural" authenticity of the sound source. Moreover, the coincident perception of sound and visual results in an impression which is qualitatively different from a product of its constituents (Basil, 1994). Therefore, the argument that natural interaction is multimodal should not encourage one-sighted imitation or modelling of real world objects.

Multimodality Provides a Perceptual Bias

As interaction with the real world always provides multimodal understanding of its objects, actions and environments, in what way, then, would that affect the "virtual world" where we are technically able to combine different presentation modalities in an artificial manner? Cinema, for example, allows theoretically endless arbitrary defined combinations of sounds and visuals. Thus, an audiovisual relationship in a multimedia presentation is not real but an illusion. Such an option to re-associate images and sounds is essential to the art of filmmaking. A single visual basically affords an infinite number of sounds. For instance, the sound of chopping wood, played in sync with a visual of hitting a baseball, is not perceived as a mistake or as two distinct events. It is heard as a baseball hit – with a particular force. Hence, when the context makes us expect a sound, it seems that the designer can use just about any sound source which produces enough believable acoustic properties for the action.

The phenomenon, where the perceiver is tricked into believing that the artificially made sound effect originates from the source indicated by the context of narration, has a long tradition of exploitation within sound design practices for radio, television and cinema. The French filmmaker and theoretician, Michel Chion (1990), talks about *synchresis* (a combination of words synchronism and synthesis) which refers to the mental fusion of a sound and a visual when these coincide. According to Chion, watching a movie, in general, involves a sort of contract in which we agree to forget that the sound is coming from loudspeakers and the picture from the screen. The spectator considers the elements of sound and image as representing the same entity or phenomenon. The result of such an audio-visual contract is that auditory and visual sensations are reciprocally influenced by each other in perception. From the perspective of the filmmaker, audio-visual contract allows the potential to provide "added value" in the perception of cinema – something bigger than the sum of the technical parts. As a perceptual bias, it induces us to perceive sound and vision as they both fuse into a natural perceptual whole. With his known assertion: "we never see the same thing when we also hear; we don't hear the same thing when we see as well", Chion (1990, p. xxvi) urges us to go beyond preoccupations such as identifying so-called redundancy between the two presentation modalities and debating which one is the more important modality.

In his account Chion (1990) also proposes the notion of transsensorial perception. He means that it is possible to achieve multimodal or cross-modal perception even with unimodal presentation and via a single sensory path. In music, for example, kinetic, tactile or visual sensations can be transmitted through a sole auditory sensory channel. In cinema, one can "infuse the soundtrack with visuality", and vice versa, images can "inject a sense of the auditory" (Chion, 1990, p. 134). By using both the visual and auditory channels, cinema can also create a wealth of other types of sensations. For example, in the case of kinetic or rhythmic sensations, the decoding should occur "in some region of the brain connected to motor functions" (Chion, 1990, p. 136).

Transsensorial perception as a concept appears to be in accordance with many theories concerning multimodal integration. The classical McGurk effect (McGurck & MacDonald, 1976), for instance, indicates the integration of different sensory modalities in a very early stage of human information processing. Interestingly, the account of audiovisual contract and especially the notions of transsensorial perception can actually be seen as apposite exemplifications of embodied multimodality and the involvement of mental imagery in perception. Next, we will focus on this embodied perspective.

Embodied Meaning

Traditional paradigms of cognitive science have had a tendency to treat the human mind in terms of functional symbol processing. The problem of these computing-oriented views is that they have not been able to explain how experiences of phenomena in the surrounding world are encoded into symbols of cognition, meaning and into concepts of thinking – not to mention how this is all linked to intentionality and how we are ultimately able to use the cognitive processes in the control of our body (Searle, 2004). As a contrasting top-down approach, scientific perspectives like phenomenology, pragmatism or ecological perception have considered meaning as being based on our interactions with the world rather than as an abstract and separated entity (Dourish, 2001). According to the ecological view of perception (Gibson, 1979), our interaction with the world is full of meanings that we can perceive rapidly without much effort. The perspective of embodied cognition continues such a line of reasoning while stressing the corporeal basis of human cognition and enactive (sensory-motor) coupling between action and thinking (Varela et. al., 1991).

Overall, the perspective of embodied cognition seeks to reveal the role of bodily experience as the core of meaning-creation, i.e., how the body is involved in our thinking. It rejects the traditional Cartesian body-mind separation altogether as the "terms body and mind are simply convenient shorthand ways of identifying aspects of ongoing organism-environment interactions" (Johnson & Rohrer, 2007). Cognition is thus seen as arising inherently from organic processes of organism-environment interaction. So, imagination, meaning, and knowledge are embodied in the way they are structured by our constant encounter and interaction with the world via our body (Lakoff & Johnson, 1999; Gallese & Lakoff, 2005). The perspective of ecological perception is closely related to the embodied one. Both perspectives draw upon the action-oriented bias of the human organism, which considers meanings as action-relevant properties of environment (i.e., affordances) that relate to our experiencing of the world.

We should now see that the embodied point of view defines a coupling between action and perception. So-called motor theories of speech perception (Liberman & Mattingly, 1985) have suggested that we understand what we hear, because we sensory-motorically "resonate" the corresponding vocal action by imaging the way the sound is produced in the vocal tract. It was later found that related motor areas of the brain indeed activate in the course of speech perception (Rizzolatti & Arbib, 1998). Several contemporary studies (see a review in Gallese & Lakoff, 2005) in neuroscience suggest that perception is coupled with action on a neural basis. Discoveries of common neural structures for motor movements and sensory perception have elevated the once speculative approach of motor theories of perception to a more plausible and appealing hypothesis. According to the studies mentioned, all sensory modalities are integrated not only with each other but also together with motor control and control of purposeful actions. As a result, doing something (e.g. grasping or seeing someone grasping) and imaging doing it activates the same parts of the brain.

Gallese and Lakoff (2005) argue that to be able to understand something, one must be able to imagine it, i.e., one understands by mentally *simulating* corresponding activity. In the case of understanding objects of the environment, one simulates how they are involved in actions, or in what way one can afford to act on them for some purpose. In other words, understanding is action-relevant mental imagery, and meaning thus equals the way something is understood in its context. Other authors have also suggested that understanding involves embodied simulation (or mental enactment) of the physical activity we perceive, predict or intend to perform. For example, the account of motor involvement in perception (Wilson & Knoblich, 2005), Godøy's (2003) account of *motor-mimetic* perception of movement in musical experience, and the theory of enactive perception (Noë, 2004) are all related to the simulation approach. According to Gallese and Lakoff (2005), action simulation seems to occur in relation to 1) motor programmes for successful interaction with objects in locations, 2) intrinsic physical features of objects and motor programmes

(i.e., manners) to act on them to achieve goals, and 3) actions and intentions of others.

The embodied simulation of motor movements of others is found to take place in so-called *mirror neurons*. It is suggested that such an innate mirror system is involved in imitation, as it indeed can explain why it is so natural for us to imitate body movements (Heiser et. al., 2003). Mirror processes, as corporeal attuning to other people, are also hypothesised to act as a basic mechanism for empathy (e.g. Gallese, 2006), i.e., perception of intentionality (mental states such as intentions and emotions). When we observe someone's action, the understanding of intentionality can be conceived as an emerging effect of the experienced motor resonances through embodied attuning (Leman, 2008). For instance a vocal utterance, as sonic vibrations in the air, "derives" cues of *corporeal intentionality* (or motor intentionality) in the vocal action. Via motor resonances, the perceived cues of the observed action relate to our personal, embodied experiences.

As we can see, the perspective of embodied cognition provides insights at least for the creation and establishing of action-related meanings and concepts. However, it is proposed that even concepts of language could be based on an embodied foundation of action-oriented understanding and multimodal sensory-motor imagery. According to the neural exploitation hypothesis, neural mechanisms originally evolved for sensory-motor integration (which are also present in nonhuman primates) have adapted to serve new roles in reasoning and language while retaining their original functions (Gallese & Lakoff, 2005; Gallese, 2008). The theory of *image schemas* (Johnson, 1987), provides maybe the most prominent perspective on how to characterise our pre-conceptual structures of embodied meaning. Johnson and Rohrer (2007) have summarised image schemas as: "

1)   recurrent patterns of bodily experience,

2) "image"-like in that they preserve the topological structure of the perceptual whole, as evidenced by pattern-completion,

3) operating dynamically in and across time,

4) realised as activation patterns/contours in and between topologic neural maps,

5) structures which link sensory-motor experience to conceptual processes of language, and

6) structures which afford "normal" pattern completions that can serve as a basis for inference."

Image schemas are *directly meaningful structures* of perception and thinking, which are grounded on recurrent experiences: bodily movements in and through space, perceptual interactions and ways of manipulating objects (Hampe, 2005). The shared corporeal nature of the human body, universal properties of the physical environment and common sensory-motor skills for everyday interactions are the providers of a range of experiential invariants.

As any recurrent experiences of everyday interaction can be considered in terms of inference patterns of image schemas, we therefore provide a couple of examples of how these image schemas can be conceived. Image schema can be based on any type of recurrently experienced action, e.g., experience of picking something up. It can thus refer to, for example, forces, kinaesthesia and intentionality experienced in picking up an object in a rising trajectory. Image schema may be based on interpersonal experiences as well. Therefore it can, for example, be related to an experience of illocutionary force (Searle, 1969; see the application in the case study in this chapter) in a vocal utterance – referring to a type of communicative intent in producing that utterance (e.g., asking something).

It is important to remember that image schemas are not equal to concepts nor are they abstract representations of recurrent experiences. Instead, they are "schematic gestalts which capture the structural contours of sensory-motor experience, integrating information from multiple modalities" (Hampe, 2005, p. 1). Such "experiential gestalts" are "principle means by which we achieve meaning structure. They generate coherence for, establish unity within, and constrain our network of meaning" (Johnson, 1987, p. 41). As such, image schemas operate in between the sensory-motor experience (acted/perceived/imagined) and the conceptualisation of it. However, they are not "lesser" level concepts. On the contrary, being close to the embodied experiences of life, the activation of image schemas in the use of concepts provides an active linkage to the richness of such experiences. Therefore when reading the sentence "Tom took the ball out of the box", in our mind we can easily complete the "picture" and "see" the instance of someone or ourselves doing just that.

We can now conclude that the creation of embodied meanings is coupled with activity. Perception and imagining of an action is proposed as involving active doing that is dependent on the embodied sensory-motor experiences and skills of a subject. It is suggested that this "internal" activity takes place in embodied simulations that use shared neural substrate with the actual performance of the similar action. It is plausible that these embodied simulations encode both mechanical characteristics of action (properties of objects in locations and motor activity) and structural characteristics of action (gestalt processes such as stream/object segregation or pattern completion) into a sensory-motor neural mechanism. Image schemas are one of the structures by which cognition is linked to sensory-motor neural mechanisms. From that position they allow the embodied experiences to be turned into distinguishable meanings and abstract concepts – and vice versa.

Multimodality is Embodied

The perspective of embodied cognition, which was discussed above, clearly has an effect on the conception of multimodality. We should now see that multimodality is a natural and inevitable property of interaction, which emerges from embodied action-oriented processes of doing and perceiving. Theory of embodied simulations suggests that in very early stages of perception, action is being simulated in a corporeal manner. Hence, the perception intrinsically involves participatory simulation of the perceived action, which activates rapidly and is an unconscious process. Such simulation is inherently multimodal because it occurs in the parts of the sensory-motor system which are shared for both perception and doing and which respond to more than one sensory modality. Embodied simulations most likely also involve perceptual gestalt-completions of "typical" and contextually "good" patterns in determining the type of the action and its meanings. Therefore, regardless of the presentation mode, mediated representations – even with limited action-specific cues of, e.g., movement, direction, force, objects properties – can make us simulate/imagine actions that make sense with the current contextual whole. The result of such (amodal) pattern completion is always multimodal experience, even when the stimulus was unimodal – e.g., in the form of written language. Embodied simulations also explain the cross-modal associations based on one presentation modality – for example, why music can create imagery of patterns of movement, body kinaesthesia, force and touch (Godøy, 2004; Tagg, 1992; Chion, 1990). In a similar way, visual presentation can evoke, for example, haptic meanings of textures.

In addition to perception, multimodality takes place in the course of interaction. This is evident in the situations in which one can effortlessly move "across" modalities, for example, when a spoken expression is continued in a subsequent hand gesture, and thus how hand gestures in general are integrated with speech and thought processes (McNeill, 2005). In the context of interaction, embodied simulations of actions should also occur prior to performed actions; when we consider how

we act and imagine what the possible outcomes would be, or when we anticipate actions of others. From the perspective of embodied cognition, it is actually impossible to keep perception fully separated from interaction. We "act out" our perception and we perceive ourselves to guide our actions (Noë, 2004). Thus, our perceptional content is unavoidably affected by the activity of perceiver.

The proposed approach questions some traditional ideas about multimodality and the underlying computational metaphor of human cognition with its separate input and output systems. The embodied view of multimodality clearly denies the existence of separate input/output modules – i.e. separate domains for all senses and motor control which do not integrate until in the presumed higher "associating area" (Gallese & Lakoff, 2005). It is thus inappropriate to consider perception as passive input to the cognitive system, or to consider sensory modalities as channels by which content could be transmitted independently from other senses. Of course, the previous paradigms concerning, for example, attentional capacity and redundant information presentation, should not be completely rejected but they could be understood in a new way in the framework of embodied cognition. Embodied cognition as an approach to design has thus potential in shifting the orientation of traditional, computation–centred models to something which could be called human-centred.

MULTIMODAL INTERACTION DESIGN

What Is "Multimodal" Design?

On the theoretical basis presented earlier in this chapter, we can formulate a new design paradigm which takes into account the natural multimodality of interaction. The proposed design paradigm seeks to support the user in the creation of perceptions that would feel natural in the sensory-motor experience of the interaction. In this sense, this paradigm is in accordance with ecological approaches of design (e.g., Norman, 1988). The essence of the multimodal design paradigm, however, lies in its conceptualisation of interaction modalities. Most importantly, this means that multimodality is not something that designers implement in applications. Rather, multimodality is the nature of interaction that designers must take into account. Hence, the multimodal design of UIs does not focus on different presentation or input modalities – nor on any communication technologies in themselves. By acknowledging this, we can begin to appreciate the modally integrated nature of our embodied sensory-motor experience – as a part of natural engagement of action and perceptual content.

What would be the main principles for multimodal design? In fact, the cornerstones of this new paradigm are those proposed already in the introduction of this chapter. Already in the discussion about the ways to conceptualise multimodality, we have concluded that "*interaction is always multimodal in nature*". As mentioned above, the embodied perspective for multimodality formulates the basic rationale for the paradigm. The second principle, "*design arises from mental images and results in mental images*", concerns the creation of embodied mental imagery in UI mediated communication. The motivation of this principle is to shift focus from communication channels and symbolic representations of information to embodied meaning-creation based on sensory-motor *attunement* between the user and the UI within the context of use. The starting point of design should be in the flow of interaction itself and the mental exploration of embodied imagery of contextually coherent action-related meanings. Being bonded to embodied experiences, such imagery is inherently multimodal. Therefore design operates on a multimodal foundation – even when its concrete implementation is technically unimodal.

The third principle; "*action-relevant mental imagery can be communicated by manifesting action-relevant attributions in the contextual appearance of UI*", asserts that a designer can actually utilise action-relevant imagery in finding propositional semantics for UI design (perceptual cues of action including, e.g., cues of movement, objects and intentionality involved in action). The aim of the exploration of mental imagery is ultimately to get ideas for physical cues that associate with the intended action-relevant imagery and thus provide support for the communication of it. For sure, the designer cannot absolutely define such couplings between certain meaning (i.e., mental imagery) and certain characteristics of UI. But these couplings are not completely coincidental either since, due to past interactions with the world, we already possess a rich history of meaningful experiences that can potentially become coupled with the contextual appearance of UI. As these couplings are intentional connections that ultimately arise in the course of interaction (Dourish, 2001), we propose that the designer must immerse herself in (actual or imaginary) participation in interaction in order to find ideas for them. The designer should possess at least a tentative mental model of application-user interaction, from which the need for and intended purpose of a certain UI element has arisen in the first place. By exploring her own relevant experiences, she should be able to pinpoint some presumably "general" action-related associations or their metaphorical extensions which would match with the communicational purpose of a UI element and would support the coherent perceptual whole in the interaction. Such meaningful imagery, as recurrent patterns of experience, is very closely related to the concept of image schema, which we discussed earlier.

The embodied attunement between user and the UI of application is the key in the communication of action-related meanings. When the user encounters the UI element, it should appear as an intentional object which indicates its purpose in the course of interaction. Action-relevant cues in UI appearances can exploit our natural tendency to attune to environment and its actions. The aim of a designer is to implement such cues that would contextually support the user in creating the intended "feel" and mental imagery. Figure 1 illustrates the UI-mediated communication process of action-relevant mental imagery. The imagery of designer is created in relation to her model of application-user interaction and to the arisen need/purpose of the certain element in UI. Correspondingly, the imagery of a user is created in relation to how she has experientially constructed (i.e., modelled) the interaction and how that certain UI element is encountered in interaction. This is quite analogous with Norman's (1988) account of system image and the related models, but in our approach, the mental construction covers the whole interaction in the context of use, not the system operation only. Unlike in traditional theory of communication (Shannon & Weaver, 1949), the process is not unidirectional. Both the user and the designer are seen as equally active in creating

mental imagery that contextually makes sense. Even though these two mental constructions would never be precisely the same, the contextual utilisation of common experiential (sensory-motor) invariants can guide the perception of the user be attuned with the designer's intention.

Mental exploration is a common practice, for instance, in film sound design (Sonnenschein, 2001). The sound designer, having immersed herself in the plot and the narration of the film, reflects her own experiential sensory-motor background in order to mentally "listen" to the sounds in order to determine acoustic invariants that would presumably relate to the preferred subjective experience.

The cues of action-relevant imagery can be encoded into various forms of physical manifestation. Therefore their appearance in UI can, for example, be realised either within a single presentation mode or in co-operative integration of multiple modes. If the propositional mental imagery involved touching a rough-surfaced object, we could equally imagine what kind of

Figure 1. The communication of action-relevant mental imagery via the processes of design and application use

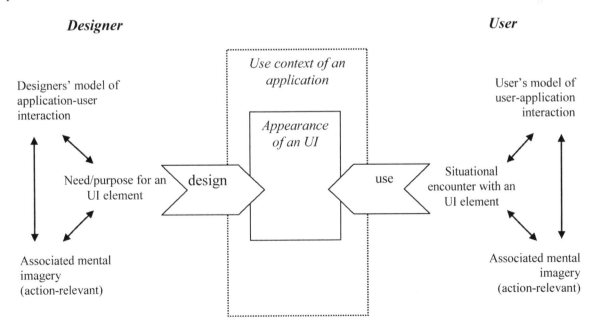

149

sounds it would produce, what it would feel like (in a haptic sense) and what it would look like. And even if the technical UI element eventually resulted in a unimodal form (e.g., audio feedback), the perception would always be engaged with the multimodal completion of perceptual content. For instance, while sliding our finger on the touch screen, associated audio feedback (indicating a rough surface) can also situationally induce pseudo-haptic perceptions of surface roughness. Radio-plays are also good examples of multimodal completion; they have been called "theatre of the mind" because they allow listeners to use their imagination solely on the basis of sound cues.

However, as in the phenomenon of synchresis, the concurrent contextual framework (e.g. UI appearance of other modes of presentation) will ultimately guide the situated perception of an UI element. Such perceptual modulation is evident, for example, in the findings of Bresciani et. al. (2005) and Guest et. al. (2002). We should therefore pursue integrated design perspectives that take into account all concurrent presentation modalities in the situational context of use of an application. The design of different UI elements should be focused on supporting the construction of coherent imagery as a perceptual whole – not on evoking competing denotations.

One of the advantages of the proposed design paradigm lies in bringing the essence of interaction – action and intentionality – to its central role in perception and meaning creation. If our perceptual content is integrated into our own actions, goal-oriented participation in interaction, we cannot distinctively talk about input and output semantics without acknowledging the embodied situation as a whole. Moreover, if our perceptual content is based on enactive simulation of activity, we should learn to conceive the semantics of HCI as embodied meanings of intentional activity – not just as representations in different channels. The fourth principle of multimodal design sums up the point: "*Acknowledging the bodily nature of*

interaction as a basis in design inevitably results in support for multimodal interaction."

The proposed design paradigm is nothing completely new; multimodal interaction design is in line with the current development in the philosophical foundation of the HCI-field, emphasising interaction design, user- and context-centred views, user experience, ecological perception, situated actions, and the ideas of tangible interaction. It is largely overlapping with the design principles for embodied interaction, introduced by Paul Dourish (2001). Recent accounts concerning the design of enactive interfaces (see e.g. Götzen et. al., 2008) are also related to the proposed design paradigm. In the following sections, we will further discuss the utilisation of mental imagery in the design of UI elements. The following design case will also demonstrate how our conceptual framework is realised in design. It should be noticed that in the following sections the discussions and examples are biased towards the design of UI presentation/feedback elements. However, it should be seen as a practical restriction of this study and not the restriction of the paradigm itself.

Designing Action-Relevant User Interface Elements

Defining Action Models

In the previous section we already suggested that the designer's mental imagery arises from her conceptual model of interaction. Imagery is also bound up with the purpose of a UI element. This *communicative function* also arises from the designer's model of interaction. Indeed, keeping interaction design in mind, we suggest that such a communicative function of a UI-element is the first and foremost factor to be considered in its design. Thus, the first questions to be asked are: Why do we need the UI element? What should be the communicational role of the UI element instance in interaction? And in what way is it used in the model/flow of interaction?

In considering the above questions, we should be able to *imagine* the flow of application-user interaction and explore different associations that are related to the role and the characteristics of the propositional UI element. This can be achieved, for example, with participatory exploration of use scenarios. These associations are what we call action-relevant mental imagery or embodied meanings, but they could as well be called design ideas or design concepts for an element. Associated mental imagery does not always have to refer to actual HCI context. For the sake of creativity, it is actually important to also explore non-HCI associations which arise from the interaction model – or from the interaction in some other, analogous, context. In the latter case, it is a question of utilising metaphors in design.

When starting to design UI elements with certain action-relevance we need to outline the activity to which this relevance refers. Such a conceptualisation of action involves a mental image of generalised and contextually plausible action, which we will call the *action model*. It is an image-like mental model of sensory-motor experience that involves the general type of activity (forces, movement, objects) and its general type of occurrence. In addition, it may involve the general type of intentionality, i.e., causality between mental states and the action. Like image schemas, action models refer to recurrent patterns of experience. Action models can be conceived as design ideas that guide the designer in attributing certain action-relevant semantics to the design. Together with the designer's model of interaction, they work as high-level mental conceptions, against which the design is reflected when considering its action-relevance.

In our design case (reported later in this chapter in more detail) we ended up using three different action models as three different approaches to the design of the same auditory feedback. These models referred to general types of actions in general types of occurrence like "connection of an object with a restraining object". It is important to notice that if more than one action model is used, they all should support the construction of the intended perceptual whole, that is, meanings that become fused into the action experience.

When considering UI presentation, it is appropriate to make a distinction between natural and artificial action-presentation relationships, similar to the action-sound distinction made by Jensenius (2007). In fact, every action-presentation relationship in UI is essentially artificial. However, like in Chion's (1990) synchresis, the relationship can be experienced as natural because of UI presentation's plausible action-relevance in the situational context on which it is mapped. Hence the instance of UI presentation can become perceptually fused with concurrent actions and result in an illusion of naturalness.

The continuum from natural to artificial operates in two dimensions. Firstly, it concerns the relationship between the action model and its context in the interaction model (see Figure 2) and secondly, the relationship between the action model and its degree of realisation in presentation (see Figure 3). These relationships in the first dimension can involve direct, indirect or arbitrary mappings. Indirect mappings involve some kind of mental mediator, for example, use of metaphors or analogical thinking. Arbitrary mappings always involve the use of some sort of code.

Let us consider the design task of creating feedback for controlling certain parameter on a touch screen. The change of parameter value needs to be done in discrete steps – either upwards or downwards. For each change operation, the feedback should be robust enough so that the user can be confident of success. One straightforward way to schematise such a control would be based on mappings between upward or downward stroking on a touch screen surface and changes in a parameter value. Based on that, we could adopt the real physical involvement of "stroking with a finger upwards or downwards" directly as an action model for feedback. But in order to support the required robustness and the feel of control, we

Figure 2. Mappings between the action model and the interaction model

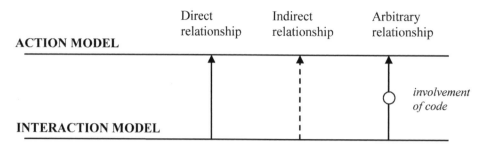

Figure 3. Relationship between the action model and its use in UI presentation

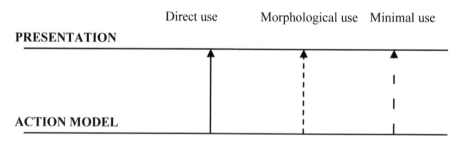

could outline the action model metaphorically as, for example "rotating a resisting wheel upwards or downwards until it snaps into a retainer that holds it". From such an action model, it would be easy to imagine visual, haptic or sonic feedback that would feel functional and natural (offering the sensations of mass, resistance and restraint) when being coupled with stroking on the screen surface. For the interaction model of discrete changes, we could also adopt a convention of pressing a "plus-button" for increase and pressing a "minus-button" for decrease. Hence the relation of the action model "pressing a button" and the parameter control would be arbitrarily defined.

As will be shown in our design case, for the design of feedback elements it may also be appropriate to explore action models from the perspective of interpersonal interaction metaphor. The active subject would be an imaginary "agent", i.e., the intentional (2nd person) counterpart of interaction. In feedback element design this can

be a very fruitful approach because it would allow the attributions of 2nd person response.

In general, the use of multiple action models is able to produce more interesting results in design than use of a single model, as it can provide multiple layers of action-based semantics. Because the degree and the means of utilisation of each model in realised presentation varies, it should be possible to implement the manifestations of more than one action model in a single presentation element. When considering multiple action models in design, they can also be allocated for use with separate presentation modalities.

The second dimension, relationships between the action model and its attributes implemented in presentation, involve quite different kinds of relations than the first dimension. Of course, before these general-level action models can be used in design, they have to be manifested in a concrete form. The second dimension operates with considerations of "how much" and "in what way" these manifested attributes of action models are utilised in the realised UI element. Roughly defined categories of the second dimension include 1) *direct use*, 2) *morphological use* and 3) *minimal use* of the action model manifestation. These categories are inspired by similar use of model appropriation in the composition of electroacoustic music (Garcia, 2000). Direct use means that the presentation element is a partial or complete reconstruction of the action model manifestation. Morphological use means that the model and its manifestation are taken as a formal structure, i.e., the designer extracts some organisational laws of the model/manifestation and utilises these features morphologically in the design. Minimal use means that only such reduced or general features or cues of the model/manifestation have been utilised in the design which do not appear as resemblances between the presentation and the model/manifestation. In the next section, we will discuss the manifestation of action models.

Manifesting Action-Relevant Mental Imagery

Godøy (2001) has proposed a model of musical imagery where our understanding of (sound-producing) actions is founded on sensory-motor images of *excitation* (imagery about what we do or imagine/mimic doing, e.g., body movement or gestures) and *resonance* (imagery about the effects, i.e. material resonances, of what we do or imagine/mimic doing). Because of the multimodal, holistic nature of such mental images, this model should be easily adaptable to the multimodal design perspective, in which it can be used as a guide to mentally *enact* motor movements and explore the reflecting material/object-related imagery. This kind of "manifesting by enacting" takes the action model as a starting point and aims to elaborate it into a more detailed scenario of action and its resonances. The result of this creative process should include tangible ideas for concrete actions; gestural imagery,

action-relevant mechanical invariants produced by certain materials and objects within interaction and, more generally, a "scheme" of the action that can be consequently articulated in physical form. Physical articulation can aim at analytic realisation of action imagery (e.g., manually or by physical modelling), or it can adopt some structures or ideas from the action imagery and interpret them in a more "artistic" manner. For example, an action model of "letting loose" might be articulated as short drum tremolo, which retains the feeling of "removing a restraint" and a decaying contour of force felt in the imagined action.

To complement Godøy's model presented above, we suggest an additional type of imagery which provides the motivational basis for the other two. Images of *intentionality* concern goals, motivations and also emotions behind the perceived/imaginary action. The exploration of intentionality behind action should be based on the designer's considerations of the functional purpose of a UI element in its context (see the earlier discussion about communicative function).

UI elements are intentional objects which, together with their contextual appearance, should indicate their functional purpose. Human perception seems to have a tendency to denote purposes or motivations behind the perceived actions. As perception involves a mirror mechanism – the attunement to the actions of other people – we thus have a corporeal apparatus for "catching" the cues of intentionality. The attribution of intentionality can also be extended to the perception of material movement, providing that we can imagine an intentional subject being involved in the action. (Leman, 2008.) Therefore UI elements can convey intentionality and can be understood as intentional objects, even if they do not directly denote the presence of another person. It is up to the designer to make sure that UI elements comprise suitable attributions of, for example, communicative intent. In other words, the designer must ensure that the imaginary action and its physical articulation is an outcome of intention-in-action (Searle, 1983) that matches the propositional communicative purpose of a UI element.

Physical articulations that spontaneously occur as the result of direct corporeal involvement in interaction can be called 2nd person descriptions of the subject's intentionality. Such articulations naturally occur, for example, in person-to-person ("from me to you") communication. As they reflect direct involvement in interaction, they are not a result of subjective interpretation but an "experience as articulated". As gestural manifestations of corporeal intentionality, they provide cues of intentionality in the form of physical features and thus offer a way to find correlates between subjective experiences and objective physical properties. (Leman, 2008.) It would be valuable, if stereotypical physical cues of specific intention to communicate could be captured in corporeal articulations. In other words, the use of such bodily determined invariants in the appearances of UI would form an intention-specific semantic basis. Intention-specific cues could be utilised to facilitate HCI – including feedback presentation and gesture recognition. In our design case, we used an action model that was based on a person-to-person interaction metaphor. The action model involved a vocal speech act with a specific communicative intent. Prior to spontaneous vocal articulations, we first explored various mental images of suitable vocal gestures in order to become situationally immersed. From recorded articulations we aimed to capture features of physical energy that would reflect the communicative intent in the speech act (i.e., the illocutionary force). In the design, these acoustic features were thus utilised as cues of corporeal intentionality. A similar type of gestural imagery and its physical articulation has already been utilised in the design of non-speech audio feedback elements (Tuuri & Eerola, 2008).

Despite the fact that the core of design process operates at a highly subjective level, the results of this process must always be transferred into a form of observable features. From the perspective of design methodology, the problem is the gap between subjective (1st person) ontology of experience and objective (3rd person) ontology of physical appearance, which makes the formulation of systematic methods very difficult. We see that at least one answer to the problem lies in spontaneous physical articulation of design ideas. As an interaction-derived, action-related imagery of the designer is manifested and expressed in physical terms, the structure and features of the physical phenomenon can be analysed and later utilised systematically in design. At the same time, the design idea can more easily be sketched and communicated within a design team.

UI Element Design as a Creative Process

To sum up the design process and utilisation of action relevant mental images at a general level, we decided to categorise the different design phases by conforming to the classic model of the creative process. Traditionally the process is described in five phases: 1) preparation, where the problem or goal is acknowledged and studied, 2) incubation, where the ideas are unconsciously processed, 3) insight, where the idea of a plausible solution emerges, 4) evaluation, where the idea is somehow concretised and exposed to criticism and 5) elaboration, where the idea is refined and implemented (Csikszentmihalyi, 1996). Our modified process includes four phases: 1) *defining the communicative function for the UI element*, which matches fairly well the classic goal defining and preparation phase, 2) *defining action models and exploring action-relevant mental images*, which comprises recursive incubation and insight phases with the aim of elaborating action-relevant imagery, 3) *articulation*, which refers to concretisation and evaluation of ideas, and finally 4) *implementation of the UI element*, which is the phase where the concretisations of action-related ideas are composed into the final appearance of a UI-element. Of course, this kind of process model is always a crude simplification of the real processes. For example, in actual design there is not always a clear boundary between mental exploration and its articulation.

In the sections above, we have introduced some conceptual tools, such as the usage of action models, which we consider essential in applying the action-oriented multimodal perspective in the UI element design. We will next expose this theoretical foundation to the creative process of real-world interaction design by presenting a case report.

Design Case: Physical Browsing Application

Background

The aim of the case was to design an appropriate sound feedback element to complement the experience of "physical selection" in the physical browsing application. Physical browsing is a means of mapping physical objects to digital information about them (Välkkynen, 2007). It is analogous to the World Wide Web: the user can physically select, or "click", links in the immediate environment by touching or pointing at them with a mobile terminal. The enabling technology for this is tags that contain the information – for example web addresses – related to the objects to which they are attached. The tags and their readers can be implemented with several technologies. In this application, we chose RFID (Radio-Frequency Identification) as the implementation technology. The tags can be placed in any objects and locations in the physical environment, creating a connection between the physical and digital entities (Want et. al., 1999).

Because of the close coupling between physical and digital, the properties and affordances (Norman, 1999) of the physical environment – namely the context, the appearance of the object and the visualisation of the tag – form a connection with the digital content of the physical hyperlink. Each part of this aggregate should complement each

other. The same holds for the interaction with the tags and the links they provide: optimally the user feels as if she is interacting with the physical objects themselves with a "magic wand". This kind of system is very close to tangible user interfaces (Holmquist et. al., 1999; Ishii et. al., 1997) even while we use a mediating device to interact with the physical objects.

While we can call the short-range selection method "touching", the reading range is not actually zero, that is, the reading device and the tag do not need to be in physical contact with each other (see Figure 4). The reading range is typically a few centimetres and because it is based on generating an electromagnetic field around the reader, it varies slightly in a way that does not seem deterministic to the user. Typically, the hardware and software combinations only report a tag proximity when the contents of the tag are already read, which means that guiding the user in bringing the reader slightly closer to the tag is not generally possible – we only know if a tag has been read or if it has not.

Because the physical action of selecting a tag is a process that happens in three-dimensional space, during tag reading the mobile device is often at an angle where it is difficult to see what is happening

Figure 4. Selecting an RFID tag with a reading device in physical browsing

on the screen, and an arm's length away from the user's eyes. Therefore, other modalities, such as sounds, are especially important in reporting the successful reading of a tag.

Design Process

The design goal was fairly straightforward, as the propositional purpose of sound was to provide feedback to the user about successful reading of an RFID tag in the physical selecting. However, in this early phase of design it was important to formulate a more detailed picture of what the actual constitutes of this *feedback function* are in the current context of interaction. Hence we created and, by imaginary participation, explored a contextually rich use scenario about the usage of a physical browsing application (for details see Pirhonen et. al., 2007) and ended up with design principles that determine some important qualitative aspects of the required feedback. Most importantly, sound should clearly illustrate, in real time, the selecting/reading of the tag. For the user, the sound would thus appear as "the" sound of that particular event of virtual touching, providing confirmation of success and a feel of control. As a consequence, the feedback sound would inform the user that physical pointing towards the tag is no longer needed. Feedback should also "answer" the user's intentions in selecting this particular tag (from many) and in wishing to peek at what kind of content the tag holds. To achieve these goals, sound design should evoke confirmatory associations in the user that are contextually suited to such intentions and the user's conception of the tag selecting action. As a secondary purpose, the sound should discriminate different tag types from each other. In this particular application there were four types of RFID tags referring to the type of media content they hold (text, picture, sound or video). However, this second purpose of the feedback sound had much lower priority in actual design considerations than the main purpose presented above.

Relating to a separate study of creative group work in design (Pirhonen et. al., 2007), we started the design by conducting three iterative panel-working sessions with six panellists. In the first panel session the aim was to immerse the panellists in the use of the physical browsing application. Such participatory experience was supported by a detailed story about a person using an application. The story was narrated as a radio-play and was based on the rich use scenario mentioned earlier. Panellists thus got the opportunity to enact contextual instances of physical selecting of RFID tags, and to explore what kind of imaginary sound they heard in association with the experience. Hence, the result of the first panel session was a list of verbally expressed ideas about the sound. These ideas fell roughly in two categories: sounds of connections (such as "clicks" and "dings") and sounds with an upward gestural motor image (such as a rising two-tone melody). By mental exploration, we interpreted the first category as sound producing actions of *"connecting an object with a restraining object"* and the second category in terms of imagery of *"picking something up (out of the container)"*. These two general types of action experiences were adopted as action models for the design. Both models have an indirect relationship with the actual doing of physical selecting, although the model of "connecting" has a much closer analogy with the physical selecting than the latter model, which is based on a gestural metaphor.

On the basis of both action models and other ideas, we made various types of sounds as recorded articulations of suitable sound-producing actions and as synthetically generated sounds. In the consequent panel sessions, these sounds were evaluated by the same panellists. The most liked one was the sound of a single metronome "click". A high-pitched two-tone melody of rapid upward 5th interval (played on a metallophone) was also liked, but because of its long decay time it was considered "a little too long".

Figure 5. Visualised pitch (F₀) contours of the selected utterance. The original segmented contour is on the left, and the contour with an interpolated gap on the right. The interpolated contour was used in the design.

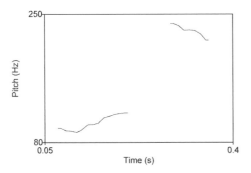 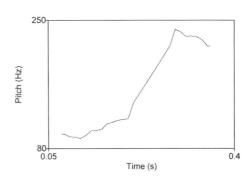

After the panel sessions, we also came up with the idea of exploring person-to-person interaction metaphors for an additional action model. We made analogies to various interpersonal communication scenarios that involve similar kind of "pointing" to an object, and the related 2ⁿᵈ person utterance as confirmatory feedback. We defined the third action model as a vocal gesture of "*asking for confirmation*" – with a polite and confidence-supporting intention such as "So, you mean this one?" The essential element of this action model was communicative intent in the speech act (i.e., the illocutionary force), not the verbal content. Suitable action relevant utterances, using a Finnish word meaning "this one?" (as a rhetorical question), were explored and spontaneously articulated the imaginary interaction scenario in mind. Articulated utterances were recorded and analysed acoustically for their prosodic features, namely fundamental frequency (F_0), intensity and first formant (F_1), to be utilised in the design. The presumption was that such prosodic features would carry cues of the corporeal intentionality of the vocal action (Leman, 2008; Banse & Scherer, 1996), allowing its intention to communicate to be implemented in design. The extraction of prosodic features was performed with Praat software (see Boersma, 2001). After some preliminary listening to the synthesised versions of F_0 pitch contours

of the utterances, one utterance was chosen as a basis for the subsequent sound design because we felt it supported the upward gestural imagery similar to the "picking something up" model. Figure 5 illustrates the original and interpolated pitch contours of the selected utterance.

We preferred the interpolated version of the pitch contour, as it provided a solid, undivided basis for sound synthesis. We then shifted all pitch values into a higher register (at the rate of 2.7), before producing renditions of the contour. Four versions were synthesised. Each version used a different type of waveform (triangle, sinusoid, sawtooth, square) resulting in different timbres. This simple feature was intended to support the distinguishing of different tag types. Next, the extracted intensity contour was applied to all four sounds in order to provide them with the dynamics of the original utterance. Likewise, all four sounds were filtered using the extracted information of the first formant (F_1) in the original utterance. It is worth noting that although formant filtering (with one formant) was used, the aim was not to reproduce speech-like sounds. However, filtering did provide the sounds with certain human-like qualities.

Finally, three different sound elements, which originated from corresponding action models, were mixed together in a multi-track audio editor.

Table 1. Summary of action models and their usage in the design of feedback sound

ACTION MODEL:	Connecting an object with a restraining object	Picking something up	Asking confirmation (speech act)
Relationship with an action of physical selecting:	Analogy of real touching instead of virtual	Based on a gestural metaphor (refers to "picking up" the content from the tag)	Based on a person-to-person interaction metaphor
Usage in the UI presentation:	Direct usage (plausible "click" sound for object-object collision)	Minimal usage (action model as a basis for musical articulation, which was partially used)	Morphological usage (acoustic structure of an utterance was used)
Propositional type of semantics:	Object-interaction resonance	Motor-mimetic image of movement (upwards)	Cues of corporeal intentionality (via motor-mimetic image)

These source elements were 1) the metronome click, which provided the proper qualities of physical connection to the beginning of the UI feedback, 2) the two-note metallophone melody, which was shortened by fading out the long-sounding decay and used in the beginning of the sound to provide damped metallic hits, and 3) the synthesised sound we produced on the basis of extracted prosodic information. The third source element provided the body element for the UI feedback sound, also providing the only difference between feedback sounds for different tag types. As a design approach, the use of the acoustic structure of a vocal act in sound design is analogous to parameterised auditory icons (Gaver, 1993). In the editing process, source elements were transformed in various ways (e.g. balanced, filtered and reverberated) in order for the result to be aesthetically pleasing and functional sound. Action models and their use in the design are summarised in Table 1.

Evaluation of the Design

So far we have discussed the physical browsing application without a context. Next we will outline a physical environment in which we evaluated the feedback sound design. We then report some usability issues of the evaluation study we conducted.

The system was used in a bicycle exhibition in a museum. The museum was situated in a former engineering works[1] thus providing a large open space for exhibition. The physical browsing system consisted of a palmtop computer (Personal Data Assistant, PDA) with wired headphones and an RFID reader integrated into a PDA. The PDA communicates with RFID tags, which are read by touching the tag with the PDA, that is, bringing the RFID reader within a few centimetres of the tag. The contents linked to the RFID tags were videos, sound files, images and short textual descriptions about the artefacts and themes in the exhibition. The software system consists of a web browser, media player, image viewer and RFID reading software. The RFID reading software is responsible for

1. detecting the proximity of the RFID tags,
2. determining the media type corresponding to the tag,
3. playing the sound feedback, and
4. presenting the media in the appropriate media display software.

Ten volunteers participated in the evaluation test. They explored the exhibition just as any other museum visitor carrying a PDA and headphones. During usage, the participants were observed and after the exhibition visit they were interviewed. The participants were asked to rate the suitability and the pleasantness of the sound feedback on a scale from 1 to 5. Then they were interviewed about the qualitative aspects of the sound. The participants were also asked whether they noticed the different types of feedback sounds (4), and whether those sounds helped to distinguish the associated media types from each other.

On a scale from 1 to 5, the average rating for suitability was 4.4 and for pleasantness 3.9. The sounds were considered especially clear and intelligible, and several subjects mentioned that the sounds were not irritating. In a positive sense, the sounds were also characterised as "not a typical sound signal" and as "friendlier than a basic beep". Other qualities of the sound mentioned were "soft" and "moving", and that the sound "corresponded well with the physical action" of reading the tag.

The sounds seemed to work very well as feedback. Most of the subjects soon stopped paying attention to the sounds as separate user interface entities and took them as a natural part of the physical user interface. After use, the participants considered the sounds to be an important response that made physical browsing easier. One subject summarised the significance of the sound as follows: "because the touching range varied slightly, and the loading of the content took some time, the sounds were important feedback for knowing when touching the tag had succeeded".

None of the participants noticed any difference between the sounds of the different media types. So, in the use context of an application, slight timbre modification for sounds of different media types evidently did not provide any difference in the feedback experience. This is an interesting observation because, although the sounds are undoubtedly similar, the difference between the sounds is clearly noticeable when comparing them directly to each other.

Illustrating the New Paradigm

The design task of the reported case was deliberately simple, even trivial. This was because a simple case made it possible to illustrate the contribution of the new paradigm at a more detailed level than a complex one. In this chapter, we are not proposing any special design method for any special need nor trying to prove its effectiveness. Rather, we are proposing a new *mental framework* to approach UI design. To illustrate that approach, the case study presented provides a clear and down-to-earth example. The fact that the design task involved a technically unimodal result was also a deliberate choice. In this way we intend to demonstrate that the proposed paradigm sees multimodality as a quality of the design process, not as multiple technical modalities in the resulting design. It can be argued that a professional designer could end up with similar results. We do not deny that. It is true that designers may more or less implicitly follow design practices which are very similar to the ones proposed. However, as mentioned at the beginning of the chapter, one of our challenges is to explicate such tacit knowledge that designers have.

The embodied perspective of multimodal design stresses the coherency of the perceptual whole which the user actively constructs in interaction. In the illustrated sound design case, the goal was to produce the sonic appearance of reading the RFID tag with a mobile device. From the viewpoint of multimodal design, the most essential issue is how this sound would be perceived within the embodied tag-reading experience. In order to achieve a design that would be coherent with the embodied experience of user, the communicative function, action models and design principles in general were built on the enactive exploration of interaction. We demonstrated the explicit usage of action models as mental tools in pinpointing action-relevant imagery to be utilised as propositional perceptual content (semantics). Hence we utilised three types of (non-linguistic) propositional semantics for a simple confirmatory feedback function. Analytic use of conceptual action models allowed these attributions/appearances to be consciously explored and explicitly implemented in the design.

Despite the unimodal presentation of the resulting design, we argue that the result is perceptually multimodal. Firstly, the situated instance of sonic appearance is intended to be coupled with motor movements of the user resulting in a situational pseudo-haptic illusion of a touch. Secondly, the feedback sound itself is designed with the intention of supporting the user in attuning to "movement in sound" and creating action-oriented sensory-motor illusions. Within the traditional conceptualisation of multimodality in HCI, a feedback event would be easily considered merely in terms of transmitting the feedback information concurrently in sonic and visual (tiny PDA screen) channels. Disembodiment from the interaction can also lead to oversimplified considerations of semantic content. The designer could think, for example, that feedback sounds can be directly recycled from application to application since they are designed to convey the same feedback "message". This kind of "absolute" view on semantics takes a model of interaction and its relation to a UI element as self-evident, and easily dismisses the communicational potential that even a simple feedback sound can have when it is designed *as* activity – and *for* activity.

CHALLENGES AHEAD

As stated already, multimodality should not be understood as a design option, but as a basic quality of human behaviour. Thus the design process is multimodal, as well as interaction between humans, and interaction between a human being and a technical device.

If human-computer interaction has always been multimodal and will always be, why even pay attention to this kind of inevitable phenomenon? Let us think about the current human-computer interaction paradigm. It was largely developed in the era when a computer meant a desktop workstation. The interaction concepts were based on the framework which was defined by the desktop

unit with its software, keyboard - later mouse - and essentially a visual display. The tradition of elaborating interaction design was inevitably connected to the existing hardware. Therefore, the conceptual framework of HCI is still largely related to past technology.

The main contribution of the current work is to provide a new conceptual basis for the multimodal design of multimodal user-interfaces. By using the ideas of Thomas Kuhn (1970), it can be said that we aim at defining a design paradigm for future needs. We argue that the concepts, or perspective, or paradigm, of the past is simply not adequate for future needs. Kuhn argued that the paradigm defines what we see and hear – it is the framework for figuring out our environment. If the paradigm of HCI is inappropriate, we as designers may fail to see something essential about the use of technical devices. Likewise, Lakoff and Johnson (1980), quite some time ago in their famous metaphor book, argued that our metaphors define what we perceive. Metaphors, in this context, mean for instance the terms which we use for conceptualising human-computer interaction. In other words, we aim at providing a conceptual framework which would serve as a highly appropriate and theoretically founded interaction design basis for future technology. The proposed conceptual framework would reveal issues which would otherwise stay hidden. It provides a perspective, which is applicable when shifting from desktop applications to mobile and ubiquitous computing, or any other novel technical setting.

A closely related design issue is the largely evolutionary nature of the development of high technology. Each new generation of technical applications inherits many of the properties of the preceding one. The way we try to interact with a new device is based on our previous experiences. Therefore it is understandable that the industry often tries to soften the introduction of new applications by making them resemble familiar products. However, evolutionary development of technology may also prevent really new ideas from being implemented. In the design of multimodal user-interfaces, an evolutionary approach implies the acceptance of the basic structure of past user interfaces. A well-known example is the so-called qwerty-keyboard as the dominating input device. Even if there is no other rationale but tradition to keep it in the form it was developed for mechanical typewriters, not too many manufacturers have dared to make any modifications to it. A similar reluctance to change the form of a technical device can be seen in numerous product categories. When digital photography freed camera designers from the technical constraints of a film roll, the whole organisation of lens, viewfinder and control devices could be rethought afresh, a "tabula rasa". Some unprejudiced studies of the digital camera were released then. Those products made brand new ways of composing and shooting possible. Unfortunately, a very conservative design has dominated ever since.

It can be argued, that conservative design dominates because consumers are conservative. We would rather not blame only consumers. Sometimes novel ideas have been commercial successes. Apple, with its innovative graphical UI and portable devices with revolutionary interaction concepts, has proved that consumers are ready to accept new ideas as well. This shows that consumers are ready to accept brand new concepts as well as modifications of existing products.

Since computation can nowadays be embedded almost anywhere, the future of computing can be expected to be truly ubiquitous by nature. In ubiquitous computing, our everyday environments are part of the user interface, and there is no clear distinction between different applications within a single physical space. This emerging setting challenges design. Interaction with different ubiquitous applications cannot rely on the models which we adopted in the era of personal computers. In so-called smart environments, technology should support more natural interaction than, for instance, typing. Understanding our naturally multimodal way of interacting with our environ-

ment would be essential in order to design usable ubiquitous applications. The physical browsing case is an early and simple example of embedding the interface – and multimodal interaction – into the environment.

When designing the products of the future, we will need to be able to take a fresh look at what people really need. The elaboration of existing technological conventions (i.e., evolutionary approach) is not necessarily adequate. By understanding the central role of bodily experience in the creation of meaning, designers would be able to get to grips with something very essential in interaction between a human being and a technical device. This, in turn, would help to create product concepts which genuinely utilise the available technology to fulfil human needs.

CONCLUDING STATEMENTS

The prevailing interaction design paradigm inherits its basic concepts from an information processing model of human cognition. In practical design, this approach has proved to be inadequate in conceptualising interaction of a human being and a technical device.

New ideas about interaction design, which challenge the legacy of the mechanical view of human cognition, have already emerged. However, the new approaches have not yet managed to formulate a concise framework. In this chapter, we have proposed a new paradigm for interaction design. It is based on the idea of the human being as an intentional actor, who constantly constructs meanings through inherently multimodal bodily experiences. The case study illustrates that the proposed approach provides a relevant conceptual framework for understanding interaction with a technical device, as well as for understanding the design process. The evaluation of the design case indicates that the implemented sounds reflected essential properties of intentional action in the given context.

The proposed approach does not contain detailed guidelines for successful design. Rather, it is a sound, theoretically founded conceptual framework. As such, it provides designers with concepts which should orientate the design process in a relevant direction in terms of human intentionality in interaction.

ACKNOWLEDGMENT

This work is funded by Finnish Funding Agency for Technology and Innovation, and the following partners: Nokia Ltd., GE Healthcare Finland Ltd., Sunit Ltd., Suunto Ltd., and Tampere City Council.

REFERENCES

Banse, R., & Scherer, K. R. (1996). Acoustic profiles in vocal emotion expression. *Journal of Personality and Social Psychology, 70*, 614–636. doi:10.1037/0022-3514.70.3.614

Basil, M. D. (1994). Multiple resource theory I: Application to television viewing. *Communication Research, 21*(2), 177–207. doi:10.1177/009365094021002003

Bernsen, N. O. (1995). A Toolbox of output modalities: representing output information in multimodal interfaces. *Esprit Basic Research Action 7040: The Amodeus Project*, document TM/WP21.

Boersma, P., & Weenink, D. (2001). Praat, a system for doing phonetics by computer. *Glot International, 5*(9/10), 341–345.

Bresciani, J.-P., Ernst, M. O., Drewing, K., Boyer, G., Maury, V., & Kheddar, A. (2005). Feeling what you hear: auditory signals can modulate tactile tap perception. *Experimental Brain Research, 162*(2), 172–180. doi:10.1007/s00221-004-2128-2

Broadbent, D. E. (1958). *Perception and communication*. London: Pergamon.

Cañadas-Quesada, F. J., & Reyes-Lecuona, A. (2006). Improvement of perceived stiffness using auditory stimuli in haptic virtual reality. In [Piscataway, NJ: IEEE Mediterranean.]. *Proceedings of Electrotechnical Conference, MELECON, 2006*, 462–465. doi:10.1109/MELCON.2006.1653138

Card, S. K., Moran, T. P., & Newell, A. (1983). *The psychology of human-computer interaction*. Hillsdale, NJ: Lawrence Erlbaum Associates.

Chion, M. (1990). *Audio-vision: Sound on screen*. New York: Columbia University Press.

Clark, A. (1997). *Being there: Putting brain, body and world together again*. Cambridge, MA: MIT Press.

Csikszentmihalyi, M. (1996). *Creativity: Flow and the psychology of discovery and invention*. New York: HarperCollins.

de Götzen, A., Mion, L., Avanzini, F., & Serafin, S. (2008). Multimodal design for enactive toys. In R. Kronland-Martinet, S. Ystad, & K. Jensen (Eds.), *CMMR 2007, LNCS 4969* (pp. 212-222). Berlin, Germany: Springer-Verlag.

de Saussure, F. (1983). *Course in general linguistics* (R. Harris, ed.). London: Duckworth. (Original work publishes in 1916).

Deutsch, J. A., & Deutsch, D. (1963). Attention: Some theoretical considerations. *Psychological Review, 70*(1), 80–90. doi:10.1037/h0039515

Dourish, P. (2001). *Where the action is: The foundations of embodied interaction*. Cambridge, MA: MIT Press.

Edwards, A. D. N. (1992). Redundancy and adaptability. In A.D.N. Edwards & S. Holland (Eds.), *Multimedia interface design in education* (pp. 145-155). Berlin, Germany: Springer-Verlag.

Fodor, J. A. (1975). *The language of thought.* Cambridge, MA: Harvard University Press.

Gallese, V. (2006). Embodied simulation: From mirror neuron systems to interpersonal relations. In G. Bock & J. Goode (Eds.), *Empathy and Fairness* (pp. 3-19). Chichester, UK: Wiley.

Gallese, V. (2008). Mirror neurons and the social nature of language: The neural exploitation hypothesis. *Social Neuroscience, 3*(3), 317–333. doi:10.1080/17470910701563608

Gallese, V., & Lakoff, G. (2005). The brain's concepts: The role of the sensory-motor system in reason and language. *Cognitive Neuropsychology, 22*, 455–479. doi:10.1080/02643290442000310

Garcia, D. (2000). Sound models, metaphor and mimesis in the composition of electroacoustic music. In *Proceedings of the 7th Brazilian Symposium on Computer Music.* Curitiba, Brasil: Universidade Federal do Paraná.

Gaver, W. W. (1993). Synthesizing auditory icons. In [New York: ACM.]. *Proceedings of INTERCHI, 93*, 228–235.

Gibson, J. J. (1979). *The ecological approach to visual perception.* Boston, MA: Houghton Mifflin.

Godøy, R. I. (2001). Imagined action, excitation, and resonance. In R. I Godøy, & H. Jørgensen (Eds.), *Musical imagery* (pp. 237-250). Lisse, The Netherlands: Swets and Zeitlinger.

Godøy, R. I. (2003). Motor-mimetic music cognition. *Leonardo, 36*(4), 317–319. doi:10.1162/002409403322258781

Godøy, R. I. (2004). Gestural imagery in the service of musical imagery. In A. Camurri & G. Volpe (Eds.), *Gesture-Based Communication in Human-Computer Interaction: 5th International Gesture Workshop, Volume LNAI 2915* (pp. 55-62). Berlin, Germany: Springer-Verlag.

Guest, S., Catmur, C., Lloyd, D., & Spence, C. (2002). Audiotactile interactions in roughness perception. *Experimental Brain Research, 146*(2), 161–171. doi:10.1007/s00221-002-1164-z

Hampe, B. (2005). Image schemas in cognitive linguistics: Introduction. In B. Hampe (Ed.), *From Perception to Meaning: Image Schemas in Cognitive Linguistics* (pp. 1-14). Berlin, Germany: Mouton de Gruyter.

Heiser, M., Iacoboni, M., Maeda, F., Marcus, J., & Mazziotta, J. C. (2003). The essential role of Broca's area in imitation. *The European Journal of Neuroscience, 17*, 1123–1128. doi:10.1046/j.1460-9568.2003.02530.x

Hoggan, E., & Brewster, S. (2007). Designing audio and tactile crossmodal icons for mobile devices. In *Proceedings of the 9th International Conference on Multimodal Interfaces* (pp. 162-169). New York: ACM.

Holmquist, L. E., Redström, J., & Ljungstrand, P. (1999). Token-based access to digital information. In *Proc. 1st International Symposium on Handheld and Ubiquitous Computing* (pp. 234-245). Berlin, Germany: Springer-Verlag.

Ishii, H., & Ullmer, B. (1997). Tangible bits: towards seamless interfaces between people, bits and atoms. In *Proc. SIGCHI Conference on Human Factors in Computing Systems* (pp. 234-241). New York: ACM.

Jensenius, A. (2007). *Action-Sound: Developing Methods and Tools to Study Music-Related Body Movement.* Ph.D. Thesis. Department of Musicology, University of Oslo.

Johnson, M. (1987). *The body in the mind: The bodily basis of meaning, imagination, and reason.* Chicago, IL: University of Chicago.

Johnson, M., & Rohrer, T. (2007). We are live creatures: Embodiment, American pragmatism and the cognitive organism. In: J. Zlatev, T. Ziemke, R. Frank, & R. Dirven (Eds.), *Body, language, and mind, vol. 1* (pp. 17-54). Berlin, Germany: Mouton de Gruyter.

Kuhn, T. S. (1970). *The structure of scientific revolutions* (2nd ed). Chicago: University of Chicago Press.

Lakoff, G., & Johnson, M. (1980) *Metaphors we live by*. Chicago: University of Chicago Press.

Lakoff, G., & Johnson, M. (1999). *Philosophy in the flesh: The embodied mind and its challenge to Western thought*. New York: Basic Books.

Lederman, S. J., Klatzky, R. L., Morgan, T., & Hamilton, C. (2002). Integrating multimodal information about surface texture via a probe: relative contributions of haptic and touch-produced sound sources. In *Proceedings of the 10th Symposium on Haptic Interfaces for Virtual Environments & Teleoperator Systems* (pp. 97-104). Piscataway, NJ: IEEE.

Leman, M. (2008). *Embodied music cognition and mediation technology*. Cambridge, MA: MIT Press.

Liberman, A. M., & Mattingly, I. G. (1985). The motor theory of speech perception revised. *Cognition, 21*, 136. doi:10.1016/0010-0277(85)90021-6

McGurk, H., & MacDonald. (1976). Hearing lips and seeing voices. *Nature, 264*, 746–748. doi:10.1038/264746a0

McLuhan, M. (1966). *Understanding media: the extensions of man*. New York: New American Library.

McNeill, D. (2005). *Gesture and thought*. Chicago: University of Chicago Press.

Noë, A. 2004. *Action in perception*. Cambridge, MA: MIT Press.

Norman, D. A. (1988). *The psychology of everyday things*. New York: Basic Books.

Norman, D. A. (1999). Affordance, conventions, and design. *Interactions (New York, N.Y.), 6*(3), 38–42. doi:10.1145/301153.301168

Oviatt, S., & Cohen, P. (2000). Multimodal interfaces that process what comes naturally. *Communications of the ACM, 43*(3), 45–53. doi:10.1145/330534.330538

Peirce, C. S. ([1894] 1998). What is a sign? In Peirce Edition Project (Ed.), *The essential Peirce: selected philosophical writings vol. 2* (pp. 4-10). Bloomington: Indiana University Press.

Pirhonen, A. (2007). Semantics of sounds and images - can they be paralleled? In W. Martens (Eds.), *Proceedings of the 13th International Conference on Auditory Display* (pp. 319-325). Montreal: Schulich School of Music, McGill University.

Pirhonen, A., & Palomäki, H. (2008). Sonification of directional and emotional content: Description of design challenges. In P. Susini & O. Warusfel (Eds.), *Proceedings of the 14th International Conference on Auditory Display*. Paris: IRCAM (Institut de Recherche et Coordination Acoustique/Musique).

Pirhonen, A., Tuuri, K., Mustonen, M., & Murphy, E. (2007). Beyond clicks and beeps: In pursuit of an effective sound design methodology. In I. Oakley & S. Brewster (Eds.), *Haptic and Audio Interaction Design: Proceedings of Second International Workshop* (pp. 133-144). Berlin, Germany: Springer-Verlag.

Rizzolatti, G., & Arbib, M. A. (1998). Language within our grasp. *Trends in Neurosciences, 21*, 188–194. doi:10.1016/S0166-2236(98)01260-0

Searle, J. R. (1969). *Speech Acts: An Essay in the Philosophy of Language*. New York, NY: Cambridge University Press.

Searle, J. R. (1983). *Intentionality: An essay in the philosophy of mind*. New York: Cambridge University Press.

Searle, J. R. (2004). *Mind: A brief introduction*. New York: Oxford University Press.

Shannon, C. E., & Weaver, W. (1949). *The mathematical theory of communication*. Urbana, IL: University of Illinois Press.

Sonnenschein, D. (2001). *Sound design: The expressive power of music, voice and sound effects in cinema*. Saline, MI: Michael Wiese Productions.

Tagg, P. (1992). Towards a sign typology of music. In R. Dalmonte & M. Baroni (Eds.), *Secondo Convegno Europeo di Analisi Musicale* (pp. 369-378). Trento, Italy: Università Degli Studi di Trento.

Tuuri, K., & Eerola, T. (2008). Could function-specific prosodic cues be used as a basis for non-speech user interface sound design? In P. Susini & O. Warusfel (Eds.), *Proceedings of the 14th International Conference on Auditory Display*. Paris: IRCAM (Institut de Recherche et Coordination Acoustique/Musique).

Välkkynen, P. (2007). *Physical selection in ubiquitous computing*. Helsinki, Edita Prima.

Varela, F., Thompson, E., & Rosch, E. (1991). *The embodied mind*. Cambridge, MA: MIT Press.

Want, R., Fishkin, K. P., Gujar, A., & Harrison, B. L. (1999). Bridging physical and virtual worlds with electronic tags. In *Proc. SIGCHI Conference on Human factors in Computing Systems* (pp. 370-377), New York: ACM Press.

Waterworth, J. A., & Chignell, M. H. (1997). Multimedia interaction. In Helander, M. G., Landauer, T. K. & Prabhu, P. V. (Eds.): *Handbook of Human-Computer Integration* (pp. 915-946). Amsterdam: Elsevier.

Wickens, C. D. (1984). Processing resources in attention. In: R. Parasuraman & D. R. Davies (Eds.), *Varieties of attention* (pp. 63-102). Orlando, FL: Academic Press.

Wilson, M., & Knoblich, G. (2005). The case for motor involvement in perceiving conspecifics. *Psychological Bulletin, 1*(3), 460473.

ENDNOTE

[1]   www.tampere.fi/english/vapriikki/

Chapter 7
Two Frameworks for the Adaptive Multimodal Presentation of Information

Yacine Bellik
Université d'Orsay, Paris-Sud, France

Christophe Jacquet
SUPELEC, France

Cyril Rousseau
Université d'Orsay, Paris-Sud, France

ABSTRACT

Our work aims at developing models and software tools that can exploit intelligently all modalities available to the system at a given moment, in order to communicate information to the user. In this chapter, we present the outcome of two research projects addressing this problem in two different areas: the first one is relative to the contextual presentation of information in a "classical" interaction situation, while the second one deals with the opportunistic presentation of information in an ambient environment. The first research work described in this chapter proposes a conceptual model for intelligent multimodal presentation of information. This model called WWHT is based on four concepts: "What," "Which," "How," and "Then." The first three concepts are about the initial presentation design while the last concept is relative to the presentation evolution. On the basis of this model, we present the ELOQUENCE software platform for the specification, the simulation and the execution of output multimodal systems. The second research work deals with the design of multimodal information systems in the framework of ambient intelligence. We propose an ubiquitous information system that is capable of providing personalized information to mobile users. Furthermore, we focus on multimodal information presentation. The proposed system architecture is based on KUP, an alternative to traditional software architecture models for human-computer interaction. The KUP model takes three logical entities into

DOI: 10.4018/978-1-60566-978-6.ch007

account: Knowledge, Users, and Presentation devices. It is accompanied by algorithms for choosing and instantiating dynamically interaction modalities. The model and the algorithms have been implemented within a platform called PRIAM (PResentation of Information in AMbient environment), with which we have performed experiments in pseudo-real scale. After comparing the results of both projects, we define the characteristics of an ideal multimodal output system and discuss some perspectives relative to the intelligent multimodal presentation of information.

INTRODUCTION

For a few years, access to computers has become possible to a large variety of users (kids, adolescents, adults, seniors, novices, experts, disabled people, etc.). At the same time, advances in the miniaturization of electronic components have allowed the development of a large variety of portable devices (laptops, mobile phones, portable media players, personnel digital assistants (PDA), etc.). New interaction situations have started to appear due to users' mobility enabled by this evolution. It is nowadays commonplace to make a phone call on the street, to work while commuting in public transportation, or to read e-mails at a fast-food. The interaction environment which was static and closed has become open and dynamic. This variety of users, systems and physical environments leads to a more complex interaction context. The interface has to adapt itself to preserve its utility and usability. Our work aims at exploiting the interaction richness allowed by multimodality as a means to adapt the interface to new interaction contexts. More precisely we focus on the output side of the interface. Our objective is to exploit intelligently all modalities available to the system at a given moment, to communicate information to the user. In this chapter, we start by presenting related work. Then we present a first framework which addresses the problem in a "classical" interaction situation. A second framework addresses the same problem in a different situation: ambient environments. After comparing the results of both projects we conclude by presenting some future research directions.

RELATED WORK

At first, multimodality was explored from the input side (user to system). The first multimodal interface was developed in 1980 by Richard Bolt (Bolt, 1980). He introduced the famous *"Put That There"* paradigm which showed some of the power of multimodal interaction. Research work on output multimodality is more recent (Elting, 2001-2003). Hence, the contextualization of interaction requires new concepts and new mechanisms to build multimodal presentations well adapted to the user, the system and the environment.

Output Multimodality Concepts

Presentation Means

When designing presentation as an output of a system, one has to choose which modalities will be used, and how they will convey information. The concept of *presentation means* represents the physical or logical system communication capacities. There are three types of presentation means: mode, modality and medium. Depending on authors, these three terms may have different meanings (Frohlich, 1991; Bernsen, 1994; Nigay, 1995; Bordegoni, 1997; Martin, 1998). In our case we adopt user-oriented definitions (Bellik, 1995; Teil, 2000). A mode refers to the human sensory system used to perceive a given presentation[1] (visual, auditory, tactile, etc.). A modality is defined by the information structure that is perceived by the user (text, ring, vibration, etc.) and not the structure used by the system[2]. Finally, a medium is a physical device which supports the expression of a modality (screen, loudspeakers, etc.). These three presentation means are dependent. A set of modalities may be associated with a given mode and a set of media may be associated with a given modality. For instance, the "Vibration" modality can be expressed through the "Vibrator" medium and invokes the "Tactile" mode.

Interaction Context

An interaction occurs in a given *context*, although the definition for what *context* means may vary depending on the research community. In our research work we adopt Dey's definition (Dey, 2000). The interaction context is considered as any information relative to a person, a place or an object considered as relevant for the interaction between the user and the system. We use a model-based approach (Arens, 1995) to specify the elements of interaction context (system model, user model, environment model, etc.). A set of dynamic or static criteria is associated to each model (media availability, user preferences, noise level, etc.). The work presented here does not propose new ways of capturing context. Instead, we suppose that we can rely on an adequate framework, as those proposed in Dey (2000) or Coutaz (2002).

Multimodal Presentation

When one uses several modalities to convey information, the presentation of information is said to be *multimodal*. A multimodal presentation is comprised of a set of (modality, medium) pairs linked by redundancy or complementarity relations according to CARE properties (Coutaz, 1995). For instance, an incoming call on a mobile phone may be expressed by a multimodal presentation composed of two (modality, medium) pairs: a first pair (Ring, Loudspeaker) indicates the call receipt while a second pair (Text, Screen) presents the caller identity (name).

Output Multimodal Models and Systems

SRM (Standard Reference Model) (Bordegoni, 1997) is one of the first conceptual models which addressed the problem of multimodal presentation. Stephanidis (Stephanidis, 1997) improved it by integrating the interaction context within the initial design of the multimodal presentation, even though this integration was incomplete. Then, Thevenin introduced the concept of plasticity (Thevenin 1999) to describe the adaptation of interfaces. At first, this concept of plasticity addressed the interface adaptation in regard to the system and environment only, while preserving interface usability. Later, it has been extended to the <user, system, environment> triplet designing the general interaction context (Calvary, 2002) (Demeure, 2003). The concept of plasticity inspired CAMELEON-RT (Balme, 2004) which is an architecture reference model that can be used to compare existing systems as well as for developing run time infrastructures for distributed, migratable, and plastic user interfaces.

Actually, we can notice that existing systems have often addressed the problem under a specific angle (Table 1). For instance, WIP (André, 1993) explored the problem of coordinating text and graphics. This system is capable of automatically generating from text and graphics user manuals for common devices. The COMET system (Feiner, 1993) also addresses the same problem in a different application domain (diagnostic, repair and maintenance). While both systems addressed the problem of coordinating visual modalities, MAGIC (Dalal, 1996) explored the coordination of visual and audio modalities. AIFresco System (Stock, 1993) addressed the problem of natural language generation in the context of an hypermedia system. PostGraphe (Fasciano, 1996) and SAGE (Kerpedjiev, 1997) have a common approach which consists in generating multimodal presentation based on the concept of presentation goal. CICERO (Arens, 1995) introduced an ap-

Table 1. Problems addressed by some existing systems. © Yacine Bellik. Used with permission.

Systems	Addressed Problems
WIP (1993), COMET (1993), AlFresco (1993)	Visual modalities coordination
MAGIC (1997)	Visual and audio modalities coordination
AlFresco (1993)	Natural language generation
CICERO (1995)	Models management
AVANTI (2001)	User model management
PostGraphe (1996), SAGE (1997)	Presentation goals management

proach based on models (media, information, task, discourse and user). AVANTI (Stephanidis, 2001) is one of the first systems, which takes into account the interaction context even though it is mainly based on user profiles. Table 1 synthesizes the contribution of these different systems.

Towards Mobile Environments: Ambient Intelligence

For a few years, computer technology has been pervading larger parts of our everyday environment. First it spread in "technological" artifacts such as cameras, mobile phones, car radios, etc. Now, researchers consider its integration into even more commonplace objects such as clothing, doors, walls, furniture, etc. This trend is referred to by terms like *pervasive* or *ubiquitous computing*, the *disappearing computer*, *mixed systems*, *ambient intelligence*, etc. All of them describe the same kind of concept, that of giving everyday objects additional capabilities in terms of computation, wireless communication and interaction with human users.

Although the basic concept dates back to the early 1990s (Weiser, 1993), its implementation was long deemed impractical because electronic devices could not be miniaturized enough. However, recent advances in the fields of miniaturization, wireless networks and interaction techniques are quickly removing these technical barriers. Moreover, until recently, researchers in the field had to master both hardware and software, which limited the develop-

ment of these systems. Now, off-the-shelf hardware platforms are readily available (Gellersen, 2004), thus software specialists can experiment with ambient systems without being hardware experts.

In consequence, more and more research groups are getting involved in the domain, and some people even think that we are at the threshold of a revolution similar to that of the 1980s when computers broke out of datacenters and spread in office environments and later at home (Lafuente-Rojo, 2007).

In 2001 the European Information Society Technologies Advisory Group (ISTAG) tried to characterize the specificities of ambient intelligence (Ducatel, 2001). It came out with three core properties:

- *Ubiquitous computing*: microprocessors can be embedded into everyday objects that traditionally lack any computing ability, such as furniture, clothing, wallpaper, etc. Some people already envision embedding RFID[3] into construction materials (concrete, paint) or furniture (Bohn 2004).
- *Ubiquitous communication*: these objects must be endowed with wireless communication abilities, rely on energy sources that provide them with good autonomy, and be capable of spontaneously interoperating with other objects, and without human intervention.
- *Intelligent user interfaces*: human users must be able to interact with these objects in a natural (using voice, gestures, etc.) and

customized way. The object must therefore take user preferences and context into account.

Ambient intelligence systems interact with users when they are not in "classical" interaction situations, i.e. sitting at one's desk or using a portable device (PDA for instance). These systems must be able to discretely and non-intrusively react to user actions. This can be significantly useful in mobile situations, when it is impractical to use a device, even a handheld one such as a mobile phone. For instance, when one is finding their way at an airport, they generally do not want to hold their mobile phone, it is much more appropriate to receive information from the background, for instance from loudspeakers or stationary display screens.

MOTIVATION FOR OUR RESEARCH WORK

Our purpose in executing the research work described in this chapter was to investigate how one can design frameworks and algorithms for mobile-based multimodal presentation of information. The work was split around two categories of systems: first those in which the user is fixed with respect to the system, but both may be mobile with respect to the environment (such as a mobile car system), and second those in which the user is mobile with respect to the system itself.

The first research work addresses the issue of creating a multimodal presentation of information that takes the current context into account. It introduces a framework that is capable of choosing a combination of modalities for a given information item (in a redundant or complementary fashion according to CARE properties). The system and the user are fixed with respect to one another. They can both be stationary (e.g. a user using their desktop telephone), or both be moving together (e.g. a traveler and their mobile phone).

In contrast, the second research work focuses on the new problems introduced by the mobility that may exist between the user and the system (like in ambient environments). It does not focus on combining modalities, but rather on using a variety of devices to provide a user with information. These devices may be stationary (fixed display screens, loudspeakers in an airport), or mobile (handheld devices), though the emphasis will be on the former. Only one modality is used at a time, but the proposed framework is nevertheless multimodal because possible modalities depend on the user and on the devices considered (exclusive multimodality (Teil, 2000)). This framework provides a mechanism for choosing a device and a modality for presenting an information item for a mobile user.

FIRST FRAMEWORK: CONTEXTUAL PRESENTATION OF INFORMATION

This research work aims at proposing an abstract model which enables one to organize and to structure the design process of a dynamic and contextual multimodal presentation. This model called WWHT (What, Which, How, Then) was implemented in a software platform (ELOQUENCE) which includes a set of tools supporting the designer/developer during the process of elaborating multimodal presentations. We have used this platform to develop two applications: a fighter cockpit simulator and an air traffic control simulator.

The WWHT Model Components

The WWHT model is based on four main components: information to present, presentation means, interaction context and the resulting multimodal presentation.

The information represents the semantic object the system has to present to the user. For instance, in mobile telephony, the receipt of a new call

Figure 1. Design process of a multimodal presentation adapted to interaction context. © Y. Bellik, C. Jacquet, C. Rousseau. Used with permission.

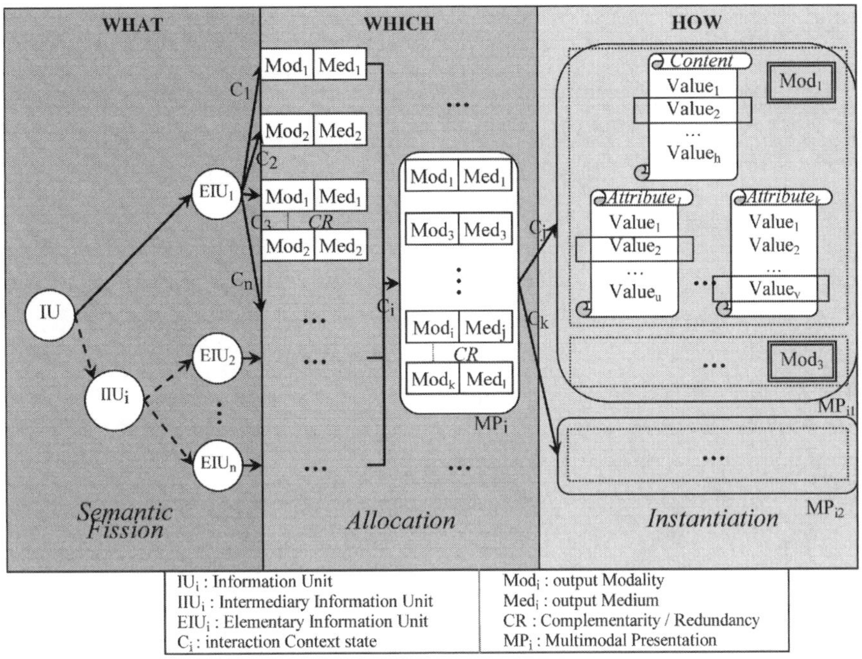

IU_i : Information Unit	Mod_i : output Modality
IIU_i : Intermediary Information Unit	Med_i : output Medium
EIU_i : Elementary Information Unit	CR : Complementarity / Redundancy
C_j : interaction Context state	MP_i : Multimodal Presentation

constitutes semantic information that the output multimodal system has to express. The presentation means, interaction context and multimodal presentation are defined as above.

The WWHT Model

The WWHT model is structured around answering four main questions:

- *What*: what is the information to present?
- *Which*: which modality (ies) should we use to present this information?
- *How*: how to present the information using this (ese) modality (ies)?
- *Then*: how to handle the evolution of the resulting presentation?

The first three questions (*What*, *Which* and *How*) refer to the initial building of a multimodal presentation while the last one (*Then*) refers to its future. Figure 1 presents the process of the initial

design. The presentation evolution is described in the sub-section entitled "Then".

Further questions could have been asked such as: "*When*", "*By Whom*", "*Where*". However we limited the model to the questions that are directly relevant for a multimodal presentation module. For instance, in our software architecture, it is not the responsibility of the presentation module to decide *when* to present information. The presentation module just waits for information to present sent by the dialog manager and when the dialog manager invokes it, it presents the information. The answer to the question "*by whom* is the presentation done and decided?" is: the system (at runtime). However the system simply complies to design rules introduced by the designer. So at last, it is the designer who determines the behavior of the system. Finally, the question "*Where*" is implicitly included inside the "*Which*" and "*How*" questions since the choice of modalities, medias and their attributes will determine the location of the presentation.

What

The starting point of the WWHT model is the semantic information (Figure 1, IU) the system has to present to the user. To reduce the complexity of the problem, we start by decomposing the initial semantic information into elementary information units[4] (Figure 1, EIUi). For instance, in the case of a phone call, the information "Call from X" may be decomposed into two elementary information units: the call event and the caller identity.

Let us underline here the fact that the term *"fission"* (Wahlster, 2003) (Nigay, 1993) is often used by opposition to the term *"fusion"* (for input multimodal interaction) to qualify the whole building process of a multimodal presentation (Figure 2). We prefer to talk about fission only during the first step of the process, which consists in splitting the initial semantic information into several elementary information units. Since the entry point of this process is semantic information, we prefer to call it semantic fission rather than multimodality fission or modality fission. The next step which consists in choosing *a modality or a combination of modalities* for each elementary information unit is then called *allocation* and is detailed in the next section.

In general, the semantic fission is done manually by the designer because an automatic semantic fission requires semantic analysis mechanisms which make the problem even harder. However, it constitutes an interesting topic for future work.

Which

When the decomposition is done, a presentation has to be allocated to the information. The allocation process consists in selecting for each elementary information unit a multimodal presentation (Figure 1, [Modi, Medj]) adapted to the current state of interaction context (Figure 1, Ci). The resulting presentation is comprised of a set of pairs (modality, medium) linked by redundancy/complementarity relations. This process may be

Figure 2. Place of semantic fission within the building process of a multimodal presentation. © Y. Bellik, C. Jacquet, C. Rousseau. Used with permission.

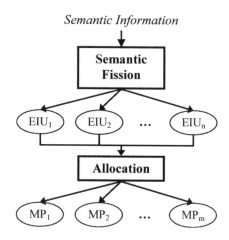

EIU: Elementary Information Unit
MP: Multimodal Presentation

complex in particular in the case of applications with several communication modalities and/or applications with a high variable interaction context. Figure 3 presents examples of possible multimodal presentations to express the semantic information "Call from X" on a mobile phone.

The selection process of presentation means is based on the use of a behavioral model. The representation of this behavioral model may vary depending on the system considered: rules (Stephanidis, 1997), matrices (Duarte, 2006), automata (Johnston, 2005), Petri Nets (Navarre, 2005), etc. In the ELOQUENCE platform we have used a rule-based representation. This representation allows an intuitive design process (If … Then…instructions). However this choice introduces problems on the scalability, the coherence and the completeness of a rule-based system. A graphical rule editor has been implemented to help the designer in the design and the modification of the rules base. Mechanisms for checking the structural coherence (two rules with equivalent premises must have coherent conclusions) are

Figure 3. Different possible presentations to express the receipt of a call on a mobile phone. © Y. Bellik, C. Jacquet, C. Rousseau. Used with permission.

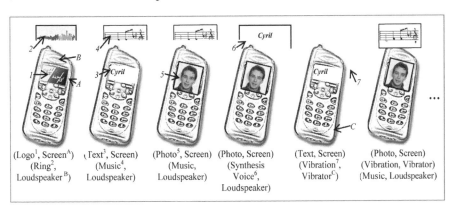

also proposed but the designer is still responsible of the completeness of the rules base.

Three types of rules are distinguished: contextual, composition and property rules. The premises of a contextual rule describe a state of the interaction context. The conclusions define contextual weights underlining the interest of the aimed interaction components (according to the context state described in the premises rule). The composition rules allow the modalities composition and so the design of multimodal presentation with several (modality, medium) pairs based on redundancy and/or complementarity criteria. Lastly, the property rules select a set of modalities using a global modality property (linguistic, analogical, confidential, etc.).

By analogy with the political world, we call our allocation process: "election". Our election process uses a rules base (voters) which add or remove points (votes) to certain modes, modalities or media (candidates), according to the current state of the interaction context (political situation, economic situation, etc.).

The application of the contextual and property rules defines the "pure" election while the application of the composition rules defines the "compound" election. The pure election elects the best modality-medium pair while the compound election enriches the presentation by selecting

new pairs redundant or complementary to the first one.

How

When the allocation is done, the resulting multimodal presentation has to be instantiated. Instantiation consists in determining the concrete lexico-syntactical content of the selected modalities and their morphological attributes[5] depending on interaction context (Figure 1, Cj). First, a concrete content to be expressed through the presentation modality has to be chosen. Then, presentation attributes (modality attributes, spatial and temporal) parameters are set. This phase of the WWHT model deals with the complex problem of multimodal generation (André, 2000; Rist, 2005). The space of possible choices for the content and attributes values of modalities may be very large and thus the choice complex.

Ideally, the content generation should be done automatically. However this is still an open problem for each considered modality such as text generation (André 2003) or gesture generation (Braffort, 2004). For us the problem is rather to select one content between n possible predetermined contents and then to determine the adequate values for the morphological attributes. Figure 4 (A) shows an example of possible contents for

Figure 4. Which content and which attributes for the photographic modality? © Yacine Bellik. © Y. Bellik, C. Jacquet, C. Rousseau. Used with permission

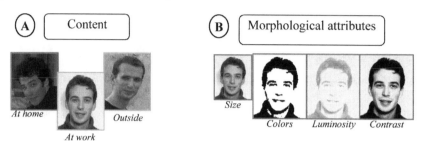

the photographic modality while Figure 4 (B) shows possible morphological attributes for the same modality.

Then

We could think that when instantiation is done, the problem of building a multimodal presentation adapted to the current interaction context is resolved. Actually, the interaction context is dynamic and thus may evolve. This reveals the problem of presentation expiration. Indeed, the presentation may be adapted when it is built but there is a risk that it becomes inadequate if the interaction context evolves. This expiration problem concerns mainly the *persistent* presentations. Thus a multimodal presentation has to remain adapted during its whole life cycle. This constraint requires the use of mechanisms that allow the presentation to evolve according to the following factors:

- information factor,
- interaction context,
- time factor,
- space factor,
- user actions.

Information factor is a common evolution factor. For instance, the presentation of a laptop battery evolves according to its power level. In the case of the interaction context, its modifications may induce presentation expiration but it is not always the case. For instance, a visual presentation will not be affected by a higher noise level. The time factor may be important in some applications. Calendar applications are a good example. An event may be presented differently at two different moments. For instance, strikethrough characters may be used to display the event when it becomes obsolete. The space factor refers to the position and space size allocated to a given presentation. For instance, the FlexClock application (Grolaux, 2002) adapts the presentation of a clock according to its window size. The clock is displayed using graphics and text when the size window is big and only text modality when it is small. Finally, user actions may also influence the presentation. For instance, when the mouse cursor hovers a given icon, an attached text tip may appear.

We define two (non-exclusive) types of presentation evolution: *refinement* and *mutation*. On the one hand, refinement doesn't change the presentation means (modalities, media) used by the presentation. It affects only their instantiations. On the other hand, mutation induces modifications in the modalities and/or media used by the presentation. This difference is important because it requires different mechanisms to handle each type of evolution. Refinement requires a back-track to the instantiation (*how*) phase only while mutation requires a back-track to the allocation (*which*) phase. For instance, let us consider a multimodal presentation for the power level of a battery.

Figure 5. Evolution types of a presentation. © Y. Bellik, C. Jacquet, C. Rousseau. Used with permission.

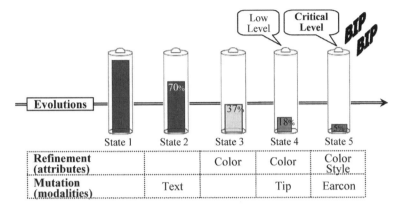

Figure 5 proposes four possible evolutions. The presentation at 70% (State 2) evolves by adding a text modality. We have a mutation here. The presentation at 37% (State 3) modifies the colors. It is a refinement. Finally, the presentation at 5% (State 5) combines both evolution types[6].

To sum up, Figure 6 shows a possible application of the different steps of the WWHT model for the mobile phone call scenario. T1, T2 and T3 show possible successive presentation evolutions.

The ELOQUENCE Platform

We have developed a software platform derived from the WWHT model to assist the designer/developer during the process of elaborating multimodal presentations. This platform, called ELOQUENCE, has been used in two different applications: a fighter cockpit simulator (Figure 7) and an air traffic control simulator (Figure 8) (Rousseau, 2006).

The ELOQUENCE platform includes two tools that respectively allow the specification and simulation of the system's outputs, and a runtime kernel which allows the execution of the final system. The specification tool (Figure 9) allows the designer to define all the elements required by the WWHT model: information units, presentation means, interaction context, and behavioral model. An analysis process must be applied to obtain these required elements. At first, it is necessary to collect a data corpus. This corpus must be composed of scenarios / storyboards (referring to nominal or degraded situations) but also of relevant knowledge on application field, system, environment, etc. Collecting this corpus must be strictly done and should produce consequent and diversified set of data. The corpus provides elementary elements needed to build the output system core (behavioral model). The quality of system outputs will highly depend on the corpus diversity. The participation and the collaboration of three actors is required: ergonomists, designers and end users (experts in the application field). Designers and users are mainly involved in the extraction of the elements while ergonomists are mainly involved in the interpretation of the extracted elements. The participation of all these actors is not an essential condition. However, the absence of an actor will be probably the source of a loss of quality in the outputs specification. Different steps should be followed to extract the required elements. The first step identifies pertinent data which can influence the output interaction (interaction context modeling). These data are interpreted to constitute context criteria and classified by models (user model, system model, environment model etc.). The next step specifies

Figure 6. Application of the WWHT model to a mobile phone call scenario. © Y. Bellik, C. Jacquet, C. Rousseau. Used with permission.

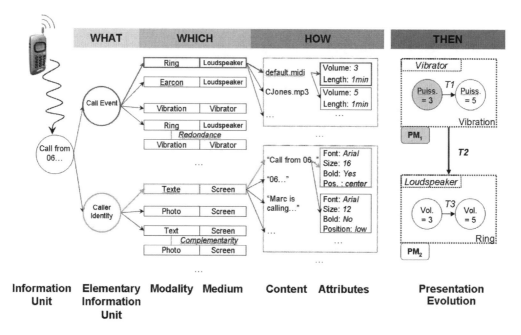

the interaction components diagram. Media are often defined first and from media it is relatively easy to identify output modes and modalities. The third step identifies semantic information which should be presented by the system. For better performance of the final system, it is recommended to decompose information into elementary semantic parts. At last, these extracted elements will allow the behavioral model definition.

The simulation tool constitutes a support for a predictive evaluation of the target application. It allows the designer to immediately check the

Figure 7. The fighter cockpit simulator. "© Thalès. Used with permission.

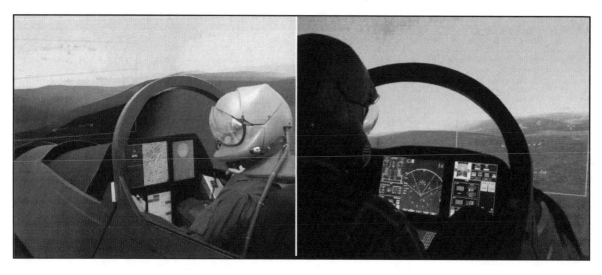

Figure 8. The air traffic control simulator. © Y. Bellik, C. Jacquet, C. Rousseau. Used with permission.

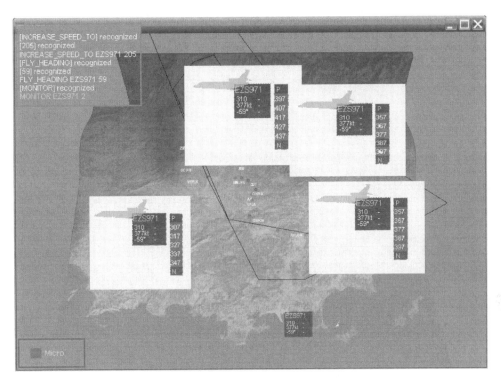

results of the specifications and thus makes the iterative design process easier. Figure 10 presents the simulation tool in the case of an incoming call on a mobile phone. It is composed of four parts. A first interface (A) simulates the dialog controller. More precisely it allows to simulate incoming information units from the dialog controller. A second interface (B) simulates a context server allowing the modification of the interaction context state. These two interfaces are generated automatically from the above specification phase. A third window (C) describes the simulation results in a textual form. Finally, a last interface (D) presents with graphics and sounds a simulation of the outputs results.

Finally, the runtime kernel is integrated in the final system. The kernel's architecture is centralized (Figure 11). The three main architecture modules (allocation, instantiation and evolution engines) implement the basic concepts of the

WWHT model. A multimodal presentation manager completes the architecture by centralizing the resources and the communications between the different modules. The ELOQUENCE platform is described in more details in (Rousseau, 2006).

In the second research work, we will see that a distributed agent architecture is more adequate for the multimodal presentation of information in an ambient environment.

SECOND FRAMEWORK: OPPORTUNISTIC PRESENTATION OF INFORMATION

The second research work is about the opportunistic and multimodal presentation of information to mobile users in an ambient environment (Jacquet, 2005). The purpose here is to provide the mobile users with information through either private

Figure 9. The specification tool. © Y. Bellik, C. Jacquet, C. Rousseau. Used with permission.

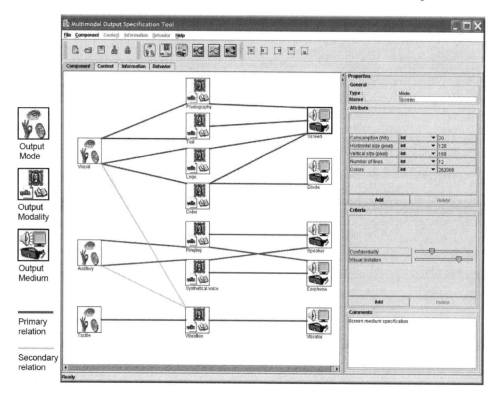

devices that they can carry with them (PDA, mobile phone, portable media player, etc.), or public devices that they may stumble upon while moving around (public display screens, loudspeakers, etc.). Let us underline three key points:

- Information relevant to users depends on the physical space where users are located. For instance, someone who gets inside a restaurant is most probably interested in the restaurant's menu; however a traveler who gets inside an airport is more likely to need to know which are his/her check-in desk and boarding gate.

- Information should be targeted to a given group of people. Indeed, it does not make sense to display an information item no one is interested in. It only confuses people and increases information lookup time. This is especially true for public displays, such

as those found in airports, which are often overloaded with information (Figure 12).

- We consider information providing and information presentation to be two distinct processes. This means that a user can (conceptually) receive information as he/she moves, but be able to *look at* (or *listen to*) it only when he/she comes close to a suitable presentation device. Information can be temporarily stored by a digital representation of the user when the person is moving.

In this way, a user can gather information items in an opportunistic way when he/she encounters them, and even if there is no suitable presentation device at that moment. Information presentation can take place at a later time, in an opportunistic fashion too, when the user is close to a suitable device. This introduces a decoupling between

Figure 10. The simulation tool. © Y. Bellik, C. Jacquet, C. Rousseau. Used with permission.

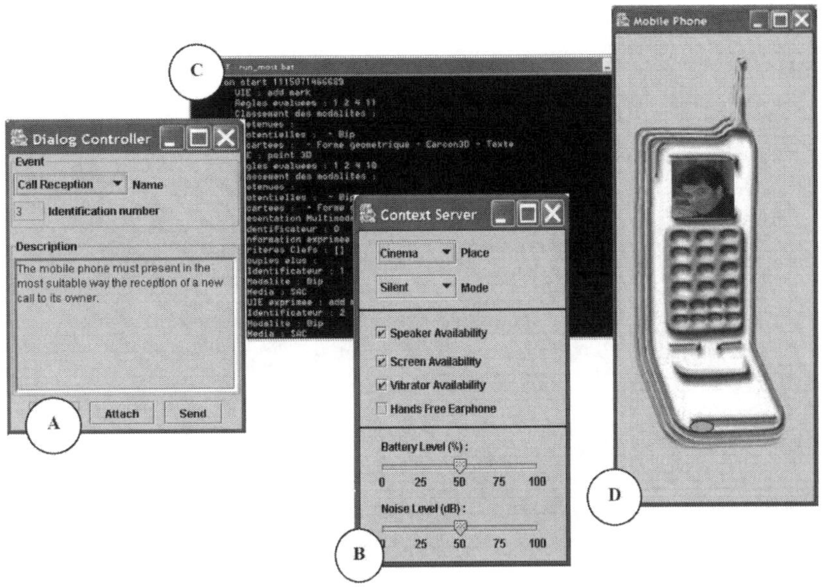

two phases (providing and presenting information), which is necessary to both opportunistic behaviors. For this decoupling to be effective, the functional core must not be linked directly with the interface, from a software architecture point of view. It is thus necessary to introduce an intermediate entity between the interface and the functional core; otherwise information providing and presentation would be linked. For this reason, we propose the KUP model, in which the aforementioned decoupling occurs through a *user entity*.

The KUP Model

In an ambient intelligence system, the co-existence of both physical and digital entities prompts for new conceptual models for interactions. We in-

Figure 11. Runtime kernel architecture. © Y. Bellik, C. Jacquet, C. Rousseau. Used with permission.

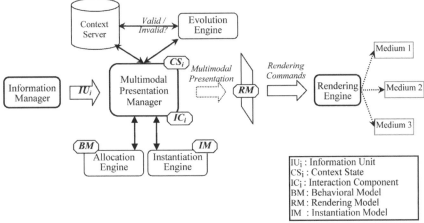

Figure 12. At Paris-Charles-de-Gaulle airport, this set of screens is permanently displaying a list of 160 flights, even when only three travelers are looking for information. © Y. Bellik, C. Jacquet, C. Rousseau. Used with permission.

troduce a model called KUP (Knowledge, User, Presentation) which is composed of three main entities, each of them having a physical facet and a digital (software) facet:

- K is the entity that represents information (or knowledge) sources. We call *semantic units* the information items produced by this entity (a semantic unit can be, for instance, the boarding gate of a given traveler; this notion is equivalent to the notion of "elementary information unit" in the WWHT model). The software facet of the K entity corresponds to the semantic component of classical architecture models (*functional core* in Seeheim (Pfaff, 1983) and ARCH (Bass, 1992), *abstraction facet* in PAC (Coutaz, 1987), *model* in MVC (Krasner, 1988), etc.).
- U is the user entity. Its physical facet corresponds to the human user. Its digital facet is active and is not limited to representing user attributes. For instance, it can store information for the user and negotiate the

presentation of information items with devices.
- P is an entity responsible for presenting information to the user. Its digital facet corresponds to the *interface* of the classical architecture models. As we only consider outputs here, the interface is therefore limited to information presentation. Its physical counterpart corresponds to the presentation device.

KUP's architecture model introduces two original features:

- it includes an active software representation of the user (U), whereas it is generally omitted or very basic in the classical models. This software representation is more than a mere description of the users with a profile or preferences;
- this software entity associated with the user lies at the center of the model, which gives an utmost importance to the user, especially because all communications

within the model are handled by this entity. Ultimately this "user" entity is responsible for decoupling information providing and information presentation.

KUP's architecture model distinguishes itself from the classical architecture models (Seeheim, ARCH, PAC, MVC, etc.) because in the latter the user is always outside the system; it is never explicitly represented by an active entity (Figure 13). Conversely, in KUP (Figure 14) the digital entity representing the user is considered as the core of the model, which allows to decouple the process of providing information (performed by K entities) from the process of presenting information (performed by P entities).

Physical space plays a significant role in an ambient intelligence system. In particular, the system is supposed to react to certain user movements in the physical space. To model these interactions and ultimately build systems able to react to the corresponding events, we introduce two concepts that we think are of highest importance in an ambient intelligence system: the *perceptual space* of an entity and the converse notion, the *radiance space* of an entity. Roughly speaking, they respectively correspond to what an entity can perceive, and to which positions it can be perceived from. Let us now define these notions properly.

First and foremost, let us define what *perception* means for the different kinds of entities considered. For a user, perception corresponds to *sensory perception*: hearing, seeing, or touching another entity. For non-human entities, perception corresponds to the *detection* of other entities. For instance, a screen or an information source can be able to detect nearby users and non-human entities. From a technological point of view, this can be achieved by a variety of means, for instance using an RFID reader.

Perceptual Space

Intuitively we wish to define the perceptual space of an entity as the set of positions in the physical space that it can *perceive*. However, *perception* depends on the input modalities that e uses, each of which has a given perceptual field. For instance, for the user entity, the visual field and the auditory field of a human being are not the same: a text displayed on a screen located at position P two meters on the back of the user cannot be seen, whereas a sound emitted by a loudspeaker located at the same place can be heard without any problems. Does P belong to the perceptual space of the user? The answer to this question depends not only on the position of P, but also on the modality considered and even on attribute values. For example, a text displayed two meters ahead of the user can be read with a point size of 72[7], but not with a point size of 8. Therefore the perceptual space depends on the physical space, on the modality space, and on the attribute modality space.

We define the notion of *multimodal space* or *m-space*, as the Cartesian product of the physical space E, the space M of available modalities, and the space of modality attributes A. (More precisely, it is the union of Cartesian products, because the space of modality attributes depends on the modality considered.) A point in the m-space is defined by a tuple containing the physical coordinates c of the point, a given modality m, and an instantiation i of this modality (set of values for the attributes of m). One can then define the perceptual space of an entity e as the set of points $X(c, m, i)$ of the m-space such that if another entity is located at the physical coordinates c and uses the modality m with the instantiation i, then it is perceived by e. Indeed, the definition of the perceptual space of an entity e includes spatial positions, but these positions are conditioned by the modalities used. Note that not all points in the m-space make sense, but this is not a problem as a perceptual space is always a *subset* of the m-space.

Figure 13. Classical HCI model: the user has no explicit representation in the interactive system. © Y. Bellik, C. Jacquet, C. Rousseau. Used with permission.

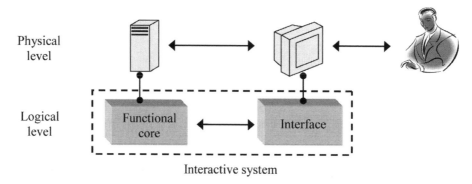

Figure 14. KUP model: the user is at the center of the information presentation system. © Y. Bellik, C. Jacquet, C. Rousseau. Used with permission.

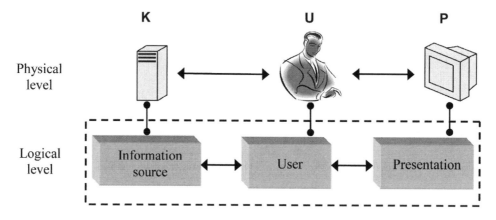

Radiance Space

We define the radiance space of an entity e, using a modality m with an instantiation i, towards an entity e', as the set of points x of the physical space E such that if e' is located in x, then e belongs to the perceptual space of e'. In other words, it is the set of points in space from which e' can perceive e. Let us notice that the radiance space of an entity e is always defined with respect to another entity e'. Indeed, the radiance space not only depends on the "emitting capabilities" of e, but also on "receiving capabilities" of e'. Thus

at a given point in space x, an entity e_1' may perceive e, whereas another entity e_2' may not. For instance, the radiance space of a loudspeaker that sends out a message at a given sound level will depend on the receiving entity (deaf user, or user with no auditory problem). Therefore it is not possible to define a radiance space in absolute terms. The situation is analogous to that of satellite telecommunication. The coverage area of a satellite (which can be termed as its radiance space), is the set of points of Earth's surface where it is possible to receive signals from the satellite, *with a dish antenna of a given diameter*. The coverage area cannot be defined independently of the receiving antenna.

The concepts of perceptual space and radiance space are respectively reminiscent of those of *nimbus* and *aura* introduced by Benford and Fahlen (Fahlen, 1992; Benford, 1993). However, the nimbus of an entity represents what is perceived by this entity, whereas the perceptual space is the spatio-modal dimension in which the nimbus can build up. Likewise, the aura of an entity represents the set of manifestations of this entity, whereas the radiance space is the spatio-modal dimension in which the aura can express itself.

Sensorial Proximity: Originating Event

In the KUP model, all interactions between entities happen following a particular event: *sensorial proximity*. This kind of event arises when an entity e_1 enters or leaves the perceptual space of another entity e_2[8]. With the above definitions for radiance and perceptual spaces, we must underline the fact that the concept of sensorial proximity spans two aspects. On the one hand, it includes spatial proximity that refers to the distance between the two entities, and their respective orientations. On the other hand it also includes the input/output capabilities of the entities[9]. For instance, a blind user coming very close to a screen will trigger no sensorial proximity event. It is the same for a sighted user coming at the same place but with his/her back towards the screen.

As our work focuses on information presentation, and therefore on *output*, we consider that changes in sensorial proximity are the only type of input events in the system. They trigger all of the system's reactions, in particular the outputs produced by the system. Up to now we do not have studied other types of input, for instance explicit user input, but this could be the subject of further investigation.

An Opportunistic Model for the Presentation of Information

Using the KUP model, one can clearly separate the process of providing information, from the process of presenting information. When a user (U) enters the radiance space of a knowledge source (K), the latter provides his/her logical entity with one or several relevant semantic units. At the moment when the user receives these semantic units, it is possible that no presentation device (P) is within his/her perceptual space. However, since users are supposedly mobile, it is possible that a presentation device enters the user's perceptual space at a later time. This will then trigger a sensorial proximity event which will initiate the process of presenting the user's semantic units on the device[10]. Figure 15 summarizes the interaction between knowledge sources, users and presentation devices.

Agent Architecture

A configuration like the one presented on Figure 18 could be hardwired, but this would not take full advantage of the very modular structure of the KUP model. For instance, if the staff changes the location of knowledge sources and presentation devices, or brings in new devices in case of a particular event, it would be cumbersome if a hardwired configuration had to be changed by hand. However, as the KUP model entirely relies on perception relationships between entities, it is possible to design implementations that auto-configure and adapt to changes without human intervention. We propose a decentralized architecture based on agents, in which each entity of the model has an agent counterpart:

- user agents (U) are an active software representation of human users,
- knowledge agents (K) provide user agents with information,
- presenter agents (P) are the software interface of the physical presentation devices.

Figure 15. A semantic unit provided by a knowledge source can be presented by several presentation devices. © Y. Bellik, C. Jacquet, C. Rousseau. Used with permission.

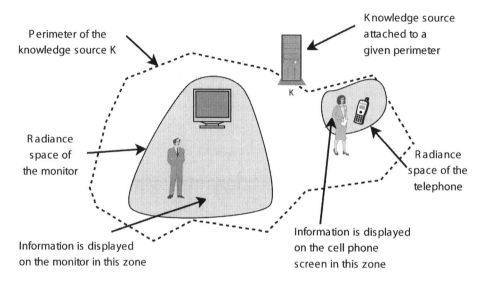

They can evaluate the cost of the presentation of an information item on a device, and perform the presentation.

This world of agents is a "mirror" of the real world, at least as far as our three types of entities of interest are concerned.

We assume that all agents can communicate with one another. Sensorial proximity relations, originating in the physical world, are mirrored in the world of agents. For instance, if a user *a* perceives a presentation device *b*, then the same relation exists between the agent counterparts. From a technical point of view, communication can use now-ubiquitous wireless networks such as Wi-Fi or mobile phone networks; sensorial proximity can be detected by various means, such as RFID tag detection, feature recognition in images, etc. Although agents have a notion of *having a location*, this does not mean that the corresponding processes have to run at this location. It is possible to have agents run on delocalized servers while retaining the flexibility of the architecture.

Agents are reactive: they are idling most of the time, and react when particular events occur. In practice, a given agent *a* can react to three types of events:

- another agent *b* has just come close to *a*[11],
- an agent *b*, that previously was close to *a*, has just gone away,
- *a* has just received a message through the network, from an agent *c*, which is not necessarily close to *a*.

Therefore, if there were only agents in the system, nothing would ever happen. Indeed, agents have reactive behaviors when the associated physical entities move. It means that the proactive properties of the system fully come from physical entities and human users: the latter usually move, and hence trigger cascades of reactions in the system.

Allocation and Instantiation in KUP

In KUP, allocating and instantiating modalities happen in a decentralized fashion. In the first framework, as interaction involved only a unique user on a unique workstation, we had preferred a centralized approach. Conversely, in this second framework, entities involved are disseminated in space, so it makes senses to resort to a decentralized architecture that matches the agent architecture mentioned above. In this way, when a U entity enters the radiance space of a P entity, the two associated agents will negotiate the most suitable modality[12] (and its instantiation) to present U's semantic units on the presentation device. This negotiation process relies on the concept of *profile*. A profile is a set of weights given to the modalities and their instances. Profiles are defined with respect to a tree-like taxonomy of modalities, common to the three types of entities.

Figure 16 gives an example of a partial taxonomic tree for output modalities.

Each entity defines a weighting tree which is superposed to the taxonomic modality tree[13]. The goal of a weighting tree is to add weights to a taxonomic tree, in order to express capabilities, preferences and constraints of users, devices and semantic units. A weight is a real number between 0 (included) and 1 (included). It can be located at two different places:

- **at node level:** the weight applies to the whole sub-tree rooted at this node. A weight of 1 means that the modalities of the sub-tree may be used, whereas a weight of 0

means that the modalities may not be used. Values in between allow one to introduce subtle variations in how much a modality is accepted or refused. This is used to express preference levels,

- **at attribute level:** the weight is a function that maps every possible attribute value to a number in the interval [0, 1]. This function indicates the weight of every possible value for the attribute. The meaning of the weights is the same as above: attributes values whose weight is close to 1 are acceptable; values whose weight is close to 0 are not.

A profile is defined as a weighting tree that spans the whole taxonomic tree. Figure 17 gives an example of a partial profile. It could correspond to an American, visually impaired user, who by far prefers auditory modalities over visual ones. The weights are given inside black ovals, just next to the nodes. Weight functions are given for some attributes. Depending on whether the attributes are of continuous or discrete nature, the functions are either continuous or discrete.

Given a user *u*, a presentation device *d* and a semantic unit *s*, the most suitable modality (along with its instantiation) to present *s* to *u* on *d* is determined by considering the *intersection* of the

Figure 16. Example of a partial taxonomy for output modalities. In this basic example we consider two kinds of output modalities, visual ones (example: text) and auditory ones (example: computer-generated spoken dialogue). © Y. Bellik, C. Jacquet, C. Rousseau. Used with permission.

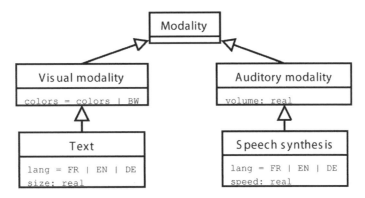

Figure 17. Example of a partial profile (weighting tree). © Y. Bellik, C. Jacquet, C. Rousseau. Used with permission.

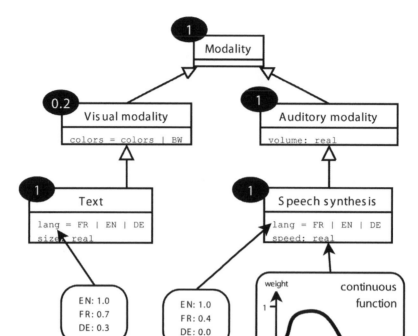

three weighting trees. This intersection method eventually produces a resulting weighting tree whose leaves are candidate modalities. Then, all that the system has to do is choose the modality with the highest weight, and instantiate it using the attribute values with the highest weights. Actually, this is the simplest of the situations, where only one semantic unit is to be presented to one user, using one presentation device. In the more general case where there are several users close to a device, or conversely if there are several devices close to one (or several) devices, more complex algorithms have been designed to have several devices collaborate with one another. They ensure a global consistency of information presentation while guaranteeing a minimal satisfaction level to each user. These algorithms are thoroughly described in (Jacquet, 2006; Jacquet, 2007).

The PRIAM Platform

In order to implement and validate the concepts introduced by the KUP model we have developed an agent platform called PRIAM (PResentation of Information in AMbient environments). Real-scale experiments being quite complex and costly to carry out, this platform includes a simulator that enables researchers to test every component of an ambient application, without being obliged to deploy it in real-scale (Figure 18). This simulator has also enabled us to validate the behavior of allocation and instantiation algorithms prior to real experiments. Every kind of situation, either simple or very complex, can be tested in this way, with the required number of presentation devices, users and knowledge sources.

We have nonetheless gone beyond the mere simulation stage. Three pseudo-real-scale experiments have been conducted. The objective

Figure 18. Simulator of the PRIAM platform. On this screenshot, three display screens and four users where simulated. The screens display information about departing trains. Screen contents appear on the simulator window, but they can also be "popped-out" like the one in the foreground. © Y. Bellik, C. Jacquet, C. Rousseau. Used with permission.

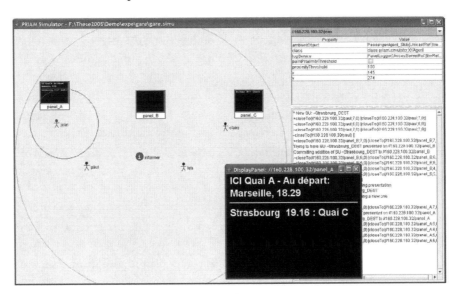

of these experiments was to prove that using a dynamic display that only displays information relevant to the people located at proximity, it was in average shorter and easier to find one's item of interest than using a long static list. Dynamic display was implemented on a computer, with an infrared detection system to detect the proximity of users. Static lists were either sheets of paper or static images on screen.

The first experiment displayed examination results for students. The second one was based on the same setting, but displayed airport information (boarding gate number). In these experiments we compared the times needed to find one's information item with static and dynamic displays. The purpose of the third experiment was to help train passengers to find a transfer, without having to walk too much. In this one, we compared the number of elementary moves needed to reach one's platform, when using static or dynamic displays. These experiments enabled us to test the platform in pseudo-real scale, and to demonstrate the benefits of displaying only information relevant

to users located in the vicinity of a presentation device. Indeed, in this case the device is far less overloaded with irrelevant items and users can lookup the items of interest more quickly. The results show that dynamic displays are superior in all cases: lookup time is respectively 50% and 25% faster in the first and second experiments, the number of elementary moves was divided by 2,4 in the third experiment (Jacquet, 2007).

The experiments have underlined some issues too, especially those related to privacy. For instance, when a passenger is alone in front of a screen, it is easy for an ill-intentioned person to know his/her destination. In one possible solution the system would introduce a few irrelevant items in the presentation so as to add noise and thus prevent third parties to gain access to private information.

In this second work, we have explored the problem of presenting multimodal information in an ambient setting. The distributed nature of information systems within the physical environment of users has led us to choose a multi-agent model in which agents representing users lie at the core. It is to be noted that the world of agents is only a mirror of the real world: the system is not proactive by itself; instead it reacts to changes and moves originating in the physical world. This means eventually that we do not really build a *world of agents*, but rather that we *agentify the real world*. This takes us back to the vision of ambient intelligence in which computerized systems monitor the actions of human beings in an unobtrusive way, so as to trigger actions when it is really relevant, and without disturbing normal user actions.

COMPARISON

Both research works presented in this chapter explored the benefits of using several output modalities to improve the way information is presented to the user. They address two different situations: a "classical" interaction situation where the user is fixed with respect to a unique interactive system and an ambient interaction situation in which the user moves and is likely to use several interactive systems.

In both cases we have proposed models for adaptive output multimodal systems. However, to not face several difficulties at the same time, we divided the issues between the two frameworks. For instance, in the second one is focused on the new constraints induced by the ambient environment. Hence, we have adopted a distributed agent architecture while the first frameworks relies on a centralized architecture. We have also insisted in the second project on the need to have an active representation of the user. However we have limited ourselves on some aspects that have already been explored in the first project. Thus we have used only exclusive[14] multimodality (while complementary or redundant multimodality was supported in the first project) and we did not

handle presentation evolution since it was already explored in the first project.

Regarding allocation and instantiation problems, the algorithms used in both frameworks are quite different. Algorithms defined in the second one are more powerful. For instance, the allocation process in the first framework is directive while it is cooperative in the second one. In the same way the instantiation process in the first framework is local while it is global in the second one.

Finally there are still some problems where we adopted the same options in both projects because these problems are still open and constitute some of our future research directions. For instance, the semantic fission process is manual in both frameworks, the content of modalities is predefined and not automatically generated in both frameworks, and the instantiation process is homogeneous in both frameworks (we will detail this problem in the next section).

Table 2 synthesizes the comparison between both frameworks.

CONCLUSION AND FUTURE RESEARCH DIRECTIONS

Thanks to the interaction richness it can offer, multimodality represents an interesting solution to the problems induced by an ever more variable interaction context. It is no longer reasonable today to continue to propose static and rigid interfaces while users, systems and environments are more and more diversified. To the dynamic character of the interaction context, the interface must also respond by a dynamic adaptation. The two frameworks described above constitute a first answer to the problem of adaptive multimodal presentation of information. However, they have also revealed some new problems that have not been addressed yet and which represent our future research directions. We summarize them in the following sections.

Table 2. Comparison between both frameworks. The last column show what an ideal adaptive output multimodal system should be. © Yacine Bellik. Used with permission.

Criterion	First framework	Second framework	Ideal system
Type of architecture	Centralized architecture	Distributed architecture	Depends
Support of CARE properties	Redundancy/Complementarity supported	Exclusive multimodality	Redundancy/Complementarity supported because a system which supports CARE properties is also capable of supporting exclusive multimodality
Content generation automaticity	No (predefined modality content)	No (predefined modality content)	Yes (generated modality content) because this will reduce the design costs.
Semantic fission automaticity	No (manual semantic fission)	No (manual semantic fission)	Yes (automatic semantic fission) because this will reduce the design costs.
Allocation strategy	Directive allocation	Cooperative allocation	Cooperative allocation because it allows to optimize the use of modalities and medias resources
Instantiation strategy	Local instantiation	Global instantiation	Global instantiation because it allows to optimize the use of modalities and medias resources
Type of instantiation	Homogeneous instantiation	Homogeneous instantiation	Heterogeneous instantiation because heterogeneous instantiation is more powerful. A system capable of heterogeneous instantiation is also capable of homogeneous instantiation

Heterogeneous instantiation

In our models we associate a unique instantiation to each elementary information unit. For instance, the string "Gate n° 15" which represent the concrete content of the elementary information unit indicating the boarding gate for a given flight, could be displayed using Arial font, 72 dots size; and white color. Hence, the instantiation of the morphological attributes is homogeneous and is applied to all elements of the modality content. However, sometimes it could be interesting to apply a particular instantiation to a part of the content. For instance, in the previous example, the number "15" could be displayed with a different color and a blinking bold style. It is not possible to specify it easily in our current models. One possible solution is to decompose again this elementary information unit into two others elementary information units, so we can instantiate each information unit independently. However this solution is not intuitive and will make the behavioral model complex. A

more interesting solution could be to define a new modality called, for instance, "2Texts" which will gather a content composed by two strings and 2 sets of morphological attributes (one set for each string). This way it becomes possible to associate to each content part a different instantiation.

Fusion in Output

Usually, fusion is a concept which is associated to input multimodality. However this concept can also be relevant to output multimodality, depending on the global software architecture (centralized/distributed) and the inconsistency detection strategy (early/late strategy). Let us take again the example seen in the first project and which is about the receipt of a phone call. We have seen that this semantic information can be decomposed into two elementary information units: the phone call event and the caller identity. Let us suppose that for each of these information units the allocation process chooses the "Ring" modality: a generic ring for the call event and a custom ring for the caller identity. This will induce inconsistence within the whole multimodal presentation and thus a back-tracking to the allocation process for a new allocation request. A possible solution could be to exploit the time factor to play both rings in a sequential way. However, if the system plays the custom ring first then the generic ring becomes useless. And if the generic ring is played first then it is likely that the user would have already answered the call before the system would have played the custom ring. A better solution would be to merge both modality contents and finally keep only the custom ring. Indeed custom ring is capable of expressing both information units, while the generic ring can only express the call event. This example shows that sometimes, a given instantiation may be attached to two different information units. This kind of relation is not yet supported in our models. We will try in future work to add this kind of relation to our models so we will be able to define output fusion algorithms.

Negotiation-Based Approach

Our current models apply their different phases in a sequential way. In the case of a blocking situation in a given phase, a back-tracking is done to the previous phase. It could be interesting to explore another approach based on a true negotiation between the different modules involved during the different phases (Figure 19). However the mechanisms of this negotiation process are still to be defined.

Influence of Inputs on Outputs

Inputs and outputs in an interactive system can be considered as dynamic interdependent flows of information. Thus, system outputs have to remain consistent with user inputs to ensure good interaction continuity. This continuity cannot be achieved unless inputs and outputs are incorporated inside the same design process. The modalities and media used in input may influence the choice of output modalities in particular in the case of lexical feedbacks. For instance, text entered on a keyboard will generally induce a visual feedback, while it could

Figure 19. Toward a negotiation-based approach. © Y. Bellik, C. Jacquet, C. Rousseau. Used with permission.

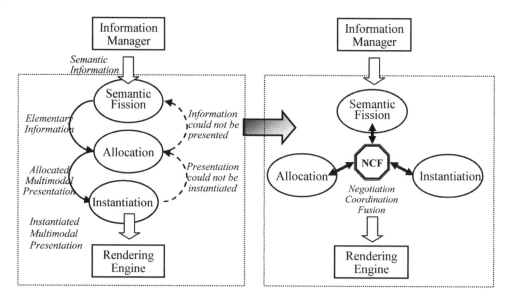

sometimes be preferable in the case of a speech command to produce a speech feedback.

As regards the semantic feedbacks, the second application developed using the first framework (air traffic control simulator) showed that the output part of the system needs to maintain an internal representation of the multimodal presentations it has provided. This allows the output part to answer requests coming from other modules, for instance about the pointed objects: when the user clicks on an (X,Y) position on the screen, only the output part of the system knows which object is located at this position since it is the output part of the system which knows which modalities and which modality attributes have been used to present application objects. This second application showed also that the output part of the system needs to know which input interaction means have been used so as to provide consistent output presentations.

Even though this application allowed us to start exploring some solutions, there are still some questions which need to be investigated, in particular those about the overall software architecture of a bidirectional (input and output) multimodal system.

REFERENCES

André, E. (2000). The generation of multimedia presentations. In R. Dale, H. Moisl & H. Somers (Eds.), *A Handbook of Natural Language Processing* (pp. 305–327). New York: CRC.

André, E. (2003). Natural language in multimedia/multimodal systems. In R. Mitkov (Ed.), *Computational Linguistics* (pp. 650-669). New York: Oxford University Press.

André, E., Finkler, W., Graf, W., Rist, T., Schauder, A., & Wahlster, W. (1993). The automatic synthesis of multimodal presentations. In M. T. Maybury, (Ed.), *Intelligent Multimedia Interfaces* (pp. 75-93). Menlo Park, CA: AAAI Press.

Arens, Y., & Hovy, E. H. (1995). The design of a model-based multimedia interaction manager. *Artificial Intelligence*, *9*(3), 167–188. doi:10.1016/0954-1810(94)00014-V

Balme, L., Demeure, A., Barralon, N., Coutaz, J., & Calvary, G. (2004). CAMELEON-RT: A software architecture reference model for distributed, migratable, and plastic user interfaces. In *Proc. of EUSAI 2004, European symposium on ambient intelligence, No. 2, vol. 3295 of Lecture Notes in Computer Science* (pp. 291-302). Eindhoven, The Netherlands: Springer.

Bass, L., Faneuf, R., Little, R., Mayer, N., Pellegrino, B., & Reed, S. (1992). A meta-model for the runtime architecture of an interactive system. *SIGCHI Bulletin*, *24*(1), 32–37. doi:10.1145/142394.142401

Bellik, Y. (1995), *Interfaces multimodales: Concepts, modèles et architectures*. Unpublished doctoral dissertation, Université Paris XI, Orsay, France.

Benford, S., & Fahlen, L. (1993). A spatial model of interaction in virtual environments. In *Proceedings of the Third European Conference on Computer Supported Cooperative Work (ECSCW'93)*.

Bernsen, N. O. (1994). Foundations of multimodal representations: a taxonomy of representational modalities. *Interacting with Computers*, *6*(4). doi:10.1016/0953-5438(94)90008-6

Bohn, J., & Mattern, F. (2004), Super-distributed RFID tag infrastructures. In *Proceedings of the 2nd European Symposium on Ambient Intelligence (EUSAI 2004)* (pp. 1–12). Berlin, Germany: Springer-Verlag.

Bolt, R. A. (1980). Put-that-there: Voice and gesture at the graphics interface. *Computer Graphics*, *14*(3), 262–270. doi:10.1145/965105.807503

Bordegoni, M., Faconti, G., Maybury, M. T., Rist, T., Ruggieri, S., Trahanias, P., & Wilson, M. (1997). A standard reference model for intelligent multimedia presentation systems. *Computer Standards & Interfaces*, *18*(6-7), 477–496. doi:10.1016/S0920-5489(97)00013-5

Braffort, A., Choisier, A., Collet, C., Dalle, P., Gianni, F., Lenseigne, B., & Segouat, J. (2004). Toward an annotation software for video of sign language, including image processing tools and signing space modeling. In *Proceedings of the Language Resources and Evaluation Conference (LREC'04)*.

Calvary, G., Coutaz, J., Thevenin, D., Limbourg, Q., Souchon, N., Bouillon, L., et al. (2002). Plasticity of user interfaces: A revised reference framework. In *TAMODIA '02: Proceedings of the First International Workshop on Task Models and Diagrams for User Interface Design* (127–134). Bucharest, Romania: INFOREC Publishing House Bucharest.

Coutaz, J. (1987). PAC, an object-oriented model for dialog design. In H.-J. Bullinger, B. Shackel (Eds.), *Proceedings of the 2nd IFIP International Conference on Human-Computer Interaction (INTERACT 87)* (pp. 431-436). Amsterdam: North-Holland.

Coutaz, J., Nigay, L., Salber, D., Blandford, A., May, J., & Young, R. M. (1995). Four easy pieces for assessing the usability of multimodal interaction: the CARE properties. In *Proceedings of the IFIP Conference on Human-Computer Interaction (INTERACT'95)*.

Coutaz, J., & Rey, G. (2002). Foundation for a theory of contextors. In [New York: ACM Press.]. *Proceedings of CADUI*, *02*, 283–302.

Dalal, M., Feiner, S., McKeown, K., Pan, S., Zhou, M., Höllerer, T., et al. (1996). Negotiation for automated generation of temporal multimedia presentations. In *Proceedings of ACM Multimedia'96* (pp. 55-64).

Demeure, A., & Calvary, G. (2003). Plasticity of user interfaces: towards an evolution model based on conceptual graphs. In *Proceedings of the 15th French-speaking Conference on Human-Computer Interaction*, Caen, France, (pp. 80-87).

Dey, A. K. (2000). *Providing architectural support for building context-aware applications*. Unpublished doctorial dissertation, Georgia Institute of Technology, College of Computing.

Duarte, C., & Carriço, L. (2006), A Conceptual Framework for Developing Adaptive Multimodal Applications. In *Proceedings of Intelligence User Interfaces (IUI'06)* (pp. 132-139).

Ducatel, K., Bogdanowicz, M., Scapolo, F., Leijten, J., & Burgelman, J.-C. (2001). *Scenarios for Ambient Intelligence in 2010, Final report*. Information Society Technologies Advisory Group (ISTAG), European Commission.

Elting, Ch., & Michelitsch, G. (2001). A multimodal presentation planner for a home entertainment environment. In *Proceedings of Perceptual User Interfaces (PUI) 2001*, Orlando, Florida.

Elting, C., Rapp, S., Möhler, G., & Strube, M. (2003). Architecture and Implementation of Multimodal Plug and Play. In *ICMI-PUI `03 Fifth International Conference on Multimodal Interfaces*, Vancouver, Canada.

Elting, Ch., Zwickel, J., & Malaka, R. (2002). Device-dependent modality selection for user-interfaces - an empirical study. In *International Conference on Intelligent User Interfaces IUI 2002*, San Francisco, CA.

Fahlen, L., & Brown, C. (1992). The use of a 3D aura metaphor for computer based conferencing and teleworking. In *Proceedings of the 4th Multi-G workshop* (pp. 69-74).

Fasciano, M., & Lapalme, G. (1996). PosGraphe: a system for the generation of statistical graphics and text. In *Proceedings of the 8th International Workshop on Natural Language Generation* (pp. 51-60).

Feiner, S. K., & McKeown, K. R. (1993). Automating the generation of coordinated multimedia explanations. In M. T. Maybury, (Ed.), *Intelligent Multimedia Interfaces* (pp. 117-139). Menlo Park, CA: AAAI Press.

Frohlich, D. M. (1991). The design space of interfaces. In L. Kjelldahl, (Ed.), *Multimedia Principles, Systems and Applications* (pp. 69-74). Berlin, Germany: Springer-Verlag.

Gellersen, H., Kortuem, G., Schmidt, A., & Beigl, M. (2004). Physical prototyping with smart-its. *IEEE Pervasive Computing / IEEE Computer Society [and] IEEE Communications Society, 3*(3), 74–82. doi:10.1109/MPRV.2004.1321032

Grolaux, D., Van Roy, P., & Vanderdonckt, J. (2002). FlexClock: A plastic clock written in Oz with the QTk toolkit. In *Proceedings of the Workshop on Task Models and Diagrams for User Interface Design (TAMODIA 2002).*

Jacquet, C. (2007). KUP: a model for the multimodal presentation of information in ambient intelligence. In *Proceedings of Intelligent Environments 2007 (IE 07)* (pp. 432-439). Herts, UK: The IET.

Jacquet, C., Bellik, Y., & Bourda, Y. (2005). An architecture for ambient computing. In H. Hagras, V. Callaghan (Eds.), *Proceedings of the IEE International Workshop on Intelligent Environments* (pp. 47-54).

Jacquet, C., Bellik, Y., & Bourda, Y. (2006), Dynamic Cooperative Information Display in Mobile Environments. In B. Gabrys, R. J. Howlett, L. C. Jain (Eds), *Proceedings of KES 2006, Knowledge-Based Intelligent Information and Engineering Systems*, (pp. 154-161). Springer.

Johnston, M., & Bangalore, S. (2005). Finite-state multimodal integration and understanding. *Natural Language Engineering, 11*(2), 159–187. doi:10.1017/S1351324904003572

Kerpedjiev, S., Carenini, G., Roth, S. F., & Moore, J. D. (1997). Integrating planning and task-based design for multimedia presentation. In *Proceedings of the International Conference on Intelligent User Interfaces* (pp. 145-152).

Krasner, G. E., & Pope, S. T. (1988). A cookbook for using the model-view controller user interface paradigm in Smalltalk-80. *Journal of Object Oriented Programming, 1*(3), 26–49.

Lafuente-Rojo, A., Abascal-González, J., & Cai, Y. (2007). Ambient intelligence: Chronicle of an announced technological revolution. *CEPIS Upgrade, 8*(4), 8–12.

Martin, J. C. (1998). TYCOON: Theoretical framework and software tools for multimodal interfaces. In J. Lee, (Ed.), *Intelligence and Multimodality in Multimedia Interfaces*. Menlo Park, CA: AAAI Press.

Navarre, D., Palanque, P., Bastide, R., Schyn, A., Winckler, M. A., Nedel, L., & Freitas, C. (2005). A formal description of multimodal interaction techniques for immersive virtual reality applications. In *Proceedings of the IFIP Conference on Human-Computer Interaction (INTERACT'05).*

Nigay, L., & Coutaz, J. (1993). Espace problème, fusion et parallélisme dans les interfaces multimodales. In *Proc. of InforMatique'93*, Montpellier (pp.67-76).

Nigay, L., & Coutaz, J. (1995). A generic platform for addressing the multimodal challenge. In *Proceedings of the Conference on Human Factors in Computing Systems (CHI'95)* (pp. 98-105).

Pfaff, G. (1983). User interface management systems. In *Proceedings of the Workshop on User Interface Management Systems.*

Rist, T. (2005). Supporting mobile users through adaptive information presentation. In O. Stock and M. Zancanaro (Eds.), *Multimodal Intelligent Information Presentation* (pp. 113–141). Amsterdam: Kluwer Academic Publishers.

Rousseau, C. (2006). *Présentation multimodale et contextuelle de l'information.* Unpublished doctoral dissertation, Paris-Sud XI University, Orsay, France.

Stephanidis, C., Karagiannidis, C., & Koumpis, A. (1997). Decision making in intelligent user interfaces. In *Proceedings of Intelligent User Interfaces (IUI'97)* (pp. 195-202).

Stephanidis, C., & Savidis, A. (2001). Universal access in the information society: Methods, tools, and interaction technologies. *UAIS Journal, 1*(1), 40–55.

Stock, O., & the ALFRESCO Project Team. (1993). ALFRESCO: Enjoying the combination of natural language processing and hypermedia for information exploration. In M. T. Maybury (Ed.), *Intelligent Multimedia Interfaces* (pp. 197-224). Menlo Park, CA: AAAI Press.

Teil, D., & Bellik, Y. (2000). Multimodal interaction interface using voice and gesture. In M. M. Taylor, F. Néel & D. G. Bouwhuis (Eds.), *The Structure of Multimodal Dialog II* (pp. 349-366).

Thevenin, D., & Coutaz, J. (1999). Plasticity of user interfaces: Framework and research agenda. In *Proceedings of the 7th IFIP Conference on Human-Computer Interaction, INTERACT'99,* Edinburgh, Scotland (pp.110-117).

Wahlster, W. (2003). Towards symmetric multimodality: Fusion and fission of speech, gesture and facial expression. In Günter, A., Kruse, R., Neumann, B. (eds.), *KI 2003: Advances in Artificial Intelligence, Proceedings of the 26th German Conference on Artificial Intelligence* (pp. 1-18). Hamburg, Germany: Springer.

Weiser, M. (1993). Some computer science issues in ubiquitous computing. *Communications of the ACM, 36*(7), 75–84. doi:10.1145/159544.159617

ENDNOTES

[1] Of course, in the case of a multimodal presentation, several modes may be used.

[2] For instance a scanned text saved by the system as a picture will be perceived by the user as a text and not as a picture.

[3] Radio-Frequency Identification

[4] This process can be done recursively as in (Wahlster, 2003) where a presentation planner recursively decomposes the presentation goal into primitive presentation tasks.

[5] The morphological attributes refer to the attributes that affect the form of a modality. For instance, font size for a visual text modality or volume for a spoken message.

[6] We observe also a refinement with respect to the internal rectangle size and the text position in steps.

[7] For a user with a normal visual acuity.

[8] This is the same as saying that entity e_2 enters or leaves the perceptual space of entity e_1.

[9] Thus the notion of perceptual proximity is not commutative in the general case.

[10] The presentation can happen as long as the semantic units are not outdated. A semantic unit can become outdated for two reasons. *Spatial outdating* may happen when the user leaves the radiance space of the knowledge source that has provided the semantic unit (but this is not always the case). *Temporal outdating* is controlled by a metadata element associated with the semantic unit.

[11] The notion of *closeness* refers to *sensorial proximity.*

[12] To make a clear distinction between problems, we have decided to restrict the second

project to *exclusive* multimodality, because complementarity and redundancy of modalities has already been studied in the first project. We have instead focussed on constraints specific to ambient environments.

[13] Except K entities that define a weighting tree for each semantic unit that they produce. Indeed, each semantic unit is supposed to be able to express itself into its own set of modalities. In consequence, weighting trees are attached to the produced semantic units, not to the K entities themselves.

[14] Exclusive multimodality allows the use of different modalities, but not in a combined way.

Chapter 8
A Usability Framework for the Design and Evaluation of Multimodal Interaction:
Application to a Multimodal Mobile Phone

Jaeseung Chang
Handmade Mobile Entertainment Ltd., UK

Marie-Luce Bourguet
University of London, UK

ABSTRACT

Currently, a lack of reliable methodologies for the design and evaluation of usable multimodal interfaces makes developing multimodal interaction systems a big challenge. In this paper, we present a usability framework to support the design and evaluation of multimodal interaction systems. First, elementary multimodal commands are elicited using traditional usability techniques. Next, based on the CARE (Complementarity, Assignment, Redundancy, and Equivalence) properties and the FSM (Finite State Machine) formalism, the original set of elementary commands is expanded to form a comprehensive set of multimodal commands. Finally, this new set of multimodal commands is evaluated in two ways: user-testing and error-robustness evaluation. This usability framework acts as a structured and general methodology both for the design and for the evaluation of multimodal interaction. We have implemented software tools and applied this methodology to the design of a multimodal mobile phone to illustrate the use and potential of the proposed framework.

INTRODUCTION

Multimodal interfaces, characterized by multiple parallel recognition-based input modes such as speech and hand gestures, have been of research interest for some years. A common claim is that they can provide greater usability than more traditional user interfaces. For example, they have the potential to be more intuitive and easily learnable because they implement interaction means that are close to the ones used in everyday human-human communication. When users are given the freedom of using the modalities of interaction of their choice, multimodal systems can also be more flexible and

DOI: 10.4018/978-1-60566-978-6.ch008

efficient. In particular, mobile devices, which generally suffer from usability problems due to their small size and typical usage in adverse and changing environments, can greatly benefit from multimodal interfaces. Moreover, the emergence of novel pervasive computing applications, which combine active interaction modes with passive modality channels based on perception, context, environment and ambience (e.g. Salber, 2000; Feki et al., 2004), raises new possibilities for the development of effective multimodal mobile devices. For example, context-aware systems can sense and incorporate data about lightning, noise level, location, time, people other than the user, as well as many other pieces of information to adjust their model of the user's environment. More robust interaction is then obtained by fusing explicit user inputs (the active modes) and implicit contextual information (the passive modes). In affective computing, sensors that can capture data about the user's physical state or behaviour, are used to gather cues which can help the system perceive users' emotions (Kapoor & Picard, 2005).

However, our lack of understanding of how recognition-based technologies can be best used and combined in the user interface often leads to interface designs with poor usability and added complexity. For designers and developers in the industry, developing multimodal interaction systems presents a number of challenges, such as how to choose optimal combinations of modalities, how to deal with uncertainty and error-prone human natural behaviour, how to integrate and interpret combinations of modalities, and how to evaluate sets of multimodal commands. These challenges are the prime motivation for our research. In particular, we aim to develop methodologies to help developers efficiently produce multimodal systems, while providing a greater user experience.

Existing related work includes the Cross-Weaver platform (Sinha & Landay, 2003), a visual prototyping tool, which allows non-programmer designers to sketch multimodal storyboards that can then be executed for quickly testing the interaction. CrossWeaver offers a practical and simple solution to building multimodal prototypes. However, design representations are let in an informal, sketched form, and analysis of the design and user test results is itself an informal process.

Flippo et al. (2003) propose an object oriented framework that enables existing applications to be equipped with a multimodal interface with relatively little effort. The framework equips existing application code with multimodal functionalities (such as modality fusion, multimodal dialog management, and ambiguity resolution), which rely on the declaration of command frames. The command frames embody the interaction design, but very little detail is provided about the process of declaring these frames and any support available for making design decisions and usability evaluation.

ICARE (Interaction CARE) (Bouchet et al., 2004) is a component-based approach to multimodal interfaces development. It defines two types of software components: elementary components for the development of pure modalities, and composition components to compose modality combinations. ICARE, however, assumes that the design of different modality combinations has been thought through prior to their implementation, and does not offer support for making design decisions.

Finally, FAME (Duarte & Carrico, 2006) is a model-based architecture for adaptive multimodal applications, together with a set of guidelines for assisting the development process. Here, the emphasis is on the introduction of adaptive capabilities in the multimodal interface to accommodate diverse users and changes in environmental conditions.

In this chapter, we present a general usability framework, which can be used in the design and evaluation phases of a large range of multimodal systems. We then illustrate its use through the design of a multimodal mobile phone, which can accept voice commands, hand writing, and on-screen keypad selections.

BACKGROUND

Despite recent progress in enabling technologies, designing and building usable multimodal systems remain difficult. Many designers and developers still lack a clear understanding and the experience about how to successfully combine multiple modalities into one stream of data. In addition, despite continuing technological progress, recognition technologies (for example speech and gesture recognition) are far from perfect and still the source of errors. In this section, we briefly discuss three main difficulties for the design of usable multimodal systems: how to choose optimal combinations of modalities, multimodal integration, and error-proneness of multimodality.

Optimal Combinations of Modalities

Some input modalities are better suited than others in some situations, so modality choice is an important design issue (Oviatt et al., 2000). An increasing number of "user-friendly" systems, whether monomodal or multimodal, try to make the best use in their interface of the sensory channels that people commonly use for natural communication. For example, in these systems, touch-screen inputs will typically replace the traditional keyboard, as people normally use their fingers to indicate or select objects in the real world. This general principle can be applied to the selection of appropriate modalities in a multimodal system. However, the following two issues remain:

1.   In a multimodal system, designers must select not only individual modalities, but optimal combinations of modalities. For instance, combining speech and hand gestures is more natural and efficient than combining drawing and hand gestures.

2.   The nature of the tasks to be accomplished with the system must also be considered when choosing combinations of modalities. For example, on a digital map or a navigation system, combining speech and finger-drawing is particularly appropriate.

Furthermore, strong user individual preferences for certain modalities or combinations of modalities must be taken into account and catered for. User cultural background can also affect the appropriateness of some combinations of modalities. To better understand user preferences and culture, comprehensive multimodal corpora would be necessary. However, such corpora are not yet available. Optimal combinations of modalities are thus combinations that are natural and adapted to the task, but that are not prescriptive. Ultimately, users should be able to use the modalities of their choice.

Modality Integration

Multimodality is only effective when multiple inputs are integrated into a single unified stream of information. Past research on multimodal integration has approached the issue in two different ways: "early fusion" and "late fusion". Whilst the former reads multiple users inputs at once and try to integrate them whenever they occur, the latter processes multiple inputs in multiple recognisers, each of which independently reads the user input and passes it to a common Integrator.

A good example of early fusion is provided by Bregler et al. (1993) who suggested a more accurate human speech recognition system that collects both auditory data (speech) and visual data (lip-movement). As this system fuses these two different data into one set of information whenever human language is uttered, it is able to improve the overall recognition accuracy. However, early fusion has a critical limitation: it can only be applied to tightly coupled modality combinations, such as speech and lip movements. For modality combinations such as speech and hand-writing, it cannot be assumed that the two modalities will always be used at the same time, and early fusion cannot be applied (Oviatt et al., 2000).

The late fusion strategy enables the integration of more complex and asynchronous combinations of modalities. In late fusion systems, each modal input is collected by an independent recogniser, as in a monomodal system. After recognition, the independent recognisers transmit data to a common "Integrator", which synthetically considers the overall context in order to integrate and synchronise multiple user inputs. Late fusion is able to flexibly cope with various modality combinations and synchronisation patterns and, based on these, to understand users' intentions.

However, how to build efficient and accurate integrators is still the subject of intense research. For example, the QuickSet system's integrator (Cohen et al., 1997), which was designed to build a list of statistically prioritised candidate interpretations derived from each independent recogniser's recommendations, implements late fusion, but is limited to the specific combination of speech and gesture modalities.

Error-Proneness of Multimodality

The third issue concerns the error-proneness of multimodality. Multimodal systems are naturally error-prone because they deal with natural human behaviour, ambiguous user inputs, and are used in rapidly changing environments (mobile systems). In addition, multimodal systems that implement recognition-based modalities of interaction (for example speech) are subject to recognition errors. The possibilities of errors and misinterpretations in context aware and pervasive computing applications, where the capture and the analysis of passive modes are key, are even greater. Not only the computing devices have become invisible, but the users may not be aware of their behaviour that

is captured by the system. They may also have a wrong understanding of what data is captured by the various devices, and how it is used. In most cases, they do not receive any feedback about the system's status and beliefs.

In the context of speech and gesture interaction, Oviatt (1999) has shown that multimodality can in fact increase recognition success rate, as different inputs mutually disambiguate each other. It remains to demonstrate whether mutual disambiguation can operate within other types of modality combinations. Bourguet (2003a) has shown that mutual disambiguation can only be achieved if it has been enabled during the design of the multimodal system.

In conclusion, general usability engineering and information technology techniques are not sufficient to guarantee effective and usable multimodal interaction systems. A structured and adapted methodology is necessary, which enables us to draw a blueprint for the design and evaluation of multimodal interactions.

THEORETICAL CORNERSTONES

The usability framework we present in the next section of the chapter is based on two theories: the CARE (Complementarity, Assignment, Redundancy, and Equivalence) properties and the FSM (Finite State Machine) formalism.

CARE Properties

To date, the most general framework for reasoning about different combinations of modalities is the CARE (Complementarity, Assignment, Redundancy, and Equivalence) framework (Coutaz et al., 1995). CARE is particularly useful for describing relationships between different multimodal inputs at the level of elementary commands (for example commands to move an object on a screen or to open a new application). The fundamental assumption in CARE is that a multimodal interaction system comprises a number of basic elements such as states, goals, agents, and modalities. On this basis, CARE describes four different properties of multimodality: Complementarity, Assignment, Redundancy, and Equivalence. *Equivalence* characterises modalities that can be used interchangeably, enabling users to use any available modality to accomplish their goal. Equivalence usually expresses the availability of choice and allows users to exercise their natural intelligence about when and how to deploy input modalities effectively. *Complementarity* describes multimodal combinations of inputs where each modal input conveys only part of the message, but the combination of all the inputs results in a complete and comprehensible message. To be complementary, two inputs must be semantically rich, i.e. they must both carry significant information. When some modalities are better suited than others for expressing certain types of information, *Assignment* guarantees that each modality is used to its natural advantage. With Assignment, only one modality can be used to change the current state to the goal state. Finally, *Redundancy* describes the simultaneous use of modal inputs that convey semantically identical information. A classical use of Redundancy is for increasing interaction robustness.

Usually, a relation between two modalities, such as the relation of equivalence between speech and handwriting, only holds for a subset of the system's commands and tasks. For example, a multimodal interface may accept both speech and handwriting to input common command names (Equivalence), but rely on hand writing only to input proper names that the speech recognition system cannot be trained to recognise (Assignment). In other words, in a multimodal system, the four CARE properties co-occur. The task of the designer is to determine, at the level of each individual command, the optimal combinations of modalities and their CARE properties, which can yield the expected usability benefits of multimodal interaction. However, although CARE provides the means to describe the fundamental properties of multimodality, it does not include any concrete methodology for the design of multimodal interaction. One of our goals is to embody the CARE properties in a practical usability framework.

FSM Formalism

FSMs (Finite State Machines) are a well-known technique for describing and controlling dialogs in graphical user interfaces (Wasserman, 1985), but FSMs are also useful for modelling multimodal commands (Bourguet, 2003b), as illustrated in Figure 1. Figure 1 represents a speech and mouse multimodal command to move an object on screen, which allows various modality combinations and synchronisation patterns (for example, it allows users to initiate the command either by speech or by mouse).

The FSM formalism provides us with a means to rapidly prototype various multimodal interaction designs. FSMs are able to represent both sequential and parallel modality synchronisation patterns (Bourguet, 2004). Furthermore, FSMs are simple to understand, logically structured, and easy to manipulate and modify. Various multimodal combinations can be easily adjusted by designers. FSMs are good late fusion integrators. FSMs can also help testing a multimodal design against possible errors (Bourguet, 2003a). Finally, FSM representations can easily be translated into program codes (e.g. Java and XML) (Van Gurp & Bosch, 1999). Accordingly, FSMs are also appropriate for the development phase of multimodal systems.

As discussed above, the concept of FSM is both effective and potentially powerful for the design of multimodal systems. However, the practical application of FSM relies on the availability of a complete system architecture such as the one proposed in (Bourguet, 2003a). This architecture comprises: (1) a "Multimodal Engine" to control and format user inputs, (2) an "Interaction Model" to integrate inputs in various sets of FSMs, (3)

Figure 1. A set of FSMs representing a "move" command and using two different modalities (Bourguet, 2004)

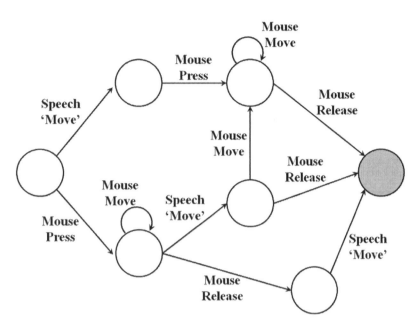

individual modality recognisers, and (4) an application to visualise and apply the results of modality integration.

The usability framework for multimodal interaction we present in the next section is based on both CARE and the FSM formalism.

USABILITY FRAMEWORK

In order to provide designers with a structured and general methodology both for the design and the evaluation of multimodal interaction, we present a usability framework, which comprises a number of steps. First, elementary multimodal commands are elicited using traditional usability techniques. Next, based on the CARE properties and the FSM formalism, the original set of elementary commands is expanded to form a comprehensive set of multimodal commands. Finally, this new set of multimodal commands is evaluated in two ways: user-testing and error-robustness evaluation. During the design process, six different stages of

multimodal commands are generated (see Figure 2): (1) Text-based Elementary Multimodal Commands, (2) FSM-based Elementary Commands, (3) Augmented Multimodal Commands, (4) User-tested Multimodal Commands, (5) Error-robust Multimodal Commands, and (6) Modelled Multimodal Commands.

Text-Based Elementary Multimodal Commands and FSM-based Elementary Multimodal Commands

The first step of the usability framework allows us to use a variety of traditional techniques such as task analysis, lo-fi prototyping, focus group discussion, and expert review, in order to elicit a number of elementary multimodal commands. The focus here is on finding and writing down possible multimodal activities for specific tasks, as would normally be done when designing monomodal interactive systems.

For example, for the task of making a call to a pre-defined speed dial number 1 with a multimodal

Figure 2. Conceptual usability framework

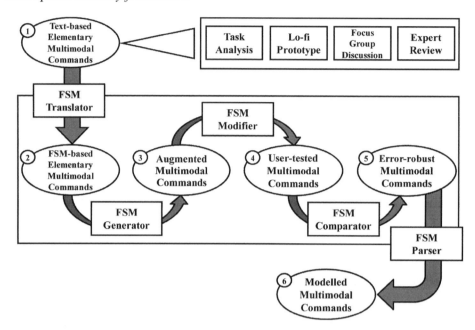

mobile phone equipped with keypads, voice recognition, and touch-screen, the task of the designer is to write down possible user action sequences. At this stage, there is no particular grammar or format for recording action sequences, and the various issues linked to multimodality design need not be considered.

Multimodal interaction modelling with the usability framework starts when these text-based elementary multimodal commands are translated into FSMs using the "FSM translator" (see Figure 2), which is a software application developed in Java for this research. The FSM Translator automatically translates the text-based elementary commands into a set of FSMs, and generates the second stage of multimodal commands: the FSM-based elementary multimodal commands.

Augmented Multimodal Commands

The purpose of the augmentation is to obtain a comprehensive set of multimodal commands based on the elementary multimodal commands elicited during the previous stage. Augmentation

is necessary because, even though classic user interaction design methodologies, such as task analysis and expert review are able to outline probable scenarios of multimodality, they cannot elicit all possible patterns of multimodal action sequences and modality combinations. The process of augmentation in the usability framework creates variations on the original set of commands and expands the size of the initial set of commands by generating new commands, which exhibit various CARE properties. Another software application, the "FSM Generator" has been developed to automate the augmenting process. The FSM Generator is also an application written in Java, which automatically augments multimodal commands and avoids inefficient manual modelling processes.

The augmentation process comprises three steps: (1) Importing elementary commands, (2) Extracting seed FSMs, and (3) Generating CARE FSMs. First, modality patterns in the original set of multimodal commands are analysed. Then, fundamental constituents of those patterns are extracted to obtain "seed FSMs". The FSM Generator

Figure 3. FSM Generator

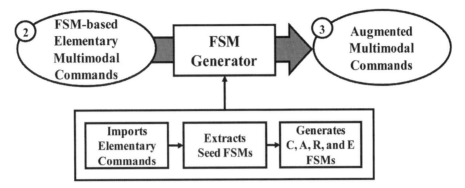

finally combines the various seed FSMs according to the rules of the CARE properties to generate additional multimodal commands. The third stage of multimodal commands is thus obtained – the augmented multimodal commands. More details on the augmentation process are provided in the Implementation section of the chapter. Figure 3 illustrates in a simple diagram the multimodality augmentation process and the role of the FSM Generator.

User-Tested Multimodal Commands

The augmented set of multimodal commands obtained in the previous step includes many possible multimodal scenarios, which need to be tested with real users' inputs. The primary aim of user testing is to filter out inappropriate or unnecessary multimodal commands from the augmented set. In addition, if it is found that users frequently activate multimodal commands that are not modelled in the augmented set of commands, these new user-generated commands are added. User-testing also allows us to elicit users' modality and modality combination preferences.

User testing can be done in a Wizard of Oz experiment (Bourguet & Chang, 2008). Firstly, a hi-fi prototype (Participant System) is required to provide participants with the virtual environment. Secondly, in order to run Wizard of Oz user testing sessions, a Facilitator System is needed that reads

and records various multimodal commands generated by the participants to the experiment. Lastly, as a separate software tool, the "FSM Modifier" (see Figure 4) is required to filter out inappropriate multimodal commands from the augmented set on the basis of collected user data, and to add new multimodal commands, which were activated by the participants during the experiment.

User testing should normally results in a much smaller and natural set of "user-tested multimodal commands". During this step, the usability framework provides a structured methodology to reflect users' multimodal behaviours and preferences upon the CARE-based augmentation.

Error-Robust Multimodal Commands

User testing in the previous step evaluates multimodal commands in isolation. Following this first evaluation, the usability framework runs another phase of evaluation to test commands in context. The second evaluation aims at making user-tested multimodal commands more robust against possible recognition errors and usability problems. Although user testing in the previous stage generated user-oriented multimodal commands, these are still error-prone. In particular, some of the user-tested multimodal combinations may be easily confusable with others. For example, if two multimodal commands contain speech inputs such as Speak "2" and Speak "to", which are highly

Figure 4. FSM modifier

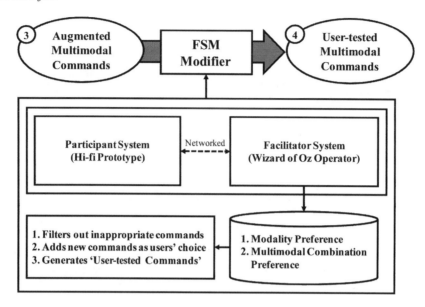

ambiguous, recognition errors are likely to occur. It is thus necessary to test multiple multimodal commands in context to detect potential conflicts and increase potential for mutual disambiguation between commands.

Because it is not possible for designers to manually test large sets of multimodal commands, we have developed another software tool, the "FSM Comparator" (see Figure 2), which exploits the FSM formalism to automate the evaluation process. Firstly, various user-tested multimodal commands across a number of multimodal tasks are merged into a single set of multimodal commands, to put them "in context". Secondly, simulated user inputs are generated to test the set of multimodal commands. Thirdly, with the help of the FSM Comparator, error robustness is evaluated and enhanced.

The FSM Comparator is able to automatically detect error-prone multimodal commands and correct them. The correction can be done in two ways: (1) Additional modality choices are added to the error-prone commands; (2) Modality combinations are modified to increase error-robustness.

Either way, the FSM Comparator updates all error-prone commands to allow them to disambiguate each other, and to prevent ambiguous or misrecognised inputs to trigger interaction problems. After completion of the error correction, the fifth stage of multimodal commands is obtained – the error-robust multimodal commands (Figure 5).

Modelled Multimodal Commands

In effect, the set of error-robust multimodal commands obtained in the previous step can be regarded as the desired outcome of the usability framework. However, if the size of the error-robust multimodal command set needs to be reduced, or if there are not clear differences between the error-robust commands and the initial set of elementary multimodal commands, iteration through the usability framework can be considered. In this case, the FSM Generator can be used to re-augment the error-robust set of multimodal commands, and further user testing and robustness evaluation must be performed.

Figure 5. FSM comparator

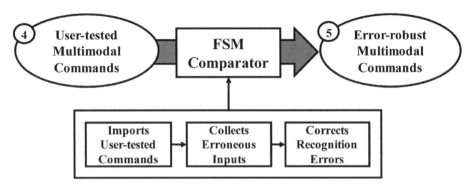

Once a satisfactory set of multimodal commands has been obtained, it is sent to the "FSM Parser" (see Figure 2), which translates the FSM-based commands into XML (eXtensible Markup Language). XML-based multimodal commands constitute the final output from the usability framework – the modelled multimodal commands, ready to be used in the development phase of the multimodal system.

IMPLEMENTATION

In order to realise the conceptual idea of the usability framework, we implemented several Java (Swing) applications such as the FSM Translator, FSM Generator, Participant System, Facilitator System, FSM Modifier, FSM Comparator, and FSM Parser (see Table 1).

FSM Translator

The main role of the FSM Translator is to translate text-based elementary multimodal commands into

Table 1. Implementation list for the usability framework

Application	Duty	Stage
FSM Translator	- Translates text-based elementary multimodal commands into FSM-based elementary multimodal commands	Stage 1 -> Stage 2
FSM Generator	- Analyses FSM-based multimodal commands - Generates augmented multimodal commands following the CARE properties	Stage 2 -> Stage 3
Participant System	- Acts as the Hi-fi prototype of a multimodal device for user testing - Collects users' multimodal behaviour and preferences	Stage 3 -> Stage 4
Facilitator System	- Receives user inputs from Participant System and gives back appropriate responses - Produces text-based user testing results	
FSM Modifier	- Filters out unnecessary augmented multimodal commands according to user-testing results and adds user generated commands - Generates user-tested multimodal commands	
FSM Comparator	- Merges several sets of user-tested multimodal commands to test them in context - Removes or corrects error-prone multimodal commands and generates error-robust multimodal commands	Stage 4 -> Stage 5
FSM Parser	- Parses error-robust multimodal commands to a XML file as the final output from the usability framework	Stage 5 -> Stage 6

Figure 6. FSM Translator. © 2009 Jaeseung Chang. Used with permission.

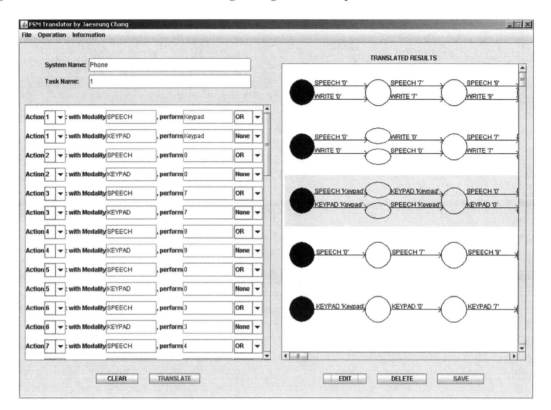

FSM-based elementary multimodal commands. This removes the burden from the designers to write FSMs by hand. The algorithm of the FSM Translator was developed to create FSM objects for each multimodal action written on the result sheet of the task analysis or expert review. As can be seen in Figure 6, action sequences can be entered on the left part of the application. Each action sequence includes four types of information: (1) The order number of the action sequence, (2) The name of the modality, (3) The input content, and (4) Relationship with adjacent inputs.

The order number of the action sequence is the ordinal number of each multimodal input. For example, it can be seen in Figure 6 that there are two rows of Action 1, which indicates that both voice utterance and keypad operation can be used for this action. The FSM Translator was developed to understand various natural expressions of modality names, such as "Say",

"Voice" and "Speech", which are all interpreted as "SPEECH". "Inputs content" contains the information as entered by the user. In Figure 6, we can see that users entered "keypad", "0", "7", and "9". Finally, the relationship with adjacent inputs indicates inputs which are equivalent, redundant or assigned. "None", "AND", and "OR" indicate Assignment, Redundancy, and Equivalence, respectively. When multimodal inputs are entered, they are translated into FSMs on the right side of the FSM Translator, where each FSM represents a different multimodal command (see example in Figure 7).

FSM Generator

The roles of the FSM Generator are to analyse the previously generated FSM-based elementary commands and to generate augmented multimodal commands according to the CARE properties. This is done in three steps: (1) Importing elementary commands, (2) Extracting seed FSMs, and (3) Augmentation.

The FSM Generator analyses the patterns of action sequences and modality combinations that were obtained from the FSM translator. In addition, it analyses the most frequent (preferred) modality combinations. The modal components of the multimodal commands ("seed FSMs") are extracted. While normal FSMs contain multiple action sequences and modalities, seed FSMs have only one pair of action sequence and modality (Figure 8). The purpose of the automatic extraction of seed FSMs is to obtain the fundamental constituents of elementary commands as the source for the augmentation process. Using these seed FSMs, the FSM Generator can generate a large number of new multimodal commands, which exhibit various CARE properties.

For the augmentation of complementary commands, the FSM Generator collects seed FSMs and builds all possible new FSMs that exhibit the Complementarity property. Firstly, it loads previously analysed patterns of action sequences.

Secondly, it generates all possible complementary commands by arranging seed FSMs onto the patterns of action sequences. The arrangement should follow the basic rule of complementary combination. For example, if one sequence uses the speech modality, the next sequence must use other modalities such as touch-screen or keypad in order to achieve complementarity.

The Assignment augmentation uses a similar algorithm, except that the FSM Generator generates multimodal commands, which only use a single modality for the entire action sequences.

For the augmentation of Redundancy and Equivalence commands, the FSM Generator uses a different algorithm. First, it merges the complementary and assigned augmented commands into one set of commands. Second, it randomly picks up some action sequences in each of the augmented commands, and applies the Redundancy or Equivalence rule. In case of redundancy, each selected action sequence is augmented with additional redundant modalities exhibiting an AND relationship. In case of equivalence, each selected action sequence is augmented with additional

Figure 7. A sample FSM-based elementary command

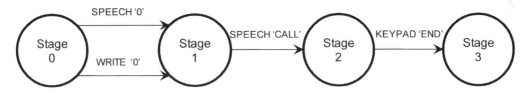

Figure 8. Four seed FSMs extracted from the FSM shown in Figure 7

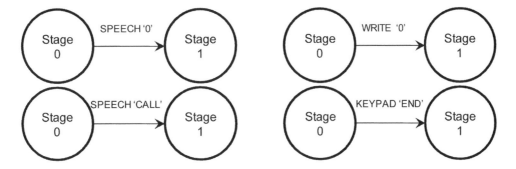

equivalent modalities exhibiting an OR relationship. The reason for randomly choosing a subset of action sequences is to avoid scalability problems. Generating both redundant and equivalent commands using all seed FSMs could potentially create several thousands of new augmented commands, and most of them would not be meaningful.

User Testing Applications and FSM Modifier

The first evaluation task of the usability framework is User Testing (UT) using the Wizard of Oz paradigm. For this purpose, two UT applications were implemented: a participant system and a facilitator system; as well as an "FSM Modifier" tool.

The participant system acts as a Hi-fi prototype, which UT participants operate during the UT sessions. As can be seen in Figure 9, the user interface to a mobile phone was implemented. This imaginary multimodal device uses three different modalities: voice command, touch-screen handwriting, and on-screen keypads. All users inputs entered on this system are sent to a facilitator (or wizard) system via a network connection.

A facilitator system was developed in order to receive users' inputs from the participant system and to send back the appropriate responses. In order to run Wizard of Oz experiment sessions, the facilitator system must be connected to the participant system via a UDP network connection. Whenever UT participants utter a voice command or operate the touch-screen, the facilitator remotely controls the participant system as if it was a working prototype with real multimodal features such as speech recognition and the detection of handwriting. In addition, the facilitator system is also in charge of collecting UT participants' multimodal behaviours and preferences and produce "user generated" text-based multimodal commands.

As mentioned earlier, the objective of the user evaluation is to flag inappropriate multimodal commands (i.e. commands that were never activated by users) and to add missing commands to

Figure 9. Participant system. © 2009 Jaeseung Chang. Used with permission.

the augmented set of commands obtained in the previous steps. To accomplish these goals, we developed an "FSM Modifier" tool, which compares the augmented commands with the user generated inputs. Prior to using the FSM Modifier, the text-based user generated multimodal commands collected by the facilitator system are translated into FSMs using the FSM Translator.

The FSM Modifier firstly analyses the patterns of action sequences and modality combinations in the user generated commands (i.e. it analyses user preferences). The FSM Modifier compares every augmented command with all patterns of action sequences and filters out the commands that do not match these patterns. For example, if we obtain an analysed pattern that corresponds to the following sequence: "1", "2", and "Send"; the augmented commands that do not follow this pattern will be flagged out (see Figure 10).

The FSM Modifier also analyses all augmented commands to check what combinations of modalities they include. For instance, if the user experiments show that users most frequently use speech and keypad in a redundant manner, the FSM Modifier will filter out some of the augmented commands that do not use speech and keypad redundantly. Depending on the original size of the augmented set of commands and on the reduction that must be achieved, the filtering process can be set to leave no more than 30% of action sequences that do not correspond to the preferred modality combinations.

Finally, the FSM Modifier adds user generated commands which were not originally included in the augmented set of commands. In other words, if a given multimodal command is commonly

used by several participants, but was not generated during the augmentation process, the FSM Modifier will automatically detect it and add it to the new set of "user tested" commands. This new set of commands should be significantly smaller than the augmented set obtained in the previous stage.

FSM Comparator for Robustness Evaluation

In order to conduct the second evaluation phase of the framework (error robustness evaluation), we implemented an "FSM Comparator" tool. The main task of the FSM Comparator is to correct error-prone multimodal commands in order to make the multimodal interaction more robust to probable recognition errors. The algorithm used in the FSM Comparator comprises four parts: (1) Importing user-tested commands, (2) Merging all user-tested commands into a single set, (3) Dispatching error-prone sample inputs to the set of commands, and (4) Error correction.

The FSM Comparator merges all user-tested commands into a single set of commands so they can be tested "in context". Error-prone inputs are

Figure 10. Filtering out inappropriate augmented multimodal commands

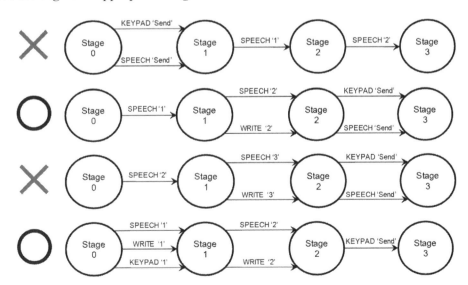

then dispatched to the set in order to observe its potential for mutual disambiguation. Example of an error-prone input might be the numeric value 1 entered using the touch-screen hand-writing feature, which is highly confusable with the alphabetic value I. The FSM Comparator checks if the set of commands will naturally be able to select the correct input interpretation via mutual disambiguation. If mutual disambiguation is not found to be operating efficiently, it can be enhanced by adding redundant modalities, or by modifying the modality combinations of some of the FSMs (Bourguet, 2003a).

While comparing all user-tested multimodal commands with the collected sample user inputs, the FSM Comparator tries to correct erroneous inputs in two ways: (1) By adding redundant modalities, and (2) By providing alternative combinations of modalities.

The first correction method (adding redundant modalities) can be applied to the commands that exhibit the Complementarity or the Assignment properties. Since these commands only have one modality available per action sequence, they cannot solve recognition error problems by themselves. For this reason, the FSM Comparator adds redundant modalities to each error-prone action sequence. For example, in the above case of writing a number 1 and a letter I, the FSM Comparator can add a redundant speech modality (Speak "one" and Speak "I") to help differentiate between the two.

The second correction method (providing alternative combinations of modalities) can be applied to commands, which exhibit the Equivalent property. Although these commands comprise more than one modality choice, the use of multiple modality remains optional (user choice). When a command proves to be error-prone, the FSM Comparator can forcibly change the Equivalent multimodal command into a Redundant multimodal command. For example, the redundant combination of speaking "1" and writing 1 can be used, instead of their equivalent combination.

FSM Parser

Finally, the FSM Parser's main task is to parse FSM-based error-robust multimodal commands into the XML (eXtensible Markup Language) format. This format, as the final output from the usability framework, is expected to be more readily usable in the implementation phase of the multimodal interaction system. While parsing the commands, the FSM Parser also automatically generates DTD (Document Type Definition) code. However, it does not produce any particular XSL (eXtensible Stylesheet Language) for the generated XML files, as we expect that the results from the usability framework will be used in various design and implementation environments.

CASE STUDY

Multimodal Mobile Phone

To illustrate the use of the usability framework, we have designed a case study using an imaginary multimodal mobile phone. As walk-up-and-use hand-held devices, mobile phones have very limited user interface capabilities. Due to their small size, mobile phones usually have small screens and keypads, which can cause various usability problems. However, when carefully designed, multimodality can help enhance the usability of mobile phones. We aim to show how the usability framework described in the previous sections can be used to design multimodal interaction on a mobile phone.

In this case-study, the mobile phone implements three different modalities of interaction: speech, hand-writing on its touch-screen, and on-screen keypads. In addition, we assume that it is mostly used in very busy and noisy environments, causing users to experience difficulties when operating the device using a single modality: speech recognition is very error-prone in noisy environments, and the touch-screen/keypad manipulations are also subject to input errors because of their small size and the users' busy circumstances. However, we hope that the interaction can be made more robust by the use of two or more modalities at the same time and the use of adequate combinations of modalities.

First of all, according to the usability framework, user interaction design must start using traditional usability techniques such as Task Analysis, Expert Review, Focus Group Discussion, Lo-fi Prototyping, etc. We conducted a Task Analysis session using a draft sketch (Lo-fi Prototype) of the multimodal mobile phone (see Figure 11) to generate text-based elementary multimodal commands.

The Lo-fi Prototype helped us visualise the probable features and form factors of the multimodal device. We then selected three user tasks

Figure 11. Lo-fi prototype

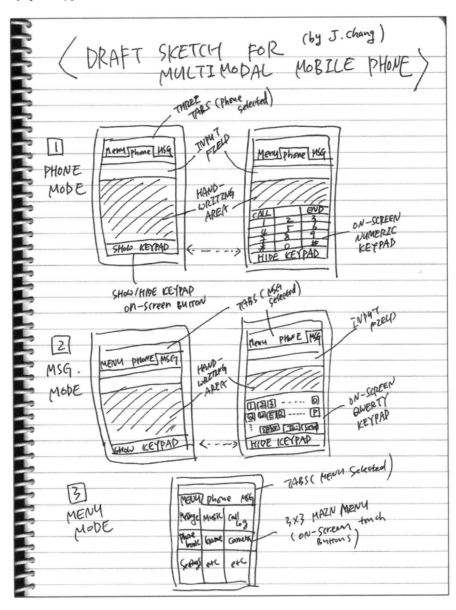

for which the necessary user-device interaction had to be designed using the usability framework: (1) Making a call to the number "07903421873"; (2) Composing the text message "hello there"; and (3) Sending a previously composed message to the number "07903421873", and checking the sent message from a "Sent Message" folder. We conducted a Task Analysis to collect possible multimodal interactions for each of the tasks. During the Task Analysis, we focused on how users would multimodally interact with the device: probable modality combinations and synchronisation patterns to complete the three tasks. As a result, we generated the text-based elementary multimodal commands (stage 1 of the usability framework) for the multimodal mobile phone. The task analysis generated a total of twenty elementary multimodal commands (six for task 1, seven for task 2, and seven for task 3), and with the use of the FSM Translator, these elementary commands were translated into twenty FSM-based representations.

The next step consists in augmenting the FSM-based elementary multimodal commands with the use of the FSM Generator. From the elementary commands, the FSM Generator extracted a total of 84 seed FSMs: 28, 20, and 36 seeds respectively for tasks 1, 2, and 3's elementary commands.

After augmentation based on the CARE properties, a total of 924 augmented multimodal commands were obtained (see Table 2). Table 2 shows how the number of multimodal commands increases after the augmentation process. As explained earlier, to avoid scalability problems, the augmentation was limited to about 300 commands for each task.

Wizard of Oz User Testing

The augmented set of multimodal commands had then to be tested through Wizard of Oz experiment, for which six participants were recruited. The hi-fi prototype shown in Figure 9 (participant system) was used for these experiments. It was connected to a facilitator system via a UDP network connection. In cases of hand-writing and keypad pressing inputs, the facilitator system automatically recorded and logged the participants' inputs. However speech commands had to be recorded and logged manually by the wizard, as the facilitator system was not able to recognise the voice commands.

In this way, all user inputs were recorded, including the following information: modality combinations for each task, concrete action sequences, and exact timings for each action sequence. At the same time, video recordings of all sessions were made. Finally, the results from all sessions of User Testing were stored in text files, which were automatically generated by the Facilitator System. As the results from User Testing were text-based, they had then to be translated into FSM-based data using the FSM Translator.

The FSM-based user-generated commands were then given to the FSM Modifier for analysis. The FSM Modifier automatically analysed the users' multimodal behaviours such as preferred action sequences and most commonly used multimodality combinations. A new set of "user-tested" commands was thus obtained, which contained 8, 9, and 15 commands for tasks 1, 2, and 3 respectively.

Table 2. Number of elementary and augmented multimodal commands

	Elementary FSMs	Seed FSMs	Augmented FSMs
Task 1	6	28	309 (C=99, A=4, R=103, E=103)
Task 2	7	20	306 (C=98, A=4, R=102, E=102)
Task 3	7	36	309 (C=99, A=4, R=103, E=103)

Robustness Evaluation

Robustness evaluation addresses probable recognition errors and usability problems in a given set of multimodal commands. The FSM Comparator was used to activate the multimodal commands using ambiguous inputs and observe the occurrences of mutual disambiguation within the set. Several ambiguous inputs were used to test the set of commands: handwritten 1 and I; spoken "o" and "0"; and spoken "2" and "to". After correction, we obtained a new set of error-robust multimodal commands, which could be parsed into XML.

CONCLUSION

Although it is well known that multimodal interaction has the potential to enhance the usability of many types of interactive systems, especially small mobile devices, the task of designing the interaction is a difficult one. The difficulties mainly consist in selecting appropriate and optimal combinations of modalities, dealing with error-prone and ambiguous user inputs, and integrating and interpreting heterogeneous data.

In response, we suggested in this paper a usability framework for the design and evaluation of multimodal interaction. The framework comprises six different stages of multimodal command design, which make the design process more easily tractable and modular. Furthermore, using the usability framework does not require any prior knowledge about multimodality design, as it incorporates traditional usability techniques such as task analysis, prototyping, and Wizard of Oz experiment. Its modular structure enables a number of iterations through the various stages of the design process. Automatic augmentation provides a rigorous and efficient way of including various CARE properties in the interaction design. The usability framework also addresses the evaluation of alternative designs, both in terms of user testing and robustness evaluation. Finally,

the use of the FSM formalism enables us to logically process various multimodal commands and to produce XML codes as the final output of the whole design/evaluation process.

FUTURE WORK

Future evolution of the work presented in this chapter will develop along three complementary axes.

First of all, we acknowledge the importance of taking the design process out of the lab. This is particularly true of context aware and mobile devices, for which a large variety of usage contexts need be considered. In the future, our usability framework will be able to encompass a full representative range of usage scenarios to better support the interaction design and evaluation tasks of mobile devices. A good starting point is provided in (de Sa & Carrico, 2008) where guidelines aimed at aiding designers on the scenario generation and selection process during design and evaluation of mobile applications are suggested.

Secondly, it is to be expected that these more realistic usage scenarios will make the use of passive modalities, i.e. implicit information which directly emanate from the context of use, more prominent. The usability framework remains to be tested, where active modalities (speech, handwriting, etc.) and passive modalities (ambient noise, location, luminosity level, etc.) can readily cooperate and be integrated. We anticipate new issues arising, such as the need for prioritisation when conflicting active and passive modality inputs will co-occur.

Finally, the very large range of possible utilisation contexts of mobile devices, combined with the unpredictability of passive modalities, has given rise to new research approaches, in which the system is more autonomous and let to choose the optimal modality or modality combination on the fly, according to some contextual parameters. In its current state, our usability framework stresses the need to increase system robustness by achieving a fine balance between modality usage flexibility (modality choice) and constraint. In the future, we will also consider alternative approaches to render the framework usable in the design of systems that make interaction modality choices at runtime.

REFERENCES

Bouchet, J., Nigay, L., & Ganille, T. (2004). ICARE software components for rapidly developing multimodal interfaces. In *Proceedings of the 6th international conference on Multimodal interfaces* (251-258).

Bourguet, M. L. (2003a). How finite state machines can be used to build error free multimodal interaction systems. In *Proceedings of the 17th British HCI Group Annual Conference* (pp. 81-84).

Bourguet, M. L. (2003b). Designing and prototyping multimodal commands. In . *Proceedings of Human-Computer Interaction INTERACT, 03,* 717–720.

Bourguet, M. L. (2004). Software design and development of multimodal interaction. In *Proceedings of IFIP 18th World Computer Congress Topical Days* (pp. 409-414).

Bourguet, M. L., & Chang, J. (2008). Design and usability evaluation of multimodal interaction with finite state machines: A conceptual framework. *Journal on Multimodal User Interfaces, 2*(1), 53–60. doi:10.1007/s12193-008-0004-2

Bregler, C., Manke, S., Hild, H., & Waibel, A. (1993). Bimodal sensor integration on the example of `speechreading'. In *Proceedings of IEEE Int'l Conference on Neural Networks* (pp. 667-671).

Cohen, P. R., Johnston, M., McGee, D., Oviatt, S., Pittman, J., Smith, I., et al. (1997). QuickSet: Multimodal interaction for simulation set-up and control. In *Proceedings of the Fifth Conference on Applied Natural Language Processing* (pp. 20-24).

Coutaz, J., Nigay, L., Salber, D., Blandford, A., May, J., & Young, R. M. (1995). Four easy pieces for assessing the usability of multimodal interaction: The CARE properties. In . *Proceedings of INTERACT, 95,* 115–120.

de Sa, M., & Carrico, L. (2008). Defining scenarios for mobile design and evaluation. In *Proceedings of CHI '08, SIGCHI Conference on Human Factors in Computing Systems* (pp. 2847-2852).

Duarte, C., & Carrico, L. (2006). A conceptual framework for developing adaptive multimodal applications. In *Proceedings of the 11th international conference on Intelligent User Interfaces,* (pp. 132-139).

Feki, M. A., Renouard, S., Abdulrazak, B., Chollet, G., & Mokhtari, M. (2004). Coupling context awareness and multimodality in smart homes concept. *Lecture Notes in Computer Science, 3118,* 906–913.

Flippo, F., Krebs, A., & Marsic, I. (2003). A framework for rapid development of multimodal interfaces. In *Proceedings of the 5th International Conference on Multimodal Interfaces* (pp. 109-116).

Kapoor, A., & Picard, R. W. (2005). Multimodal affect recognition in learning environments. In *Proceedings of the ACM MM '05* (pp. 6–11).

Oviatt, S. (1999). Mutual disambiguation of recognition errors in a multimodal architecture. *Proceedings of the SIGCHI Conference on Human Factors in Computing Systems: the CHI is the Limit* (pp. 576-583).

Oviatt, S., Cohen, P., Wu, L., Vergo, J., Duncan, L., & Suhm, B. (2000). Designing the user interface for multimodal speech and pen-based gesture applications: State-of-the-art systems and future research directions. *Human-Computer Interaction, 15*, 263–322. doi:10.1207/S15327051HCI1504_1

Salber, D. (2000). Context-awareness and multimodality. In *Proceedings of First Workshop on Multimodal User Interfaces.*

Sinha, A. K., & Landay, J. A. (2003). Capturing user tests in a multimodal, multidevice informal prototyping tool. In *5th International Conference on Multimodal Interfaces* (pp. 117-124).

Van Gurp, J., & Bosch, J. (1999). On the implementation of finite state machines. In *Proceedings of the 3rd Annual IASTED International Conference on Software Engineering and Applications* (pp. 172-178).

Wasserman, A. (1985). Extending state transition diagrams for the specification of human-computer interaction. *IEEE Transactions on Software Engineering, 11*(8), 699–713. doi:10.1109/TSE.1985.232519

Section 4
Applications and Field Reports

Chapter 9
Exploiting Multimodality for Intelligent Mobile Access to Pervasive Services in Cultural Heritage Sites

Antonio Gentile
Università di Palermo, Italy

Antonella Santangelo
Università di Palermo, Italy

Salvatore Sorce
Università di Palermo, Italy

Agnese Augello
Università di Palermo, Italy

Giovanni Pilato
Consiglio Nazionale delle Ricerche, Italy

Alessandro Genco
Università di Palermo, Italy

Salvatore Gaglio
Consiglio Nazionale delle Ricerche, Italy

ABSTRACT

In this chapter the role of multimodality in intelligent, mobile guides for cultural heritage environments is discussed. Multimodal access to information contents enables the creation of systems with a higher degree of accessibility and usability. A multimodal interaction may involve several human interaction modes, such as sight, touch and voice to navigate contents, or gestures to activate controls. We first start our discussion by presenting a timeline of cultural heritage system evolution, spanning from 2001 to 2008, which highlights design issues such as intelligence and context-awareness in providing information. Then, multimodal access to contents is discussed, along with problems and corresponding solutions; an evaluation of several reviewed systems is also presented. Lastly, a case study multimodal framework termed MAGA is described, which combines intelligent conversational agents with speech recognition/ synthesis technology in a framework employing RFID based location and Wi-Fi based data exchange.

DOI: 10.4018/978-1-60566-978-6.ch009

INTRODUCTION

Imagine your last visit at a popular art museum. You approach the cashier to buy your ticket and the lady offers, at an extra cost, to rent an audio guide. You take it and move to the entry gate where a guard check your ticket and let you in. Your visit then begins, and here you are, fiddling with the buttons to find the beginning of your story. As you wade between halls and paintings you keep punching buttons, following the numbered path, and hearing a fascinating voice narrating the wonder behind each piece of art. "What did she just say?" Not a problem, you simply push that button again and here is she untiringly repeating the story. "So she is really referring to that other painting I saw two or three halls ago" – you find yourself thinking – "what was the name of the artist?" You can't remember and look for a clue in your keypad to only find anonymous numbers. You end up putting your question away and continue the visit. Can technology advances help us envision better scenarios in the next future? We believe the answer is yes, and multimodal mobile devices will play a key role. As this chapter will attempt to show, in a non-distant future the same you will now enter the museum with your smartphone in your pocket and wearing a headset. The entry gate will now open as you pass through, activated via a Bluetooth exchange that checks your e-ticket, activating your virtual guide. Your visit begins, and you move through the halls following a path in the museum map, which you previously set while planning your visit from home. A wi-fi connection updates your screen as you wade, and the your guide, as before, is telling about what you see. This time, when you ask, an answer is given and further details materialize on your screen if so you desire…Seems too far-away? Well all of the technologies involved are already here, as we will see in the rest of the chapter.

Such technologies, both hardware and software, made the Mark Weiser's vision of Ubiquitous Computing real and more and more available for users in their in everyday life. The Ubiquitous Computing paradigm relies on a framework of smart devices that are thoroughly integrated into common objects and activities. Such a framework implements what is otherwise called a pervasive system, which main goal is to provide people with useful services for everyday activities.

As a consequence, the environment in which a pervasive system is operating becomes augmented, that is enriched by the possibility to access additional information and/or resources on a per-needed basis. Augmented environments can be seen as the composition of two parts: a visible part populated by animated (visitors, operators) or inanimate (artificial intelligence) entities interacting through digital devices in a real landscape, and an invisible part made of software objects performing specific tasks within a underlying framework. People would perceive the system as a whole entity in which personal mobile devices are used as human-environment adaptable interfaces.

The exponential diffusion of small and mobile devices, third-generation wireless communication devices, as well as location technologies, has led to a growing interest towards the development of pervasive and context-aware services. There are many domains where pervasive systems are suitably exploited. One of the most recent and interesting applications of pervasive technology is the provision of advanced information services within public places, such as cultural heritage sites or schools and university campuses. In such contexts, concurrent technologies exploited in smart mobile devices can be used to satisfy the mobility need of users allowing them to access relevant resources in a context-dependent manner. Of course, most of the constraints to be taken into account when designing a pervasive information providing system are given by the actual domain where they are deployed.

Cultural Heritage applications pose tremendous challenges to designers under different aspects. Firstly, because of the large variety of visitors they have to deal with, each with specific needs and expectations about the visit. Secondly, no two sites are the same, and pretty much you need a framework that can easily produce a new installation given the site characteristics (indoor versus outdoor, distributed versus centralized, individual centered versus group centered, etc.). Lastly, the technologies involved must be robust to failures, redundant and, above all, easy and intuitive to use. These are the reasons why there are several research groups that are focusing their attention on this applicative domain. It is a good test bed to validate almost all models and design choices.

As said elsewhere in this book, "software applications or computing systems that combine multiple modalities of input and output are referred to as multimodal." Being free to choose among multiple modes to interact with such augmented environments and systems is crucial to their wider acceptance by a vast variety of users. This is often the case when such systems are intended for mass fruition at museums or cultural heritage sites. Multimodality is the capability of a system to allow for multiple modes of interaction, from traditional point-and-click to voice navigation or gesture activation of controls. A pervasive system targeted to cultural heritage fruition, therefore, can no longer be designed without first addressing how it will handle interaction with users. Additionally, multimodality relies on redundant information, resulting on more dependable systems that can adapt to the needs of large and diverse groups of users, under many usage contexts.

In this chapter, we will first offer an overview of the evolution of hardware and software technologies relevant to the field of Cultural Heritage fruition, reviewing some of the most relevant projects. The review will start from basic web-based interfaces, and then move to second-generation applications that exploit virtual reality. This give users immersive experiences, such as archaeological site navigation, time and space travel, and 3D reconstruction of ancient artifacts. The review

will end with a discussion of some recent third-generation systems, which aim at integration of human factors and computer science. Such systems provide users with a wider range of context-related services tailored on users' profiles.

The second part will focus on multimodality as a key enabler for a more natural interaction with the virtual guide and its surrounding environment. In this second part, we also evaluate the approaches examined in the chapter in terms of multimodality, presence of pervasive access to contents, and intelligent handling of interaction with the user. Discussions above will give the readers an adequate background on issues and problems researchers deal with when designing multimodal systems for information provision in cultural heritage sites.

In the third part, we will discuss how our research group intended this envisions, presenting our designs and their implementations. In particular, we will step into the development of a pervasive and multimodal virtual guide that integrates different cooperating technologies. Our main goal is to offer a more natural interaction between users and information access services compared to traditional pre-recorded, audio-visual guides. In more detail, we exploit hardware and software technologies, such as personal mobile devices (PDAs and cellular phones), speech recognition/synthesis and intelligent conversational agents, by integrating them into a single, easy-to-use interface. In a single, design the system overcomes limitations of PDAs and Smartphones (due to small screens and dimensions) integrating voice activated commands and navigation by means of Automatic Speech Recognition (ASR) and Text-To-Speech (TTS) embedded technologies. The system maintains context sensitiveness by means of RFID technology, which allows a mobile device to estimate its position within the environment and to detect context changes occurring where the user currently is. This way the system behavior is effectively adapted to user's needs.

The chapter will end with a discussion on the lessons learnt and the challenges facing large scale development of pervasive multimodal systems for cultural heritage applications.

CULTURAL HERITAGE FRUITION IN THE NEW MILLENNIUM: A TIMELINE

During the last decade, several workgroups focused their research activities on the definition of models to be implemented with the aim at providing users of a cultural heritage site with some kind of useful services. In this section we will introduce some of the resulting projects we consider relevant.

We start our review of applications for cultural heritage fruition at the onset of the new millennium, for no better reason than the availability of the first systems integrating multiple hardware/ software technologies onto a single mobile tour-guide with some interaction capability. As a matter of fact, trivial automatic guide systems proliferated in archaeological sites during the last decade of twentieth century. They were mainly based on pre-recorded audio files played by fixed kiosk-style devices or portable headsets. Such systems were completely unaware of the context, thus forcing users to follow timing and flow of the narration (often users found themselves looking around to identify items the guide had been talking about).

In the following sub-sections we present the evolution of enabling technologies in the field of cultural heritage from the year 2001 to nowadays. Specifically, we will focus on technologies connected to access devices, positioning/location, human-environment interfaces, and ambient intelligence. In order to support our review, we will discuss a selection of that we deem best representing the evolution of systems for fruition of cultural heritage sites. Figure 1 depicts the timeline we will follow, along with the projects discussed in this chapter.

2001-2002

One of the first systems that integrates different technologies to provide visitors at cultural heritage site with brand-new services, is Archeoguide (Augmented Reality based Cultural Heritage On-site Guide) (Vlahadis, 2001; Vlahadis, 2002). Its main objective is to offer visitors a feeling of how a given site was at different points in time. Archeoguide is a client-server system for personalized mobile touring and navigation in archaeological sites, which uses both augmented and virtual reality technologies on different location-sensitive devices.

The advent of wireless connections plays an important role in content fruition: it allows for dynamic contents access and adaptation to both the actual context and the user needs. For example, in Hyperaudio and HIPS (Petrelli, 2001)

Figure 1. Timeline of discussed projects

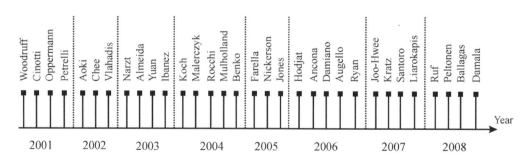

presentations composed of audio, images and hyper textual links, are dynamically generated on mobile devices according to the analysis of the context based on user interaction.

In order to give users more immersive experiences, Cinotti (2001) presented MUSE (MUseums and Sites Explorer), an experimental mobile multimedia system for information retrieval in cultural heritage sites, offering virtual reality and multimedia tours on e-book terminals.

In the same year, Oppermann (2001) proposes Hippie, a nomadic museum guide. It is an information system in which the user accesses his or her personal information space anywhere and independently from the particular access device.

As far as human-computer interaction is concerned, Woodruff (2001) and Aoki (2002) addressed the issue of social interaction during museum tours. They enabled social interactions providing visitors with an electronic guidebook. In its first release, the electronic guidebook delivered audio contents through speakers in an open air modality. The evolution of this system, called Sotto Voce, enables visitors to share audio information using a audio eavesdropping mechanism. The aim is to augment user satisfaction using museum assistant devices, overcoming the hindrances of traditional pre-recorded museum audio-guide.

In this period, the key enabling technology is undoubtedly the wireless connection among devices. This allowed designers to start thinking about dynamic content fruition in mobility. First implementations provided that users must carry a large set of hardware, such as a laptop in a backpack, a long GPS antenna above their heads, and some handy I/O interface (Vlahadis, 2001; Vlahadis, 2002). This opened a new research trend, focused on the set up of visual interfaces to be visualized on a portable device. Furthermore, more interaction modes than point-and-click were hoped, thus giving a stimulus to implement new techniques to this end.

2003-2004

The main lack of pre-recorder audio/video guides, is that they provide users with static contents, often triggered by the users themselves, apart from their profiles. As a consequence, one of the main goal of guide systems is to take into account the user profile, in terms of *who* is him, *what* he is looking for, *where* is him and *how* to present information to him. In other words, researchers had to define elements of the *context* to be considered for information composition and presentation. Since the user position within a site is a key data to define the context, most of systems proposed in this period start to show basic context awareness by means of location technologies.

In 2003, Narzt et al. (2003) propose a framework for mobile navigation systems that pervasively finds out both position and orientation of the user from any sensory source. It improves the association to the real world by merging video techniques and 3D-graphics. Even if virtual environments are extremely realistic, it is desirable to link objects and locations to specific narrations so to make the visit more interesting and amusing for the user. Along this line, Ibanez et al. (2003) presented intelligent storytellers for virtual environments that are able to narrate stories and to describe items, each with its own perspective. Rocchi et al. in (2004) addressed the issue of the seamless interleaving of interaction between mobile and fixed devices. In particular, they focus on a museum tour as application domain, offering a dynamic presentation of contents, while assuring its coherence throughout the visit.

During this period, studies on conversational agents begin to be used to handle the interaction process in a more natural way. The use of this technology leads to interactive virtual guides that allow users to ask questions during the visit. In (Yuan, 2003) the user interacts with a conversational agent that guides him during his virtual tour inside an art gallery. The agent tries to identify the intentions of the user through the analysis of his input. If this goal is reached, the reply of the agent is more accurate; otherwise it is less satisfactory if user queries are expressed using several keywords.

This period witnessed also the appearance of cellular phones as terminals of choice for their widespread use and increasing computing power, display and memory capabilities. Koch and Sonenberg (2004) used the standard Multimedia Messaging Service (MMS), along with other cooperating technologies to create a pervasive museum guide system. In (Mulholland, 2004), an intelligent support for cultural heritage story research and exploration is presented. Their "Story Fountain" system allows the user to obtain information trough the research and the exploration of stories about the cultural heritage site.

This is the age when personal mobile devices, such ad PDAs and smartphones, are more and more accepted among common people. Almost everyone has a small wearable computer in his pocket, wirelessly connected with the surrounding digital world, and with an easy-to-use interface. This allowed designers to start thinking about them as the enabling media both for *explicit* and *implicit* interaction. In the first case, personal mobile devices are used as I/O devices where ad-hoc user interfaces can be deployed. In the second one, such devices are exploited to build the context and to infer what users expect next. This is the case of indoor and short-range positioning systems based on RF measurements, and of user behavior tracking systems based on wireless radio MAC address.

2005-2006

The growing need of information personalization, resulted in the need of a more accurate context definition and detection. This led to the consideration of more than one technology at a time, concurrently exploited to implement a multi-channel interaction. As a matter of fact, during this timeframe we witnessed the spreading of several technologies

(Wi-Fi, Bluetooth, RFID, GPS, voice recognition and synthesis, image recognition, conversational agents) and smart access devices (such as PDA and smartphones) that more and more get integrated into enhanced systems.

Farella et al. (2005) presented a work that exploits widely available personal mobile devices (PDAs and cellular phones) and software environments (usually based on Java) to create highly interactive Virtual Heritage applications based on popular, low-cost, wireless terminals with highly optimized interfaces. In 2006 Ancona et al. (2006) proposed a system called AGAMEMNON that uses mobile phones equipped with embedded cameras to enhance the effectiveness of tours of both archaeological sites and museums.

The evolution of speech synthesis techniques allow for a new multimedia content delivery way, based on text-to-speech technology. This allows providing the user with new services accessible also by telephone. This solution tries to overcome the well-known limits of static pre-recorded audio guides, by vocally presenting a text that can be dynamically composed. Using this technology, Nickerson (2005) proposes the system named History Calls. He developed an experimental system using VoiceXML technology that is capable to deliver an automated audio museum tour directly to cell phones.

The main target of some research group is to make the visit within a cultural heritage site as more natural as possible for the largest part of users, avoiding the use of ad-hoc or complex devices. This goal can be achieved by integrating different well-known technologies in a different way.

In 2006, authors of (Augello, 2006) proposed a multimodal approach for virtual guides in cultural heritage sites that enables more natural interaction modes using off-the-shelf devices. The system, termed MAGA, integrates intelligent conversational agents (based on chatbots) and speech recognition/synthesis in a RFID-based location framework with Wi-Fi based data exchange. Moving along the line of a more natural interaction,

Jones et al. (2005) proposed a system in which multiple virtual guides interact with visitors, each characterized by its own personality. In their system, it is then possible to interact with two intelligent agents, with divergent personalities, in an augmented reality environment. Another example of system that is capable to adapt itself to the user behavior is CRUSE (Context Reactive User Experience) proposed by Hodjat in (2006). It is a user interface framework that enables the delivery of applications and services to mobile users.

In Dramatour (Damiano, 2006), Carletto, a virtual guide for the Savoy apartments of Palazzo Chiablese (in piazza Castello - Torino) provides users with information on portable devices presenting them in a dramatized form (being himself a small spider, long time inhabitant of Palazzo Chiablese).

2007-2008

The past two years see the advent of ambient integrated guides that are based on mixed reality, where real physical elements interact with virtual ones. In (Liarokapis, 2007) authors focus their work on the design issues of high-level user-centered Mixed Reality (MR) interfaces. They propose a framework of a tangible MR interface that contains Augmented Reality, Virtual Reality and Cyber Reality rendering modes. This framework can be used to design effective MR environments where participants can dynamically switch among the available rendering modes to achieve the best possible visualization. In (Santoro, 2007) Santoro et al. propose a multimodal museum guide that provide user with gesture interaction mode: using a PDA equipped with a 2D accelerometer, they control interface and content navigation by means of hand tilt gestures.

Damala et al. in (2008) present a prototype of an augmented reality mobile multimedia museum guide, also addressing the full development cycle of the guide, from conception, to implementation,

testing and assessment of the final system. They use last-generation ultra-mobile PCs, which allow designers to exploit some of the most powerful software technologies, such as OpenCV for video acquisition, ARToolkitPlus for tracking the paintings, OGRE3D for the insertion of the virtual objects, Open AL for the audio output and XERCES for XML document parsing.

In these last years attention of researchers in the field of pervasive service provision moved to an higher level of abstraction, mainly a middleware level. Software technologies have been exploited to give systems more adaptability, usability and, above all, intelligence. These features are the more promising ones in order to achieve the naturalness needed to get the largest acceptance of virtual guide systems among common people. As a consequence, the current research interests in the field of service provision in cultural heritage sites are mainly aimed at reaching these features. A common line researchers are following is the concurrent use of multiple hardware and software technologies at a time. The goal is to give users the possibility to interact in multiple modes, according to their habits, skills and capabilities. This will have the desired side effect to expand the number of prospective users, allowing also disabled ones to exploit such guide systems with their residual capabilities.

EXPLOITING MULTIMODALITY: INTERACTING IN MULTIPLE WAYS

The simplest way to support users during their tours of cultural heritage sites is to provide them with pre-arranged contents delivered through kiosk-style or headset devices in audio/video format. This solution, however, has several limits, despite being currently the most largely adopted inside museums. In the first place, it restricts visitor mobility to the fixed standing area in front of the exhibit, where information can be accessed through the kiosk. Secondly, contents is statically pre-arranged, no matter who is the actual recipient and his/her specific preferences and needs. Lastly, users are allowed limited or no interaction at all with such systems, often unable to even pause the record being played.

Multimodal and context-aware interaction allows to overcome these drawbacks. As a matter of fact, during the last decade, the advent of many technologies (such as wireless communications or identification with radio frequency) as anticipated in the timeline section, has contributed to make more real the Mark Weiser's vision of ubiquitous computing. From a practical point of view, we have witnessed the development of a wealth of applications, enabled by some of these technologies, all aimed at simplifying human-system interaction and improving the user's experience and satisfaction.

The overall goal of designing a modern visitor guide for cultural heritage sites is to provide users with a personalized fruition experience, co-involving as many of the human interaction modes as possible, while offering in-depth, context based contents delivery. Seamless integration of different technologies, both hardware and software, allows for the creation of personal virtual assistants guiding people during the visit. Such virtual guides may tell enchanting tales about the artifacts on display, or even entertain people with amusing comments to cover for small network or power failures.

Mobility within the site is another key aspect to address, either by using locally-available, dedicated mobile devices, or by exploiting visitor's own personal ones. Integrating these devices into the surrounding augmented environment will allow a system for a non-invasive pervasivity, and users for a pleasant and natural visit experience.

After decades of research, input/output devices for kiosk-based interaction have converged to keyboard, mouse and monitor. While these devices suffice for interacting with 2D and simple 3D applications, they are no longer effective for mobile access around immersive virtual environments encapsulating cultural heritage sites. A whole new set of I/O modes are sought, such as spatial tracking, force feedback, voice recognition, voice synthesis, position/orientation detection, etc. These multiple, concurrent modes of interaction need to be properly orchestrated to provide the visitor with a natural visit experience, without burdening him/her with cumbersome or confusing stimuli.

In the following sub-sections we will discuss how multiple senses and interaction modes have been concurrently involved, both for input and for output purposes, in some of the above presented systems. In particular, we will consider touch and sight, hearing and speaking, position and orientation, gesture recognition, and image recognition.

Touch and Sight

Touch and sight are the most used modes for human-environment interaction, as they were involved in the trivial keyboard-and-monitor or point-and-click interfaces. Some recent implementation focused on new ways to exploit these interaction modes.

One of the simplest ways to give touch new possibilities, is to exploit it for remote application control. Thanks to the features of personal mobile devices, this solution is easy to be implemented and used by non-skilled users, and it gives a good feedback in terms of user experience.

In this field, researchers focused their work on the interaction between user and device, by allowing the first to use his or her own palmtop or mobile phone, without having to become familiar with a new I/O interface. In fact, users can control remotely running applications with their own device by the well-known drag, click and double-click operations. Such systems rely on wireless networks for data exchange among devices, thus allowing designers to manage concurrent multiuser interaction (Woodroof, 2001).

Besides the sheer information about the site, the data flow can be enriched with some additional content in order to promote and support the fruition of works of art using information technologies. An existing implementation is the MUSE system (Cinotti, 2001), that is based on hand-held computers connected to a site control center by a wireless link. This system provides users with interactive learning and entertainment services, supplies site monitoring and control functions, thematic visits, virtual access to unavailable items, logistic information.

In our multimodal virtual guide we included the traditional touch and sight interaction as a backup mode, in order to let people use their own personal mobile devices. This allows us to make our guide accessible even by means of devices with a reduced hardware equipment compared with those holding all that is necessary for the full featured interaction (Augello, 2006).

In the above described implementations, the sight sense is involved in a simple way, by means of traditional displays for the final output to the user. In order to improve the user's experience, some existing solution makes a more extensive use of the sight sense, exploiting immersive displays (such as cylindrical screens, multiple projectors, and active stereo shutter glasses) (Farella, 2005). This way users feel to be active part of a virtual reality immersive environment.

This is the case of Archeoguide (Vlahadis, 2001; Vlahadis, 2002), which exploited first developments of mobile computing, augmented reality, 3D visualization, networking, and archiving, to provide users with on-site and online tours of a physical site according to their profile and behavior during the tour. Users of Archeoguide are equipped with an Head-Mounted Display (HMD) and a mobile unit (a laptop in a backpack, or a PDA that allows a reduced set of features). These devices are wirelessly networked to a local server that acts as the central repository of data related to the cultural heritage site. Information related to the user position and orientation are refined by an optical tracking algorithm, and then they are used to render 3D models of the monuments of the site and display the images to the user's HMD as a superimposed image on the real one.

Augmented reality representations were used over the years by several different research groups, with significant advances in modeling. Among the latest results, authors of (Damala, 2008) proposed a three layer navigation scheme, which allows users to navigate into 2D and 3D multimedia presentations related to a detected painting.

In last years also multi-touch technology use has significantly increased. As an example, in (Peltonen, 2008) multi-touch displays have been used in a central location in Helsinki, Finland. The system is named CityWall and provides a zoomable timeline used to arrange public images of the city. Pictures, downloaded from Flickr, can be resized, rotated, and moved with simple one- or two-handed gestures.

Hearing and Speaking

The presentation of contents in the audio format evolved over the years, starting from first approaches with the traditional pre-recorded audio guides. This solution provides users with pre-recorded audio contents by means of ad-hoc handheld devices, sometimes related to the user's position by means of a number or a letter the user must manually input. Since there is no awareness about the actual position of the user, information are presented in a abstract way, that is often also confusing. The user is a passive actor in such systems, and the resulting content fruition is boring, despite its simplicity.

The use of audio as output channel had a significant spin off with the advent of text-to-speech technology. It allows a system to convert normal language text into synthesized (computer generated) speech. Using this technology, researchers overcame limits of static pre-recorded audio, by dynamic text generation to be audio reproduced using a text-to-speech engine. Among them,

Nickerson (Nickerson, 2005) proposed the History Calls system, based on the VoiceXML technology. It dynamically generates contents about a tour within a museum in the form of text to be spoken and delivered to cell phones.

In 2006 we started our experiments on the use of audio as input channel for the information provision system (Augello, 2006). We exploit the Automatic Speech Recognition technology to translate an audio stream in a sequence of words, to which a semantic value is attached by means of conversational agents. The guide vocally answers by means of the Text-to-Speech technology, thus implementing the full-duplex vocal interaction. More details about our solution and achieved results will be described in the "Case study" section.

Vocal interaction, both for input and output purposes, is probably the more natural interaction mode, making a virtual guide more human-like. As it can be easily guessed, it is more suitable for not-so-much crowded environments, particularly when used as input mode. In fact, whereas vocal output could be made suitable by means of earphones, accuracy of vocal input can be improved by talking loud and/or near the microphone. This could lower the naturalness of interaction, as well as the accuracy.

Position and Orientation

An input channel that is more and more used is the user's position along with his spatial orientation. It can be used to start building a location-related content, and to trigger its delivery too. This was the first way to take into account some context-related factor, thus giving systems the context-awareness. Over the years, more context factors have been considered, such as the user behavior (based on the followed path, or on the requests list), as well as the user profile (based on user preferences or skills), but the position of users within a site is a key data for retrieving context-based information. As a consequence, there are several researches

whose main goal is to set up positioning or location systems as more accurate as possible, according to different kind of constraints (cost, devices, type of users, environment).

In HIPS (Petrelli, 2001) both the content and the linguistic form of presentations are dynamically generated on a mobile device depending on constantly updated variables like the user position and orientation, the history of interaction, including places visited and received information, and the user model.

The Global Positioning System (GPS) is the common basis for suitable positioning within outdoor environments. In order to improve the GPS native accuracy, research groups make use of some enhanced solution such as the D-GPS (Differential GPS). As an example, in the Archeoguide system (Vlahadis, 2001; Vlahadis, 2002) the user position within the site is detected by means of a D-GPS relying on a local network of reference antennas integrated with the well-known constellation of GPS satellites. This allows the system to achieve an accuracy of less than 1 meter. The system also detects users orientation by a digital compass that allows an accuracy of $0.5° \div 5°$.

The actual position of users is not the only information used to detect the context in order to tune suitable contents. For example, Jones et al. (2005) propose the use of narrative agents, accessible by means of a PDA in which multisensory systems are integrated. The agents are able to invoke empathy in the user, mould to his/her behavior in the environment. The system relies on a wireless framework and the GPS is used to detect the user position. As a further example, the CRUSE system (Hodjat, 2006) exploits the context, composed of user preferences, behavior and position, in order to display dynamically generated information. CRUSE detects the topics of interest by user actions (coming from command lines or GUI manipulation), or by contextual events (such as the proximity to a specific position). CRUSE also plans next actions analyzing the interaction history and the topics of interest.

GPS is a good solution for cheap and easy position detection outdoors since involved protocols are part of a standard and devices are commonly available, but it is not suitable for indoors, where obstacles like walls, roof, etc. block out satellite signals. In this case, a wireless framework suitably exploited can come in help to address this issue. Koch et al. (2004) use a Bluetooth-based location system that relies on a proximity detection framework implemented by Bluetooth-enabled receivers acting as Points of Presence (PoP). When the PoP detects a device in its range, it transmits an event to a location server informing the device's ID and its distance. The device position is then determined by combining information from several PoPs. Once the position is detected, a context solver checks the user's profile and infers if he would be interested in receiving information about his current position. In positive cases, the representation system of museum portfolio is queried in order to compose and deliver multimedia contents to the user's device.

In our MAGA system, we exploit the RFID technology to detect the user's position within the site (Augello, 2006). In more detail, we trigger the interaction between the user and the system by the detection of a RFID tag attached to an exhibit or a point of interest. As it will be discussed in the "Case study" section, RFID allowed us to set up a positioning framework that is easily scalable and cost-effective. Of course, this solution is suitable for indoor sites, where points of interests are near one to each other, such as our target sites. In such situations, RFID enables a raw compass detection too, that could be accurate enough for common applications. It has to be noticed that the needed hardware is very cheap and easy to be programmed, particularly if compared with other compass detection systems, such as those based on magnetic trackers or accelerometers.

Despite most of the used technologies were at their beginning, Oppermann and Specht proposed an interesting model: HIPPIE (Oppermann, 2001) supports users at all stages of the visit to a museum, including the preparation, the actual tour, and the evaluation after a tour, aiming at the improvement of the comprehension and amusement of exhibits. The mobile support is just a part of the whole system. As a matter of fact, information can be accessed at home through the Internet for the preparation and evaluation of a tour. Inside the museum, information is obtained using wireless technologies. The system detects the followed path in the physical space and provides individual visualization adapting the interface to the current context. As a result, it proposes exhibit tours tailored to the estimated interests of the individual user. The system can be based on a sub-notebook or on PDAs equipped with a simplified interface.

Gesture Activation of Controls

The research towards the achievement of a natural interaction has recently involved technologies aimed at detecting user gestures, such as accelerometers, magnetic tracking sensors, and motion detection by video analysis. In (Santoro, 2007) the user can control and navigate the interface with gestures through a paradigm called "Scan & Tilt". This is obtained using a 2D acceleration sensor, which allows the detection of small movements of the handheld device, where the direction and speed of movements is translated to actions/events of the application, the same way the Nintendo WII remote control does.

Malerczyk (2004) exploits a pointing gesture recognition system through which the user can establish his own exhibit selecting distinct painters, artistic topics or images. The approach integrates Mixed Reality technologies into a traditional museum environment. The system is particularly useful in small-spaced museums, where it is difficult to show paintings in an adequate manner. Once selected the painting, the user can easily interact with it by just pointing at the interaction canvas.

Besides that, the user can look at the details of the paintings without using a magnifying glass.

Ballagas et al (Ballagas, R.,2008; Kratz, S., 2007) presented REXplorer, a game designed to inform visitors about history in a funny way. The player can communicate with spirits associated to significant buildings of a town. The system makes use of Nokia N70 mobile phones linked with Bluetooth GPS receiver. The game exploits a novel mobile interaction mechanism, named "casting a spell", i.e. the user makes a gesture by waving a mobile phone through the air. This allows the user to evoke and interact with a character, named as "spirit", to continue playing the game. In order to reach this goal, a gesture recognition system for camera-based motion has been developed. The algorithm makes use of state machines, modeled from a gesture rule set, that parse the motion data and decode the action the user has depicted.

A multi-user, multitouch, projected table surface, large wall displays and tracked hand-held display have also been used in a Collaborative Mixed Reality Visualization of an Archaeological Excavation presented in 2004 by Benko et al. (2004). Through these tools, users can choose their own view preference, look at all the archaeological site, select objects, display additional multimedia information. The system is capable to recognize three different gestures: point, grab, and thumbs-up. Besides that, the table is capable to discern among simultaneous interactions by multiple users. This way several users can interact with the system at the same time.

The use of gestures as input is one of the more challenging thread in the research on natural interaction modes, since it involves two parallel tasks, both computing-bound: recognition and interpretation. The first one aims at the translation of gestures in a digital representation. This task is mainly accomplished by means of wearable devices, such as accelerometers and magnetic trackers, or by means of the motion detection from a (multiple) video streaming. Wearable devices have the advantage to allow for a more accurate detection, but they need to be used within a dedicated hardware framework. The wearable feature allow for the contemporary capture of multiple gesture sources (visitors of a site, each wearing their own tracking device), but not all people could accepted to wear some device during a visit.

The accuracy of motion detection from a video stream depends both on hardware and software capabilities. High-definition cameras along with a fast and effective image processing algorithm, allow for good recognition performances. This solution is non-intrusive, but it could be not suitable for application within a public place with a large number of visitors, such a cultural heritage site is.

Once gestures are digitized, they need to be correctly interpreted. This is a very hard task, because the semantic of the same gesture may vary according to a two-dimensional (al least) input state: the user profile (habits, skills, knowledge), and the context in which he is. For instance, the straight horizontal movement of the index finger may represent the will to point at an item, or to push in the same item. The solution of this problem requires an accurate context reconstruction (based on an accurate context representation), tightly coupled with a correct analysis of the user's profile. This moves the focus to a wealth of problems in different fields that have to be taken into account in order to correctly infer the semantic of a gesture.

Discussions above may explain why gesture recognition is mainly used in small spaces with only one user (e.g. CAVE-like systems), or in situations where a reduced set of actions are classified for a given application and all gestures are mapped in one of the possible actions (e.g. the Nintendo Wii remote control).

Image Recognition

Due to the increasing availability of low-cost and high-resolution cameras, often embedded in personal mobile devices, image analysis-based techniques are more and more used to detect the user context, sometimes in cooperation with other techniques. For example, the AGAMEMNON system (Ancona, 2006) makes use of mobile phones equipped with embedded cameras to support visitors in archaeological sites and museums. The system usage is twofold: based on an image recognition service, tourists can enjoy their visit with a customized and detailed multimedia guide, and site personnel can use images provided by tourists devices to monitor the state of the monuments and preserve them from vandalism and time injuries.

Image recognition has been also successfully used in Snap2Tell (Joo-Hwee, 2007), a system for tourist information access. Snap2Tell has been developed with the aim to provide a multimodal scene description based on an image of the site captured and sent by a camera phone. In particular, the system uses a Nokia N80 client, a unique database STOIC 101 (Singapore Tourist Object Identification Collection of 101 scenes), and a discriminative pattern discovery algorithm in order to learn and recognize scene-specific local features.

In (Ryan, 2006) a museum guide has been presented. In this case, a webcam is exploited in order to acquire images, which are processed by a tablet PC. When the user desires more information about an exhibit, he presses the information button on the tablet pc as soon as the exhibit is shown on the screen. The exhibit is automatically identified through an image recognition algorithm and the corresponding information is displayed.

In more recent times camera phones have been used in museum guide systems to recognize paintings in art galleries (Ruf, 2008). The system uses Scale-Invariant Feature Transform (SIFT) and Speeded Up Robust Features (SURF) algorithms to reach this goal. The benefit is that objects are selected by simply taking a picture of them.

This kind of interaction gives users the possibility to select a portion of the surrounding space (by simply taking a photo) and to submit it to ask for related information. The counterpart is that the system must have the capability to recognize items or, more generally, contexts among a potential infinite manifold of possible pictures. This task is often accomplished by reducing the number of items about which information can be requested, or by involving users in the item selection process.

MULTIMODAL MOBILE ACCESS TO SERVICES AND CONTENTS IN CULTURAL HERITAGE SITES

The cultural heritage target domain is extremely challenging for the development of systems that assist users during their visit according to their mobility requirements. Moreover these systems should attract and involve users by showing an easy and friendly access. Most of these requirements are fulfilled by means of personal mobile devices, suitably exploited as adaptive human-system interfaces.

We start our discussion by summarizing the main features of a sub-set of projects, among those previously discussed, that make use of hand-held devices (Table 1). Systems are listed by year of presentation, along with their features under four main categories: compass/position detection, context awareness, intelligent interaction, output modes and input modes.

The *compass/position detection* column shows technologies used to detect the user position and orientation within the cultural heritage site. All positioning-based systems exploit information about the proximity of the user to a particular item or point of interest. With no further information, such systems are not able to detect or estimate what the user is looking at, thus making this solution unsuitable for sites with items that are close one to each other. To disambiguate such situations, information about the spatial orientation of the user within the environment is needed in addition to his position. To this end, some systems use

Table 1. System features

Year	System	Compass / position detection	Context Awareness	Intelligent Interaction	Input Modes	Output Modes
2001	Woodruff	-	-	-	keyboard, touch pen or touch-screen	text, pre-recorded audio
2001	Cinotti (MUSE)	-	-	-	keyboard, touch pen or touch-screen	text, pre-recorded audio, video clip, images
2001	Oppermann (Hippie)	infrared, electronic compass	location-based, profile-based	-	keyboard, touch pen or touch-screen, user position and/or compass	text, pre-recorded audio, images
2001	Petrelli	-	location-based, state-based, profile-based	proactivity	keyboard, touch pen or touch-screen, user position and/or compass	pre-recorded audio, images
2002	Aoki (SottoVoce)	-	-	-	keyboard, touch pen or touch-screen	pre-recorded audio, images
2002	Vlahadis (Archeoguide)	GPS	location-based, user-based	natural language	keyboard, touch pen or touch-screen, user position and/or compass	text, pre-recorded audio, video clip, images
2004	Koch	-	location-based, state-based, profile-based	proactivity	user position and/or compass	text, pre-recorded audio, video clip, images
2004	Rocchi	infrared, accelerometers	location-based, state-based, profile-based	proactivity, character	keyboard, touch pen or touch-screen,	pre-recorded audio, video clip, images
2005	Farella	-	-	-	keyboard, touch pen or touch-screen,	text, pre-recorded audio, video clip, images, virtual reality
2005	Nickerson	-	-	-	vocal interface	pre-recorded audio, synthesized narrations
2005	Jones	GPS	location-based, state-based, profile-based	natural language, proactivity, character	keyboard, touch pen or touch-screen, user position and/or compass	text, pre-recorded audio, images
2006	Hodjat (CRUSE)	GPS	location-based, state-based, profile-based	natural language, proactivity	keyboard, touch pen or touch-screen, user position and/or compass	text
2006	Ancona	-	state-based, profile-based	proactivity	keyboard, touch pen or touch-screen, vocal interface, image Recognition	text, pre-recorded audio, synthesized narrations, video clip, images
2006	Damiano (DramaTour)	-	state-based	proactivity, character	keyboard, touch pen or touch-screen, user position and/or compass	video clip
2006	Augello (MAGA)	RFID	location-based, state-based	natural language, inference, proactivity, character	keyboard, touch pen or touch-screen, vocal interface, user position and/or compass	text, synthesized narrations, images
2007	Santoro	RFID	location-based, state-based	proactivity	keyboard, touch pen or touch-screen, gesture caption, user position and/or compass	text, pre-recorded audio, video clip, images
2007	Joo-Hwee	-	-	-	image Recognition	text, audio, images
2008	Ruf	-	location-based	-	image Recognition	text, audio, images, virtual reality

different combinations of technologies and techniques, such as infrared combined with electronic compass or with accelerometers embedded into the hand-held devices, thus providing users with fine-tuned contents. Simpler, cheaper and less intrusive location frameworks are made available by means of the RFID technology. In fact, the short tag detection distance can be exploited to estimate the user interest for a specific item with a good accuracy.

The *Context-awareness* column reports the different techniques used to build context-related contents. Systems that miss this feature provide users with information that is neither position- nor profile- based. Three different methods have been identified for contents composition to make the human-system interaction more natural and interesting:

- *location-based*: users are provided with information related to items that are located near their current position;
- *profile-based*: delivered contents are generated according to the user's profile, such as preferences, skills, age, etc;
- *state-based*: information are generated taking into account different context factors, such as the user position, the user's profile, the interaction flow, the followed path, etc. Systems may use a sub-set of these elements to detect the current state (e.g., the history of inputs).

Research results in the field of artificial intelligence suggest new tools to make the interaction more natural. (Jones, 2005; Ibanez, 2003; Almeida, 2003). To take this trend into account, in the *Intelligent Interaction* column we then report the capability of a system to implement methodologies that are typical in the field of artificial intelligence. Specifically, the *proactivity* feature shows that the system can spontaneously initiate the interaction with the user without his explicit request, according to the detected context.

Systems that support *natural language* interaction accept user input in natural language (e.g.: "give me information about marble statues dating from the 5th century BC"), and/or provide user with spoken information. Systems with the *character* feature embed a life-like tour virtual assistant with a specific personality. *Inference* means that a system has the capability to make inferences on a domain ontology to update its knowledge-base and to generate ad-hoc contents.

The *Input Mode* column lists the input modes available for each system, whereas the *Output Modes* column lists the content delivery modes a system is enabled to use.

In the following we will discuss how issues and problems of multiple concurrent interaction modes have been faced in the systems and approaches examined so far. In particular we now focus our attention on concurrent input modes, as most of researchers we cited did. Actually, the processing of multiple inputs simultaneously coming from different channels present several constraints, for instance in terms of synchronization and concurrency, whereas the management of contemporary output modes is less compelling. To this end, in Figure 2 we depict the temporal evolution of such systems, comparing them under three dimensions: degree of multimodality, pervasive access to contents, and intelligence in interaction management.

Any given system is therefore classified as either *none, intelligent, pervasive,* or *both* according to the performed interaction, as illustrated in the figure with the bar filling pattern. Systems classified as *none* do support neither intelligent interaction nor pervasive access. A system offers *pervasive* access to cultural heritage contents if it relies on a framework of smart devices where data are stored and processed. The retrieved information is then suitably formatted to fit the specific access device and transmitted over a wireless connection. We classify as *intelligent* in the interaction a system in which contents generation and/ or delivery are managed with techniques that are

Figure 2. Evolution of multimodal input and performed interaction

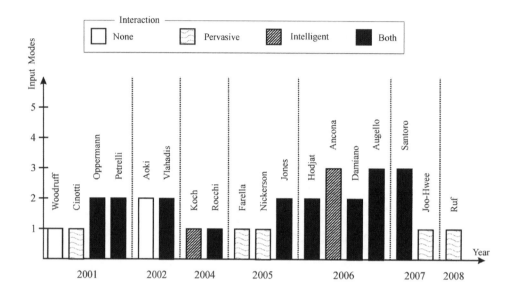

inherited from the artificial intelligence research field. In particular, from the contents generation point of view, the system should use one of the following methods: state-based, location-based, profile-based and inference. From the contents delivery point of view, the system should either interact using natural language, or exhibit proactive behavior, or show a life-like character, or any combination of the three.

The degree of *multimodality* of each system is presented on the vertical axis, and it is evaluated by counting the number of input modes by which users gain access to information.

Figure 2 shows that in the last decade the pervasive approach in service provision within cultural heritage sites has been largely adopted. This is mainly due to the distributed nature of data and to the need of accessing those data any where at any time. The figure also shows that the use of algorithms and techniques of the Artificial Intelligence field is considered a key feature for the success of a system, particularly when the main goal is the naturalness of the interaction. Pervasive systems accessed by mobile devices substituted the pre-recorded audio guides, providing data mining

algorithms with higher computational power and with more data sources.

In order to better highlight the evolution of interaction, in Table 2 we present the detailed list of input modes implemented by each of the discussed systems. This allow readers to realize that, despite new interaction modes are studied and made available, most of existing systems keep showing the traditional point-and-click interface, although being implemented by up-to-date devices.

Another commonly used input mode is the user position. In fact, this allows designers to give their systems the context-awareness by taking into account an important context element. Furthermore, the needed framework can be relatively cheap, and often may partly rely on some existing implementation (e.g. the GPS).

Different input modes, such as voice, gesture and image recognition, are commonly considered as useful in order to improve the naturalness of interaction, but they have been not so largely adopted mainly due to their complexity. Advances of both hardware and software technologies in these fields, are giving an important stimulus to

Table 2. Input modes implemented by systems

Year	System	Keyboard, touch pen or touch-screen	Vocal interface	User position and/or compass	Gesture recognition	Image recognition
2001	Woodruff	X				
2001	Cinotti	X				
2001	Oppermann	X		X		
2001	Petrelli	X		X		
2002	Aoki	X				
2002	Vlahadis	X		X		
2004	Koch			X		
2004	Rocchi	X				
2005	Farella	X				
2005	Nickerson		X			
2005	Jones	X		X		
2006	Hodjat	X		X		
2006	Ancona	X	X			X
2006	Damiano	X		X		
2006	Augello	X	X	X		
2007	Santoro	X		X	X	
2007	Joo-Hwee					X
2008	Ruf					X

their adoption, and several research groups are currently working in the field of integration of these technologies.

In the following section, we will discuss how we intended the multimodality for mobile access to an intelligent service provision system in a museum. The presented case study will show a system that implements three concurrent input modes (touch, voice, position) and two output modes (text, voice) for interaction with humans.

MAGA: A CASE STUDY ON MULTIMODAL INTERACTION

In (Augello, 2006) we proposed a multimodal and pervasive approach for virtual guides in cultural heritage sites, named MAGA (Multimodal Archaeological Guide in Agrigento), which inte-grates intelligent conversational agents and speech recognition/synthesis technologies to a RFID based location framework and a Wi-Fi based data exchange framework. The application is accessible using a PDA equipped with an RFID-based, auto-location module, while the information retrieval service is provided by means of a spoken, natural language interaction with a conversational agent, enhanced by integrating reasoning capabilities into its Knowledge Representation module. The system architecture is shown in Figure 3.

The system is composed by two main areas: the Interactive Area and the Rational Area. The first one provides a multimodal access to the system services. This area embeds the vocal interaction, the traditional point-and-click interface, and the automatic self-location. It allows to acquire user inputs in multimodal form and it feeds the Rational Area with simple questions in textual form.

After the Information Retrieval (IR) process, the Rational Area returns text, images and links to the Interactive Area for the output process.

The Rational Area is made of a Cyc (from "encyclopedia") ontology (CYC ontology documentation) and a conversational agent, which is usually called *chatbot*. The conversational agent is provided with inferential capabilities thanks to the use of a Java porting of the CyN project (Cyc + Program N = CyN, where Program N is an interpreter of the Alice chatbot) (Coursey, 2004). It allows the chatbot to link its knowledge base, composed of AIML rules (Artificial Intelligence Markup Language) (AIML, 2009), to the information stored in the common sense Cyc ontology.

The interaction occurs through the loading of X+V (XHTML + Voice) pages, which can be triggered by user vocal and visual command or RFID detection. In every page the user can have a vocal dialogue with the chatbot. Whenever the user pronounces his request, the multimodal browser looks for a match in the grammar file. If a match is found the application submits the recognized query to the chatbot awaiting the answer. The interaction between the application running on the PDA and the system is also started by the detection of a RFID tag, which is used to estimate the PDA position within the environment. According to this feature, people can go on asking questions about the current object to the chatbot with vocal queries, or they can discard the information and continue their tour.

The Rational Area

In the Rational Area (see Figure 4), the chatbot is provided with inferential capabilities. In particular an "ad hoc" microtheory for the specific context, properly defining the collections of concepts and facts regarding the analyzed domain, has been created into the Cyc ontology. It is possible to exploit the information already present in Cyc, by hooking up the created microtheory to some of the existing ones. During the conversation the

Figure 3. The MAGA system

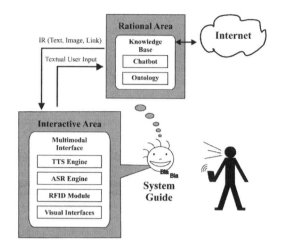

chatbot looks for a lexical matching between the user query and the set of modules stored in its knowledge base.

The chatbot searches for the best match rule in its knowledge base. The Aiml rule can directly produce an answer, or it might be necessary to query Cyc in order to construct a more suitable answer to the user's request. The Cyc responses are embedded in a natural language sentence according to the rules of the template. The chatbot answer could also be the result of a query to standard search engines, which will search for local or remote documents related to the user query. The choice of using Cyc allows the system developers to exploit the large amount of data already organized and described in this ontology. This makes the system more adaptable to domain changes, as it is not necessary to write the entire set of knowledge every time, but only the most specific.

The Interactive Area

The Interactive Area manages synchronization between different kind of user inputs and provides IR results to the user. As shown in Figure 5, the

Figure 4. The Rational Area

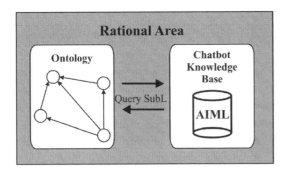

system is accessible through a web page in a multimodal browser from the handheld device. The interaction occurs by means of the loading of X+V (XHTML + Voice) pages, that can be triggered by user vocal and visual command or RFID detection.

Whenever the user submits his request visually, vocally or through RFID detection, the multimodal browser loads a new page built with the Rational Area reply.

X+V is a mark-up language for web applications, which implements the vocal part of the multimodal interface with a subset of W3C standard VoiceXML and the visual part with XHTML (Axelsson, 2004).

The speech recognition process is carried out through an "ad-hoc" built-in speech grammar, including a set of rules which specify utterances that a user may say. This allows the improvement of the automatic speech recognition process: in fact, a small domain context allows for the building of a richer and more thorough set of pre-defined sentences. Furthermore, a particular domain-restricted grammar improves the user's freedom of expression, enhancing the naturalness of the dialogue. The grammar has been built using the XML Form language of the SRGS (Speech Recognition Grammar Specification) from a set of key-words about the domain. These key-words, which are names related to the application domain, are classified in categories and stored in a database.

An "ad-hoc" algorithm gets the key-words from the database and puts them into the grammar to build spoken user utterances. The vocal interface is easily adaptable to application domain changes. The grammar is easy-fitting and can be improved with minimal effort; it is sufficient to change the key-words database to induce a grammar for the new context domain.

When the speech matching occurs, the application submits the recognized query to the chatbot and waits for the answer. In the same way the application submits the query associated to the element selected with the touch-pen on the PDA screen.

The interaction between the application running on the PDA and the system is also triggered by the detection of a RFID tag, which is used to estimate the PDA position within the environment. RFID module in the Interactive Area allows the system to sense changes in the environment in which the user is located and to automatically adapt itself and act accordingly to these changes based on user needs. This technology can be exploited to improve the users experience within augmented environments, thus allowing the access to relevant resources in a context-dependent fashion. We have used passive RFID tags, which can work without a separate external power source, obtaining operating power generated from the reader. We have passive tags each one holding an unique ID, and each one attached to a point of interest. The information related to the items of interest can be obtained by means of a compact flash reader module hooked into the PDA.

Once the RFID reader on the PDA detects a tag, its unique ID is passed to the server application over the wireless network. The link between the information sent to the PDA and the user's position is given by the detected ID. This allows the chatbot to start the position-related interaction with the user, who can go on asking the chatbot with further questions or discard the information.

Figure 5. The Interactive Area

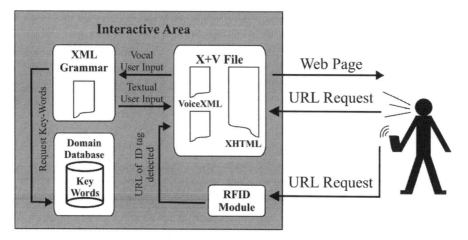

The Multimodal Interaction

Users can request and access information through spoken commands and traditional visual and keyboard (or stylus) based commands by means of Automatic Speech Recognition (ASR) and Text-To-Speech (TTS) technologies embedded in a multimodal browser on a PDA. The chatbot processes user questions and gives consistent replies using its KB enriched with ontology conceptualization and internet services too.

The Interaction Area synchronizes multimodal inputs during the user navigation. Figure 6 shows an example of an interaction between the user and the proposed system, in which all three available access modalities are involved.

The first step is to tag any point of interest within a given environment. People walk across such an environment with a PDA running our client application. When an RFID tag is detected by the reader module on the PDA, data read from the tag is passed to the PDA to be processed. Then the client application running on the PDA sends the detected tag ID to the server via the wireless connection. This allows the chatbot to start the interaction with the PDA by providing it with some basic information related to its current position.

When a tag is detected, the RFID Module sends a query to the chatbot about the selected item specifying that the input channel is the RFID one. The chatbot does not reply with information about the item but asks the user to choose between starting an interaction on the detected item or continuing his previous dialogue. In the interaction shown in Figure 6 the user accepts the first proposal of the system using traditional point-and-click visual interface. Hence, the chatbot starts interaction about the detected item which, in the specific example, is the room 3 inside the museum. The system provides a short description of the room and a list of available information. The interaction goes on by means of the voice channel: the user asks the chatbot for more information about the showcase list. After every user request, the system builds web pages with contents retrieved by the Rational Area. Such contents can be visually and vocally navigated according with the user preference.

Discussion

We implemented a trial version of MAGA within the "Museo Archeologico Regionale di Agrigento" (Regional Archaeological Museum of Agrigento – Sicily, Italy). The museum, located in the "Valle dei

Figure 6. An example of multimodal interaction in MAGA

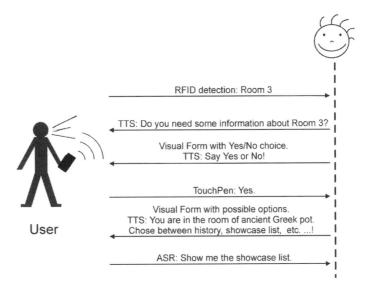

Templi" (Temples Valley) area, contains artifacts that are historical evidences of Agrigento. Visitors of the museum can use the guide exploiting voice, position (RFID-driven) and point-and-click interaction by means of a PDA. The Human-System interaction takes place by means of speech-to-text recognition (ASR) and text-to-speech synthesis (TTS) in English. Our proposed system makes use of a speaker independent voice recognition engine. The implemented information provision service has been tested during a period that lasted 30 days by 135 visitors.

Studies on feedbacks we collected from test users, show that 69% of them had a satisfactory interaction experience. In addition, our observations revealed that the system behaved according to its specifications in 87% of the cases, with a speech recognition success of 80%, over a total of about 1200 vocal items.

Performance limitations, associated with slower processing speeds and memory decline, are mostly referred to users that speak a bad English. However, it has to be noticed that users with bad English pronunciation, after a short practice with the guide by means of the on-line vocal help,

improve their own pronunciation achieving better results. During our tests, we also observed that users never changed the subject of the conversation before obtaining the desired information, and that the maximum number of trials with the same question was four. This shows that the spoken interaction establishes a sort of "feeling" between the human users and the system, thus making users more patient and favorably disposed towards the system speech recognition failures.

The environment where we carried out our trials was often crowded and noisy, with some point of interest placed outdoors, so we collected interesting indications by observing the system response in those situations. When the noise source is not related with the human speech, the system performances decrease almost linearly with the drop in S/N ratio. We obtained different results when the noise, intended as non-desired signal, is given by human talks in the vicinity of the microphone. With a crowded situation of two or three people talking each other within the microphone range, the system performance decays dramatically. We expected this result since the system is not fine-tuned for a specific voice, therefore it tries

Table 3. System evaluation summary

	Not useful	A little useful	Quite useful	Very useful
How do you rate the possibility to vocally interact with a virtual guide?	3	16	24	92
How do you rate the accuracy of voice recognition?	4	9	74	41
How do you rate the use of a PDA to access information?	6	23	59	47
How do you rate the system in comparison with the usual kiosk-style ones?	5	9	60	61
How do you rate the friendliness of the interface?	4	10	47	74
How do you rate the overall response time of the system?	5	42	62	26
What is your overall rate for the system?	3	18	21	93

to recognize all words contained in the signal captured by the microphone. In this cases, users spontaneously and instinctively talked closer to the microphone, thus exploiting the Automatic Gain Control (AGC) feature present on almost all audio acquisition devices.

We received positive feedback from visitors who found this way to access information more natural than traditional ones, such as info-points on fixed kiosks in which information are requested and presented by means of a Web browser. They reported that they felt the same experience as if they talked with human employee, with the advantage that they have not to reach the information desk, and they have not to wait any line to be served. Table 3 summarizes the visitors feedback by showing their answers to our questionnaire.

Table 3 shows that the worst result is given by the system response time. Actually, users had to wait from 6 to 14 seconds before the system answers something (such as the requested information, or the request to repeat the question). This is partially due to the wireless connection we used, that was Bluetooth based. In order to overcome this problem, we plan to use only Wi-Fi based connections, or to narrow the band of the audio channel (we used the CD-quality - single channel - settings for the output) when using Bluetooth.

Another useful indication we collected is related to the use of a PDA as interaction media. Users are quite satisfied, but they would prefer the use of smartphone since they has a more familiar interface. Actually, smartphones and cellular phones are still more diffuse among people than PDAs.

The proposed system is a first prototype intended as a smart replacement of the outdated traditional audio/visual pre-recorded guides. Future work will regard the enhancement of the system improving the interaction naturalness, with particular attention to the time response.

CONCLUSIONS AND FUTURE WORK

Cultural heritage fruition and communication are an exciting new arena in where to exercise many enabling technologies and study novel, more natural interaction schema, all aimed at engaging visitors with multisensorial, memorable visit experiences. We have thus explored systems for cultural heritage fruition from the onset of the new millennium, looking at their capability to engage visitors with pervasive, multimodal, intelligent access to information contents. We have particularly focused on multimodality as it is a key enabler for natural, unobtrusive interaction with exhibits and site virtual environment. In addition we have also looked at how pervasive access to information contents is deployed, and to what degree some intelligence is embodied to resemble human tracts in system responses to visitors' queries and preferences.

It appears that in the past decade or so, applications for cultural heritage fruition have focused on locating visitor position inside the environment, as this piece of information is key to contextualize dynamic contents and offer a natural interaction, specifically geared for the exhibit/artifact at hand. Voice-based techniques are used to a lesser extent, due to the difficult tuning that vocal interface often require. It is still open research the development of robust speech and speaker recognition systems that operate reliably in crowded, noisy environments. An interaction mode that is receiving a growing interest is movement and gesture recognition. Many exemplary projects are available in literature that propose immersive fruition of virtual world and augmented reality, in which gesture based commands are intuitive for the visitor.

Work is in progress to develop augmented reality systems in which object tracking freely depend on specific features of the physical objects in sight, such as edges and contours. Further intelligence may also be conferred to the virtual guides by adding semantic representations and semantic analysis capabilities, which facilitate smart information retrieval from large repositories (even on the web) and enable spontaneous, friendly and engaging dialogue generation.

REFERENCES

AIML. *(Artificial Intelligence Mark-up Language) Documentation.* Retrieved January 28, 2009, from http://www.alicebot.org/aiml.html.

Almeida, & P., Yokoi, S. (2003). Interactive character as a virtual tour guide to an online museum exhibition. In *Proceedings of Museum and the Web 2003.*

Ancona, M., Cappello, M., Casamassima, M., Cazzola, W., Conte, D., Pittore, M., et al. (2006). Mobile vision and cultural heritage: the AGAMEMNON project. In *Proceedings of the 1st International Workshop on Mobile Vision*, Austria.

Aoki, P. M., Grinter, R. E., Hurst, A., Szymanski, M. H., Thornton, J. D., & Woodruff, A. (2002). Sotto voce: Exploring the interplay of conversation and mobile audio spaces. In *Proceedings of ACM Conference on Human Factors in Computing Systems* (pp. 431-438). New York: ACM Press.

Augello, A., Santangelo, A., Sorce, S., Pilato, G., Gentile, A., Genco, A., & Gaglio, S. (2006). MAGA: A mobile archaeological guide at Agrigento. *Proceedings of the workshop Giornata nazionale su Guide Mobili Virtuali 2006, ACM-SIGCHI.* Retrieved from http://hcilab.uniud.it/sigchi/doc/Virtuality06/index.html.

Axelsson, J., Cross, C., Ferrans, J., McCobb, G., Raman, T., & Wilson, L. (2004). Xhtml+voice profile 1.2. *W3C.*

Ballagas, R., Kuntze, A., & Walz, S. P. (2008). Gaming tourism: Lessons from evaluating REXplorer, a pervasive game for tourists. In *Pervasive '08: Proceedings of the 6th International Conference on Pervasive Computing.* Berlin, Germany: Springer-Verlag.

Benko, H., Ishak, E. W., & Feiner, S. (2004). Collaborative Mixed Reality Visualization of an Archaeological Excavation. *The International Symposium on Mixed and Augmented Reality (ISMAR 2004)*, November 2004.

Chee, Y. S., & Hooi, C. M. (2002). C-VISions: socialized learning through collaborative, virtual, interactive simulations. *Proceedings of CSCL 2002 Conference on Computer Support for Collaborative Learning*, Boulder, CO, USA, 2002, pp. 687-696

Coursey, K. Living in cyn: Mating aiml and cyc together with program n, 2004. R. Wallace. The anatomy of a.l.i.c.e. Resources avalaible at: http://alice.sunlitsurf.com/alice/about.html.

CYC ontology documentation. Retrieved January 28, 2009, from http://www.cyc.com/.

Damala, A., Cubaud, P., Bationo, A., Houlier, P., & Marchal, I. (2008). Bridging the gap between the digital and the physical: design and evaluation of a mobile augmented reality guide for the museum visit, *Proceedings of the 3rd international conference on Digital Interactive Media in Entertainment and Arts* (DIMEA '08), Athens, Greece, Pages 120-127

Damiano, R., Galia, C., & Lombardo, V. (2006). Virtual tours across different media in DramaTour project, Workshop Intelligent Technologies for Cultural Heritage *Exploitation at the 17th European Conference on Artificial Intelligence* (ECAI 2006), Riva del Garda, 2006, pp. 21-25.

Farella, E., Brunelli, D., Benini, L., Ricco, B., & Bonfigli, M. E. (2005). Computing for interactive virtual heritage. *IEEE MultiMedia, 12*(3), 46–58. doi:10.1109/MMUL.2005.54

Hodjat, B., Hodjat, S., Treadgold, N., & Jonsson, I. (2006). CRUSE: a context reactive natural language mobile interface. In *Proceedings of the 2nd Annual international workshop on wireless internet,* Boston, MA (pp. 20). New York: ACM.

Ibanez, J., Aylett, R., & Ruiz-Rodarte, R. (2003). Storytelling in virtual environments from a virtual guide perspective. *Virtual Reality (Waltham Cross), 7*(1). doi:10.1007/s10055-003-0112-y

Jones, C. M., Lim, M. Y., & Aylett, R. (2005). Empathic interaction with a virtual guide. *AISB Symposium on Empathic Interaction*, University of Hertfordshire, UK.

Joo-Hwee Lim, Yiqun Li, Yilun You, & Chevallet, J.-P. (2007). Scene recognition with camera phones for tourist information access. In *2007 IEEE International Conference on Multimedia and Expo* (pp. 100-103).

Koch, F., & Sonenberg, L. (2004). Using multimedia content in intelligent mobile services. In *Proceedings of the WebMedia & LA-Web Joint Conference 10th Brazilian Symposium on Multimedia and the Web 2nd Latin American Web Congress* (pp.1-43).

Kratz, S., & Ballagas, R. (2007). Gesture recognition using motion estimation on mobile phones. In *Proceedings of 3rd Intl. Workshop on Pervasive Mobile Interaction Devices at Pervasive 2007.*

Liarokapis F., & Newman,Design, R. M. (2007). Experiences of multimodal mixed reality interfaces. In *Proceedings of the 25th annual ACM international conference on Design of communication* (pp. 34-41).

Malerczyk, C. (2004). Interactive museum exhibit using pointing gesture recognition. In *WSCG'2004*, Plzen, Czech Republic.

Mulholland, P., Collins, T., & Zdrahal, Z. (2004). Story fountain: intelligent support for story research and exploration. In *Proceedings of the 9th international conference on intelligent user interfaces*, Madeira, Portugal (pp. 62-69). New York: ACM.

Narzt, W., Pomberger, G., Ferscha, A., Kolb, D., Muller, R., Wieghardt, J., et al. (2003). Pervasive information acquisition for mobile AR-navigation systems. In *Proceedings of the Fifth IEEE Workshop on Mobile Computing Systems & Applications* (pp. 13-20).

Nickerson, M. (2005). All the world is a museum: Access to cultural heritage information anytime, anywhere. *Proceedings of International Cultural Heritage Informatics Meeting, ICHIM05.*

Oppermann, R., & Specht, M. (2001). Contextualised Information Systems for an Information Society for All. In *Proceedings of HCI International 2001. Universal Access in HCI: Towards an Information Society for All* (pp. 850 – 853).

Peltonen, P., Kurvinen, E., Salovaara, A., Jacucci, G., Ilmonen, T., Evans, J., et al. (2008). It's mine, don't touch: Interactions at a large multitouch display in a city center. In *Proc. of the SIGCHI conference on human factors in computing systems (CHI'08)* (pp. 1285-1294). New York: ACM.

Petrelli, D., Not, E., Zancanaro, M., Strapparava, C., & Stock, O. (2001). Modeling and adapting to context. *Personal and Ubiquitous Computing, 5*(1), 20–24. doi:10.1007/s007790170023

Raffa, G., Mohr, P. H., Ryan, N., Manzaroli, D., Pettinari, M., Roffia, L., et al. (2007). Cimad - a framework for the development of context-aware and multi-channel cultural heritage services. In *Proc. International Cultural Heritage Informatics Meeting (ICHIM07)*.

Rocchi, C., Stock, O., & Zancanaro, M. (2004). The museum visit: Generating seamless personalized presentations on multiple devices. In *Proceedings of the International Conference on Intelligent User Interfaces*.

Ruf, B., Kokiopoulou, E., & Detyniecki, M. (2008). Mobile museum guide based on fast SIFT recognition. In *6th International Workshop on Adaptive Multimedia Retrieval*.

Ryan, N. S., Raffa, G., Mohr, P. H., Manzaroli, D., Roffia, L., Pettinari, M., et al. (2006). A smart museum installation in the Stadsmuseum in Stockholm - From visitor guides to museum management. In *EPOCH Workshop on the Integration of Location Based Systems in Tourism and Cultural Heritage*.

Salmon Cinotti, T., Summa, S., Malavasi, M., Romagnoli, E., & Sforza, F. (2001). MUSE: An integrated system for mobile fruition and site management. In *Proceedings of International Cultural Heritage Informatics Meeting*. Toronto, Canada: A&MI.

Santoro, C., Paternò, F., Ricci, G., & Leporini, B. (2007). A multimodal mobile museum guide for all. In *Proceedings of the Mobile Interaction with the Real World* (MIRW 2007). New York: ACM.

Vlahakis, V., Ioannidis, M., Karigiannis, J., Tsotros, M., Gounaris, M., & Stricker, D. (2002). Archeoguide: an augmented reality guide for archaeological sites. *IEEE Computer Graphics and Applications, 22*(5), 52–60. doi:10.1109/MCG.2002.1028726

Vlahakis, V., Karigiannis, J., Tsotros, M., Gounaris, M., Almeida, L., Stricker, D., et al. (2001). Archeoguide: first results of an augmented reality, mobile computing system in cultural heritage sites. In *Proceedings of the 2001 conference on Virtual reality, archeology, and cultural heritage* (pp. 131-140). New York: ACM.

Woodruff, A., et al. (2001). Electronic guidebooks and visitor attention. *Proceedings of the 6th Int'l Cultural Heritage Informatics Meeting* (pp. 623-637). Pittsburgh: A&MI.

Yuan, X., & Chee, Y. S. (2003). Embodied tour guide in an interactive virtual art gallery. In *Proceedings of the 2003 international Conference on Cyberworlds* (pp. 432). Washington, DC: IEEE Computer Society

Chapter 10
Multimodal Search on Mobile Devices:
Exploring Innovative Query Modalities for Mobile Search

Xin Fan
University of Sheffield, UK

Mark Sanderson
University of Sheffield, UK

Xing Xie
Microsoft Research Asia, China

ABSTRACT

The increasingly popularity of powerful mobile devices, such as smart phones and PDAs, enables users to search for information on the move. However, text is still the main input modality in most current mobile search services although some providers are attempting to provide voice-based mobile search solutions. In this chapter, we explore the innovative query modalities to enable mobile devices to support queries such as text, voice, image, location, and their combinations. We propose a solution to support mobile users to perform visual queries. The queries by captured pictures and text information are studied in depth. For example, the user can simply take a photo of an unfamiliar flower or surrounding buildings to find related information from the Web. A set of indexing schemes are designed to achieve accurate results and fast search through large volumes of data. Experimental results show that our prototype system achieved satisfactory performance. Also, we briefly introduce a prospective mobile search solution based on our ongoing research, which supports multimodal queries including location information, captured pictures and text information.

INTRODUCTION

As search technology has proliferated on the desktop, search service providers are competing to develop innovative technologies and offer new services in order to attract more users and generate more revenue from advertisements. Nowadays, the competition is moving to mobile platforms since mobile phone, smart phone and PDA users are a

DOI: 10.4018/978-1-60566-978-6.ch010

larger community than computer users. Major Internet search providers, such as Google, Yahoo!, Microsoft or Ask.com are actively promoting new mobile services, including traditional information search and more mobile specific services such as product search and local search. Many such services use text-based interfaces, which, due to the constrained input/output modalities make the services inconvenient to use. However, mobile phones can support rich multimodal queries, composed of both text combined with image, or audio, or video, or location information.

Research into technology and user interaction in modalities other than text is becoming increasingly important. Since the basic function of mobile phones is to enable voice communications, voice is the most natural way on a mobile to input search queries. The ready availability of the microphone makes non-speech audio queries a natural option; embedded cameras offer the possibility of visual input; also the increasing prevalence of built-in GPS provides location information as well. Not only can GPS in a mobile offer the user a chance to navigate around, it can provide the search engine useful geographical information based on the user's current location. Therefore, with such technologies, the mobile phone can be turned into a powerful tool to carry out multimodal search and acquire information on the go.

In this chapter, we will examine mobile search based on multimodal queries composed of text, local information, image and audio. In the next section, we will investigate existing work on location based mobile search, visual mobile search and audio mobile search from both an industrial and academic perspective, followed by typical system design of a multimodal mobile search service.

A mobile search system by visual query is introduced in detail as an example system. As audio queries based on audio matching and speech recognition are relatively mature in the research perspective, we will mainly emphasise on visual and location modalities when introducing the implementation details. Finally, we propose a

prospective mobile search solution to process multimodal queries including location information, captured pictures and text information, followed by a conclusion.

STATE OF ART

Initially, mobile search services were simple counterpart versions of web search ones and the content was mainly limited to news, weather, sports, etc. In recent years, remarkable efforts have been made on both interface and content of mobile search services in the research community. The search providers are also shifting their focus on multimodal interfaces and richer content types. A location-aware system (e.g. GPS), embedded camera and micro-phone can be successfully exploited to collect multimodal information and generate multimodal queries for better search experiences on mobile devices. In this section, we will specifically study the present research work and commercial search services on mobile audio search, mobile visual search and location based mobile search.

Mobile Audio Search

A considerable number of mobile audio search services have been described by the major search engine companies in recent years; the use scenarios include but are not limited to the following aspects (Xie, et al., 2008):

- To search for local information, such as restaurants, travel destinations etc, through a spoken keyword or dialogue;
- To search for text information on the Web that is relevant to the corresponding text of an audio query;
- To search for similar audio materials, such as songs or ringtones, to the audio query;

Google has released an experimental voice-activated local search system Goog411(Goog411, 2008) which uses voice recognition to enable users to simply speak their query to find the needed information. There are other mobile search services, which also use speech recognition. Yahoo! oneSearch (oneSearch, 2008) and Microsoft's Tellme (Tellme, 2008) both aim to provide Web information search including local information query, directory assistance, etc. Similar search products like PromptU (PromptU, 2008) and V-ENABLE allow mobile users to obtain information with spoken queries from mobile handsets. ChaCha (ChaCha, 2008) is another search service that supports voice queries employing call centre staff to answer those queries the automated system cannot handle.

Technologies of audio fingerprinting (Burges, Platt, & Jana, 2003; Cano, Batle, Kalker, & Haitsma, 2002) and query by humming (Ghias, Logan, Chamberlin, & Smith, 1995; Lu, You, & Zhang, 2001; Zhu & Shasha, 2003) are applied in audio material search, such as music search, from mobile devices. An audio fingerprint is a compact signature extracted from an audio signal, which can be used to identify an audio sample or quickly find similar items in an audio database. For instance, the search service 411-SONG (411-SONG, 2008) allows users the ability to identify an unknown song simply by recording a 15 second segment of it. The user will receive a text message with the artist and title and a link to download/purchase the CD or the ringtone.

Query by humming is a music search solution to allow users to find a song by humming part of the tune when a melody is just stuck in the mind. It's generally suitable for retrieval of songs or other music with a distinct single theme or melody. The search service NayioMedia (NayioMedia, 2008) in South Korea can identify tunes that users hum into microphone. It can recognise the tune and show where to buy it or share with others.

Mobile Visual Search

Fan et al (Fan, Xie, Li, Li, & Ma, 2005) broke visual mobile search services into three categories based on the information need:

- To search for certain characters or signs for the purpose of character recognition like OCR. For example, barcode and QR code (QRCode.com, 2006).
- To search for the result with similar visual features or the same concept to the query. For example, the query image is a certain flower or an animal;
- To search for a result that is the same one or a copy of the query image. For example, a famous painting, a certain product or a landmark building;

The research and applications developed to address the above three needs are described as follows.

Some effective character recognition techniques have been developed specifically for camera phones in recent years (M. Smith, Duncan, & Howard, 2003; Watanabe, Sono, Yokoizo, & Okad, 2003). For example, some mobile ISPs in Japan and China offer search by QR code. This kind of service began to be widely used to query the information of various products and even Web navigation. A mobile user captured the QR code of a product and were provided with related reviews or were navigated to the official website to learn more about the product.

The challenge of searching for visually similar objects is related to traditional Content Based Image Retrieval (CBIR) technology (Flickner, Sawhney, Niblack, Ashley, & Huang, 1995; J. R. Smith & Chang, 1996) which has been widely investigated in the last decade. Early work on CBIR is mostly based on global descriptors, while in recent work, much effort has been drawn to research on local descriptors and their associated semantics.

Identifying an identical object/scene by local descriptors is now coming to some maturity and has achieved satisfying results in object recognition and duplicate image detection. Local feature detectors (K. Mikolajczyk, et al., 2005; Sebe, Tian, Loupias, Lew, & Huang, 2003) are employed to detect salient points in a image, which are usually distinctive from neighbouring area. The specific image features around the salient points are extracted to be represented by local descriptors, which are generally high-dimensional vectors (Krystian Mikolajczyk & Schmid, 2005). More details on salient points and local features will be presented in the "Example System" section. The local descriptors have a degree of invariance and can handle image scaling, cropping, shearing, rotation, partial occlusions, brightness and contrast change etc. In addition, they have some invariance to viewpoint and illumination. Therefore, the local descriptors are particularly suitable to represent low-quality and transformed pictures captured by embedded cameras. The descriptors are computed as feature vectors for all images in a dataset. Consequently, the identical images/scenes are assumed to be proportional to the similarity of the feature vectors, followed by some verification such as spatial constraints of salient points.

Most research on searching by pictures captured on mobile cameras cast their tasks into a traditional CBIR context (Hare & Lewis, 2005; Noda, Sonobe, Takagi, & Yoshimoto, 2002; Sonobe, Takagi, & Yoshimoto, 2004; Yeh, Tollmar, & Darrell, 2004; Yeh, Tollmar, Grauman, & Darrell, 2005). Considering the computational cost and the limited accuracy of the CBIR method, the tasks are generally designed to be specific and based on a relatively small local image database, though (Yeh, et al., 2004) used text-based Web image search engine to locate image data sets on the fly.

Some commercial mobile visual search systems have recently emerged from the research. For instance, Mobot (Mobot, 2008) offers a service to search for relevant items within certain products based on a picture taken with a mobile device. It can recognise a picture of a printed advertisement or a real product and provide the links to real-time product information.

Location Based Mobile Search

Nowadays, GPS-enabled devices are being widely adopted. Mobile users cited GPS capability more frequently than Internet access as a desired feature on their mobile devices, according to a survey by market researcher Leo J. Shapiro and Associates (LJS) (GPSWorld.com, 2007). From a usability point of view, Location-awareness technology enables users to search for information without the effort of explicitly entering the current location. When searching for the needed information, such as restaurants or entertainment services, mobile users are most likely to be interested in finding businesses near their current location. For example, a search engine can use a person's location to filter retrieved Web pages by proximity. Location information used as query input in conjunction with relevant keywords will offer the mobile user more relevant search results. The geographical information derived from the user's location can also be used to learn preferences or contexts to generate and improve personalised search results. Choi (2007) used this combination of query with a user's physical location and distance from user's physical location, to generate more personalised and locally targeted search results. GPS data from mobile devices are adopted to search for personalised traffic information and route planning (Balke, Kiebling, & Unbehend, 2003; Zheng, Liu, Wang, & Xie, 2008) in a convenient way.

Location based search services can operate without integrated GPS: Google launched the "My Location" technology for mobile application based on its Google Maps in 2007, which uses mobile phone tower information to approximate the user's location and find nearby services or attractions, without the need of a GPS unit. The "My Location" technology was added to its mobile search service (GoogleMobile, 2008) to offer Google mobile search users the ability to locate the nearest restaurants, petrol stations or hospitals from their mobile devices. Also, this technology is incorporated in the Google Mobile App for iPhone. Search results can be automatically adapted based on user's current location.

Other major mobile search engines also launched local search based on GPS information recently, such as Microsoft Live search mobile (live.com, 2008) and Ask Mobile GPS (ask.com, 2008). The newly emerging location based mobile search solution, Nuance Voice Control (Nuance, 2008) is distinguished from other similar ones by providing a voice driven operation interface. Also, the GPS data are used to locate the user's current position in the map. For instance, it enables the user to ask from a GPS enabled phone for a near destination with much less interaction when driving.

Summary

Multimodal mobile search is still at an early stage and facing many challenges for both commercial companies and research communities. The present practical search services are mainly focused on mobile audio search using technologies based on audio matching and speech recognition.

We believe ample opportunities exist in location based mobile search due to the increasing prevalence of GPS-enabled mobile phones. Also, mobile visual search is promising as embedded cameras are becoming more powerful and more people are getting used to capturing pictures on mobiles. Therefore, after introducing the typical architecture of a multimodal mobile search system in the next section, we will propose two example systems which respectively adopt visual modality and location modality as main information input to perform mobile search.

SYSTEM DESIGN

Since the computation capability of mobile handsets is limited, a multimodal mobile search system is generally designed using a client-server architecture. A typical system structure was proposed for visual and audio search on mobile devices (Xie, et al., 2008). We extend this framework to a broader context to cover location based mobile search as well.

The architecture mainly consists of four parts as shown in Figure 1. A detailed explanation of each part is given below.

Figure 1. General structure of a multimodal search system

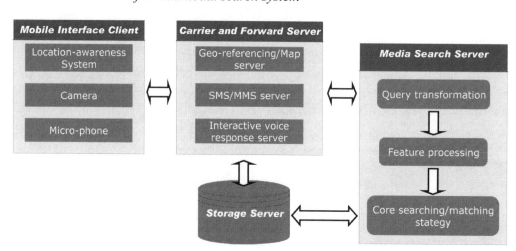

- **Mobile client interface:** A client application running on the mobile device generates multimodal queries with built-in GPS, microphone, camera, and other sensors through audio recording, photo taking etc. The client receives and displays search results as well.
- **Carrier and forward server:** An intermediate service designed to transmit multimodal queries generated by mobile devices; it also returns the search results in a format compatible for mobiles.
- **Storage server:** Multi-media content, such as web pages, audio and images, is collected here for user access. Certain features are extracted and indexed offline for the searching/matching process.
- **Media search server:** Multimodal queries are transformed here. For instance, GPS information will be geo-referenced so that it can be used by a geographic information system. The audio will be decoded into temporal signals. The exact features to be extracted will be determined by the content in the storage server. Search results will be returned to the mobile via the Carrier and forward server.

EXAMPLE SYSTEM: MOBILE SEARCH BY VISUAL QUERIES

In this section, an example system is presented to illustrate the working of the system design described above. In this system, both the text and visual queries are used and the search mechanism mainly relies on the visual information.

The search process can be described in a specific scenario. A mobile user travels to an unfamiliar place and needs some information of the current area. He/she can simply take some pictures of a surrounding landmark building and send the pictures with a couple of words, such as "building Beijing", as queries.

The attached text is used to match the keywords and select a predefined image dataset. An image matching solution is designed to find images containing the identical building. It is based on identifying semi-local visual parts, which are grouped from local salient points. The processing flowchart is illustrated in Figure 2. As the captured pictures often contain cluttered background and other objects, users are encouraged to capture an image sequence to generate visual parts. This solution can significantly reduce the computation time of salient point matching and improve the accuracy of retrieved results.

Background Knowledge

Scale-Invariant Feature Transform (SIFT)

In the existing methods of image matching estimation, local feature based schemes (Baumberg, 2000; Schmid & Mohr, 1997) have shown the advantages in robustness and accuracy over the traditional global measurements. Recently, there have been considerable efforts (Lowe, 2004; Krystian Mikolajczyk & Schmid, 2004; Schaffalitzky & Zisserman, 2002) on extending local features to be invariant and robust to image transformations, such as rotation, scale, affine and intensity transformation. Stable feature points should be detectable regardless of these transformations. Especially, 2D affine transformations in Figure 3 can well approximate the image deformations of planar objects or patches under a range of viewpoint changes in a 3D scene.

Therefore, invariant local descriptors are extensively used to recognise different views of the same object as well as different instances of the same object class. It is empirically (Krystian Mikolajczyk & Schmid, 2005) found that the SIFT descriptor (Lowe, 2004) shows very good performance and invariance to image rotation, scale, intensity change, and to moderate affine transformations. The local descriptor for each

Figure 2. Flowchart of visual search by photos on camera phones for identical objects

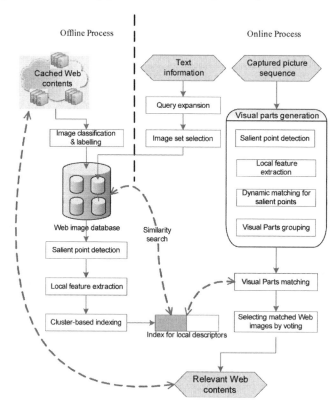

Figure 3. Image changes concerned in invariant local features

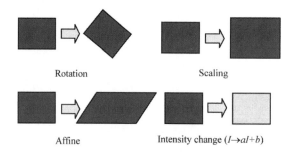

salient point can be represented by a vector with scale, orientation and location in SIFT algorithm. The salient regions the descriptors indicate can be shown as a series of circular regions with different radiuses within the image. Though some other affine invariant detectors and descriptors (Krystian Mikolajczyk & Schmid, 2004) can provide more stable results under larger viewpoint changes,

considering the computational cost, we adopt the SIFT based method in our solution.

High-Dimensional Indexing and Similar Search

The similarity of two images is defined to be proportional to the similarity of local salient points shared between the images. Therefore, the problem of searching similar images is converted to searching the matched salient points between the images. In a large-scale dataset, indexing techniques are critical to speed up the search over a simple sequential scan. The retrieval of similar points in high dimensional space is a K-nearest neighbour search or ε-range search. Since the value of ε is hard to determine in different datasets and application, we focus on K-nearest neighbour search in our experiments. There is already much work on this aspect, but when the number of dimensions is greater than 20, almost all suffer from the "curse of dimensionality", i.e. the performance degrades to linear search with higher dimensions. The phenomenon is particularly prevalent in the exact nearest neighbour search, thus approximate nearest neighbour is popularly used in this case.

Traditional high dimensional indexing methods include:

1) Tree-based methods, such as R-tree, SS-tree, and SR-tree etc. These methods are based on space partitioning and work well with lower dimensions;

2) Hash-based methods, such as locality sensitive hashing (LSH). The retrieval performance is good for higher dimensions, but it requires too much storage space and suffers from the slower I/O speed due to random disk access;

3) Scan-based methods, such as VA-file, VA+-file. They also suffer from random disk access. When the dimension is higher, too many bits are required to quantise points. Otherwise more irrelevant points have to been scanned;

4) Cluster-based methods. An efficient high dimensional cluster algorithm will work well and does not require too much storage space.

Overview Design

In our system, index structures are based on unsupervised clustering with Growing Cell Structures (GCS) artificial neural networks (Fritzke, 1994), which has good performance in the higher dimensional space.

As shown in Figure 2, in an offline process, images are cached and classified into a series of image sets by the keywords extracted from surrounding text. Local features (SIFT features) in each image are extracted and indexed based on the GCS method.

In the online query process shown in Figure 4, an image sequence is captured and sent with words. The attached text are use to match the keywords of each predefined image set. The semi-local visual parts are generated from the query images before searching in the specific image set and finding images with identical objects. The images with the

Figure 4. Visual query process from camera phones based on identical object detection

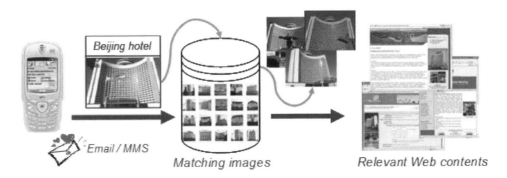

same prominent objects or scenes are returned by calculating voting results of matched visual parts. Finally, the web pages in which these images are located can be returned for reference.

The implementation detail of the online query process is as follows.

First, query expansion is performed based on the attached words in the query, which is sent from the camera phone. The expanded query words are matched with the key words of a given image set. The image set is selected as the database in current task and further identical image/object detection will be carried out on this set.

The local configurations of salient points are often stable across different images of an identical object (Lazebnik, Schmid, & Ponce, 2004). Therefore, the spatial configurations with similar appearance are potential visual correspondences. Inspired by this, we first find the correspondences from multiple query images, and then group the stable neighbouring salient points to compose visual parts (Figure 5 (c)).

A large proportion of local feature matches in the captured photos are false matches, which are caused by features from background clutter and other objects. Consequently we often need to identify the correct matches among over 90% outliers (Lowe, 2004). By introducing the visual parts instead of using the point matches directly, the precision of the retrieval results is considerably improved. In addition, the number of query points is greatly reduced, as is the computational cost of searching for similar images in the database.

The visual part generation and matching algorithms are summarised in Figure 6 and described in detail in the following two subsections.

Visual Part Generation

For the query input, the user simply needs to capture multiple pictures continuously. The visual parts can be extracted by grouping the corresponding salient points between the input images.

The algorithms of visual part generation can be summarised in three steps as shown in Figure 5.

First, the SIFT algorithm is used to extract local features from each query image. Each feature is represented by a 128-dimensional vector.

Second, the matched features between neighbouring query image pairs are identified respectively. The features are matched corresponding to the nearest neighbour (NN) relationship in the 128-dimensional vector space. Here the L2 distance is used to measure the NN relationship. As the nearest feature vector is not always the correct correspondence in the large amount of feature vectors, two nearest neighbours are chosen as candidate matches. The matches are supposed to be rejected when the ratio of the distance from the closest neighbour to the distance from the second closest is larger than 0.8. In our experiments, it was proved that this method eliminated more than 90% of false matches while discarding less than 5% of correct matches.

Third, triples of matched salient points which are neighbours in terms of position are selected in an image. Assuming that we build up N corresponding salient point pairs, for any point A in these N pairs which belongs to the first image, we can define its local neighbourhood region as a circle. The circle's centre is the point A and its radius is a constant factor (for example 3 in our experiments). Any two points B, C in matched pairs whose positions locate in the local neighbourhood region within the first image are grouped as a visual part with the point A.

Thus, a visual part can be regarded as a triple of neighbouring salient points. All the triples in the image are enumerated in this neighbourhood as the visual parts. This method can effectively suppress the noisy salient points since they are often on smaller scales and there are no other salient points nearby.

For multiple query images, we add the visual parts extracted from each image pair to form a visual part set. In this case, more query images require longer computation time. Considering the

Figure 5. Generation of semi-local visual parts according to local features from salient points

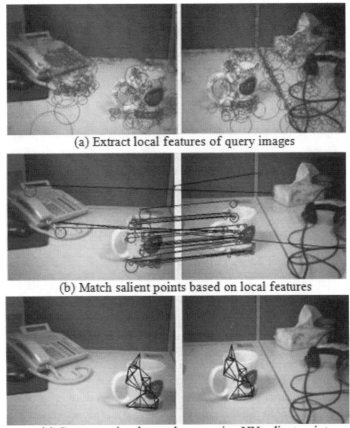

(a) Extract local features of query images

(b) Match salient points based on local features

(c) Generate visual parts by grouping NN salient points

Figure 6. Summary of main algorithms about visual parts

<u>**Visual parts generation:**</u>

Step 1: Extract local descriptors based on SIFT algorithm

Step 2: Find matched salient points in query image pair

Step 3: Select triples of salient points and group parts from the matched points that are neighbours in position.

Visual parts matching:

Step 1: Find nearest-neighbour points of individual points that belong to these query parts from indexes of GCS

Step 2: Rank images from the database by the number of matched points

Step 3: For each returned image

    Determine whether there are parts matching the query parts
    If yes, validate spatial constraints between matched parts
    Discard if no valid constraints

Step 4: Return web pages which contain the matched images in step 3.

trade-off between the computation time and the performance, we usually use two query images in our system.

Visual Part Matching in Database

The generated visual parts will be used as inputs to find matched counterparts from the database.

The first step is to get K-NNs from the database for each point in a query visual part. Then the returned points are grouped if they belong to the same image in the database.

Second, for a query image, we choose those images in the database with the largest number matched points. This procedure is as a voting mechanism. Thus, according to the matched relationship between salient points, it is easy to construct matched parts in each candidate images from the database.

The third step is to validate spatial constraints between these matched parts between the candidate images and the query parts. The candidate triple parts are rejected if they cannot meet affine transform of query parts. The simple voting count is performed based on the number of matched parts for each candidate image from the database. The spatial validation based on visual parts is actually a kind of semi-local spatial constraint which allows a greater range of global shape deformations and local deformations.

Fourth, an image succeeding in the spatial validation is judged as the one containing an identical object in the query images. Accordingly, the web pages in which these matched images locate can be processed and the relevant content can be returned to mobile users.

Experimental Results

A number of experiments were carried out to evaluate our system in the aspects of efficiency and effectiveness. We analysed the performance of our solution and compared it with the conven-

tional visual based search algorithm adopted by most of previous mobile search systems (Hare & Lewis, 2005; Yeh, et al., 2005).

In the conventional algorithm, only one image is used as query image. Then all the extracted feature vectors are input into the database to search NN as matched features. Matched results are returned based on the number of features matched between the query image and each database image. Finally, the transformation verification based on RANdom SAmple Consensus (RANSAC, Lowe, 2004) is used to validate spatial constraints of these matched points.

Datasets and Settings

To evaluate our solution, we built two image databases. One included 2,557 images of book covers that were collected from online book shops. The other consisted of 1,259 pictures of famous buildings captured by volunteers in Beijing, China. All the images were labelled with correct names.

The query sets consisted of pictures of the book covers and buildings which appeared in the above two databases. All the query images in our experiments were captured by camera phones. Their dimensions were limited to 320x240 pixels to meet the actual mobile transmission environment. For the building dataset, we captured 25 pairs of query images and on average 1,187 salient points per query image. For the book cover dataset, we captured 18 pairs of query images, on average 546 salient points per query image.

Query Time Analysis

The computational cost of our method mainly consisted of two parts: 1) generation of visual parts from query images; 2) searching matched parts in the database.

As illustrated in Table 1, our method showed better query performance than the conventional method. This is because the searching time in the

Table 1. Average query time statistics and comparison

Datasets	Visual part based method (ms)			Conventional method (ms)
	Parts extraction	Query in database	Average query time	Average query time
Building	487.4	875.8	1,363.2	3695.9
Book cover	143.1	350.7	493.8	1108.1

Table 2. Comparison of the average number of query salient points

Datasets	Visual part based method		Conventional method
	Average number of query parts	Average number of query points	Average number of query points
Building	1848	262	1187
Book cover	1428	171	546

database was proportional to the number of input query points. We greatly reduced the number of query points by generating visual parts from query images as shown in Table 2.

Precision and Recall Analysis

When the same object appeared in both the query image and the result image, we defined it as a correct match. Considering that the display area of a mobile device was limited, a mobile search system was required to maximise the precision and to ensure an acceptable recall in the top retrieval results.

Assuming that the Web pages containing Top-N matched images were presented to the mobile user, we compared the performances by choosing the different values of N from 1 to 10. As shown in Figure 7, our solution outperformed the conventional approach in most cases of the top-10 results. Several examples are given in Figure 8.

Figure 7. The precision and recall comparison on different datasets

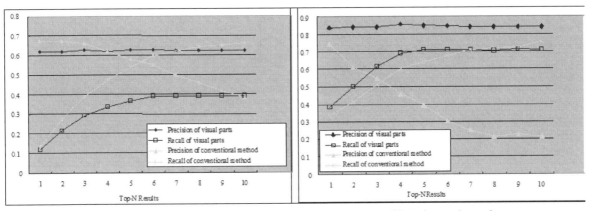

(a) Building image dataset   (b) Book cover image dataset

Figure 8. Query examples on the datasets of buildings and book covers

Figure 9. Flowchart of the location based multimodal search solution

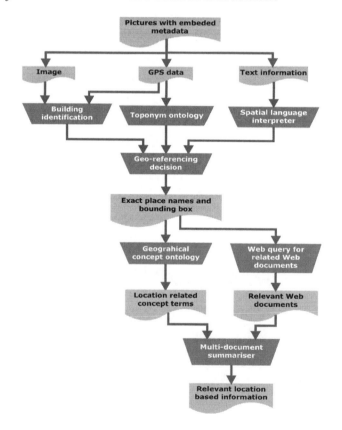

PROSPECTIVE CASE STUDY: MOBILE SEARCH FROM LOCATION

In this section, we propose a prospective multi-modal mobile search system based on an ongoing research project TRIPOD (TRIPOD, 2008) to illustrate how to obtain location related information from mobiles by multimodal search (i.e. text, GPS data and picture). The information query process can be described in a scenario: a tourist travels to

Figure 10. Building discovery by matching in the 3D spatial model

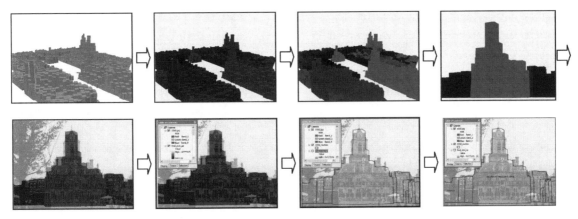

Figure 11. An example of query process and returned result

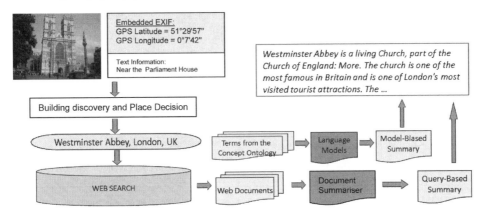

an unfamiliar city and needs some information on the surrounding place; he/she can simply take a picture (captured with GPS metadata) of a nearby landmark building and then send it with some hint text information, such as "near the Parliament House", to the proposed system. The multimodal search flowchart is illustrated in Figure 9.

For the above scenario, GPS data can be mapped into possible place names by a toponym ontology service (Abdelmoty, Smart, & Jones, 2007). The hint text is examined by a spatial language interpreter (Fu, Jones, & Abdelmoty, 2005) to help the toponym ontology service assign an accurate place name and footprint (generally a bounding box for a certain place name).

Also, with the GPS coordinates, the captured picture can be matched in a detailed 3D spatial building model in urban areas according to the shape, height etc (De Boer, Dias, & Verbree, 2008). Therefore, an exact building name can be discovered as shown in Figure 10.

We take advantage of spatial data from a Geographic Information System (GIS) to derive more semantic information about the image location. In terms of the validated place name, a comprehensive set of keywords can be extracted from a geographical concept ontology service (Purves, Edwardes, & Sanderson, 2008) to enrich the image semantics.

A number of Web documents are retrieved by the place name. Different levels of text descriptions can be summarised from the text content of these Web documents with regard to both the place name and the extracted geographical keywords (Aker, Fan, Sanderson, & Gaizauskas, 2008). The location related descriptions can be returned to the mobile user to help them know better about the surrounding areas as depicted in Figure 11.

CONCLUSION

In this chapter, an in-depth study on innovative query modalities was presented in the context of mobile search. We described how the emerging mobile voice search technology is drawing much attention and has been deployed in some commercial mobile search systems. Also, we investigated both the technical and research state of the art of promising visual and location query modalities. In terms of the discussion of the above three query modalities, a general architecture of a multimodal search system was then introduced. We concentrated on the relatively overlooked visual query modality and discussed more implementation details in a prototype mobile visual search system. A further mobile search system was proposed based on an ongoing research project, which aims to incorporate text, visual and location input modalities. In our proposed solutions, the audio/ voice modality has not yet been fully exploited. Integration of this modality will provide more natural interaction methods and probably help in refining search results.

Multimodal mobile search is highly promising from both commercial and academic perspectives. Meanwhile, there are also many interesting challenges with regard to modality incorporation, search result relevance and search speed on mobile devices.

1) Information coming from different modalities may be contradictory. Different modalities may have different reliabilities in different contexts. For example, GPS information may be erroneous or even not be available in narrow lanes and voice may be incorrectly recognised when recorded in a noisy restaurant. It's still a technical difficulty to integrate the interpretations of multiple modalities into a single semantic representation. We would plan to borrow ideas from multimodal information fusion techniques, which have been investigated for many years for the computer-human interaction systems.

2) As mobile devices only have small display screens, the amount that can be displayed is very limited. Result relevance is critical in this situation. It could be improved by introducing more multimedia features, using effective content matching algorithms and exploiting incorporation of multimodal inputs.

3) The multimodal inputs generally require searching in multimedia database. The waiting time is a more important factor for mobile users than that for desktop users. Better indexing structures and distributed search systems can offer faster search speed for mobile search systems.

It's our hope that the readers can be intrigued and reflective when they attempt to work on more effective mobile search systems.

ACKNOWLEDGMENT

This material is based upon work partly supported by the European Community in the TRIPOD project (FP6 045335).

REFERENCES

Abdelmoty, A. I., Smart, P., & Jones, C. B. (2007). Building place ontologies for the semantic web: Issues and approaches. In *Proceedings of the 4th ACM workshop on Geographical information retrieval*.

Aker, A., Fan, X., Sanderson, M., & Gaizauskas, R. (2008). *Automatic image captioning from web documents by image location*. Paper presented submitted to the 31st European Conference on Information Retrieval.

Ask.com. (2008). *Ask mobile GPS: The next level of local*. Retrieved from http://www.ask.com

Balke, W.-T., Kiebling, W., & Unbehend, C. (2003). *Personalized services for mobile route planning: a demonstration*. Paper presented at the Proceedings of 19th International Conference on Data Engineering.

Baumberg, A. (2000, June). *Reliable feature matching across widely separated views*. Paper presented at the Proceedings of IEEE Conference on Computer Vision and Pattern Recognition, Hilton Head Island, USA.

Burges, C. J. C., Platt, J. C., & Jana, S. (2003). Distortion discriminant analysis for audio fingerprinting. *IEEE Transactions on Speech and Audio Processing, 11*(3), 165–174. doi:10.1109/TSA.2003.811538

Cano, P., Batle, E., Kalker, T., & Haitsma, J. (2002). *A review of algorithms for audio fingerprinting*. Paper presented at the IEEE Workshop on Multimedia Signal Processing 2002.

ChaCha. (2008). *ChaCha. Ask away*. Retrieved from http://www.chacha.com/

Choi, D.-Y. (2007). Personalized local internet in the location-based mobile web search. *Decision Support Systems, 43*(1), 31–45. doi:10.1016/j.dss.2005.05.005

De Boer, A., Dias, E., & Verbree, E. (2008). *Processing 3D geo-information for augmenting georeferenced and oriented photographs with text labels*. Paper presented at the GIM Spatial Data Handling Conference 2008.

Fan, X., Xie, X., Li, Z., Li, M., & Ma, W.-Y. (2005, Nov.). *Photo-to-Search: Using Multimodal Queries to Search the Web from Mobile Devices*. Paper presented at the 7th ACM SIGMM International Workshop on Multimedia Information Retrieval in conjunction with ACM Multimedia 2005, Singapore.

Flickner, M., Sawhney, H., Niblack, W., Ashley, J., & Huang, Q. (1995). Query by image and video content: the QBIC system. *IEEE Computer Special Issue on Content-Based Retrieval, 28*(9), 23–32.

Fritzke, B. (1994). Growing cell structures - a self-organizing network for unsupervised and supervised learning. *Neural Networks, 7*(9), 1441–1460. doi:10.1016/0893-6080(94)90091-4

Fu, G., Jones, C. B., & Abdelmoty, A. I. (2005). *Ontology-Based Spatial Query Expansion in Information Retrieval*. Paper presented at the On the Move to Meaningful Internet Systems 2005: CoopIS, DOA, and ODBASE, Agia Napa, Cyprus.

Ghias, A., Logan, J., Chamberlin, D., & Smith, B. C. (1995). Query by humming - musical information retrieval in an audio database. *ACM Multimedia*, 231-236.

Goog411. (2008). *Find and connect with local businesses for free from your phone*. Retrieved from http://www.google.com/goog411/

GoogleMobile. (2008). *Google mobile search with my location*. Retrieved from http://m.google.com/

GPSWorld.com. (2007). *GPS Wins Out over Internet for Mobile Phone Users*. Retrieved from http://lbs.gpsworld.com/gpslbs/LBS+News/GPS-Wins-Out-over-Internet-for-Mobile-Phone-Users/ArticleStandard/Article/detail/480477

Hare, J. S., & Lewis, P. H. (2005, January). *Content-based image retrieval using a mobile device as a novel interface*. Paper presented at the Proceedings of SPIE Storage and Retrieval Methods and Applications for Multimedia 2005, San Jose, USA.

Lazebnik, S., Schmid, C., & Ponce, J. (2004). *Semi-local affine parts for object recognition*. Paper presented at the Proceedings of British Machine Vision Conference 2004, Kingston, UK.

Live.com. (2008). *Windows Live Search for Mobile*. Retrieved from http://wls.live.com/

Lowe, D. G. (2004). Distinctive image features from scale-invariant keypoints. *International Journal of Computer Vision, 60*(2), 91–110. doi:10.1023/B:VISI.0000029664.99615.94

Lu, L., You, H., & Zhang, H.-J. (2001). *A new approach to query by humming in music retrieval*. Paper presented at the Proc. IEEE Int. Conf. Multimedia and Expo 2001.

Mikolajczyk, K., & Schmid, C. (2004). Scale and affine invariant interest point detectors. *International Journal of Computer Vision, 60*(1), 63–86. doi:10.1023/B:VISI.0000027790.02288.f2

Mikolajczyk, K., & Schmid, C. (2005). A performance evaluation of local descriptors. *IEEE Transactions on Pattern Analysis and Machine Intelligence, 27*(10), 1615–1630. doi:10.1109/TPAMI.2005.188

Mikolajczyk, K., Tuytelaars, T., Schmid, C., Zisserman, A., Matas, J., & Schaffalitzky, F. (2005). A comparison of affine region detectors. *International Journal of Computer Vision, 65*(1-2), 43–72. doi:10.1007/s11263-005-3848-x

Mobot. (2008). *Mobile visual search*. Retrieved from http://www.mobot.com/

NayioMedia. (2008). Retrieved from http://www.nayio.com/

Noda, M., Sonobe, H., Takagi, S., & Yoshimoto, F. (2002, Sep.). *Cosmos: convenient image retrieval system of flowers for mobile computing situations*. Paper presented at the Proeedings of the IASTED Conf. on Information Systems and Databases 2002, Tokyo, Japan.

Nuance (2008). *Nuance voice control*. Retrieved from http://www.nuance.com/voicecontrol/

oneSearch (2008). *Get answers with search designed for mobile*. Retrieved from http://mobile.yahoo.com/onesearch

Prompt, U. (2008). Retrieved from http://www.promptu.com/

Purves, R. S., Edwardes, A., & Sanderson, M. (2008). *Describing the where - improving image annotation and search through geography*. Paper presented at the proceedings of the workshop on Metadata Mining for Image Understanding 2008.

QRCode.com. (2006). *A homepage about QR code*. Retrieved from http://www.qrcode.com/

Schaffalitzky, F., & Zisserman, A. (2002, May). *Multi-view matching for unordered image sets*. Paper presented at the Proceedings of the 7th European Conference on Computer Vision, Copenhagen, Denmark.

Schmid, C., & Mohr, R. (1997). Local grayvalue invariants for image retrieval. *IEEE Transactions on Pattern Analysis and Machine Intelligence, 19*(5), 530–535. doi:10.1109/34.589215

Sebe, N., Tian, Q., Loupias, E., Lew, M., & Huang, T. (2003). Evaluation of salient point techniques. *Image and Vision Computing, 17*(13-14), 1087–1095. doi:10.1016/j.imavis.2003.08.012

Smith, J. R., & Chang, S.-F. (1996, Nov.). *Visu- alSEEk: a fully automated content-based image query system.* Paper presented at the Proceedings of the 4th ACM International Conference on Multimedia, Boston, USA.

Smith, M., Duncan, D., & Howard, H. (2003). *AURA: A mobile platform for object and location annotation.* Paper presented at the 5th Int. Conf. Ubiquitous Computing.

411SONG. (2008). *ID songs anywhere.* Retrieved from http://www.411song.com/

Sonobe, H., Takagi, S., & Yoshimoto, F. (2004, Jan.). *Image retrieval system of fishes using a mobile device.* Paper presented at the Proceedings of International Workshop on Advanced Image Technology 2004, Singapore.

Tellme (2008). *Tellme: Say it. Get it.* Retrieved from http://www.tellme.com/

TRIPOD. (2008). *Project Tripod - Automatically captioning photographs.* Retrieved from http:// www.projecttripod.org/

Watanabe, Y., Sono, K., Yokoizo, K., & Okad, Y. (2003). *Translation camera on mobile phone.* Paper presented at the IEEE Int. Conf. Multimedia and Expo 2003.

Xie, X., Lu, L., Jia, M., Li, H., Seide, F., & Ma, W.-Y. (2008). Mobile Search With Multimodal Queries. *Proceedings of the IEEE, 96*(4), 589–601. doi:10.1109/JPROC.2008.916351

Yeh, T., Tollmar, K., & Darrell, T. (2004, June). *Searching the Web with mobile images for location recognition.* Paper presented at the Proceedings of IEEE Conference on Computer Vision and Pattern Recognition 2004 Washington D.C., USA.

Yeh, T., Tollmar, K., Grauman, K., & Darrell, T. (2005, April). *A picture is worth a thousand keywords: Image-based object search on a mobile platform.* Paper presented at the Proceedings of the 2005 Conference on Human Factors in Computing Systems, Portland, USA.

Zheng, Y., Liu, L., Wang, L.-H., & Xie, X. (2008). *Learning transportation mode from raw GPS data for geographic applications on the Web.* Paper presented at the 17th International World Wide Web Conference Beijing, China.

Zhu, Y., & Shasha, D. (2003). *Warping indexes with envelope transforms for query by humming.* Paper presented at the Proceedings of the 2003 ACM SIGMOD, San Diego, California, USA

Chapter 11
Simplifying the Multimodal Mobile User Experience

Keith Waters
Orange Labs Boston, USA

ABSTRACT

Multimodality presents challenges within a mobile cellular network. Variable connectivity, coupled with a wide variety of handset capabilities, present significant constraints that are difficult to overcome. As a result, commercial mobile multimodal implementations have yet to reach the consumer mass market, and are considered niche services. This chapter describes multimodality with handsets in cellular mobile networks that are coupled to new opportunities in targeted Web services. Such Web services aim to simplify and speed up interactions through new user experiences. This chapter highlights some key components with respect to a few existing approaches. While the most common forms of multimodality use voice and graphics, new modes of interaction are enabled via simple access to device properties, call the Delivery Context: Client Interfaces (DCCI).

INTRODUCTION

Many initial developments in multimodality focused on systems with unconstrained set of resources, such as those found in desktops, kiosks and rooms with high performance processors, large displays and hi-fidelity audio capabilities. In contrast, mobile cellular handsets are highly constrained, characterized by small screens, restrictive keyboards and intermittent network connectivity. Despite these

DOI: 10.4018/978-1-60566-978-6.ch011

restrictions, it has been suggested that multimodality should in fact be *more* useful in a wireless mobile environment (Kernchen and Tafazolli 05). This chapter endorses this view and further suggests that a simplified approach to mobile multimodality can be achieved through the incorporation of mobile device modes.

The recent emergence of the mobile Web and well-defined Web standards, allows the development of rich Web applications and services. Such data services are likely to reshape how users interact with their mobile cellular phones in the next few years.

For example, today's commercially available smart phones, such as the iPhone and handsets running the Google Android platform, are fully capable of rendering Web compliant content that go well beyond traditional limits of the Wireless Application Protocol (WAP) and mobile specific specifications of the Open Mobile Alliance (Alliance 07). Furthermore, Web standard compliant mobile browsers will reshape how mobile Web applications can be presented on mid-tier mobile cellular handsets. As a result, there are emerging opportunities to integrate novel and simplified mobile multimodal modes within a Web-based interaction.

When accessing Web services, multimodality is well suited to impoverished mobile interactions, especially when the screen is small and the inputs and outputs are both awkward and cumbersome. In such situations, interactions that follow a *path-of-least-resistance* are appropriate. For example, speech recognition can simply replace keyboard input when the user requires hands-free operation. Likewise, while some text fields can be filled via speech input, selecting items from a list using a stylus is often quicker. Larger tasks can thus be completed faster with multimodal interactions than with single modes alone. In addressing common multimodal tasks such as form filling, one must be sensitive to users' needs and leverage good design principles. Speech interactions, for example, tend to be cumbersome for spatial tasks such as identifying regions on maps (Oviatt 00). In addition, offering both modalities all the time demands careful interaction design because it can be confusing to the user as to what modes to use when. Nevertheless, it is possible to demonstrate mobile multimodal systems that complete tasks faster than a single mode alone.

Mobility introduces additional dimensions to the multimodal user experience. Users who are in motion walking, on a loading dock engaged in moving items, or driving in an automobile are usually focusing on those tasks. In such situations, a multi-input interface is not feasible. The ability to switch between modes is an important multimodal capability, especially when single-handed or hands-free operation is demanded.

It has been recognized that commercial mobile multimodality has reached a crossroads, and the challenge of mobile multimodal services have direct dependencies on the specific capabilities of mobile devices (YamaKami 07). This chapter similarly concludes that combining multiple human modes of interaction coherently into a single well understood and well defined mobile standard is a challenge. Nevertheless, an alternative approach to multimodality is presented in this chapter that can simplify the user experience through the novel use of device-based modalities. This approach is both realistic and practical.

Mobile presence is one example of a mobile device modality. Undoubtedly mobile presence coupled to location will spur new types of mobile location-based services, and it is clear that multimodal systems will be able to capitalize on these newly available modes. Importantly, on board device sensors will be able to provide unambiguous environmental status as inputs to multimodal systems. For example, *What is the status of my network?* Event notifications can represent dynamically changing properties such as *What is my device's current location?* An application's changing patterns can also be represented, for example, *Can my application automatically adapt from quiet to noisy street?* Exposing these new forms of a device's system level status at the level of Web markup, facilitate the process of integrating novel device properties which in turn can simplify mobile multimodal user experience.

The concept of exposing mobile device properties for use in multimodal systems was first sketched out within the World Wide Web Consortium (W3C) Multimodal Interaction Group (MMI) (Larson et al. 03, Waters et al. 07), first as a system and environment framework and later as a candidate recommendation *Delivery Context: Client Interfaces (DCCI)* within the Ubiquitous

Web Applications (UWA) group. The underlying concept leverages system and environmental variables to help shape the user's experience, for example by brightening the screen in dim light situations, reducing an application's power consumption based on battery levels, or modifying the interaction mode depending on network bandwidth. These new areas of multimodal content adaptation are just starting to be explored and may well prove valuable in customizing and adapting content or even an individual user experience.

Another hurdle preventing mobile multimodality development is the lack of authoring tools. Existing systems are tend to be carefully crafted by expert embedded mobile handset programmers (B'Far 05). As a result, creating new interaction scenarios is resource intensive. Providing Web markup APIs is one approach to expand the reach to Web development authors rather than programmers, thereby simplifying the mobile multimodal authoring procedure. The simple rationale is that there are an order of magnitude more Web developers than embedded mobile handset programmers.

This chapter starts by highlighting the real-world constraints that shape mobile multimodal systems. The network and mobile cellular handset components described in this chapter are two major constraints that have little to do with the core of many multimodal platforms. However, they directly influence what kind of interaction can take place. Understanding such technical limitations is necessary when developing multimodal systems for large scale mobile cellular network deployments.

Rather than presenting a new form of mobile multimodality, presents an overview of several scenarios that have commercial potential. The chapter also provides insight to key constraints which are hard to overcome, need to be addressed in multimodal systems before consumer-based mass market adoption of multimodality becomes a reality. Finally, this chapter suggests that a simplified access to modes, and especially device property status, has the potential to become an additional mode in mobile multimodality. The chapter is broken up as follows:

The *Background* section provides a context for the chapter and simple overview of some multimodal systems to date, while the following section describes some of the related work with respect to deployed mobile cellular technologies. *Multimodality for Mobile Systems* details the need for thin or fat mobile clients, while the following sections describe *Network Constraints* and *Mobile Cellular Handset Constraints* in the context of the market and operating systems. *Mobile Device Properties* and *Delivery Context: Client Interfaces* describe how device properties can be exposed to Web developers via simple APIs that can be considered modes in a multimodal system. The *Future Trends* section suggests that mobile multimodality can be enhanced though the simple addition of device modes in a mobile system. Finally, the *Conclusion* summarizes the chapter.

BACKGROUND

Human interaction is inherently multimodal, so constructing computer systems based on a taxonomy of human input and output modalities is often hypothesized to be key to successful multimodal systems (Schomaker et al. 95). Over the years, many multimodal systems have been built on this assumption through logical combinations of voice, visual, gesture and tactile modalities. Further research has revealed important misconceptions when combining modes in multimodal interaction systems; for example, a successful system is unlikely to be created from a naive combination of modes (Oviatt 99). Such research has subsequently helped shape multimodal system design.

One of the earliest multimodal implementations was MIT's *Put-that-there* system, in which a user's spoken command combined with pointing gestures could control the placement of objects on a large projection screen (Bolt 80). The system supported the concept of simultaneous modes of speech and gesture capable of evaluating commands such as *that* and *there*.

Rather than creating a whole-room environment, a more obvious multimodal system involves sitting in front of a large display with a pointing device, such as an electronic pen, while wearing a headset to pick up speech utterances used in a speech recognition engine (Oviatt and Kuhn 97). Such desktop systems are capable of synchronized voice and gesture commands such as *"How far from here to here"* while pointing to items on display.

These early systems often displayed the problems of ambiguous inputs, with voice commands alone providing insufficient information to disambiguate the command. However, with a pen input discretely pointing to a location on the screen while uttering *"here"* provides enough context to disambiguate the command and resolve the request. Logically, such a command is reduced to *"How far from (pen point(x,y))***here***to (pen point(x,y))***here***"*. The patterns of multimodal integration with spatial and temporal data present challenges to multimodal system designers.

The term *mutual disambiguation* refers to situations in which one modal input, such as speech, can be resolved with another, such as pen input (Oviatt 00). Further evidence suggests that walking, talking, and interacting simultaneously reduces the need for multiple-input modalities. While an important consideration in a multimodal system, several performance issues affect real-world mobile interactions, well before disambiguation need be considered.

Many systems collect input primarily from the user in the loop, where ambiguous input is almost guaranteed from one or more modes. As a result, many multimodal research systems concern themselves with *fusion* to create an interaction loop (B'Far 05). For example, many speech and gestures systems require one or more components of parallel input processing, signal processing, a statistical ranking to unify semantic interpreta-

tions, language interpretation and a multimodal integrator.

Such a complex set of components is built with the intent of producing a coherent set of results in a multimodal user interface. However, some components generate non-deterministic behavior and can therefore produce different results from similar input. For a large-scale commercial system designers point of view, is problematic. Generating inconsistent results from a complex system are unlikely to be reliable or robust enough in demanding real world situations. Some form of simplification is required for commercialization.

The resolution of ambiguity in a multimodal system remains an active research topic. Within industrial consumer solutions it is important that ambiguous input and output be removed thereby simplifying the overall system design and reliability. A machine mode lacks such ambiguity, for example the location of the device, or the ambient brightness of the environment, or the status of the cellular radio network. Such modes are likely to become first class modes in next-generation multimodal systems, because they generate reliable data.

Mobile multimodal systems have been created from laptops in backpacks with portable displays to investigate mobility issues (Oviatt 00). More recently, street tests have employed a portable device with a pen in a more compact format (Ehlen et al. 02). These systems essentially inherited the desktop metaphor and, literally, made them portable, while ignoring important criteria implied by their mobility. This chapter explores several reasons why this is the case. Multimodal developers have tried to adapt the familiar desktop metaphor, and have not addressed the constraints of mobile users.

Today's mobile cellular handset has a significantly different form factor than the desktop, laptop, or personal digital assistant (PDA), which presents new challenges for mobile multimodal systems. Mobile devices are designed primarily to be carried in your pocket, so they have much smaller screens than their desktop counterparts. While a mobile cellular handset's form factor is a key issue in designing multimodality, other physical device constraints are discussed in the following section. Multimodal applications designed for a desktop do not necessarily translate well to a mobile context in which the user may be walking or driving a vehicle (Tarasewich 03).

RELATED WORK

As described in the previous section, multimodality may map well into a mobile cellular environment, however to date mobile multimodality has yet to become a mass market solution. Why? Despite years of research, multimodal interfaces remain difficult to construct, author, and publish content for mobile cellular handsets. In addition, developing mobile multimodal interfaces becomes even more challenging in today's highly varied networks. To be clear this chapter focuses on mobile cellular multimodality in a mass-market mobile network. There are, of course, high bandwidth WiFi, WiMax and other localized mobile networks, but they lack consistent widespread coverage. Therefore, the network and mobile cellular handset common denominators can impose significant real-world constraints that render many multimodal systems useless. Despite the sever constraints of the mobile cellular industry, there are component technologies that are relevant and are briefly discussed below.

Device independence is a common approach to map content to the device. Authoring content for multiple devices becomes a logical way to approach device independence (Simon et al. 05). Operators use well-defined data bases of device properties to provide adaptation to extend the reach of mobile content.

Content adaptation based on device profiles are in commercial use today. User Agent Profile (UAProf) is a well understood standard for accessing static properties such as screen width and

height amongst other things. Mobile content can then be re-purposed to better suit the mobile cellular handset capabilities. Content adaptation is a common technique used by mobile operators to widen the range of devices the content can reach. The Open Mobile Alliance (OMA) published User Agent Profiles (UAProf), which provides information about the device characteristics that can be stored in a database (Wireless Application Forum 01). However, properties that change overtime are not incorporated.

Context adaptation attempt to create personalized interactions and learn new profiles, based on a prior interaction history (Salden et al. 05, Paternò and Giammarino 06). This adaptation occurs in the network in conjunction with other components of a multimodal system.

Migratory interfaces are those that allow an interface to be transposed onto another surface or device platform (Berti and Paternò 05). The concept is similar in nature to content adaptation with the added functionality of determining how to hand off a task in mid-flow to other devices. A task mapping engine essentially re-organizes graphical presentation into an audio X+V presentation for the target device.

The techniques described above use static information provided by the device, or accessed through well-defined databases. Properties such as the orientation of the device, the fact that the screen was closed in the middle of the interaction for example, are not considered in the interactions. Research in this area has yet to leverage simple properties that are modified over time. These dynamic properties can be a valuable mode in mobile multimodal system design.

MULTIMODALITY FOR MOBILE SYSTEMS

Figure 1 illustrates a simplified view of the components in a mobile multimodal system delivering Web content(6) to a mobile cellular handset(4).

Figure 1. A simple multimodal system collecting voice input and delivering data output from a Web document server

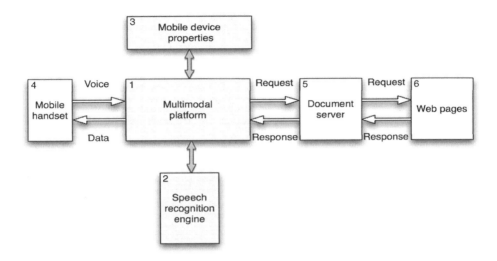

Such a system could be presenting a Web page's search results based on the voice request: *"Find me Starbucks in Boston"*. The core of the system is the multimodal platform(1) comprising a speech recognition engine(2), and mobile device properties(3). The network components are deliberately absent to illustrate key components. The sequence of events starts with speech inputs collected on the mobile cellular handset, encoded and transmitted to the multimodal platform, then decoded and passed to a speech recognition system. The results of the speech recognition are formulated into a Web request that is sent to the document server(5), which generates Web pages(6) that are passed back as data to the multimodal platform(1) which finally be shaped and rendered by a device data based stored in system properties(3) to the mobile cellular handset(4).

The multimodal platform is responsible for several tasks, but essentially it handles voice inputs and generates a graphical output response shaped by the device properties database. Inputs can be expanded to include a combination of voice or graphic commands, for example stylus/ pen or keypad/DTMF, while outputs can be voice-generated from a text-to-speech engine or graphical data generated from a Web server.

There are several ways to construct such a multimodal example. Within a Web-based context, the open standard VoiceXML (McGlashan et al. 04) is typically used for the voice channel, and XHTML (Pemberton et al. 00) for the graphical channel. Many multimodal systems consider multimodality as the integration of voice and graphics more commonly referred to as X+V, to provide a visual and voice markup combined with grammar rules (Axelsson et al. 04, Niklfeld et al. 01, Nardelli et al. 04).

Today's mobile cellular smart phones are sufficiently powerful to run most applications and services, albeit on a small screen. The following section describes wireless network and device form factor constraints in more detail.

Such an architecture poses the question of where the multimodal interaction and integration is to occur: in the network or on the device? What about the speech recognition and text-to-speech components? Clearly these are architectural

Figure 2. A typical client/server relationship in a mobile multimodal system. Some components can only be reside on a client or the server, whereas others marked with an asterisk, can be on either depending on the application.

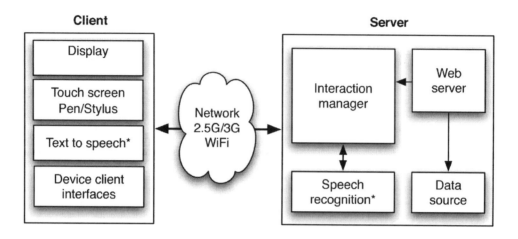

decisions that depend on how a mobile multimodal system needs to function. In some cases where device resources are limited, there may be no option but to have the primary interaction management take place on the network server. In other cases where interactions involve input with a device using a pen or stylus on a touch screen, interaction management can be performed locally. A mostly server-based interaction management entails increased network data traffic, with audio exchanged back and forth between the client and the server. Conversely, a system with client-based speech recognition and text-to-speech would work well in low bandwidth situations, and would not require constant network connectivity. Figure 2 illustrates an architecture relying on a network. Devices that rely heavily on their relationship with a server are considered "thin" clients, while "fat" clients can integrate more computationally intensive multimodal functions. A thin client may only be responsible for rendering Web content and a small DSR component (Pearce 00, Zaykovskiy 06), whereas a fat client may well contain a complete speech recognition system. Such decisions are highly dependent on the mobile cellular handset

deployment and the mobile cellular network in which the system operates.

NETWORK CONSTRAINTS

It is important to consider what type of wireless network provisioning is available for mobile multimodal systems, because it dictates what kind of interaction can take place. Global System for Mobile (GSM) represents the majority of the global wireless market, but many parts of Europe, Asia, and North America provide faster mobile networks, thereby increasing the complexity of the data service provisioning. While GSM is more ubiquitous, it cannot reasonably support the data traffic necessary for mobile multimodality. (Of course, areas only covered by GSM do not have wide access to fat clients, either.) While describing all the potential wireless provisioning configurations is beyond the scope of this chapter, a brief outline helps illustrate the trade-offs of each environment.

General Packet Data Radio Service (GPRS) is a packet-based data service supporting Short Messaging Service (SMS) and Multimedia Messaging Service (MMS). 2.5G combines 2G (based on GSM) cellular technology with GPRS data transfer, while 3G will eventually supersede 2.5G to allow broadband wireless data, including High Speed Packet Access (HSPA), with speeds up to 14.4Mbits/sec down-link and 5.8Mbits/sec on the up-link. While not immediately obvious to mobile multimodal system developers, handsets and networks that can support *simultaneous* voice and data are critical constraints to multimodality.

Mobile multimodality is usually built around the concepts of voice and data interaction. In one scenario the voice channel can be encoded as data, so the radio network need only deal with data packets within the network, feasible with a GPRS or 2.5G network. But a 3G wireless network allows the data channel to be open at the same time as the voice channel, supporting a multimodal architecture in which a control channel can be used at the same time as voice commands.

The constraints of mobile provisioning indicate different types of multimodal interaction, as follows:

- *Sequential multimodality*, often referred to as switched multimodality, was the first system developed for common usage on GSM networks. It is widely implemented on mobile cellular handsets supporting Short Messaging Service (SMS). Because no device installations are required for a functioning multimodal application, this technology has a considerable advantage over more sophisticated technologies still waiting for networks capable of supporting simultaneous voice and data. Note that in a sequential, or switched application, the user typically has to switch between voice and data modes of interaction.
- *Synchronous multimodality* goes one step beyond simple SMS applications, providing a multimodal experience that can also work with many of today's networks and more advanced mobile cellular handsets.

In a synchronous multimodal system, both speech and data are used in separate sessions. The session switching is handled automatically by the application and wireless network software. In addition, specific mobile cellular handsets with Class B radio modules need to be present. Class B mobile phones can be attached to both GPRS and GSM services, using only one service at a time. Class C mobile phones are attached to either GPRS or GSM voice service, requiring the user switch manually between services.

- *Simultaneous multimodality* systems require all network traffic to be data and operate best on 2.5G or 3G networks. Typically this requires encoding of voice data on the mobile cellular handset prior to transmission. Once encoded, all transactions can be IP packets offering mixed modes of operation on the mobile cellular handsets. They provide much higher data transfer rates than existing mobile networks, typically with several Mbits/sec up-link and down-link performance, and reduce latencies and switching artifacts introduced in earlier multimodal solutions. For example: *"Reschedule this (pointing to a calendar item) to Tuesday"* can be implemented within the network.

MOBILE CELLULAR HANDSET CONSTRAINTS

The limitations of current mobile devices impose a key constraint on multimodal systems. Mobiles come in a variety of different form factors, with a broad range of displays, keyboards, touchscreens and peripherals, running many different combinations of software on various operating systems. While smart phones offer capabilities comparable to personal computers and thus a rich environment for multimodal systems, they represent a small percentage of the global mobile cellular handset market. Today's mobile multimodal systems tend to target these mobile cellular handsets, especially when fat clients are required.

Operating systems provide the software interface between the hardware and software applications. The commercially available worldwide mobile smart phone operating systems according to Gartner (Gartner, Inc., 08) indicate that Symbian (~50%) represents the largest worldwide mobile operating system, RIM (~16%), Windows Mobile (~12%), Linux (~8%), and iPhone (~12%). Each operating system shapes what software can be run on a specific mobile cellular handset. As a result, presenting a single point of reference for a majority of commercially available mobile cellular handsets is a significant challenge; applications compiled for one target architecture may not execute on another. For Web content, however, it is essential that each device can render it faithfully, a task traditionally relegated to Web browsers. Web rendering engines such as WebKit currently provide compliance to a baseline of well understood Web markup standards such as XHTML, CSS, and ECMAScript that make rich mobile Web applications and services possible.

Mapping Web content to a large number of devices is a routine task for Operators, especially when the content is rendered within a browser. It is possible to detect the type and version of the browser, as well as the device itself that can be subsequently used to adapt the content. The goal of such techniques is to be device-independent, abstracting away the issues addressing the plethora of devices in an operator's portfolio, an approach known as *Write Once, Publish Many*. Successful content adaptation techniques depend on accuracy and maintenance of a large portfolio of device characteristics in databases that rely on User Agent Profiles (Wireless Application Forum 01). Developers have yet to full explore using content adaptation to target multimodal interactions for specific devices. Instead, mobile multimodal systems typically rely on simple switched mul-

timodality with a thin client with an SMS push, requiring little device adaptation. On the other hand, fat client devices with large color displays and on-device speech recognition may well benefit from content adaptation.

Thin clients generally have higher network demands than fat clients, because fat clients can have localized device interactions and exchange only small updates with the server. While seemingly advantageous, fat clients are usually compiled for the target architecture, and are thus less portable among devices. Thin clients rely on independent functionality resident in the server: speech recognition and text-to-speech, for example. To perform remote speech recognition, audio has to be packetized and transmitted to the server. A common way to reduce the resulting network traffic is to support distributed speech recognition (DSR), in which voice cepstral coefficients are generated on the client, then transmitted to the server for use in the speech recognition system (Pearce 00). Regardless of the capacity of the client, no clear universal solution has emerged, and the industry can expect both client/server scenarios to flourish.

MOBILE DEVICE PROPERTIES

Device sensor resources such as battery levels, signal strength, orientation, ambient brightness, temperature, proximity and location are often overlooked as first-class inputs to multimodal systems. Instead, the ambiguous human modes of speech, touch, vision are considered first in many multimodal systems, which require complex processing to create an interaction. In contrast, the machine-level modes are unambiguous and can provide precise information about the users status and mode of interaction. It is suggested that these will become the dominate modes in commercial consumer-based systems because of their simple usage and integration.

Such properties and their associated states can provide valuable input within a multimodal system to adapt and control input and output behaviors. Location based services (LBS), for example, can determine the device's approximate position in time and space. This is possible not only with Global Positioning System (GPS) data, but can also be calculated through base-station triangulation or WiFi positioning that explicitly know the location of an access point. A multimodal system can leverage these coordinates by customizing speech recognition to a narrow set of street names in that area. Localized device system properties, such as light sensors, can adapt the interaction to ensure the brightness of the display is appropriate for outdoor or indoor use.

To date, multimodal interactions primarily combine speech and graphics through XHTML and VXML markup languages; alternative modes are rarely envisioned in multimodal systems (Niklfeld et al. 01). However, mobile device-based sensors – detecting physical location, orientation, motion, ambient light, body heat, battery life, mobile signal strength and audio levels – can provide valuable input and output modes to enrich and simplify a mobile user experience.

Client-side access to device properties is in the process of being standardized though the World Wide Web Consortium (W3C) with the candidate recommendation of Delivery Context: Client Interfaces (DCCI) (Waters et al. 07). This standard exposes programming interfaces through markup, so device properties can be accessed uniformly across all devices that run Web browsers. Once the interfaces are exposed, property values can be used to adapt the multimodal interaction.

DELIVERY CONTEXT: CLIENT INTERFACES

Allowing content to adapt to the specific context of the device is the essence of Delivery Context: Client Interfaces. These client-side interfaces provide access to properties and are seen as an important component of multimodal interaction frameworks (Larson et al. 03). DCCI provides a multimodal system's Interaction Manager with easy access to a hierarchy of properties representing the current device capabilities, device configuration, user preferences and environmental conditions. These properties can indicate which modes are supported, or currently active, as well as a means to enable or disable particular modal functions and can be notified when users make their own changes. While DCCI is a client-side framework, it does not itself define any properties.

A variety of different properties of a mobile cellular handset can be exposed programmatically. For example, most devices provide some API access to the current battery level and network signal strength. Once exposed, it is useful to distinguish between persistent and transient properties. Static properties, also known as persistent properties, are those that remain constant for the duration of the session, for example a user's language preference for prompt playback. Dynamic properties, also known as transient properties, refer to notifications and events that occur during a session, for example Global Positioning System (GPS) notifications updating the location of a mobile device.

It is natural to express device properties as a hierarchy, for which the Document Object Model (DOM) provides a familiar way to express such hierarchies. The DOM also provides an eventing model, which is required for dynamically changing properties. Eventing is a powerful technique to allow asynchronous signaling to the application that a property value has changed without blocking the program's execution. Typical environmental properties involve dynamic information, such as the remaining battery life, mobile signal strength, ambient brightness, location, and display orientation. These properties vary during a session, and need to signal the application to adapt to the new environmental conditions.

Static properties are typically defined by user preferences and device context, such as the

Figure 3. The complete system on the device accessing DCCI properties

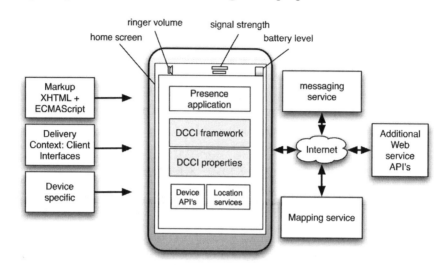

selected voice for speech synthesis, a preference for stylus input over the keypad, or preferences for font size and color. Such properties usually remain persistent for the duration of a session, and can be configured at the start of the session.

Figure 3 illustrates a prototypical architecture demonstrating an example social presence messaging application. From a user's point of view,

a map displays a friend's location, coupled to a lightweight overlay messaging interface. The ringer volume on/off settings of the friend's device determines his availability. If the volume is off, a grayed-out image indicates that the friend does not want to be disturbed.

There are several components resident on the device: the application itself written in Web

Figure 4. An example social presence messaging application constructed from DCCI properties, illustrating how device location and ringer notifications can be reflected into a Web application

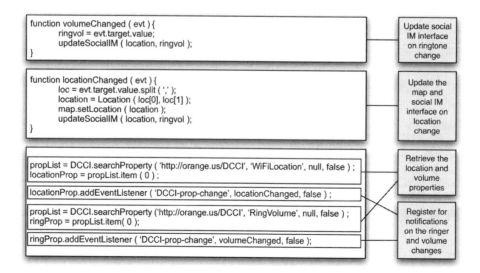

markup, the DCCI framework and the device-specific API code. In this example, the device code accesses the device ringer volume status. It is easy to see that additional device-specific APIs could be generated, for example the network signal strength or the battery level, but for this application they are not needed. The mapping and messaging services are not local to the device; instead they are accessible over the network.

Figure 4 illustrates the basic ECMAScript required to use the DCCI framework and to set up properties and notifications that can be used by the application. First the location and device ringer properties are retrieved, then registered for notifications when they change. In this application, when the ringer volume is turned off the **volumeChanged** function is called, displaying a *do-not-disturb* message to the friend. Likewise when the position of the device changes, the **locationChanged** function is called, invoking an update to the individual's location on the map. Figure 5 illustrates how this simple application appears on a mobile device. Overlaying the map is a simple visual messaging service with an icon marked to indicate that the user does not want to be disturbed.

The example illustrates how DCCI can be used in a particular social presence and mapping application. There are of course many such mapping services that use local device GPS and instant messenger clients. Nevertheless, providing a framework for accessing device properties and exposing them to the Web developer is a key asset. This is especially true when considering how important XHTML and VoiceXML have been to multimodal development to date.

FUTURE TRENDS

Mobile multimodality is fraught with constraints and barriers to entry. This chapter has described two key constraints: the mobile cellular handset and the mobile network. While they inhibit the

Figure 5. An example screen of a mobile social presence messaging application. The marker on the map indicates a friend's location, while the grayed-out image indicates that the friend does not want to be disturbed.

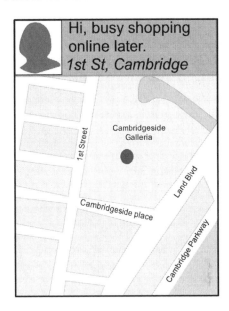

large-scale adoption of multimodal technologies for mobile, there are clear opportunities.

The advantage of multimodal interactions is their redundancy and fall-back. A mobile screen in bright sunlight can barely be visible, in which case speech can be used. When users are mobile and engaged in other activities, there's an opportunity for alternative modes of operation to address a task. These supplementary modes of interaction are often overlooked in multimodal development in favor of more sweeping, wholesale replacements of existing user interactions. This is perhaps a misguided impulse, especially considering the trend towards simplification of tasks for mobile services.

Pervasive computing presupposes that anytime and anyplace information services are always available from devices and sensors embedded in the environment (Weiser 99). In many ways the mobile phone embodies the first widely accessible mass commercial embodiment in this direction, especially considering that they are rarely further than a few meters always from us at any time. Mobile multimodality can be seen as a key enabler for pervasive computing where context-aware services can be adapted to these small devices.

Multimodal content adaptation may well prove valuable in customizing and simplifying users' mobile experience. Simple system variables such as battery level, signal strength, and ambient brightness serve as cues to change the mobile's behavior. A simple ambient noise detector may prove valuable to a user in a noisy train station, by prompting the system to request alternatives to audio input. As described with a specific example of the DCCI framework, the state of a device can be used effectively to adapt content to the mobile. In a larger sense the DCCI framework concept can extend to devices and environmental sensors that may have single-purpose functionality. A permanent room motion sensor integrated into a home system is not usually considered a mode in a multimodal system, however combine that sensor with a person in the room using voice command and control and it can be considered multimodal.

Both static and dynamic properties should provide context sensitivity for individual users and can also be considered a companion to pervasive computing, where a variety of factors influence how an application behaves ((Tarasewich 03). It is easy to imagine multimodal inputs and outputs for systems that are embedded into the fabric of the environment responding to environmental and human inputs.

CONCLUSION

This chapter highlights two key constraints that developers of commercial-grade mobile multimodal Web services must confront. These constraints have little to do with the multimodal platform in which speech and graphics are integrated, and

output generated. Instead, they are based on network characteristics and performance, the mobile cellular handset, and interaction design. Complex mobile multimodality, featuring full synchronous integration of interaction modes, is typically fragile. System designers need to pay close attention to the reliability of overall task completion before considering commercial deployments.

Multimodality's advantage over graphic- or voice-only systems is its reliance on redundant modes. If a speech recognition system fails due to a noisy environment, a graphical mode should be available to complete the task, or else Dual-Tone Multi-Frequency (DTMF) as a last resort to respond to a spoken set of options. Multimodal systems should use fallback to present the most reliable device modality, as appropriate. Speech is considered a key modality on small mobile devices with limited display capabilities. However, it is difficult to integrate reliably under typical mobile conditions when there is noise from the street, transit station, bars or cafes. High levels of city street noise can inhibit even the best speech recognition systems, even those featuring noise cancellation. Under these conditions, fallback modalities are key to successful deployments.

Further observations on mobile multimodal systems are as follows:

- Simple mobile multimodal interactions can be enhanced with device properties, especially dynamic properties that vary over time, simplifying the user's experience and expediting completion of tasks.
- Because Web technologies are standard across devices and their operating systems, device properties need to be exposed at a Web markup level, as proposed in this chapter through the use of DCCI. The binding of hardware to Web markup provides Web developers the easiest path to implement mobile multimodal interaction.
- Web-based content authoring offers a common denominator for multimodality.

Just as VoiceXML has been successful in consolidated commercial speech recognition systems, multimodal recommendations should provide a basis for designing mixed-modality systems. While markup suggests a logical way forward, there remains the task of clearly defining and communicating a common language for Web developers.

- There is little evidence to suggest that mobile users want to use more than one mode at once. Observed behaviors indicate that users tend to stay in one mode or another, at best switching among modes. There is little evidence yet that full synchronous mobile multimodal behavior is even necessary.

- Mobile device form factors vary enormously, not only their screen sizes, keyboards and processors, but also the quality of audio pick-up through embedded microphones and Bluetooth headsets. This factor alone presents a significant impediment to the large-scale adoption of multimodality. Only the simplest forms of switched multimodality on GSM networks so far offer commercial-grade reliability. A deployment strategy that considers only a limited range of high-performance mobile cellular smart phones needs to be carefully considered.

- Mobile operator network constraints and next-generation service opportunities for mobile multimodality should leverage networks' ability to carry simultaneous voice and data interactions. So far there have been few examples of this type of system.

It is feasible to simplify mobile user interactions using multimodality. Relying on common Web markup technology facilitates content and interactions that go beyond the sort of custom interfaces that for many years have dominated the mobile application space. Easy access to device properties for Web developers will be an important feature for the next generation of mobile development.

The premise that mobile multimodality can be simple remains a challenge for operators, handset manufacturers, and content authors. There needs to be a great deal of investigation of the problems discussed here before considering deployments. Mobile multimodality is clearly much different from more traditional fixed systems. The unique, pressing need to simplify the mobile multimodal experience is what elevates it to a mass-market application.

ACKNOWLEDGMENT

The work reported in this chapter is the result of Orange colleges over several years on a variety of specific multimodal systems, specific field tests and DCCI implementations. Their contributions are duly acknowledged: Benoit Simon, Mark Mersiol, and Raul Oak for early mobile multimodal systems, in-lab tests, field trials and analysis. Keith Rosenblatt, Amar Bogani and Mark Boudreau for work on DCCI. Finally, Mike Sierra for valuable editorial input.

REFERENCES

Axelsson, J., Cross, C., Ferrans, J., McCobb, G., Raman, T. V., & Wilson, L. (2004). XHTML+Voice Profile 1.2. *W3C*. Retrieved from http://www.voicexml.org/specs/multimodal/x+v/12/

B'Far. R. (2005). *Mobile computing principles*. Cambridge, UK: Cambridge University Press.

Berti, S., & Paternò, F. (2005). Migratory multimodal interfaces in multidevice environments. In *ICMI '05: Proceedings of the 7th International Conference on Multimodal Interfaces* (pp. 92–99). New York: ACM.

Bolt, R. A. (1980). 'Put-that-there': Voice and gesture at the graphics interface. *In SIGGRAPH '80: Proceedings of the 7th annual conference on Computer Graphics and Interactive Techniques* (pp. 262–270). New York: ACM. Ehlen, P., Johnston, M., & Vasireddy, G. (n.d.). Collecting mobile multimodal data for MATCH. In *Proceedings of ICSLP*, Denver, Colorado, (pp. 2557–2560).

Gartner, Inc. (2008). *Press Release: Worldwide SmartPhone Sales, 2008*. Retrieved from http://www.gartner.com.

Kernchen, K., Mossner, R., & Tafazolli, R. (2005). Adaptivity for multimodal user interfaces in mobile situations, autonomous decentralized systems. In *Autonomous Decentralized Systems, ISADS* (pp. 469–472). Washington, DC: IEEE.

Larson, J., Raman, T. V., & Raggett, D. (2003). Multimodal interaction framework. *W3C*. Retrieved from http://www.w3.org/TR/mmi-framework/

McGlashan, S., Burnett, D., Carter, J., Danielsen, P., Ferrans, F., Hunt, A., et al. (2004). Voice extensible markup language (VoiceXML) version 2.0. *W3C*. Retrieved from http://www.w3.org/TR/voicexml20/

Nardelli, L., Orlandi, M., & Falavigna, D. (2004). A multi-modal architecture for cellular phones. In *ICMI '04: Proceedings of the 6th International Conference on Multimodal Interfaces* (pp. 323–324). New York: ACM.

Niklfeld, G. Finan, R., & Pucher, M. (2001). Multimodal interface architecture for mobile data services. In *Proceedings of TCMC2001 Workshop on Wearable Computing*, Graz, Austria.

Open Mobile Alliance. (2007). *OMA Mobile Profiles*. Retrieved from http://www.openmobilealliance.org/

Oviatt, A., DeAngeli, S., & Kuhn, K. (1997). Integration and synchronization of input modes during multimodal human-computer interaction. *CHI, 97*, 415–422.

Oviatt, S. (1999). Ten myths of multimodal interaction. *Communications of the ACM, 42*(11), 74–81. doi:10.1145/319382.319398

Oviatt, S. (2000). Multimodal system processing in mobile environments. In *UIST '00: Proceedings of the 13th annual ACM symposium on User Interface Software and Technology* (pp. 21–30). New York: ACM.

Paternò, F., & Giammarino, F. (2006). Authoring interfaces with combined use of graphics and voice for both stationary and mobile devices. In *AVI '06: Proceedings of the working conference on Advanced Visual Interfaces* (pp. 329–335). New York: ACM.

Pearce, D. (2000). Enabling new speech driven services for mobile devices: An overview of the ETSI standards activities for distributed speech recognition front-ends. *Technical Report AVIOS 2000, Aurora DSR Working Group*.

Pemberton, S., Austin, D., Axelsson, J., Celik, T., Dominiak, D., Elenbaas, H., et al. (2000). XHTML 1.0 - The extensible hypertext markup language (2nd ed.) 1.0. *W3C*. Retrieved from http://www.w3.org/TR/2002/REC-xhtml1-20020801.

Salden, A., Poortinga, R., Bouzid, M., Picault, J., Droegehorn, O., Sutterer, M., et al. (2005). Contextual personalization of a mobile multimodal application. In *Proceedings of the International Conference on Internet Computing, ICOMP*.

Schomaker, L., Nijtmans, J., Camurri, A., Lavagetto, F., Morasso, P., Benoit, C., et al. &. Blauert, J. (1995). A taxonomy of multimodal interaction in the human information processing system. *Technical Report 8579, Esprit Project*.

Simon, R., Wegscheider, F., & Tolar, K. (2005). Tool-supported single authoring for device independence and multimodality. In *MobileHCI '05: Proceedings of the 7th international conference on Human computer interaction with mobile devices & services* (pp. 91–98). New York: ACM.

Tarasewich, P. (2003). Designing mobile commerce applications. *Communications of the ACM, 46*(12), 57–60. doi:10.1145/953460.953489

Waters, K., Hosn, R., Raggett, D., Sathish, S., Womer, M., Froumentin, M., et al. (2007). Delivery Context: Client Interfaces (DCCI) 1.0. *W3C.* Retrieved from http://www.w3.org/TR/DPF

Weiser, M. (1999). The computer for the 21st century. *SIGMOBILE Mobile Computing and Communications Review, 3*(3), 3–11. doi:10.1145/329124.329126

Wireless Application Forum. (2001). *User Agent Profile.* Retrieved from http://www.openmobilealliance.org/tech/affiliates/wap/wap-248-uaprof-20011020-a.pdf

Yamakami, T. (2007). Challenges in mobile multimodal application architecture. In *Workshop on W3C's Multimodal Architecture and Interfaces.*

Zaykovskiy, D. (2006). Survey of the speech recognition techniques for mobile devices. In *Proceedings of SPECOM* (pp. 88–93).

KEY TERMS AND DEFINITIONS

Context Aware: Any system or device that reacts to changes in the environment.

DTMF: Dual-Tone Multi-Frequency. This technology uses in in-band voice-frequencies as a series of tones, often referred to as touch-tones. These tones can be recognized to trigger particular behaviors.

Fall-Back Mobile Modalities: The ability of a Multimodal system to have redundancy build into the design.

GSM: Global System for Mobile. The most popular standard for mobile phones.

GPS: Global Positioning System. A system based on orbiting satellites.

GPRS: General Packet Data Radio Service. A standard that focuses on data for mobile networks.

Location Based Services: The location of a device can be used to deliver targets services. The location of a device can be calculated through a variety of means; base-station triangulation, GPS or WiFi access points.

Mobile Phone: Also known as a cellular phone used primarily as a voice communication device with limited data capabilities.

Mobile Device-Base Sensors: A collection of specific hardware components capable of determining the status of a device.

Multimodal Content Adaptation: Content that is shaped based on the user, the constraints of the device and the network.

Mutual Disambiguation: Resolving input modalities where one or more inputs are ambiguous.

Pervasive Computing: Seamless integrations into everyday devices that communicate with one-another removing the need for explicit user interaction.

Persistent Properties: Static properties.

Smart Phone: A mobile cellular phone with advanced features that are similar to a personal computer.

Transient Properties: Dynamic properties.

Section 5
New Directions

Chapter 12
Multimodal Cues:
Exploring Pause Intervals between Haptic/Audio Cues and Subsequent Speech Information

Aidan Kehoe
University College Cork, Ireland

Flaithri Neff
University College Cork, Ireland

Ian Pitt
University College Cork, Ireland

ABSTRACT

There are numerous challenges to accessing user assistance information in mobile and ubiquitous computing scenarios. For example, there may be little-or-no display real estate on which to present information visually, the user's eyes may be busy with another task (e.g., driving), it can be difficult to read text while moving, etc. Speech, together with non-speech sounds and haptic feedback can be used to make assistance information available to users in these situations. Non-speech sounds and haptic feedback can be used to cue information that is to be presented to users via speech, ensuring that the listener is prepared and that leading words are not missed. In this chapter, we report on two studies that examine user perception of the duration of a pause between a cue (which may be a variety of non-speech sounds, haptic effects or combined non-speech sound plus haptic effects) and the subsequent delivery of assistance information using speech. Based on these user studies, recommendations for use of cue pause intervals in the range of 600 ms to 800 ms are made.

INTRODUCTION

The proliferation of mobile computing devices is moving us towards a ubiquitous computing scenario of people and environments that are augmented with

DOI: 10.4018/978-1-60566-978-6.ch012

computational resources (Abowd et al., 2002). To accomplish tasks, users operate a variety of network-enabled devices such as Smart Phones, PDAs (Personal Digital Assistants) and hybrid devices that are increasingly powerful and sophisticated. Despite the best efforts of product designers to design for usability, there are still situations in which users need assistance to operate a product, access a service, or accomplish a task. In such situations, ubiquitous online assistance should be available to support the users in completing their goals.

In literature, this type of support is typically referred to as "online help" or "user assistance". Traditionally, the term "online help" refers to the documentation available to support users in their usage of software applications, e.g., "brief, task-oriented modules of information" (Harris & Hosier, 1991). This type of material, and much more, is needed to support use of the broad range of interactive products and services available to users today.

In recent years, the term "online help" has been gradually replaced in technical literature by the broader term "user assistance". One definition of user assistance is "the information channels that help users evaluate, learn, and use software tools" (BCS, 2001). This broader definition includes "other forms of online documentation, such as quick tours, online manuals, tutorials, and other collections of information that help people use and understand products" (Gelernter, 1998).

To date, much of the research relating to user assistance has been focused on the users of software packages in a desktop/laptop usage scenario, i.e., the user has a mouse, keyboard and large monitor. Studies in these usage scenarios have shown that mainstream user assistance approaches can be effective (Grayling, 2002; Hackos & Stevens, 1997; Horton, 1994), but there are also many documented difficulties and limitations associated with these approaches (Carroll & Rosson 1987; Delisle & Moulin 2002; Rosenbaum & Kantner, 2005).

User assistance systems that evolved for use in supporting desktop/laptop software applications have been adapted and are now used on mobile handheld devices, even though these devices have significantly different capabilities, form factors and usage scenarios. As a result, many of the problems associated with desktop/laptop user assistance functionality also exist on smart mobile devices. In some cases the usability issues on these platforms are even more severe.

To date, most user assistance material has been developed under the assumption that the material will be read, either on a visual display or in print format. However, display of assistance material on portable devices with small display sizes, limited resolution and fonts is difficult. There is typically very little space available to display assistance information in the context of the application user interface. Small amounts of pop-up text can be displayed, but this also risks obscuring important information in the application user interface itself. On many handheld platforms, the user must switch to an independent "help viewer" program to view assistance material, i.e., they move away from the associated application window. Such a context switch has been shown to be problematic in desktop applications (Kearsley, 1988; Hackos & Stevens 1997), and similar problems can be expected in mobile usage scenarios too. Reading on small form factor devices also presents numerous challenges (Marshall, 2002).

Speech technology can be used to enable access to user assistance material in a variety of scenarios which are problematic for traditional user assistance access methods, e.g., when the user's hands/eyes are busy, or where there is limited or no visual display, etc.

Today's mobile computing platforms frequently support audio input and output, and often have some limited haptic feedback capability. These audio and haptic capabilities can also be utilized, either with or without the use of a visual display, to support the delivery of assistance information using speech. The term multimodal user assistance is used to refer to this type of functionality.

Audio, and to a lesser extent haptic, signals are also commonly used in mobile HCI for a wide variety of purposes including: attention-grabbing alerts; providing system status information to a user (e.g., sounds are often used to signify the success or failure of an operation); and in telephony applications for branding and land-marking purposes, i.e., letting the user know where in the application they are. Such signals are typically shorter in duration than the equivalent spoken language message and they can be successfully used in conjunction with speech output.

This chapter is concerned with the use of non-speech audio and haptic signals as a cue, in supporting the delivery of assistance information to a user via speech. In the course of some initial pilot studies with auditory assistance (Kehoe et al., 2007), a number of subjects commented that there was an inadequate gap between the non-speech sound cue and the subsequent speech information. In these studies the cue was modeled as being the equivalent of a printed semi-colon, and used a value of 400 milliseconds (ms) as recommended by Cohen (2004). There is a lot of information to guide developers in the design of earcons and auditory icons (Blattner et al., 1989; Brewster, 1997; Edworthy & Hellier, 2006; Gaver, 1997), but little to guide them with respect to the subsequent pause duration that should follow an audio or haptic cue. Cohen suggested 400 ms as a starting point, and then "increase or decrease from there, letting your ear be your guide". If some users considered the 400 ms pause duration to be too short, then this raises the question: what is the optimal pause duration?

This chapter provides some background to the issues associated with the use of speech to present user assistance information, the role of audio and haptic cues, and a discussion on the range of pause intervals that occur in human speech communications. This chapter then reports the findings from two user studies that examine user perception of the duration of a pause between a cue (which may be non-speech sound, haptic, or combined non-speech sound plus haptic) and the subsequent delivery of user assistance material using speech, i.e., the assistance system presents information using speech following a cue. Based on the subject rating data collected in the study, a recommendation is made for pause intervals following audio and haptic cues.

BACKGROUND

The field of multimodal HCI has continued to expand and evolve since Bolt's (1980) "Put-that-there" demonstration in 1980. There is a significant volume of literature available to support and guide a developer working in this field. General principles underlying the development of multimodal HCI have been described (Benoit et al., 2000; Oviatt, 2002). Guidelines (Raman, 2003; Reeves et al., 2004) and frameworks (Flippo et al., 2003; MMIF, 2008) to support overall high-level system design have been proposed. Issues associated with usability evaluation of multimodal systems have also been considered (Beringer et al., 2002; Dybkjær et al., 2004).

The design and implementation of non-speech audio and haptics in mobile and ubiquitous HCI is an active area of research, and some of the related work is highlighted later in this section of the chapter. This chapter attempts to complement existing research by reporting findings from two user studies that focus on one very specific topic: the exploration of pause intervals between cues (which may be non-speech sound, haptic, or combined non-speech sound plus haptics) and subsequent information presented using speech. The studies investigated the role of a pause in allowing temporal separation of user assistance information to facilitate comprehension.

The presence and spatial distribution of blank space in a visual document is important in enabling a reader to discriminate words, sentences, paragraphs, headings, etc. When information is presented temporally, either through audio or haptics, pauses in delivery play a similar role in allowing a person to efficiently comprehend the information. A number of researchers have explored the role of pauses in speech, and the appropriate selection of pause durations has been found to improve the accuracy of detection and recall of lists of digits and letters (Duez, 1982; Goldman-Eisler, 1972; Reich, 1980).

Accessing Information via Speech

In the past, handheld mobile devices have been widely used to access voice-enabled applications and services via voice browsers that reside on the server, e.g., VoiceXML. More recent, but much less widely deployed developments, facilitate usage of voice input as a component in a multimodal interaction scenario, e.g., XHTML+Voice and SALT. These technologies are primarily server-based. Ward (Ward & Novick, 2003) makes a distinction between general information access using speech, and information access when the user requires assistance in completing a task. For example, when a user interacts with a VoiceXML application to check their bank balance, checking their bank balance is their primary task. However, in user assistance scenarios the user is already occupied with a primary task, i.e., the task for which they need assistance.

Speech is already widely used to assist users in a number of different narrow application domains. For example, speech is increasingly used in automotive applications for driving directions, and vehicle equipment control while the user's hands are busy (Geutner et al., 2000). Researchers have also explored the use of speech as part of a multimodal interface in home entertainment and public information terminals (Kirste et al., 2001). A number of consumer electronics devices, including printers and digital cameras, can now provide guided instruction to the user via speech output.

In recent years, researchers have begun to explore the issues associated with the use of speech technology to access technical and product documentation in mobile environments (Block et al., 2004; Ward, 2003). Providing access to user assistance material via speech raises a number of challenges with respect to material authoring, storage and interaction. To date, most user assistance material has been developed with the assumption that the material will be read, either on a display or in print format. However, there are significant differences between written and spoken versions of a language, and material should be authored with consideration for effective delivery using speech. Speech itself has some inherent limitations as a channel for communicating information, e.g., listening time, linear access. Kehoe and Pitt proposed a number of guidelines for authoring of assistance topics which are suitable for presentation using speech (Kehoe & Pitt, 2006).

Accessing technical product documentation can be significantly different from today's common speech assistance interactions, i.e., receiving directions while driving, directory assistance applications, etc. The information supplied using speech in these scenarios is typically very constrained. The information included in the majority of user assistance information repositories is more complex, more variable and larger in volume.

Range of Pause Durations Explored

While presenting auditory assistance material to a user on a mobile device it is important that the speech is cued so that the listener is prepared for it, otherwise some of the leading words in the assistance topic might be missed. If the pause between the cue and the subsequent speech is too short then the user may not be ready to receive the information. If the pause following the cue is too long then it wastes time and the listener may become annoyed and frustrated at the delay.

The range of pause intervals explored in the context of this study was guided by research with respect to the use of pauses in human speech (Apel et al., 2004; Campione & Véronis, 2002; Zellner, 1994), in telephony applications (Cohen et al., 2004; Larson, 2002) and sonification (Bregman, 1990).

Zellner (1994) stated that around 200 - 250 ms appears to be the lower auditory threshold for the reliable perception of pauses. Following analysis of a large database of human speech records, Campione & Véronis (2002) observed that the distribution of pauses appears as trimodal, suggesting a categorization of brief (< 200 ms), medium (200-1000 ms) and long (> 1000 ms) pauses. Long pauses occurred mainly in spontaneous speech, and were often indicative of cognitive activity, e.g., a person trying to clarify thoughts, or uncertainty.

Pauses are also frequently used as prosodic cues for turn-taking, both in human speech conversations and in speech dialogue systems. For example, a pause in a Spoken Dialogue System's output may be used to facilitate an easier "barge-in" by the user (Larson, 2002; McTear, 2004). A pause in speech input may indicate that the user is in difficulty and needs some additional information or guidance. Pause intervals in the range from 500 to 2000 ms are typically used for such conversational cue purposes (Ferrer, et al., 2002; Shriberg & Stolcke, 2004).

Based on this literature review, the range of pauses explored in the study was 300 to 1000 ms. Shorter pauses could be problematic for listeners to reliably distinguish, and longer pauses are typically only encountered in spontaneous speech and can be suggestive of uncertainty.

Using Non-Speech Sound and Haptic Cues

Speech, non-speech sound and haptics have important roles in making user assistance material accessible in a broader range of usage scenarios. Research has shown that large amounts of information can be presented using earcons and auditory icons (Blattner et al., 1989; Brewster, 1997; Edworthy & Hellier, 2006; Gaver, 1997). An earcon is defined as "non-verbal audio messages that are used in the computer/user interface to provide information to the user about some computer object, operation or interaction" (Blattner, 1989). There is no intuitive link between an earcon and what it represents, and earcons have to be learnt by the user. Auditory icons are defined as "natural, everyday sounds that are recorded in the environment and mapped to system events

by analogy. The advantage of auditory icons is that only a minimal learning effort is required to understand the connection between sound and the to-be-represented object" (Vilimek, 2005).

They can be used for a wide variety of purposes including alarms, positional markers, presenting routine messages and summary information. Papers have discussed haptics, used alone or in conjunction with non-speech sounds, being utilized for similar purposes (Brewster & Brown, 2004; Chang & O'Sullivan, 2005). There are also social, environmental and user preference issues to be considered in relation to the use of cues in mobile environments. For example, audio played through a speaker on a mobile device may not be effective in a noisy environment or socially acceptable in a meeting; haptics cues require that the user is in physical contact with their mobile device in order for the effects to be felt.

While audio is frequently used as a cue or an alert in HCI, haptics is much less so. Much of the research in the field of haptics has been in virtual reality and simulation applications using high-end equipment while striving for tactile fidelity. However, some studies have been performed that explore the use of simple haptic cues for spatial orientation, navigation, and situational awareness in driving and flying activities (Gilson & Fenton, 1974; Ho et al., 2005; Tan et al., 2001). These studies demonstrated that additional stimulation via a simple haptic cue is an effective method to enhance human attention and performance.

The role of audio and haptic cues in these studies is significantly different from their common usage as attention grabbing alerts in mobile HCI today. When used as alerts, the audio and haptic signals are typically designed to immediately attract attention (Hansson et al., 2001). These are designed to be noticeable even in a very noisy sound environment, and when the recipient is busy with another task. However, the studies described in these chapters explore the use of audio and haptic cues in supporting speech information delivery in a multimodal user assistance scenario.

Figure 1. Auditory user interface model overview

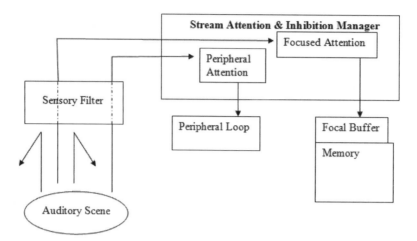

Auditory Perceptual System Model

In light of some of the perceptual issues raised in other studies, we developed a theoretical model (Figure 1) outlining the various stages of the auditory perceptual system (Neff et al., 2007). The model is based on contemporary perceptual principles such as Auditory Scene Analysis (Bregman, 1990) and the Changing-State Hypothesis (Jones et al., 1999). Rules based on these principles allow us to consider the acoustic structure of cues in relation to auditory streaming, attention and memory. The consideration of these perceptual issues in relation to subtle sonic elements, such as cue structure and cue/speech time gaps, is especially important in mobile device interface design given the visual constraints involved and the reliance on effective conveyance of auditory information.

All auditory content presented to the user in the user assistance system (all non-speech cues and speech strings) are collectively interpreted as an Auditory Scene. From this auditory scene, a primitive perceptual process organizes the scene into simpler components called streams (Bregman, 1990). The acoustic makeup of all the auditory components in the scene is the determining fac-

tor in how sounds are organized into streams. Therefore, depending on particular acoustic construction, one sound element will either join with another or segregate from it, allowing the perceptual system to identify sound sources in space and the relationship between them.

For example, a sound element consisting of a repetitive high pitch tone and a low pitch tone played sequentially at approximately walking pace will perceptually consolidate into one stream (a two pitch sequence). However, if this sequence is sped up, the two tones perceptually segregate into separate streams (a repeating low pitch tone and repeating high pitch tone). The behavior of each scenario (one stream or two streams) could potentially be very different further on in the auditory perceptual pathway (or in latter stages of the model; see Figure 1). The perceptual processing of any subsequent auditory stream, such as speech, would also be affected.

The primary concern with regard to streaming is the interaction and interference that several streams can create at all stages of the model. The initial design of our auditory scene is therefore critical to the success of relaying the correct information to the user. This usually implies the simplification of the auditory scene by removing redundant auditory information but yet efficiently transmitting as much relevant information as possible. Most concerns with respect to stream interaction and interference relate to parallel streaming, where more than one stream is concurrently presented to the user. Removing redundant auditory information from the scene greatly reduces the chances of such stream interaction. Interface designers incorporating sonic elements in mobile devices are often concerned with aesthetic rather than functional issues, but such decisions have serious ramifications in relation to the efficiency of the auditory data presented to the user.

In relation to a user assistance system interface in a mobile environment, we must be clear about the purpose of the auditory information we produce. Therefore, instead of parallel streaming, sequential streaming is the preferred option where a cue is perceptually interpreted as one stream that is temporally separated from the following speech string (the second stream). Although sequential stream presentation greatly reduces the possibilities of stream interaction, it does not mean that this approach is immune to other perceptual related pitfalls. The purpose of the cue is to notify the user that they need to pay attention to information that is on its way. It must be attention grabbing but designed and presented in a manner that does not distract from the upcoming speech stream. Other properties determined by the rules of the model are 'when' to play the cue and 'when' speech should follow. If a cue is played too soon, it loses its perceptual connection with the following speech stream and therefore loses its purpose. If it is too late, it runs the risk of occupying attention while the initial onset of the speech stream is lost.

Furthermore, a single tone cue structure is interpreted as a single stream, but potentially so too are complex cue structures depending on their acoustic makeup. Principles of aesthetics and of using complex real-world sounds come into

play, and it is possible to have more appealing sounds that fulfill the same function as simple sine tones. There is a fine line between complex cue structures that are interpreted as single streams and those that are interpreted as several competing streams. In our model, cues that break into several competing streams and those that retain their single stream structure have different knock-on effects on subsequent perceptual mechanisms such as attention and memory. In particular, we are interested in the influence of auditory/multimodal cue structure on the attention mechanism of the auditory perceptual system. The attention mechanism requires time to readjust between sequential streams (Wrigley & Brown, 2000), and in this case that consists of a non-speech or multimodal cue followed by a speech stream. Even though this scenario does not involve competing concurrent streams, which have more obvious perceptual effects, the attention mechanism remains sensitive to the acoustic structure of the initial cue stream. If the attention mechanism requires some recalibration time between events, then does the acoustic structure of the initial cue stream have an influence on the amount of time required by the attention mechanism to fully focus on the following speech stream? Based on the principle of the Changing State Hypothesis (Jones et al., 1999) where an acoustically changing cue (i.e. complex) interferes with memory rehearsal of a concurrent speech stream, we believe that a complex cue in a sequential scenario will also interfere with the speech stream, but in relation to the attention mechanism. As a result, we expect to find that longer pause times between complex cue and speech stream will be required in order for the attention mechanism to properly recalibrate and focus on the following speech content after the complex cue has been presented. In order to examine this theory, we therefore investigate the influence of cue structures (simple, complex and multimodal) on pause times.

Model Structure

In order to evaluate the effect certain non-speech cues have on subsequent speech content, we need to understand the flow and behavior of information in the human auditory system. Figure 1 presents a general overview of the model we have developed for this purpose.

One or more streams are identified from the auditory scene. For the perceptual system to cope, some of these streams will be deemed important and allowed to pass, while others will be deemed unimportant and blocked. This is the first stage of filtering, where the perceptual system simplifies the environment in order to reduce cognitive overload. In our model, the Sensory Filter performs this task. A schema hosted in memory allows the Sensory Filter to decipher which streams are important and which streams to block. This schema is a user-experience template which the user acquires over time dealing with similar scenarios. Therefore, if the user is inexperienced with the particular scenario, a less appropriate schema will be allocated to the Sensory Filter and mistakes may be made by the user (i.e. important streams may be blocked and unimportant streams pass). However, the simplification of the scene and the utilization of intuitive methods of presentation reduce the need for the user to acquire a specific schema through practice (such as using the attention grabbing 'whistle' auditory icon in our system).

After unimportant streams have been successfully blocked, a second type of filtering takes place in the form of stream attention and inhibition. Only one stream is allowed into Focused Attention, while all others reside in the Peripheral Loop. Focused Attention is attached to the Focal Buffer which has access to higher processes such as memory encoding etc. The Peripheral Loop, however, does not have such privileges but it can interfere with the processes associated with the Focal Buffer. The acoustic makeup of streams plays a significant role in this regard, and the rules pertaining to this stage of the model are based on the Changing State Hypothesis (Jones et al., 1999). Here, a stream that changes acoustically over time will interfere with a stream being encoded in memory. However, our cues are not concurrent with the speech stream and so we have theoretically avoided interferences with the speech stream in this regard. What is very significant in terms of our system design is the time it takes the Stream Attention and Inhibition Manger to drop one stream from Focused Attention when another is about to enter the cycle. We are interested in the time frame between the cue being dropped from Focused Attention and the onset of the speech stream entering Focused Attention. Given the impact that different cue structures have on concurrent speech streams, we are interested to see if there is a significant difference between unchanging cue streams, changing cue streams and multimodal cue streams when occurring sequentially with speech.

USER STUDY ONE: PAUSE DURATION AND SIMPLE CUES

The purpose of this first study was to explore the user's perception of a range of pauses between a cue and the subsequent speech information. In terms of our user model, this study would give us an indication as to the time required by the Stream Attention & Inhibition Manager to successfully terminate the cue stream in Focused Attention and uptake the onset of the speech stream. Another issue to be explored was whether the preferred pause duration varied between audio and haptic cues. The data from this study is then used as the basis for proposing guidelines for pause durations in user assistance information delivery.

Speech is temporal, and a rather slow presentation format as compared with visual alternatives. Regular users of screen reader software frequently increase the software speaking rate as they adapt to the system. Similar user adaptive behaviour

Figure 2. System overview for evaluation of multimodal cues. © 2009 Aidan Kehoe. Used with permission.

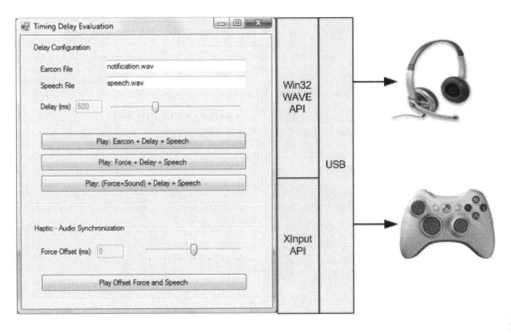

Test Platform

was observed in the previous user study conducted involving speech-enabled user assistance (Kehoe, 2007), i.e., while performing the initial task some subjects commented that the speaking rate was too fast, but this was not reported in subsequent tasks.

Our user model predicts a dynamic build-up of user experience which determines the accuracy of the initial schema template applied to the Sensory Filter (Figure 1). However, we are unsure whether this pertained to user experience and progressive exposure to pause intervals. As a result, this study was also designed to determine if there was a noticeable change in a user's perception of the pause durations over time with usage, i.e., as the subjects used the system more would they prefer shorter pauses between the cue and the subsequent speech information? If this were the case, it would be possible for a system to dynamically modify pause intervals based on a subject's use of the system.

The test hardware was a Windows Vista PC, a wired Xbox 360 gamepad (used to generate haptic cues), and a stereo USB headset (for delivery of non-speech sound cues and subsequent speech assistance topic information) as shown in Figure 2.

This study was performed using a standalone program (screenshot included in Figure 2) that was developed specifically for these studies, i.e., the test was not performed in a computer game-play context. This standalone program was required to explore the issues associated with generating accurate synchronized audio and haptic cues. The results have implications beyond the specific hardware configuration used for this study.

Synchronizing Audio and Haptic Output

People are sensitive to haptic-audio asynchrony (Adelstein et al., 2003). A number of initial experiments were performed to understand the issues associated with generation of accurate synchronized audio and haptic cues using the hardware and software components of this test platform.

Hi-fidelity haptic equipment frequently used in research (e.g., Sensable PHANTOM), and some technologies available in the mobile device space (Chang & O'Sullivan, 2005; Vibetonz, 2008), support much more sophisticated audio and haptic synchronization capabilities. For example, multi-function transducers, as used in products such as Motorola mobile phones e380, e398, and e680, can produce synchronized audible and vibrotactile output from the same audio signal. However, the test equipment in this study involved the use of off-the-shelf components (Windows Vista, Xbox 360 gamepad, USB headset) that are widely available and available at low-cost. Generation of haptic effects on battery-operated mobile devices introduces a number of challenges. These include the limited capabilities of the haptic hardware to generate effects, limited controls options available to a developer via the platform APIs, and battery life management. Researchers have explored trade-offs between the perceived magnitude of haptic effects and the impact on power consumption (Jung & Choi, 2007). However, it is also important to understand and control potential latencies in the cue generation that could have a significant impact on the results.

For example, audio output is transferred from the test application, to the Windows AudioClass driver, to the USB driver, over the USB cable to the headset microprocessor, and is finally available as an audio signal via the headset DAC outputs.

The Xbox 360 controller is typical of inertial tactile feedback controllers which use one or more motors to vibrate the controller plastic housing, and thus generate haptic effects that can be felt by a user. This is achieved through the use of an eccentric rotating mass motor that typically only offers very limited software control, i.e., usually the controlling software application can only specify a motor rotation speed. For example, the Symbian S60 3rd Edition CHWRMVibra API offers the following methods for a developer to control vibration:

- StartVibraL (TInt aDuration): Starts the device vibration feature with the product specific default intensity.
- StartVibraL (TInt aDuration, TInt aIntensity): Starts the device vibration feature. This API call allows the developer to specify an intensity level; however this functionality is not supported on all mobile devices.
- StopVibraL (): Stops the device vibration feature.

Figure 3. Spectrograph of cue and motor ramp up/down

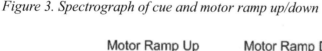

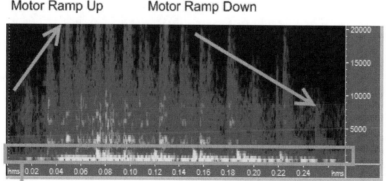

While using these devices the motor does not start up immediately; there is a ramp up time as the motor reaches the desired speed. The motor also requires some time to stop rotation, i.e., even after the motor is turned off by the software the motor vibration effects will still be generated for a period of time as the motor spins down, as shown in the spectrograph in Figure 3.

The test software was tuned to cater for the motor ramp up and ramp down times, so as to allow proper synchronization of the haptic output and audio output. A separate computer was used to examine the motor PWM signals (using the National Instruments NI USB-6008 Multifunction DAQ) and record the sounds generated by both an audio cue and the sound of the motor rotating. A sample recording is shown in Figure 3. In this figure the audio cue is a 440Hz sine wave, and the broad higher frequency spectral energy is the audio recording of the motor, with visible ramp up and ramp down times. Following these experiments the test software was modified to cater for the observed motor ramp up (40-50 ms) and motor ramp down (120-130 ms) times.

Test Process

Fourteen subjects, ranging in age from 18 to 50 years of age, participated in the study (4 female, 10 male). None of the subjects had participated in this type of study previously. The test subjects were presented with user assistance topics structured as shown in Figure 4; a cue of fixed duration, followed by a variable length pause, followed by the spoken assistance topic text. For example, if a user had requested help on a specific topic then the topic would be presented to the user as a cue

Figure 4. Topic presentation sequence

1. Cue (300ms)	2. Pause (300 - 1000ms)	3. Speech Assistance Topic

of duration 300 ms, followed by a pause, followed the spoken assistance topic.

Only simple audio and haptic cues were used in this first study; they are described in more detail later in the chapter. As part of the introduction to the study, the cues were played for the subjects, so that they were familiar with them prior to their evaluation. During the study the subjects were not occupied with other tasks. The subjects were told that they were going to be presented with an assistance topic. Thus the cue which preceded the speech assistance topic delivery was not performing the role of an attention-grabbing alert.

3Cues and User Ratings

The cue duration was 300 milliseconds. The cue could be a non-speech sound (440Hz sine wave); haptic output (single motor driven at 70% of max scale output); or combined non-speech sound plus haptic. The pause following the cue ranged in duration from 300 ms to 1000 ms, in intervals of 100 ms.

The first two pause intervals in each test sequence were at 900 ms and 500 ms, to allow the subject to become familiar with the cues. This was followed by eight more pause durations at 100 ms intervals in the range 300 to 1000 ms that were selected at random by the test software. In total, each subject evaluated 30 pause intervals (i.e., ten pause intervals repeated for the three differ-

Table 1. 7-point Likert scale used in test

1. Much Too Short	2. Too Short	3. A Little Too Short	4. About Right	5. A Little Too Long	6. Too Long	7. Much Too Long

ent cue types). For each of the 30 pause intervals the subjects were asked to rate the duration of the pause on a 7-point Likert scale (Table 1).

Since Likert scales usually do not completely satisfy the requirements for an interval scale, it was decided to use primarily non-parametric tests for the subsequent analyses. Statistical analysis was performed using Minitab software version 13.32

Results and Discussion

Figure 5 shows box-and-whisker plots of the user ratings for each of the three different cue types. One subject, who described himself as a very impatient person, rated all of the non-speech sound cues over 500 ms as being "much too long" and this is very noticeable in the charts.

At various times during the test, five of the subjects reported that they were uncertain how to classify a cue, e.g., should it be "too long" or "a little too long". One subject reported that she felt that her rankings might be inconsistent. This occurred when users were presented (at random) with pause intervals that were very close together, e.g., if a 600 ms pause was followed by 500 ms pause then the subject might express concern that they could not effectively distinguish between both durations. There are many cases in the collected data in which user rankings of adjacent time intervals were not consistent with the actual elapsed time (e.g., user ranks a 600 ms interval as just right, but ranks a 500 ms as a little too long). However, in spite of these occurrences the overall rankings of time intervals was consistent as can be seen in box-and-whisker plots in Figure 5, i.e., as X-axis time interval increases, the median user rating also increases.

A visual comparison of the graphs in Figure 5 suggests that the subjects rated non-speech sound cues and haptic cues differently, with the subjects possibly having a preference for longer pause intervals for haptic cues. Performing a two-sample Wilcoxon rank sum test of equality for both cue

interval medians (N=10) using a 95% confidence interval resulted in W=131.0 and p-value= 0.0411 supporting the hypothesis that there is a difference between the population median ratings for non-speech sound cues and haptic cues.

During a previous pilot study, and again in this study, it was observed that several subjects changed their ratings for identical pause intervals. There seemed to be a tendency for subjects to prefer shorter time intervals as they progressed through the test sequence. In this study the first pause interval to be rated by subjects was always 900 ms (to ensure subjects were familiar with the cue), and that same 900 ms interval would be rated again by each subject in the course of the nine additional intervals that were presented to the subjects at random.

Figure 6 shows data relating to the non-speech sound cue. It shows each subject ratings for the initial 900 ms pause interval, and for the second rating of that same pause interval. The median subject rating for the initial 900 ms pause interval is lower than their second rating of that same interval, i.e., as observed in an earlier pilot study.

However, performing a two-sample Wilcoxon rank sum for equality of medians for both data sets (N=14) using a 95% confidence interval resulted in the W and p-values as shown in Table 2. Therefore, the limited amount of data collected in this study does not support the hypothesis that there is a difference between the population medians.

Summary

This first study explored the user perception of a range of pause intervals, for a variety of different cue types, in the context of a user assistance scenario. The subjects in the study preferred slightly longer duration pauses for haptic cues than they did for non-speech sound cues.

In a previous pilot study it was observed that, following extended use of the system, some subjects preferred shorter pauses. If this were the case it would be possible for a mobile user assistance

Figure 5. Subject ratings of pause intervals for different cue types

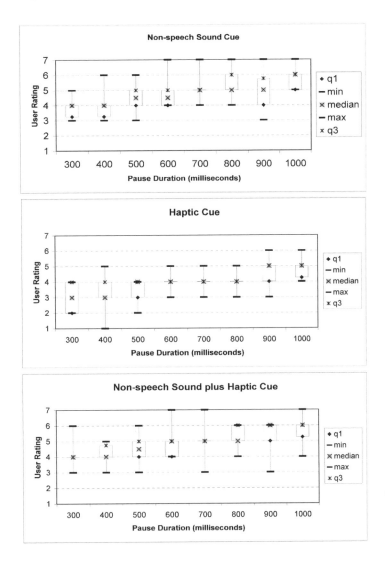

Figure 6. First and subsequent rating of 900 ms non-speech sound cue

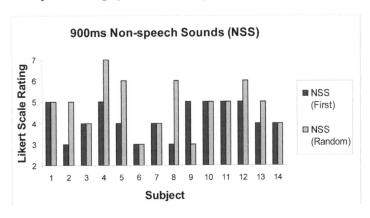

Table 2. Initial and later rating for 900 ms pause duration

Cue Type	Median (1st Rating)	Median (Later Rating)	W	p-value
Non-speech	4.5	5	180.0	0.1393
Haptic	4	5	189.0	0.2555
Non-speech+Haptic	5	6	170.0	0.0677

system to dynamically modify pause intervals based on a subject's use of the system, i.e., after a period of extended or regular use the pause duration could be shortened. However, the limited amount of data that was collected and analyzed in this study did not support this hypothesis.

USER STUDY TWO: PAUSE DURATION AND MORE COMPLEX CUES

While there has been a significant amount of research on the design and use of auditory icons and earcons, the reality is that the majority of HCI implementations continue to use rather simple and short duration audio constructs for cues. The first user study involved only very simple cues with a fixed duration of 300 milliseconds. The audio cue was a 440Hz sine wave; the haptic cue was a single motor driven at 70% of max scale output. The purpose of this subsequent study was to explore the user rating of pause duration ratings with a much wider variety of audio and haptic cues. The study also explored whether subjects would prefer longer pauses following more complex cues.

The earcons are one element and compound types using simple motives (Blattner, 1989). The earcons are constructed from simple motives that are short, rhythmic sequences of pitches (Sumikawa et al., 1986) arranged in such a way as to produce a "tonal pattern sufficiently distinct to allow it to function as an individual recognisable entity". E1 is a simple earcon using a 2kHz tone (Figure 7). E2 is a two part earcon (Figure 8). The first half of the earcon is a 2kHz tone, the second

half is 2.5kHz. E3 consists of five different tones (Figure 9).

The auditory icon is a human whistle (Figure 10). This is a sound that humans use to attract attention, i.e., ecological sound.

The section of the model, from the Stream Attention & Inhibition Manager to Memory Encoding, contains rules associated with the Changing State Hypothesis (Jones et al., 1999). This relates to how an unimportant, acoustically-changing, auditory stream (whether speech or non-speech), which is relayed concurrently with an important speech stream, can interfere with the memory rehearsal processes involved. In the model, this is specifically associated with streams in the Peripheral Loop and the Focal Buffer (the entity that supplies sonic information to memory). In our adaptation of the Changing State Hypothesis, a speech stream has two segments of data – its lexical component and its acoustic component. In the model, the lexical component is not interfered with by non-speech content, however, its acoustic component is. Because both lexical and acoustic components are not separated at this level of the perceptual system, the acoustic traits of another non-speech stream can easily interfere with the acoustic component of a speech stream and, as a result, its important lexical message. Hence, the morphing of the term Irrelevant Speech Effect to the Irrelevant Sound Effect.

Therefore, with regard to concurrent speech and non-speech streams, the more acoustically complex the non-speech stream is, the higher the chance that it will interfere with the speech stream during the rehearsal process of memory encoding. Because our cues are not concurrent,

Figure 7. Illustration of earcon E1

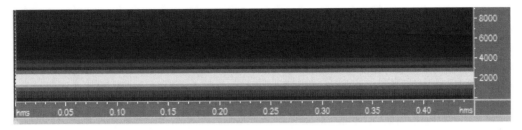

Figure 8. Illustration of earcon E2

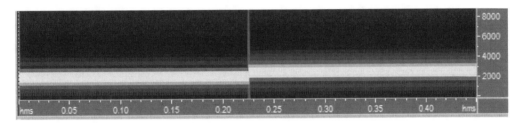

Figure 9. Illustration of earcon E3

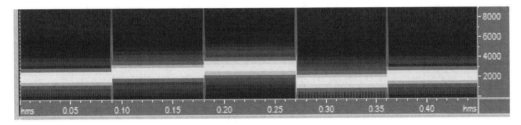

Figure 10. Illustration of auditory icon, a human whistle

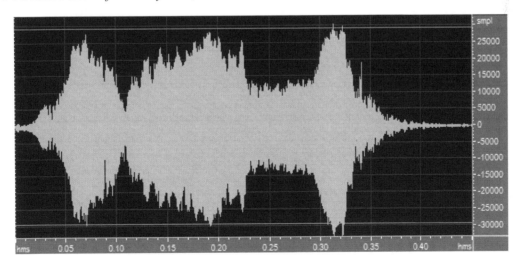

the model predicts that the chance of interference is reduced. However, it does not yet consider the possibility of attention and memory residue, where the cue is not entirely wiped from the Focal Buffer before the onset of the speech stream. Therefore, if there is such a possibility, how would a simple cue compare with a complex cue if played prior to a speech stream? If attention and memory residue is indeed a factor, perhaps more relevant is the question, how much of a pause is required to ensure that a complex cue does not result in the issues encountered by Jones et al. (Jones et al., 1999)?

Today's mobile devices are capable of rendering complex audio cues. A wide range of earcons and auditory icons are commonly used to represent events such as arrival of new mail, successful completion of an operation, the occurrence of an error, etc. Studies have demonstrated that the typical audio working environment can impact a person's performance at a variety of tasks.

Using mobile devices in a very loud environment can obviously be a challenge based on sound level alone. However, much lower sound levels in an environment can also be problematic. For example, acoustic variation (Colle & Welsh, 1976), and variation in frequency (Jones et al., 1999) and rate of change (Macken et al., 2003) have been shown to influence task performance.

Test Process

The test subjects in this study had not participated in the previous study. The subjects ranged in age from 25 to 55 years of age (6 female, 8 male). The test platform was the same as in the first study, i.e., Windows Vista PC, a wired Xbox 360 gamepad and a stereo USB headset. This test also used more complex auditory and haptic cues; these are described in more detail later in the chapter. Differences in user ratings of simple and complex cues were also explored.

As in the previous study, the test subjects were presented with user assistance topics structured

as shown in Figure 4; a cue of fixed duration, followed by a variable length pause, followed by the spoken assistance topic text. As part of the introduction to the study, all of the cues were played for the subjects so that they were familiar with them prior to their evaluation.

3Cues and User Ratings

Only pause durations of 300 ms, 600 ms and 900 ms were evaluated in the study. The cue duration was a fixed 450 ms for all cue types. Four auditory cues (three earcons and one auditory icon) were used, and two haptic cues. In addition to a simple auditory cue, more complex earcons and an auditory icon were created using the Audacity software package (audacity.sourceforge.net).

While mobile interaction has the potential to be enhanced with well-designed haptic effects, the reality is that the limitations of current generation of haptic technology available in mobile devices can be a significant barrier to implementation (Luk et al., 2006). Generating a variety of distinguishable haptics cues on the current generation of mobile phones is a challenge given the limitations of the hardware, and the APIs available to developers. However, even in situations where only a simple on/off API is supported, more complex haptic cues can be created through the use of software pulse width modulation techniques (Li et al., 2008). In the future, the availability of more sophisticated actuators (e.g., Vibetonz, 2008), and multiple actuators, will offer even more possibilities to developers on mobile platforms.

The Xbox 360 controller used in this study can also be used to generate more complex haptic cues than those used in the first study. The controller has two motors. Each of the two motor speeds can be controlled independently; both motors have different eccentric rotating masses that result in significantly different haptic effects; and the fact that the motors are located in the left and right controller handle allows for spatial variation in generated effects. H1 in this study is a simple haptic cue in which the left hand motor is driven 70% of the max scale output, i.e., the same type of cue as was used in the first study. H2 is a more complex haptic cue involving both left and right motors driven at 70% of max output. The cue starts in the left hand motor only. After 100 ms the effect starts playing on the right hand motor. After 240 ms the effect is stopped on the left hand motor (Figure 11).

For each of the 18 test cases (6 cues, 3 pause durations) the subjects were asked to rate the duration of the pause on the same 7-point Likert scale as used in the previous study.

Results and Discussion

Figure 12 shows a box-and-whisker plot of data collected in the study. In the previous study, some subjects had expressed a doubt about their ability to correctly rate pause durations. In that study the pause intervals were 100 ms. In this study, only 300/600/900 ms pause durations were used and the subjects made no such comments. In general (even though there were a few exceptions), the subjects were consistent with their rating of pauses, as can be see from the summed average ratings in Figure 13, i.e., broadly linear correlation between the pause duration and the user rating.

The data suggests that, for this broad range of cue types, the optimally rated pause duration is

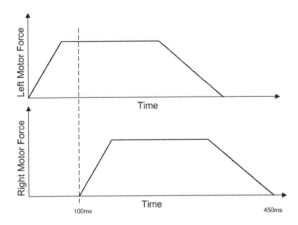

Figure 11. Illustration of haptic cue H2

between 600 ms and 800 ms. This is even clearer when viewing Figure 14, which adds linear trend lines for cues E2, AI and H2.

When a speech synthesizer is rendering text it aims to generate appropriate pauses for punctuation marks. Detail with respect to implementation of pause durations in many commercial synthesizers is not publically available, but can be estimated by making recordings of the synthesized speech and analyzing the pause durations generated for punctuation marks. Table 3 lists the estimated pause duration associated with a colon and a full-stop for a number of speech synthesizers. The recommendation for use of pause durations in the range of 600 – 800 ms results in a pause much longer than that typically associated with a colon; and probably longer than a full stop.

Figure 12. User ratings for different cues and pause durations

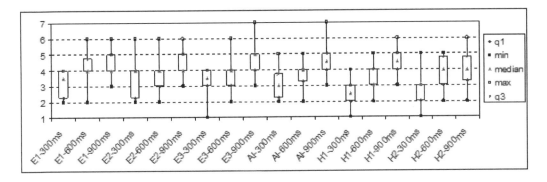

Another issue to be explored in this study was whether there was any significant difference in user rating of the pause duration for simple and complex cues. Complex cues can be more attention-grabbing, but perhaps the changing nature of these cues would require more processing by the listener, and they might prefer a longer pause duration following such cues.

For the 600 ms pause duration, several pairs of simple-complex cues were analyzed. The data in

Table 4 shows the average and median ratings for the selected simple-complex cue pairs: earcon 1 user ratings were compared with the more complex earcon 3; earcon 1 ratings were compared with the more complex auditory icon; and simple haptic H1 cue was compared with the more complex H2 cue. As shown in Table 4, in both of these cases, a two-sample Wilcoxon rank sum test (N=14) using a 95% confidence interval could not reject the hypothesis that the medians for both cues were different.

Figure 13. User ratings for different cues and pause durations

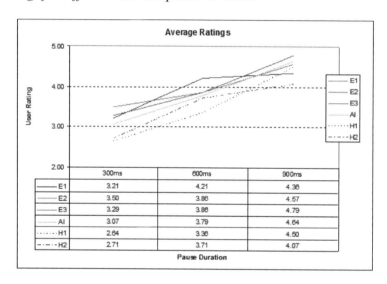

Figure 14. User ratings for different cues and pause durations

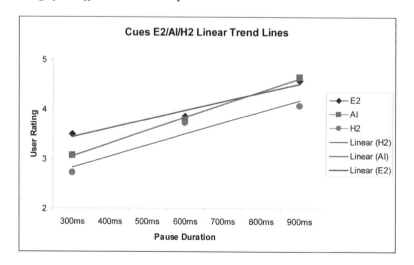

Table 3. Estimated pause interval for speech synthesizers

Synthesizer	Colon	Full stop
ATT DNV1.4 Mike	180 ms	690 ms
L&H Michael	190 ms	510 ms
Microsoft Anna (Vista)	330 ms	620 ms
DecTalk 5.0	160 ms	640 ms

Table 4. Comparison of simple-complex cue pairs

Cue Pair	Avg. 1	Avg. 2	Median 1	Median 2	p-value
Earcon E1 > Earcon E3	4.21	3.86	4.00	4.00	0.1191
Earcon E1 > Auditory Icon	4.21	3.79	4.00	4.00	0.1351
Haptic H1= Haptic H2	3.36	3.71	3.00	4.00	0.3779

Table 5. Comparison of audio (E1) and haptic (H1) cue ratings

Pause Duration	Avg. E1	Avg. H1	Median E1	Median H1	p-value
300 ms	3.21	2.64	3.50	2.50	0.1200
600 ms	4.21	3.36	4.00	3.00	0.0248
900 ms	4.36	4.50	4.00	4.50	0.7006

The first study suggested that the users preferred longer pauses for haptic cues, as compared with auditory cues. However, this finding was not repeated in all cases in this study. Table 5 lists the p-values for comparison of user ratings for earcon E1 and haptic cue H1 for N=14 and using a 95% confidence interval (i.e., these cues are the most similar to the cues used in the first study). Only the ratings for the 600 ms interval support the hypothesis that there is a difference in population medians.

FUTURE RESEARCH DIRECTIONS

Both studies were limited in the number of subjects who participated. Higher precision test equipment (with respect to the ability to control generation of haptic cues), together with a larger number of subjects, would be required to perform a more complete evaluation of users perceptions of pause intervals between cues (non-speech sound, haptic, and non-speech sound plus haptic) and subsequent speech information.

It is possible that the duration of the pause interval following the cue, or the cue type, could impact the successful completion of the subsequent task. For example, if the pause interval was very short then perhaps this could result in a lower successful task completion rate. However, there are a large number of additional factors that can impact task completion, and as a result the impact of the pause interval and cue type on successful task completion rates was not explored in these studies.

The studies have raised some important issues with regard to our user model. Further tests are required to determine the impact of user exposure to pause time intervals and whether the schema template theory in the model translates to cue-pause-speech scenarios. Also, further studies are required to determine differences between cue types (simple versus complex) and their influence on subsequent speech streams, and how pause time intervals play a role.

User assistance on mobile devices is typically integrated with, or closely associated with, a software application. However, smart mobile devices also have potential to be used as a portal to access user assistance information that is not tied to a specific application running on the device, e.g., the mobile device could be used to access assistance material on how to change the spark-plug on a lawnmower, replace a printer cartridge, check a car engine oil level, etc. We are currently exploring the use of smart mobile devices for such purposes.

The studies outlined in this chapter have been concerned specifically with user assistance scenarios. Audio, and to a lesser extent haptic, signals are also commonly used in mobile HCI as alerts and as elements in spoken dialogue systems. A more detailed exploration of optimal pause intervals following the use of audio and haptic signals in these usage scenarios is also desirable, and is being planned as future work.

CONCLUSION

The technological challenges raised by ubiquitous computing environments create serious performance and versatility issues for current user assistance system functionality. User assistance on smart mobile devices needs to employ a more sophisticated approach to effectively support users. Detailed and comprehensive assistance information is especially important for applications running on these devices because, given the typical usage scenarios, it seems unlikely the users would carry manuals with them.

Speech technology can be used to enable access to user assistance material in a variety of mobile usage scenarios which are otherwise problematic, e.g., when the user's hands/eyes are busy, or where there is limited or no visual display. Many of today's mobile computing platforms and consumer electronics devices have the ability to generate a variety of audio and haptic cues that can be used to complement delivery of user assistance information via speech.

In practice, the actual type of cue utilized could vary depending on the usage scenario, equipment capability and user preferences. Numerous studies have demonstrated improved performance (Dennerlein et al., 2000) and higher rates of user satisfaction (Brewster et al., 2007; Chang & O'Sullivan, 2005) when haptic feedback is incorporated as part of a user interface. In order to use cues effectively in user assistance it is important to understand the range of acceptable pause intervals for non-speech sound, haptic, and combined non-speech sound plus haptic cues.

Some user ratings, and the averages of all the user ratings, in the studies seemed to suggest that as they heard more topics, they would prefer shorter pause durations. However, analysis of the data using a two-sample Wilcoxon rank sum test did not support this theory. Results from the first study indicated that subjects preferred longer pause durations for haptic cues, as compared with audio cues. But this result was not replicated in all cases in the second study.

These studies suggest that, for a broad range of cues, pause durations in the range of 600 to 800 ms between the cue and the subsequent speech information would be acceptable for most users. This is at the upper range of values used for the pause duration generated by a full stop in commercial speech synthesizers. The data collected in the tests did not support the hypothesis that users prefer longer pause durations for more complex cues, e.g., single frequency earcon versus multi-frequency earcon.

The haptic capabilities of most mobile devices today are limited. However, even within the constraints of current technology, it is possible to generate a range of distinguishable haptic effects (Li et al., 2008). It is important for system implementers to be aware of potential latencies when attempting to synchronize audio output with haptic feedback generated using these low-end vibrotactile devices. In some system configurations these latencies can result in haptic effects not being felt by the user as soon as expected; and the effects may continue to be felt by the user longer than expected due to motor spin down time. These system-specific latencies must be understood and taken into consideration while generating multimodal cues on mobile devices.

REFERENCES

Abowd, G. D., Mynatt, E. D., & Rodden, T. (2002). The human experience. *IEEE Pervasive Computing / IEEE Computer Society [and] IEEE Communications Society, 1*(1), 48–57. doi:10.1109/MPRV.2002.993144

Adelstein, B. D., Begault, D. R., Anderson, M. R., & Wenzel, E. M. (2003). Sensitivity to haptic-audio asynchrony. In *Proceedings of 5th international Conference on Multimodal interfaces ICMI '03* (pp. 73-76). New York: ACM.

Apel, J., Neubarth, F., Pirker, H., & Trost, H. (2004). Have a break! Modelling pauses in German speech. In *Proceedings of KONVENS 2004, Österreichische Gesellschaft für Artificial Intelligence OEGAI* (pp. 5-12, 2004).

BCS Specialist Interest Group in Software Testing. (2001). Retrieved September 12, 2008, from http://www.bcs.org.uk/siggroup/sigist/index.htm

Benoit, C., Martin, J. C., Pelachaud, C., Schomaker, L., & Suhm, B. (2000). Audio-visual and multimodal speech systems. In D. Gibbon (Ed.), *Handbook of Standards and Resources for Spoken Language Systems.* Amsterdam: Kluwer.

Beringer, N., Kartal, U., Louka, K., Schiel, F., & Turk, U. (2002). PROMISE: A procedure for multimodal interactive system evaluation. In *Proceedings of LREC Workshop on Multimodal Resources and Multimodal Systems Evaluation* (pp. 77–80).

Blattner, M. M., Sumikawa, D. A., & Greenberg, R. M. (1989). Earcons and icons: Their structure and common design principles. *SIGCHI Bull, 21*(1), 123–124. doi:10.1145/67880.1046599

Block, H. U., Caspari, R., & Schachtl, S. (2004). Callable manuals - access to product documentation via voice. *Information Technology, 46*(6), 299–304.

Bolt, R. A. (1980). "Put-that-there": Voice and gesture at the graphics interface. In *Proceedings of 7th Annual Conference on Computer Graphics and interactive Techniques SIGGRAPH '80* (pp. 262-270). New York: ACM

Bregman, A. S. (1990). *Auditory scene analysis.* Cambridge, MA: MIT Press.

Brewster, S., Faraz, C., & Brown, L. (2007). Tactile Feedback for Mobile Interactions. In [New York: ACM.]. *Proceedings of SIGCHI Conference on Human Factors in Computing Systems CHI, 2007,* 159–162. doi:10.1145/1240624.1240649

Brewster, S. A. (1997). Navigating Telephone-Based Interfaces with Earcons. In *Proceedings of HCI on People and Computers XII H* (pp. 39-56). London: Springer-Verlag.

Brewster, S. A., & Brown, L. M. (2004). Non-visual information display using tactons. In *Proceedings of CHI '04 Extended Abstracts on Human Factors in Computing Systems* (pp. 787-788). New York: ACM.

Campione, E., & Véronis, J. (2002). A Large-Scale Multilingual Study of Silent Pause Duration. In *Proceedings of Speech Prosody 2002 Conference* (pp. 199-202).

Carroll, J. M., & Rosson, M. B. (1987). *The paradox of the active user. Interfacing Thought: Cognitive Aspects of Human-Computer Interaction.* Cambridge, MA: MIT Press.

Chang, A., & O'Sullivan, C. (2005). Audio-haptic feedback in mobile phones. In *Proceedings of CHI '05 Extended Abstracts on Human Factors in Computing Systems* (pp. 1264-1267). New York: ACM.

Cohen, M. C., Giangola, J., & Balogh, J. (2004). *Voice User Interface Design.* Boston, MA: Addison-Wesley.

Colle, H. A., & Welsh, A. (1976). Acoustic masking in primary memory. *Journal of Verbal Learning and Verbal Behavior, 15*, 17–31. doi:10.1016/S0022-5371(76)90003-7

Delisle, S., & Moulin, B. (2002). User interfaces and help systems: from helplessness to intelligent assistance. *Artificial Intelligence Review, 18*(2), 117–157. doi:10.1023/A:1015179704819

Dennerlein, J., Martin, D., & Hasser, C. (2000). Force-feedback improves performance for steering and combined steering-targeting tasks. *CHI Letters, 2*(1), 423–429.

Deuz, D. (1982). Silent and non-silent pauses in three speech styles. *Language and Speech, 25*(1), 11–28.

Dybkjær, L., Bernsen, N. O., & Minker, W. (2004). Evaluation and usability of multimodal spoken language dialogue. *Speech Communication, 43*(2), 33–54. doi:10.1016/j.specom.2004.02.001

Edworthy, J., & Hellier, E. (2006) Complex auditory signals and speech. In Wogalter, M. (Ed.), *Handbook of Warnings* (pp.199–220). Philadelphia, PA: Lawrence Erlbaum.

Ferrer, L., Shriberg, E., & Stolcke, A. (2002). Is the speaker done yet? Faster and more accurate end-of-utterance detection using prosody in human-computer dialog. In . *Proceedings of LICSLP, 2002*, 2061–2064.

Flippo, F., Krebs, A., & Marsic, I. (2003). A framework for rapid development of multimodal interfaces. In *Proceedings of 5th international Conference on Multimodal interfaces ICMI '03* (pp. 109-116). New York: ACM.

Gaver, W. (1997). Auditory interfaces. In M. Helander. T Landauer, & P. Prabhu (Eds.), *Handbook of human-computer interaction (2nd ed.)* (pp. 1003-1042). Amsterdam: Elsevier.

Gelernter, B. (1998). Help design challenges in network computing. In *Proceedings of 16th Annual international Conference on Computer Documentation SIGDOC '98* (pp. 184-193). New York: ACM.

Geutner, P., Arevalo, L., & Breuninger, J. (2000). VODIS - voice-operated driver information systems: a usability study on advanced speech technologies for car environments. In . *Proceedings of, ICSLP-2000*, 378–382.

Gilson, R. D., & Fenton, R. E. (1974). Kinesthetic-tactual information presentations - inflight studies. *IEEE Transactions on Systems, Man, and Cybernetics, 4*(6), 531–535. doi:10.1109/TSMC.1974.4309361

Goldman-Eisler, F. (1972). Pauses, clauses, sentences. *Language and Speech, 15*, 103–113.

Grayling, T. (2002). If we build it, will they come? A usability test of two browser-based embedded help systems. *Technical Communication, 49*(2), 193–209.

Hackos, J. T., & Stevens, D. M. (1997). *Standards for online communication*. New York, NY: John Wiley & Sons, Inc.

Hansson, R., Ljungstrand, P., & Redström, J. (2001). Subtle and public notification cues for mobile devices. In *Proceedings of 3rd international Conference on Ubiquitous Computing* (pp. 240-246). London: Springer-Verlag.

Harris, R. A., & Hosier, W. J. (1991). A taxonomy of online information. *Technical Communication, 38*(2), 197–210.

Ho, C., Tan, H. Z., & Spence, C. (2005). Using spatial vibrotactile cues to direct visual attention in driving scenes. *Transportation Research Part F: Traffic Psychology and Behaviour, 8*, 397–412. doi:10.1016/j.trf.2005.05.002

Horton, W. (1994). *Designing and Writing Online Documentation*. Wiley: New York, NY.

Jones, D. M., Saint-Aubin, J., & Tremblay, S. (1999). Modulation of the irrelevant sound effect by organizational factors: Further evidence from streaming by location. *The Quarterly Journal of Experimental Psychology, 52*(3), 545–554. doi:10.1080/027249899390954

Jung, J., & Choi, S. (2007). Perceived magnitude and power consumption of vibration feedback in mobile devices. In [Berlin, Germany: Springer-Verlag.]. *Proceedings of Human-Computer Interaction HCII, 2007*, 354–363.

Kearsley, G. (1988). *Online help systems: design and implementation*, Norwood, NJ: Ablex Publishing.

Kehoe, A., Neff, F., Pitt, I., & Russell, G. (2007). Improvements to a speech-enabled user assistance system based on pilot study results. In *Proceedings of 25th Annual ACM international Conference on Design of Communication SIGDOC '07* (pp. 42-47). New York: ACM.

Kehoe, A., & Pitt, I. (2006). Designing help topics for use with text-to-speech. In *Proceedings of 24th Annual ACM international Conference on Design of Communication SIGDOC '06* (pp. 157-163). New York: ACM.

Kirste, T., Herfet, T., & Schnaider, M. (2001). EMBASSI: multimodal assistance for universal access to infotainment and service infrastructures. In *Proceedings of 2001 EC/NSF Workshop on Universal Accessibility of Ubiquitous Computing: Providing For the Elderly WUAUC'01* (pp. 41-50). New York: ACM.

Larson, J. A. (2002). *Voicexml: Introduction to developing speech applications*. Upper Saddle River, NJ: Prentice Hall.

Li, K. A., Sohn, T. Y., Huang, S., & Griswold, W. G. (2008). Peopletones: a system for the detection and notification of buddy proximity on mobile phones. In *Proceedings of 6th international Conference on Mobile Systems, Applications, and Services Mobi-Sys '08* (pp. 160-173). New York: ACM.

Luk, J., Pasquero, J., Little, S., MacLean, K., Levesque, V., & Hayward, V. (2006). A role for haptics in mobile interaction: initial design using a handheld tactile display prototype. In [New York: ACM.]. *Proceedings of SIGCHI Conference on Human Factors in Computing Systems CHI, 06*, 171–180. doi:10.1145/1124772.1124800

Macken, W. J., Tremblay, S., Houghton, R. J., Nicholls, A. P., & Jones, D. M. (2003). Does auditory streaming require attention? Evidence from attentional selectivity in short-term memory. *Journal of Experimental Psychology. Human Perception and Performance, 29*(1), 43–51. doi:10.1037/0096-1523.29.1.43

Marshall, C. C., & Ruotolo, C. (2002). Reading-in-the-small: a study of reading on small form factor devices. In *Proceedings of 2nd ACM/IEEE-CS Joint Conference on Digital Libraries JCDL '02* (pp. 56-64). New York: ACM.

McTear, M. F. (2002). Spoken dialogue technology: Enabling the conversational user interface. In *Proceedings of 1-85 ACM Computing Surveys (CSUR)*. New York: ACM Press.

MMIF. (2008). *W3C multimodal interaction framework*. Retrieved September 12, 2008, from http://www.w3.org/TR/2008/WD-mmi-arch-20080414/

Neff, F., Kehoe, A., & Pitt, I. (2007). User modeling to support the development of an auditory help system. In *Lecture Notes in Computer Science. Text, Speech and Dialogue*. Berlin, Germany: Springer.

Oviatt, S. L. (2002). Breaking the robustness barrier: recent progress on the design of robust multimodal systems. In M. Zelkowitz (Ed.), *Advances in Computers* (pp. 56). San Diego, CA: Academic Press.

Raman, T. V. (2003). User Interface Principles for Multimodal Interaction. In *Proceedings of MMI Workshop CHI 2003*.

Reeves, L. M., Lai, J., Larson, J. A., Oviatt, S., Balaji, T. S., & Buisine, S. (2004). Guidelines for multimodal user interface design. *Communications of the ACM, 47*(1), 57–59. doi:10.1145/962081.962106

Reich, S. S. (1980). Significance of pauses for speech perception. *Journal of Psycholinguistic Research, 9*(4), 379–389. doi:10.1007/BF01067450

Rosenbaum, S., & Kantner, L. (2005). Helping users to use help: Results from two international conference workshops. In *Proceedings from Professional Communication Conference IPCC 2005* (pp. 181–187).

Shriberg, E., & Stolcke, A. (2004). Direct Modeling of Prosody: An Overview of Applications in Automatic Speech Processing. *Speech Prosody, 2004*, 575–582.

Sumikawa, D., Blattner, M., & Greenberg, R. (1986). *Earcons: Structured audio messages.* Unpublished paper.

Tan, H. Z., Gray, R., Young, J. J., & Irawan, P. (2001). Haptic cueing of a visual change-detection task: Implications for multimodal interfaces. In *Usability Evaluation and Interface Design: Cognitive Engineering, Intelligent Agents and Virtual Reality* (pp. 678-682). Philadelphia, PA: Lawrence Erlbaum Associates.

VibeTonz System. (2008). *Immersion Corporation.* Retrieved September 12, 2008, from http://www.immersion.com/mobility

Vilimek, R., & Hempel, T. (2005). Effects of speech and non-speech sounds on short-term memory and possible implications for in-vehicle use. In *Proceedings of ICAD 2005.*

Ward, K., & Novick, D. G. (2003). Hands-free documentation. In *Proceedings of 21st Annual international Conference on Documentation (SIGDOC '03)* (pp. 147-154.). New York: ACM.

Wrigley, S. N., & Brown, G. J. (2000). A model of auditory attention. *Technical Report CS-00-07, Speech and Hearing Research Group.* University of Sheffield, UK.

Zellner, B. (1994). Pauses and the temporal structure of speech. In E. Keller (Ed.), *Fundamentals of Speech Synthesis and Speech Recognition: Basic Concepts, State-of-the-Art and Future Challenges* (pp. 41-62). Chichester, UK: John Wiley and Sons Ltd.

Chapter 13
Towards Multimodal Mobile GIS for the Elderly

Julie Doyle
University College Dublin, Ireland

Michela Bertolotto
School of Computer Science and Informatics, Ireland

David Wilson
University of North Carolina at Charlotte, USA

ABSTRACT

Information technology can play an important role in helping the elderly to live full, healthy and independent lives. However, elders are often overlooked as a potential user group of many technologies. In particular, we are concerned with the lack of GIS applications which might be useful to the elderly population. The main underlying reasons which make it difficult to design usable applications for elders are threefold. The first concerns a lack of digital literacy within this cohort, the second involves physical and cognitive age-related impairments while the third involves a lack of knowledge on improving usability in interactive geovisualisation and spatial systems. As such, in this chapter we analyse existing literature in the fields of mobile multimodal interfaces with emphasis on GIS and the specific requirements of the elderly in relation to the use of such technologies. We also examine the potential benefits that the elderly could gain through using such technology, as well as the shortcomings that current systems have, with the aim to ensure full potential for this diverse, user group. In particular, we identify specific requirements for the design of multimodal GIS through a usage example of a system we have developed. Such a system produced very good evaluation results in terms of usability and effectiveness when tested by a different user group. However, a number of changes are necessary to ensure usability and acceptability by an elderly cohort. A discussion of these concludes the chapter.

INTRODUCTION

New information technology (ICT) can play a very important role in the quality of life of elderly people who wish to continue living autonomously as they age. For example, mobile phones enable people to stay connected, in addition to hosting services such as location-based applications that aid people navigating and locating points of interest.

DOI: 10.4018/978-1-60566-978-6.ch013

Furthermore, using such applications encourages cognitive activity among its users, which is important in keeping an active mind. However, technology, particularly mobile technology, can be difficult for elderly users to accept and adopt for a number of reasons. Firstly, the vast majority of the current, and near future, elderly cohort are not digitally literate as they did not grow up during this technological age. Secondly, elders often suffer from age-related impairments. Cognitive disabilities resulting from age degenerative processes, can significantly increase the learning curve of an elder, making it difficult for them to learn new skills. Additional cognitive problems include limited short-term memory, lower co-ordination capacity, lower sensory capability and slower ability to react (Dong et al, 2002). Physical impairments, related to sensory loss, are another obvious effect of ageing (Hawthorn, 2000). Such impairments affect visual, auditory and tactile capabilities, as well as speech intelligibility, further distancing the elderly from technology. Another significant hindrance to adopting technology is the elderly's' sometimes negative attitudes towards technology, which they perceive as unfamiliar or unnecessary. Finally, many systems themselves are often poorly designed and hence complex to use, even for the average user. Therefore, it is critical that designers of systems consider the requirements of the elderly, who have vastly different needs to the average-aged user, if such systems are to become more widely useful to a larger context of users.

Multimodal interfaces can potentially enable elder users to benefit from technology. Verbal communication between humans is often supplemented with additional sensory input, such as gestures, gaze and facial expressions, to convey emotions. Multimodal systems that process two or more naturally co-occurring modalities, aim to emulate such communication between humans and computers. The rationale for multimodal HCI is that such interaction can provide increased naturalness, intuitiveness, flexibility and efficiency for users, in addition to being easy to learn and use (Oviatt et al, 2000). The naturalness and intuitiveness of multimodal HCI are important factors in decreasing the complexity of applications that employ multimodality. This in turn may contribute to decreasing the learning curve of the elderly, when learning to use such applications. Furthermore, it may reduce the user's apprehension towards technology as multimodal interaction, using speech and gestures for example, may seem a more familiar means of interacting with a computer, than the traditional mouse and keyboard. Motor and sensory issues relating to elderly users can be overcome through the ability for them to choose the mode of interaction that best suits their capabilities. For example, an elderly user with arthritic problems might find speech a preferable interaction mode to handwriting or touch. On the other hand, a user whose speech might be unclear due to a lack of articulation for example, or who has an auditory impairment might prefer to interact through touch or gesture. Each of these potential benefits, coupled with the fact that multimodality has been shown to reduce complexity of human computer interaction in many application domains, makes multimodality an ideal paradigm with which to help elderly users begin to embrace technology.

In addition to intuitive input modalities, the large range of relatively inexpensive mobile devices currently available, ensure applications supporting multimodality are available to a broad range of diverse users in society, including the elderly. As such, multimodal interfaces are now incorporated into various applications contexts, including healthcare (Keskin et al, 2007), applications for vision-impaired users (Jacobson, 2002), independent living for the elderly (Sainz Salces et al, 2006) and mobile GIS to name but a few. This latter application area represents the focus of our research interest and in this chapter we examine how integrating multimodality into mobile GIS can contribute to systems that are usable and useful for the elderly.

Mobile geospatial applications have recently surged in popularity among a wide range of non-professional user groups. Such applications are now being used by a diverse range of users to carry out a vast range of activities. Such activities include navigating an area, getting directions, locating points of interest (POI) and discovering information regarding these POIs. The elderly are a large group of potential users who should be afforded the ability to partake in and benefit from such activities, if they wish to do so. Furthermore, mobile GIS supports many recreational activities that may be enjoyed by elderly users, such as fishing or Geocaching. Activities such as these help to promote social connection among the elderly, as well as providing them with a cognitive task to tackle. For example, navigating to an annotated location on a map which another fisherman has noted is a good spot for fishing, or working as part of a team to find 'treasure' while Geocaching.

Designing multimodal interfaces for mobile geospatial applications will help to ensure increased access to spatial information for the elderly and as such, helps to promote universal access. For those users with little computing experience of using buttons and menus to navigate and query, voice commands and pointing gestures may provide a more natural and intuitive means of communication with a GIS. In addition, information presented through different modalities enables a user to adapt to the format of information display that suits their own cognitive learning style, or that they have sensory access to (Wickens & Baker, 1995). For example, non-visual interfaces, using tactile, haptic and auditory modalities, can potentially increase the use of spatial applications to vision-impaired users. A further benefit of mobile GIS to the elderly population is the ability for such mobile applications to act as a tracking device for family members. The presence of GPS may enable a son or daughter to track their elderly parent, providing peace of mind as to where they are and whether they are keeping active. As such, mobile geospatial applications supporting multi-modal interfaces can significantly aid in fostering independent living for the elderly, irrespective of their age, skill level or impairments.

Despite the potential benefits of designing multimodal mobile GIS with elderly users in mind, little research within the field considers these users. The primary goal of this chapter is to explore how the body of work concerning the elderly and mobile and multimodal applications can be used to support research into ensuring GIS applications are designed with the elderly as a potential user group in mind. We analyse existing literature in the fields of mobile computing and multimodal interfaces for the elderly, and examine how such research can be applied to increasing the usability of GIS for the elderly. While we are aware that many elders may have no interest in using any form of ICT, including mobile GIS, we feel it is important to facilitate interaction for those who do. Our wish is not to 'push' technology usage on elders, but rather to assist those who are interested in using it. As such, we believe it is vital to design applications and interaction with this cohort of users in mind. The primary contribution of this chapter, therefore, concerns an overview of what we can learn from the current literature as well as a discussion of the benefits, issues and challenges, in addition to an identification of what still needs to be done, to foster the design of mobile multimodal GIS for the elderly.

The remainder of this chapter is organised as follows. The following section sets the background for our chapter, highlighting the characteristics of the elderly population which must be taken into account when developing technological applications for such users. It focuses on related literature in the areas of both mobile and multimodal applications for the elderly, highlighting the benefits and challenges of designing such systems for this user group. The third section examines how this research can be applied within the field of GIS. It also examines the complexities of GIS that need to be addressed to ensure such applications both appeal to the elderly, in addition to being easy to use. Section 4 introduces the multimodal mobile GIS we have developed, which to date has targeted younger users. Using this system as a case study, we outline how it might be adapted to be suitable for elderly usage. Finally, we conclude with a practitioner's 'take-away' for designing and developing multimodal mobile GIS for the elderly, and outline some areas of future work.

BACKGROUND AND RELATED WORK

This section provides some background research on the elderly as a cohort, in addition to current research examining aspects of mobile and multimodal application use by the elderly. Mobility and multimodality are two very important characteristics of geospatial applications which increase their usage context and usability for many diverse user groups.

The Elderly Cohort

The ageing population is rapidly increasing and people over the age of 65 now represent a large section of society. As such, efforts should be made to integrate the elderly into society as much as possible, ensuring that they enjoy a good standard of living, well into old age. One means of achieving this is to ensure new technology is created with the elderly cohort as potential end users in mind, enabling them to reap the benefits of many different types of technological applications. Within the context of this chapter, we clarify 'technology' to mean ICT. While we are aware that elders may be technically skilled in other fields, ICT is a relatively young discipline, to which younger adults and children have had more exposure. As such, it requires more effort to ensure technological applications are usable for elders. It is also important to mention that a large proportion of today's population under the age of 50 are mostly technologically adept. As such,

these users will expect to be able to continue using information technology in the future and this must be possible even in the midst of age-related physical or cognitive impairments.

An important point to note is that the elderly population is an extremely heterogeneous group, representing a very broad and diverse set of characteristics and different degrees of disabilities and impairments (Jorge, 2001). Firstly, there may be large differences in age, for example between a 65 year old and an 85 year old and such differences may manifest themselves in different levels of impairments, mobility, capabilities etc. The mind-set of elders is another distinguishing factor. Many may want to be independent and self-sufficient and hence may be very open to adopting new technologies that may help them achieve this state of living. Others, however, may live a closed life and may not be willing to try something new. These, combined with a multitude of additional characteristics of ageing, may affect how elders can interact with technology. As adults age, they begin to experience a wide range of age-related physical impairments, such as hearing and vision loss and reduced mobility, as well as cognitive impairments such as memory loss and difficulty in absorbing information. Such impairments vary from individual to individual and also within individuals from day to day. The effects of these impairments include a slower learning curve, a loss of confidence, a fear to try something new and difficulties in spatial orientation. Sainz Salces et al (2006) provide a good discussion on the physical and cognitive affects of ageing.

Apprehension towards technology and a reluctance to accept it, is a further issue which may hinder an elder's uptake of technological applications. However, as computers and technology play an increasing role in everyday life, it becomes necessary to identify and address those issues that cause such fear or reluctance, as a thorough understanding of the elderly's misgivings, as well as their requirements, is essential to increase acceptability. It is necessary that elders perceive the usefulness of the technology in their lives as a benefit great enough to outweigh the effort required to learn it.

One way in which technology may prove beneficial to elders is in facilitating social connectedness. As people grow older, their peer network shrinks and they may become inclined to spend more time in their home and less inclined to perform activities outdoors. In addition, friends and family may live far away, or be busy, adding to an elder's isolation. A number of other factors may also contribute to social isolation, including a fear of falling (if this is something an elder has already experienced), which may lead to a fear of going outside or a feeling of not wanting to burden family and friends. While technological intervention cannot take the place of human contact, it can help to facilitate interactions and social activity among elders. Technology usage may facilitate interaction in certain situations (such as with applications of GIS, which we will discuss later). However, involving elders in the design process is also another very effective way to connect them socially to their peers. For example, involving a number of elders in a recurring focus group during an iterative design and testing process may facilitate new friendships and may provide the elder with a feel of actively contributing to a 'worthy cause'.

While it is generally agreed that older people are capable of learning new skills (Gardner & Helmes, 1999), it is necessary for designers of applications that target elders as a user group, to ensure applications are designed with their unique needs in mind. The following two subsections address related research within the fields of mobile application usage by the elderly and multimodal interaction issues for the elderly.

The Elderly and the Mobile Context

Within the context of this chapter, mobility refers to both mobile devices and the issues involved in interacting whilst moving. While the availability and usage of mobile devices has increased significantly in recent years, alongside advances in device technology, there are still many limitations associated with such devices which can have negative effects on the usability of mobile applications. A significant problem with mobile devices is that they attempt to give people access to powerful computing services through small interfaces, which typically have extremely small visual displays, poor input techniques and limited audio interaction facilities. Furthermore, use in motion affects the usability of mobile applications as it may be difficult to interact with such an application with the limited input methods typically supported. Such limitations may affect all users of mobile applications, irrespective of age, technical know-how or physical capabilities. However, when we consider the physical and cognitive issues associated with ageing, it becomes even more critical to address human computer interaction challenges associated with mobile device technology for the elderly population.

There are many application areas of mobile technology which may offer numerous benefits to the elderly. These include mobile phones, mobile medical applications (Lorenz et al, 2007; Nischelwitzer et al, 2007) and GIS ((Kawamura, Umezu & Ohsuga, 2008)). Carrying a mobile phone enables an elder to reach, and be reached by, family and friends wherever they may be. Indeed, for the vast majority of people of all ages, a mobile phone is considered an essential accessory of everyday life. While today's generation of mobile phones have a wide range of functionalities, the elderly population use mobile phones mostly for spoken communication. Thus, ensuring a large enough text size and keypad, in addition to a loud enough ring volume, may help disabled users with vision and hearing impairments to use a mobile phone. A user study carried out in (Lorenz et al, 2007) evaluates a number of different interface layouts, on different mobile hardware platforms (handhelds), to determine the most desirable design for a mobile health monitoring application. Throughout the user centred design process outlined in such a paper, the authors gain valuable feedback from elders regarding design requirements. For example, users with sight problems preferred a text size of up to 36pt. One user with a hand tremor requested a button size of at least two finger widths. Further requests included little, or no, animation, no auto-scrolling text and a loud acoustic sound and strong vibration when alarming the user in the event of a distinctive value.

A recent case, outlined in an online article, demonstrates the difficulty in convincing mobile phone retailers that the elderly constitute a large and growing market (Vnunet.com, 2007). A mobile phone aimed specifically at the elderly has been manufactured by an Austrian company. This phone features a large screen display, oversized and easy to use buttons, a louder than usual speaker and ring volume and a powerful vibration alert. In addition, the phone has an emergency button which can be programmed to call up to 5 people for help. Despite the many benefits of such a design for the large elderly population, the manufacturer cannot convince any mobile networks in the UK to retail their product, as the networks are driven very much by the youth market. As such, a further challenge arises to ensure mobile applications for the elderly, once designed and implemented, actually make it into the retail market. Plos and Buisine (2006) describe a case study of universal design applied to mobile phones which, while catering to the unique needs of elderly or disabled users, resembles a regular mobile phone and hence avoids the stigmatising effect of being labelled solely for the elderly or disabled. This is an important issue to keep in mind when designing mobile applications for the elderly, as if a particular product conveys the sense of 'disability', it is highly likely to be rejected by the cohort it is intended for. Furthermore, if a mobile phone is targeted for the general population, while being usable by groups such as the elderly, there is a higher likelihood that it will make it onto the retail market.

The limitations associated with mobile device usage, coupled with the dynamic diversity of the elderly population, give rise to a new set of challenges when designing mobile applications to be used by this cohort. For example, scrolling is difficult for many users and as such an application free of scrolling may be more important than an application displaying a large font, but which requires the user to scroll. Furthermore, it is necessary to correctly balance weight (of the device) with screen size. A relatively large mobile device with a large amount of screen real estate might accommodate users with visual impairments. However, such a device may be too heavy, or awkward to carry, for prolonged usage. We hope that as the future brings more advanced mobile technology, size and weight will become less of an issue. One possible means to overcome screen space and interaction issues during elderly usage is to introduce multimodal interfaces for mobile applications for the elderly.

The Elderly and Multimodal Interaction

The elderly population is a very heterogeneous group, displaying a wide variety of characteristics and abilities (Hawthorn, 2000). For example, many older people are physically fit and may share similar 'user characteristics' as younger adults. On the other hand, and particularly as people reach old age, some elders become affected by vision, hearing and memory loss, in addition to dexterity problems caused by arthritis, for example. This ageing process affects an individual's ability to interact successfully with a standard graphical user interface (Zajicek, 2001). The process of interacting with mobile user interfaces is even more difficult, as diminished visual acuity makes it very difficult to see items on a small interface and cognitive impairment increases the difficulty of conceptualising and navigating an interface. This makes it difficult to design technological applications that suit this entire, diverse user group. Multimodal interfaces, however, offer a flexible design, allowing people to choose between a combination of modalities, or to switch to a better-suited modality, depending on the specifics of their abilities, the task and the usage conditions. This flexibility is a key feature which makes multimodal interfaces ideal for mobile computing for the elderly.

Interest in multimodal interface design is motivated by the objective to support more efficient, transparent, flexible and expressive means of human computer interaction (Oviatt et al, 2000). Multimodal interaction allows users to interact in a manner to what they are used to when interacting with humans. Using speech, gestures, touch and gaze tracking, for example, constitutes 'natural' interactions. It is believed that multimodality can improve user friendliness, the impact of the message, the entertaining value of the system and improves learning of the system (Sainz Salces et al, 2006). As such, multimodal interfaces support the intake and delivery of information for a diverse segment of society, including the elderly.

Multimodal output can also prove beneficial for the elderly. The ability to present information redundantly through different output modalities has the potential to compensate for the failing senses of the elderly. Zajicek & Morrissey (2003) performed a set of experiments with elderly subjects, using a voice Web browser which aims to compensate for the age-associated impairments of memory and vision loss. The aim of the experiments was to examine how well older adults absorb information through different output modalities (speech and text) and to determine whether mixing output modes might help elders with failing senses. The authors found that more information was retained when voice output messages were shorter, and so inferred that, for older adults, speech output should be broken down into small chunks. They also found that older adults found synthetic speech output confusing to listen to and as such, inferred that natural, pre-recorded speech would be preferable for older users.

There are many benefits of multimodal interfaces for the elderly cohort. Continued use of one particular mode of input or output might lead to physical or mental discomfort or tiredness for users with certain disabilities. For example, an elder with arthritis of the hands may find it difficult to use a touch screen for a prolonged amount of time. Given that manual dexterity becomes an issue with age, continual interaction with a pen or small buttons may also be particularly cumbersome for older users. Multimodal interaction should reduce the heavy load on working memory. Computing applications require some aspect of mental modelling to 'solve' a task, which may be difficult for elderly users suffering from cognitive decline. However, multimodal interfaces, given their flexibility and naturalness, can increase the usability of an application for non-specialist users who have little or no computing experience.

When designing a multimodal interface that includes speech, the designer must be aware not only of the advantages of speech technology which may enhance the user's interaction, but also of its limitations. There are a number of factors that can contribute to the failure of a speech applications, the most notable being that of background noise which can disrupt a speech signal, making it more difficult to process and decreasing recognition rates. Indeed, reducing the error rate of a speech recognition system, under all environmental contexts, remains one of the greatest challenges facing speech recognition manufacturers (Deng & Huang, 2004). Furthermore, the speech of elderly individuals, particularly those over the age of 70, is subject to recognition error rates that may be double that of younger or middle-aged adults (Wilpon & Jacobson, 1996). This may be due to a lack of articulation and a slower rate of speech production.

Given the potential user- and system-centred errors that can result from a multimodal interface, it is necessary to have specific methods of error correction in such systems. Multimodal error correction has been suggested as a solution to avoiding repeated errors during tasks with multimodal interfaces (Suhm, Meyers & Weibel, 2001). While there are many advantages of incorporating multimodality into an interface, these advantages can significantly diminish if interactive and efficient methods are not provided to correct errors resulting from one modality. Multimodal error correction – allowing the user to switch modality to correct an error made through usage of a different modality – can be more accurate and efficient than repeating the task through the input modality which caused the error. Providing alternative modes of correction increases the user's satisfaction with the system's error handling facilities, leading to an overall augmentation in system usability and increased user satisfaction (Suhm, Meyers & Weibel, 2001). This is important for elderly users who may experience problems with a particular input modality. A multimodal system that can efficiently, and seamlessly, switch interaction modes to fix an error can greatly reduce the cognitive complexity of such a task for an elderly user.

Summary

We provide here an overview of the issues, benefits and challenges of designing and fostering usage of mobile, multimodal applications for elders.

- Aim for universal design rather than design for elders. Within the elderly cohort itself there exists a large range of diversity with regard to capabilities and attitudes towards technology. Furthermore, universal design helps to eliminate the stigmatising effect caused by labelling a piece of technology as specifically for elders.
- There are a number of design challenges which arise when developing mobile applications for the elderly. For example, buttons must be large enough to be pressed accurately by elders with motor impairments; text must be readable by those with vision impairments; strong vibrations may

be required for alerting those with hearing disabilities; it is generally agreed that scrolling should be avoided, etc. These issues need to be reconciled with the necessity of universal design.

- It is necessary to achieve a suitable balance between weight (of the mobile device) and screen size.
- Multimodal interfaces may help overcome some of these design issues, by providing a choice in how to interact etc. Redundant modes of interaction at an interface, such as speech, pen, touch, gesture, audio etc. allow the elder to choose the mode of interaction that best suits their physical and cognitive capabilities. As these capabilities change through time, so can their chosen mode of interaction.
- Supporting and efficiently executing multimodal error correction facilities is essential for dealing with errors resulting from a particular modality. It is important that the user perceives no overhead in switching between modalities.
- Dealing with speech interaction errors, whether system- or user- initiated is critical. For example, what is the best means to alert the user to the fact that the word they have spoken is not in the grammar and is therefore incorrect? Alternatively, the speech recognition software may not understand the user's command, even if it is correct, due to an inarticulation for example. Other common speech recognition errors result from the user speaking before the system is ready to listen, the user pauses, the system recognises a mis-fire, such as background speech which triggers recognition, or a non-speech sound such as a cough. It is necessary to ensure the system is designed correctly and can deal appropriately and efficiently with such errors. This will increase the system's perceived usability for elders.

We believe mobile GIS is a further area where elderly users can enjoy many benefits. However, research in this area is relatively unexplored. This application area is the focus of the next section of this chapter. However, first we summarise the benefits, issues and challenges involved in designing mobile multimodal applications for elders.

GIS FOR THE ELDERLY: ISSUES AND REQUIREMENTS

A Geographic Information System is a computer system for capturing, storing, querying, analysing and displaying geographic data (Chang, 2002). While many traditional GIS applications are desktop-based, the real benefits of such systems can be gained through field use. Mobile GIS, then, refers to the acquisition and use of GIS functionality through mobile devices and in a mobile context. In recent years, mobile GIS usage has moved from exclusively professional spatial applications (such as cartography and geography) to wider contexts. As such, users of mobile GIS now include professional GIS experts such as cartographers and surveyors in addition to non-expert users such as tourists, outdoor enthusiasts such as fishermen and non-expert professionals such as electricians who may use mobile GIS frequently for their work. However, very little research within the field of mobile GIS considers the elderly population as possible end users. As such, there is a need to address the potential benefits and challenges in designing mobile GIS for this user group.

The first question to address is what types of geospatial applications the elderly population are likely to use, and how such applications might benefit them. As previously mentioned, it is important that the elder perceives the potential benefit of an application as greater than the effort required to learn it. We perceive a number of useful geospatial applications for the elderly, the most obvious being a location-based system (LBS).

An LBS is a tourist-oriented, context-aware application which generally provides tour-guide or route-finding services based on the user's current geographical location. LBS allow users to navigate a map and search for and locate points of interest (POIs) (e.g. all Italian restaurants within 1 km of the user's current position), retrieve additional non-spatial data describing POIs (e.g. restaurant opening times) and plan routes to their desired location. Examples of such systems include Deep Map (Malaka, 2000) and CRUMPET (Poslad, 2001). However, the research described in these, and the majority of similar, applications does not account for elderly users, despite the fact that elders constitute a large cohort of potential tourists. An LBS that provides clear visual and auditory navigation instructions can aid users in finding POIs near their current location. Furthermore, a mobile navigation system that alerts the user to obstacles such as steps or slopes in addition to rest spots such as benches, or public washrooms would provide a great sense of independence to elders in their everyday lives.

Other application areas of mobile GIS for the elderly might include support for outdoor activities, such as fishing or Geocaching. For an elder interested in fishing, a geospatial application might provide information on 'hot spots' with plenty of fish, based on other users' annotations describing their experiences over the past number of days. An activity such as Geocaching might be performed in groups, whereby elders use a GPS-based geospatial application to solve clues and ultimately navigate to a hidden 'treasure'. This can benefit the elder in a number of ways. Primarily, such activities encourage social behaviour which is important as social isolation is a problem facing many elders as their social circle decreases as they age. Secondly, providing the elder with a task to solve encourages cognitive function, but in the context of a fun task, which may help to overcome an elder's feelings of fear, inability or inadequacy when interacting with a technological application. Finally, an elder using a GPS-enabled mobile device might allow a family member to track them, providing peace of mind as to where the user is and whether they are keeping active.

While the potential advantages of mobile GIS for the elderly are evident, there are a number of issues concerning such applications which may inhibit their usability by elders. Mobile geospatial applications are inherently complex, even for the general population of non-expert users, and as such it is difficult for such users to exploit geospatial applications effectively. Such complexities involve querying and annotating spatial data. Furthermore, given that users need access to information and services provided by geospatial applications while on the move, many geospatial applications are dependent on mobility. As such, a characteristic of mobile spatial applications is the continually changing environmental context, and changing noise levels which may make it particularly difficult, or inappropriate, to interact using certain modalities in certain situations. For example, whilst walking around, speech might be a preferable interaction modality over pen input, as it is easier to walk and speak than walk and point to small interface buttons. However, speech input and output may not be an ideal modality in a particularly noisy environment, or in a situation where a user is visiting a museum or church. This raises an issue relating to the complexity of switching between modalities, for an elderly user, as they move through changing contexts. It is necessary to ensure that minimum overhead is required when switching modalities and, further, that it is evident to the user, at the interface level, how to turn on or off certain input or output modalities.

Mobility introduces further complexities when interacting with a GIS, such as a small user interface footprint which can result in information overload if too much spatial detail is presented at once, limited use in motion, in addition to limited computational power, limited memory and unreliable bandwidth. While much research has addressed this issue of complexity embedded

within mobile geospatial applications (Doyle, Bertolotto & Wilson, 2008b; Oviatt, 1996), it remains to be seen whether such research applies to the elderly population.

The majority of research focusing on GIS for the elderly examines aiding such users in navigating a route (Pittarello & De Faveri, 2006; Kawamura, Umezu & Ohsuga, 2008). Applications that aid elder users in navigating a route are an important aspect of any GIS system for the elderly, and necessary for a number of reasons. Geo-spatial tasks involve navigation and some problem-solving strategy, which is difficult for older users. Indeed, research has shown that older people are less able to retrace and navigate a route than younger people (Wilkness et al, 1997). Furthermore, it is difficult for elderly people with disabilities to move around outside their homes. Despite this, it is important to get elderly people out into society, to engage them in social activities and to try to increase their willingness to leave their homes in an effort to prevent social isolation.

A mobile navigation system to help the elderly is described in (Kawamura, Umezu & Ohsuga, 2008). The aim of such a system is to detect barriers in an urban location, based on a user's location and profile. Such barriers may include obstacles which make it difficult for the elderly to navigate, such as steps, slopes or a lack of footpaths. Prior to system design, the authors undertook a survey with a number of elders, to determine the requirements for a mobile navigation system. Critical findings included an unwillingness to leave the home (due to environmental barriers), a desire to be independent (as most elders are reluctant to accept help from family and friends because they don't want to be a burden) and a difficulty in using new technology (due to age-related impairments or an unfamiliarity with technology). Based on these requirements, Kawamura et al identified three major system requirements, including barrier and useful information notification, an easy to use interface with appropriately sized icons and buttons and autonomous system behaviour to allow users to be passive. The resulting system was deployed on a mobile phone with GPS. Alerts (of barriers) to users consisted of both auditory and visual cues. Following notification, the user can press a large button on the phone to view a map of the immediate area with the location of the barrier highlighted. Users also have to opportunity to annotate the system by adding a barrier's location to the map.

The application described by Kawamura et al was evaluated over a number of days with a cohort of 10 elderly users. Results indicated that half of the subjects felt the alerts occurred at the wrong time, but they felt barrier and places-of-interest notifications were a useful concept. Furthermore, none of the users submitted annotated content, although 70% expressed a desire to submit content. This suggests that novel input modalities should be considered to encourage elders to interact with technology. A number of users commented that while the application was not very difficult to use, they found the map display to be too small. This may suggest that other devices should be examined to deploy such an application. Overall, however, 90% of the test subjects found the application fun, which is a critical part of encouraging elders to use a technological system.

In addition to displaying barrier information to aid elders with navigation, research has also shown that landmarks are the most important navigational cue as they can help the user to orient themselves in the environment (May et al, 2003). In a similar vein, Pittarello & De Faveri (2006) describe a semantic based approach to improve access of elderly people, with visual and cognitive deficiencies, to real environments. In addition to the timely identification of barriers, the authors identify a number of user requirements for elderly pedestrian navigation. These include a description of semantic objects in the immediate location, the availability of output through visual and speech interfaces, the availability to input information through the interface or a speech recognition system and the possibility to repeat relevant infor-

mation, given that elderly people have a reduced capacity to retain information.

While aiding navigation is definitely an important first step for many geospatial applications, more is required. A full GIS provides its users with the capability to query (such as finding the distance between two points, locating a feature on a map, finding the best route to reach a destination etc.), to annotate and to manipulate map feature content. Tasks such as these can easily increase the complexity of GIS, particularly for an elder. Research has shown that multimodal interfaces can considerably help to reduce the complexity involved in interacting with GIS interfaces (Doyle et al, 2007; Fuhrmann et al, 2005; Oviatt, 1996). From the research described in Section 2 above, we know that multimodal interfaces can also provide many benefits for elderly users. As such, integrating multimodality into mobile GIS seems an ideal solution to make such applications usable and accessible by this large section of the population. In the following section we outline a system we have developed, which provides mobile mapping services for users, on mobile devices and supporting multimodal interaction. Such an application has produced good results in terms of ease of use and efficiency for younger users. Here, we examine how we might adapt this application for the elderly cohort.

CASE STUDY: ADAPTING COMPASS FOR ELDERLY USAGE

CoMPASS (Combining Mobile Personalised Applications with Spatial Services) is a mobile, multimodal GIS that we have developed to provide mobile mapping services to both novice and professional users within the field of GIS. CoMPASS incorporates the delivery of vector map information using personalization (Weakliam, Bertolotto & Wilson, 2005) as well as interactive augmented reality for media annotation and retrieval (Lynch, Bertolotto & Wilson, 2005), which is relevant to the user's immediate spatial location. The front end of CoMPASS consists of a Graphical User Interface which supports geospatial visualization and interaction using multiple input modalities for mobile users (Doyle, Bertolotto & Wilson, 2008b). While CoMPASS targets, and caters for, users who may have little computing experience, our initial prototype did not consider the elderly as a potential user group, largely due to time restrictions in developing the first iteration. However, we are now interested in how CoMPASS might be adapted to both appeal to elderly users and to ensure such users can easily operate the application. As such, we present an overview in this section of the current functionality of CoMPASS, those features that might work for the elderly and those aspects of the system that might need to be changed. Such changes relate to hardware, interface layout and interaction modalities. Having devised a first draft of these changes, we then aim to perform a user-centred design with a group of elderly users to determine what else is required.

System Architecture

We begin by describing the system architecture of the CoMPASS multimodal mobile GIS, to provide readers with an overview of the design of such systems. The following architecture is proposed for delivering personalised geo-spatial data using non-proprietary software and to efficiently represent maps in either Web or mobile environments. The scope of the system's functionality ranges from geospatial information handling and transmission over slow communication links to personalisation techniques and human-computer interaction (HCI). In order to fulfil the requirements of the system in a distributed computing environment, the system architecture comprises an n-tier client-server application structure (Figure 2). The architecture encompasses three aspects: (1) a map server, (2) a service provider and (3) a data deployment component. While not common to all GIS architectures, CoMPASS supports user profil-

ing. This is a benefit for elders, as their interests can be stored and maps representing these interests can be downloaded on request. This, coupled with the user's location obtained via GPS, helps alleviate the information overload problem, which may be very problematic for elders, particularly those with impaired vision or cognitive decline. However, of most interest to us for this chapter are the HCI handling service of the Services Layer and the Deployment Layer.

The HCI Handling Service is depicted in Figure 2. The client can handle and process interactions in the form of gesture, speech and handwriting. Separate recognisers are responsible for recognition of each mode of input. The output of each recognition system is then interpreted and translated into feedback, which the user sees in the form of actions being carried out, results of their queries etc.

The system architecture must be efficient for elders. I.e. there must be a quick turnaround between the elders executing an action and them seeing the results on the interface. While this depends on many (sometimes uncontrollable) factors, such as wireless connectivity, we should ensure the best possible system architecture is in place.

CoMPASS Design and Interaction

CoMPASS has been developed on a Tablet PC with a 10.4 inch display and weighing 3 pounds. While this is considered a light-weight device, prolonged usage may not be possible with elderly users who may suffer from dexterity problems, or arthritis. Furthermore, carrying a device of such weight may strain an elderly user's arm and neck, which is less than ideal. As such, an alternative device will be considered for elderly usage. We have developed a scaled-down version of CoMPASS on a PDA, which may be preferable over a Tablet PC. Further devices will also be considered, such as the Epee PC, which weighs a total of 2 pounds and has a display size ranging from 7 to

Figure 1. CoMPASS Architecture

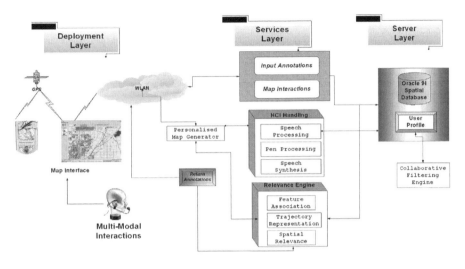

Figure 2. Structure of the Client Module

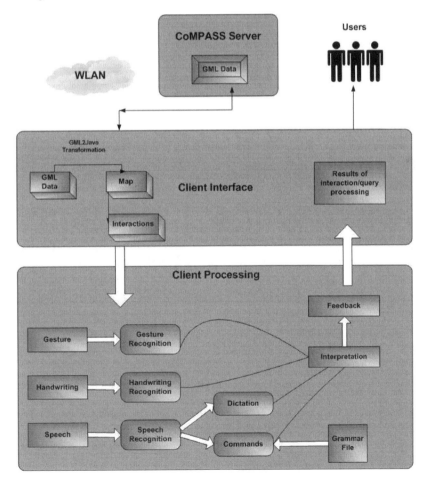

10.2 inches. Research is required to find a mobile device with an ideal balance between weight, processing capability and screen size.

CoMPASS allows users to connect to a remote server and download interactive, personalised, vector maps over wireless connections to mobile devices. It is then possible for users to dynamically interact with their map. Available interactions include navigation through zooming and panning. This is possible through a series of buttons on the interface and also by drawing an area on the map where the user would like to focus. It is also possible to re-centre the map on a particular location by simply clicking on that map location. Manipulation of map features is possible through turning on/off map features (such as highways, restaurants etc.) and changing the colour and size of map features for individual personalisation of map feature content. CoMPASS also supports feature querying including highlighting specific features in an area defined by a user, highlighting a specific number of features closest to a given point, finding the distance between two points on the map and multimedia annotation and retrieval. All of this functionality can be executed using pen input, speech input or a combination of both. A more detailed description of the CoMPASS mobile GIS is presented in (Wilson et al, 2007; Doyle, Bertolotto & Wilson, 2008a).

A screenshot of the CoMPASS interface is shown in Figure 3. As can be seen, the Tablet PC affords a relatively large screen space. Our aim was to give most of the screen space over to map display, and ensure there was minimal clutter to enhance usability for users. However, changes would be necessary to adapt such a display for elderly usage. Firstly, given that elderly users will likely use a smaller mobile device, with less screen real estate, the area for map display will be significantly reduced, requiring increased usage of zooming and panning functionalities to navigate the map. To ensure that the map display does not become too overloaded with spatial information, it is vital to ensure that no unnecessary spatial

content is displayed. This is possible with CoMPASS through the process of map content personalisation, displaying only those map features which are relevant to the user's current location and preferences (Weakliam, Bertolotto & Wilson, 2005). Furthermore, despite a decrease in screen space, it will be necessary to increase the size of the available navigation buttons considerably. Past evaluations of the CoMPASS system with young users (between the ages of 20 and 35) indicated that the navigation buttons were too small to point to accurately with the Tablet PC's pen. Given the problems of visual acuity associated with ageing, in addition to the possibility of hand tremors, it is vital that buttons are large enough to be clearly seen and accurately pointed too. It will also be necessary to increase the font size of all text that is displayed to users, such as on menu items, within the annotation box and in the information bar at the bottom of the screen. It has been suggested that a font size of 36pt would be preferred for elderly users of handheld devices with visual impairments (Lorenz et al, 2007). Whether such large fonts are feasible on mobile devices, without introducing large amounts of scrolling, is an issue requiring further investigation.

With regard to interaction, CoMPASS currently recognises voice input through voice commands and dictation, and pen input in the form of intra-gestures and handwriting. While we intend to retain the multimodality aspect of CoMPASS for elderly users, given the benefits such interfaces can afford elderly users as outlined above, there are a number of changes we may need to make before CoMPASS is usable by the elderly. Firstly, as Zajicek & Morrissey's research (2003) showed, the elderly prefer natural speech output over computerised output. Given that CoMPASS currently incorporates the latter form, we will re-record any speech output in natural speech. We believe, due to results from previous evaluations of the CoMPASS system, that voice commands issued by the elderly will be recognisable by a speech recognition system, such as IBM's Via

Figure 3. Screenshot of tourist viewing an annotation regarding an art exhibition in a local park. ©
2009 Julie Doyle. Used with permission.

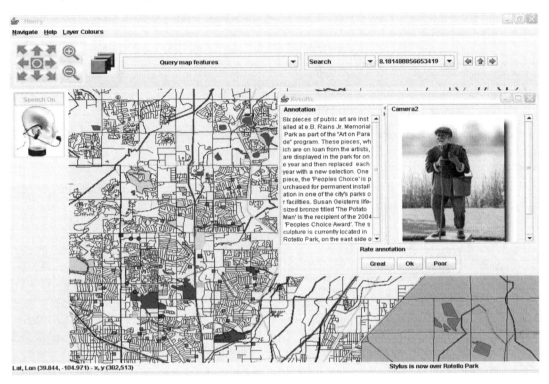

Voice™ which is used within CoMPASS. Such evaluations demonstrated that, when a speech recognition grammar is employed, recognition rates for non-native English speakers were as high as those for native English speakers (Doyle, Bertolotto & Wilson, 2008a). While specific testing with elders is required, we believe keeping voice commands brief will not only ensure they are easy to learn, but will support more robust recognition. Furthermore, elder-specific speech recognition engines are a topic of current interest, with research in (Mueller & Wasinger, 2003) describing use of a corpus of elderly speech from Scan Soft as their speech model.

One area which will require much investigation is that of dictation input by elders. Dictation is essentially free-form speech, and so enforces fewer restrictions on what a user can say. Such input is not matched against a rule grammar, but rather a dictation grammar. Dictation is used within CoMPASS for annotating map features. Once a user issues the command 'Annotate', the rule grammar is disabled and the dictation grammar becomes active. The user then delivers their voice annotation and an interaction ensues with the system to determine whether the dictation recognised is correct, or whether multimodal error correction (through one of pen and keyboard input, voice memo, handwriting or repeated dictation) should be invoked. However, in evaluating the CoMPASS system previously, dictation accuracy was very poor and proved frustrating for many users (Doyle, Bertolotto & Wilson, 2008a). Multimodal error correction through voice memo input, whereby a recording of the user's message is stored alongside their annotation, was preferred by 60% of users. As such, only providing this option for elders who would like to input annotations, will likely be significantly more efficient than if they were to input a dictation and have to correct this afterwards.

In addition to voice commands and dictation, the CoMPASS multimodal interface also recognises and processes interactions in the form of gestures. Gestures can take the form of 'intra-gestures' i.e. pointing or selecting with the stylus/pen to locations or objects on the Tablet PC screen. Such gestures can take two forms within CoMPASS: pointing and dragging. Users can point at objects to re-centre the map at this point, to discover the name and type of objects, to specify what feature they would like to query or what feature they would like to annotate. Dragging gestures specify a 'zoom-in' on the area over which the pen is dragged, or, when used in conjunction with a query, to specify the area of interest for that query. In addition to gesture input with a pen/stylus, it is also possible for the user to interact with their finger, as the Tablet PC screen represents a touch-screen. Previous interaction studies with CoMPASS have indicated that users feel more control when using the pen, as such an input device allows for more accurate and precise input. However an interesting interaction study might investigate whether this is also the case for elderly users.

A further issue in designing multimodal interfaces is that strategies are required for coordinating input modalities and for resolving issues related to integration and synchronisation of input modes. There are three main classes of modality integration:

- **Sequential,** whereby only one modality is available at a time. With sequential integration, there is no requirement for inputs to be active simultaneously. In any given state, only one input mode will be active when the user takes some action. Furthermore, inputs from different modalities are interpreted separately by the system's inference engine.

- **Simultaneous Uncoordinated,** whereby several modalities are available but only one modality is interpreted at any given time. While input modalities are simultaneously active, there is no requirement that these modalities are interpreted together. This type of integration is implemented within CoMPASS.

- **Simultaneous Coordinated,** whereby several modalities are coordinated and are interpreted together. This is the most advanced type of multimodal integration.

Although a common belief, multimodal signals will not always occur co-temporally. Research has shown that for map based tasks using speech and pen input, as few as 25% of user's multimodal commands actually contain overlapping inputs (Oviatt, DeAngeli & Kuhn, 1997). While there is no specific research on whether this is the case for elderly users, we believe it is likely. That is, given the process of cognitive impairment experienced by many elders, we believe that it would be easier for elderly users to perform one action followed by another (e.g. speak, then point), as is currently the case within CoMPASS, rather than to perform two actions at the same time (e.g. speak whilst pointing). However, an experimental process is necessary to confirm this.

While we have begun the process of adapting the CoMPASS system for elderly usage, much work has still to be done. Indeed reviewing the appropriate literature has greatly aided us in making certain design decisions. However, we believe that a user-centred design process with the elderly is required. As such, designing CoMPASS for elderly users will be an iterative process, tightly coupled with feedback from elderly users. We currently have prototypes of CoMPASS on both a PDA and a Tablet PC, and these will be presented to elders at focus groups over the coming months to determine what redesigns or additional features might be necessary to ensure usability.

CONCLUSION AND FUTURE TRENDS

This chapter has presented a detailed analysis of the issues involved in developing multimodal mobile GIS for the elderly. Our reporting has emerged primarily from a review of current literature and a consideration of how our own multimodal mobile GIS might need to be adapted to address such issues for the elderly. We summarise the most important points here, in the form of a 'take-away' for other practitioners in the field.

- Demographics are changing – the elderly population is rapidly increasing and as such, it is time to start considering the unique needs of such users with regard to usage of technology.
- The elderly population needs to understand the benefits a particular technology can afford them and it is necessary that they perceive this benefit as being greater than the effort required to learn the new technology.
- Mobile devices for the elderly should be small and light enough so as not to cause strain to the elder. However, they must also contain a large enough screen to effectively display information in a large font, ideally without requiring the user to scroll extensively.
- Multimodal systems have the capability to expand the accessibility of mobile computing to the elderly cohort, given the flexibility they afford users in interacting through modalities that best suit the elder's capabilities or preferences. However, seamless switching between modalities is necessary to minimise the complexity of multimodal interaction for older users and efficient and effective error correction facilities must be in place.
- Currently, very little research exists on the challenges and potential benefits of mobile

GIS for elderly users. Numerous types of GIS applications could be utilised by elders to aid them in carrying out day to day activities, including LBS, navigation alerts to 'obstacles' in the elder's environment, such as steps, or places to rest, outdoor activities (which can encourage group interaction and social connection with peers). Furthermore, family tracking is possible using GPS. Such an application may provide peace of mind to both the elder and their family/caregiver.

- Using techniques such as user profiling and personalisation can reduce information overload on a mobile device and hence increase the usability of mobile GIS for elders. Furthermore, supporting multimodal interaction offers further ease of use, given the flexibility they can provide a user when interacting with a mobile mapping application.

Next Steps:

- It is necessary to understand the ethical issues involved in designing any technological applications for the elderly. This means understanding not only how (geospatial) technology might benefit the elderly, but also the misgivings, or apprehensions of the elderly, so that these issues can be addressed. While many elders may have no interest in adopting technology in their daily lives, we feel it is important to facilitate easy access and interaction for those who are interested. While technology for elders is often motivated by the idea of independent living, it is important to realise that it cannot take the place of human-human interaction. As such, technology that can bring people together, such as mobile GIS, may provide many benefits for elders.
- The design process should be user-based and should begin with ethnographic

research to determine how elders live and what they really need. Too often, technology research is driven by technologists. An idea is developed and deployed as it is perceived to solve a problem. However, when designing technology for elders, it is necessary to start by understanding older people's needs, then design and test the technology to determine whether it is beneficial to the elder, and modify the design if necessary. As such, the design process begins and ends with the elder. Such research involves collaboration between ethnographers and technologists to find areas where technology could make a positive impact in the everyday lives of elders.

- Furthermore, involving elders themselves in the design process of technology may prove beneficial. Focus groups, for example, foster social interaction. Elders can become active contributors and feel they are making a difference. Such activities help elders in maintaining social interaction with their peers, which may contribute to better physical and mental health.

- We plan to adhere to this design process and to test our first adapted prototype for elders at a focus group in the coming months.

- We will eventually undertake a comprehensive evaluation to determine the acceptance level by elders of our multimodal, mobile GIS. We will examine their learning curve, interaction preferences and usage patterns over a 6 week period. We will also examine motivation for use, to better help us understand the factors behind elderly adoption of ICT in the context of multimodal mobile GIS.

We hope this work will encourage others in the field of GIS to consider the elderly as a potential user group. Future work will require much research and evaluation to confirm exactly how geospatial interfaces for the elderly should be designed, to ensure maximum usability and acceptability by such users.

ACKNOWLEDGMENT

Research presented in this paper was funded by a Strategic Research Cluster grant (07/SRC/I1168) by Science Foundation Ireland under the National Development Plan. The authors gratefully acknowledge this support.

REFERENCES

Deng, L., & Huang, X. (2004). Challenges in adopting speech recognition. *Communications of the ACM*, *47*(1), 69–75. doi:10.1145/962081.962108

Dong, H., Keates, S., & Clarkson, P. J. (2002). Accommodating older user's functional capabilities. In S. Brewster & M. Zajicek (Eds.), *HCI BCS*.

Doyle, J., Bertolotto, M., & Wilson, D. (2007). A survey of multimodal interfaces for mobile mapping applications. In L. Meng, A. Zipf, & S. Winter (Eds.), *Map-based Mobile Services – Interactivity and Usability*. Berlin, Germany: Springer Verlag

Doyle, J., Bertolotto, M., & Wilson, D. (2008a). (in press). Evaluating the benefits of multimodal interface design for CoMPASS – a mobile GIS. *GeoInformatica*.

Doyle, J., Bertolotto, M., & Wilson, D. (2008b). Multimodal interaction – improving usability and efficiency in a mobile GIS context. In *Advances in Computer Human Interaction* (pp. 63-68). Washington, DC: IEEE Press.

Fuhrmann, S., MacEachren, A., Dou, J., Wang, K., & Cox, A. (2005). Gesture and speech-based maps to support use of GIS for crisis management: A user study. In *Proceedings of AutoCarto*. Bethesda, MD: Cartography and Geographic Information Society.

Gardner, D. K., & Helmes, E. (1999). Locus of control and self-directed learning as predictors of well-being in the elderly. *Australian Psychologist, 34*(2). doi:10.1080/00050069908257436

Hawthorn, D. (2000). Possible implications of aging for interface designers. *Interacting with Computers, 12*, 507–528. doi:10.1016/S0953-5438(99)00021-1

Jacobson, R. D. (2002). Representing spatial information through multimodal interfaces. In *Proceedings of the 6th International Conference on Information Visualisation* (pp. 730-734). Washington, DC: IEEE Press.

Jorge, J. (2001). Adaptive tools for the elderly: New devices to cope with age-induced cognitive disabilities. In *Workshop on Universal Accessibility of Ubiquitous Computing,* (pp. 66-70). New York: ACM Press.

Kawamura, T., Umezu, K., & Ohsuga, A. (2008). Mobile navigation system for the elderly – preliminary experiment and evaluation. In *Ubiquitous Intelligence and Computing* (pp. 579-590). Berlin, Germany: Springer LNCS.

Keskin, C., Balci, K., Aran, O., Sankar, B., & Akarun, L. (2007). A multimodal 3D healthcare communication system. In *3DTV Conference*. Washington, DC: IEEE Press.

Lorenz, A., Mielke, D., Oppermann, R., & Zahl, L. (2007). Personalized Mobile Health Monitoring for Elderly. In *Mobile* [New York: ACM Press.]. *Human-Computer Interaction, 07*, 297–304.

Lynch, D., Bertolotto, M., & Wilson, D. (2005). Spatial annotations in a mapping environment. In *GIS Research UK* (pp. 524-528).

Malaka, M., & Zipf, A. (2000). Deep map – Challenging IT research in the framework of a tourist information system. In *Proceedings of the 7th International Congress on Tourism and Communication Technologies in Tourism (Enter 2000)* (pp. 15-27). Berlin, Germany: Springer LNCS.

May, A. J., Ross, T., Bayer, S. H., & Tarkiainen, M. J. (2003). Pedestrian navigation aids: Information requirements and design implications. *Personal and Ubiquitous Computing, 7*(6), 331–338. doi:10.1007/s00779-003-0248-5

Mueller, C., & Wasinger, R. (2002). Adapting multimodal dialog for the elderly. In *ABIS-Workshop on Personalization for the Mobile World* (pp. 31-34).

Nischelwitzer, A., Pintoffl, K., Loss, C., & Holzinger, A. (2007). Design and development of a mobile medical application for the management of chronic diseases: Methods of improved data input for older people. In *USAB 07 – Usability & HCI for Medicine and Health Care* (pp. 119-132). Berlin, Germany: Springer LNCS.

Oviatt, S. (1996). Multimodal interfaces for dynamic interactive maps. In *SIGCHI Conference on Human Factors in Computing Systems* (pp. 95-102). New York: ACM Press.

Oviatt, S., Cohen, P., Wu, L., Vergo, J., Duncan, L., & Suhm, B. (2000). Designing the user interface for multimodal speech and pen-based gesture applications: State-of-the-art systems and future research directions. *Human-Computer Interaction, 15*(4), 263–322. doi:10.1207/S15327051HCI1504_1

Oviatt, S., DeAngeli, A., & Kuhn, K. (1997). Integration and synchronization of input modes during multimodal human computer interaction. In *Conference on Human Factors in Computing Systems (CHI '97)* (pp. 415-422). New York: ACM Press.

Pittarello, F., & De Faveri, A. (2006). Improving access of elderly people to real environments: A semantic based approach. In *Advanced Visual Interfaces,* (pp. 364-368). New York: ACM Press.

Plos, O., & Buisine, S. (2006). Universal design for mobile phones: A case study. In *CHI Extended Abstracts on Human Factors in Computing Systems* (pp. 1229-1234). New York: ACM Press.

Poslad, S., Laamanen, H., Malaka, R., Nick, A., Buckle, P., & Zipf, A. (2001). CRUMPET: Creation of user-friendly mobile services personalised for tourism. In *Conference on 3G Mobile Communication Technologies* (pp. 26-29).

Sainz Salces, F., Baskett, M., Llewelyn-Jones, D., & England, D. (2006). Ambient interfaces for elderly people at home. In Y. Cai & J. Abascal (Eds.), *Ambient Intelligence in Everyday Life* (pp. 256-284). Berlin, Germany: Springer LNAI.

Suhm, B., Meyers, B., & Weibel, A. (2001). Multimodal error correction for speech user interfaces. [New York: ACM.]. *ACM Transactions on Computer-Human Interaction, 8*(1), 60–98. doi:10.1145/371127.371166

Vnunet.com. (2007). *UK operators shun mobile for the elderly*. Retrieved October 9th 2008 from http://www.vnunet.com/vnunet/news/2194636/uk-mobile-operators-shun-phone-elderly.

Weakliam, J., Bertolotto, M., & Wilson, D. (2005). Implicit interaction profiling for recommending spatial content. In *ACMGIS '05,* (pp. 285-294). New York: ACM Press.

Wickens, C. D., & Baker, P. (1995). Cognitive issues in virtual reality. In T.A. Furness & W. Barfield (Eds.), *Virtual Environments and Advanced Interface Design* (pp. 514-541). Oxford, UK: Oxford University Press.

Wilkniss, S. M., Jones, M., Korel, D., Gold, P., & Manning, C. (1997). Age-related differences in an ecologically based study of route learning. *Psychology and Aging, 12*(2), 372–375. doi:10.1037/0882-7974.12.2.372

Wilpon, J., & Jacobsen, C. (1996). A study of speech recognition for children and the elderly. In *International Conference on Acoustics, Speech and Signal Processing,* (pp. 349-352). Washington, DC: IEEE Press.

Wilson, D., Doyle, J., Weakliam, J., Bertolotto, M., & Lynch, D. (2007). Personalised maps in multimodal mobile GIS. *International Journal of Web Engineering and Technology, Special issue on Web and Wireless GIS, 3(2),* 196-216.

Zajicek, M. (2001). Interface design for older adults. In *Workshop on Universal Accessibility of Ubiquitous Computing* (pp. 60-65). New York: ACM.

Zajicek, M. (2006). Aspects of HCI research for older people. *Universal Access in the Information Society, 5*(3), 279–286. doi:10.1007/s10209-006-0046-8

Zajicek, M., & Morrissey, W. (2003). Multimodality and interactional differences in older adults. *Universal Access in the Information Society, Special Issue on Multimodality: A Step Towards Universal Access, 2(2),* 125-133.

Chapter 14
Automatic Signature Verification on Handheld Devices

Marcos Martinez-Diaz
Universidad Autonoma de Madrid, Spain

Julian Fierrez
Universidad Autonoma de Madrid, Spain

Javier Ortega-Garcia
Universidad Autonoma de Madrid, Spain

ABSTRACT

Automatic signature verification on handheld devices can be seen as a means to improve usability in consumer applications and a way to reduce costs in corporate environments. It can be easily integrated in touchscreen devices, for example, as a part of combined handwriting and keypad-based multimodal interfaces. In the last few decades, several approaches to the problem of signature verification have been proposed. However, most research has been carried out considering signatures captured with digitizing tables, in which the quality of the captured data is much higher than in handheld devices. Signature verification on handheld devices represents a new scenario both for researchers and vendors. In this chapter, we introduce automatic signature verification as a component of multimodal interfaces; we analyze the applications and challenges of signature verification and overview available resources and research directions. A case study is also given, in which a state-of-the-art signature verification system adapted to handheld devices is presented.

INTRODUCTION

The current proliferation and ubiquity of electronic applications and services has motivated the need for user authentication means, which must be convenient and reliable at the same time. Nowadays, access control and user authentication are common tasks that are usually performed with tokens or passwords. In the case of handheld devices, user authentication is performed in most cases with passwords that are provided through keypad-based monomodal interfaces. Biometric-based systems represent an alternative to traditional user validation methods, as they rely on anatomical (e.g. fingerprint, iris) or behavioral (e.g. voice, signature) traits to authenticate a user (Jain, Ross & Pankanti, 2006).

DOI: 10.4018/978-1-60566-978-6.ch014

These traits cannot be in general forgotten or stolen without severe consequences for the user. As an example, it is now common to observe fingerprint verification systems in handheld and portable electronic devices (e.g. laptops). Due to the enhanced convenience provided to the users compared to the use of passwords, biometrics such as voice, fingerprint or signature can improve the usability of mobile user authentication applications.

Among biometric traits, signature is one of the most socially accepted as it has been used in financial and legal transactions for centuries. Despite its acceptance, automatic signature verification is a challenging task per se, as it must face at the same time the variability among genuine signatures from an individual (high *intra-class* variability) and the possibility of skilled forgers, which can imitate a signature with high accuracy (low *inter-class* variability). Moreover, forgers are not always possible to model during the design of a verification system due to their unknown nature. Consequently, reliable automatic signature verification is still an open issue, which can be corroborated by the notable variety of research works conducted on this subject in the last decades (Plamondon & Lorette, 1989; Fierrez & Ortega-Garcia, 2007).

Two main types of signature verification systems exist, depending on the nature of the information they use to perform verification. *Off-line* systems use static signature images, which may have been scanned or acquired using a camera. *On-line* or *dynamic* systems use captured signature time-functions, extracted from the pen motion. These functions can be obtained using digitizer tablets or touchscreen-enabled devices such as Tablet-PCs and smartphones. Traditionally, dynamic systems have presented a better performance than off-line systems as more information levels (e.g. pen speed, pen pressure, etc.) than the signature static image are available to perform verification (Plamondon & Lorette, 1989).

In this context, touchscreen-enabled handheld devices represent an appropriate computing plat-form for the deployment of dynamic signature verification systems as they provide both a pen-based input and reasonable computing power. These devices are already prepared for multimodal interaction, usually based on the combination of keypad, handwriting or speech modalities. As a matter of fact, commercial devices already provide handwritten character recognition as a text input alternative for years. Touchscreen-enabled smartphones have recently experienced an outstanding technological evolution, representing many new promising scenarios and applications for ubiquitous user interfaces and, specifically, for signature verification. They have gathered an increasing interest among the scientific and industrial communities as they provide multimodal capabilities (Oviatt & Lunsford, 2003) and a convenient way of interfacing with other systems; being thus able to host a wide range of user-centric applications (Ballagas, Borchers, Rohs, Sheridan, 2006). Verification of signatures or graphical passwords on handheld devices provides a convenient method for ubiquitous user authentication in commercial payments or financial transactions among other applications.

In this chapter we consider dynamic signature verification on handheld devices and overview its major applications and challenges. The chapter is focused on the handwritten signature modality, which can be integrated in multimodal interfaces for mobile devices based on touchscreens and keypads or speech. Signature may be used as an optional (and more natural) validation means instead of typed passwords or as a mandatory step in electronic transactions. The chapter is organized as follows. First, the typical architecture of a dynamic signature verification system is outlined. Applications, a market overview and challenges are then presented, which are followed by a summary of available resources and standards. We present as a case study a state-of-the-art signature verification system based on local and global features that has been specifically adapted to handheld devices. The system is tested using a signature database captured with a PDA device under realistic usage conditions. Finally, future research directions are given and conclusions are drawn.

BACKGROUND

Architecture of a Dynamic Signature Verification System

Automatic signature verification systems generally share a common architecture, which is depicted in Figure 1. The following steps are usually performed (Fierrez & Ortega-Garcia, 2007):

1. **Data Acquisition**. Signature signals are captured using a digitizing tablet or touchscreen. The signature signal is sampled and stored as discrete time series. While some digitizing tablets provide pressure or pen angle information, these are not commonly available in handheld devices. The sampling rate is usually equal to or above 100 Hz, which is an acceptable rate, since the maximum frequencies of the signature time functions are around 30 Hz (Plamondon & Lorette, 1989). Preprocessing steps may be carried out after data acquisition, such as noise filtering, resampling, or interpolation of missing samples.

2. **Feature Extraction**. Two main approaches have been followed by most implementations: *feature-based* or global systems describe each signature by a holistic vector consisting of global features (e.g.

signature duration, number of pen-ups, average velocity) from the signature (Lee, Berger & Aviczer, 1996). On the other hand, *function-based* or local systems use the sampled signature time functions (e.g. position, pressure) for verification. The use of pen orientation features such as azimuth or altitude has been reported to provide good results (Muramatsu and Matsumoto, 2007); although it has been discussed by other authors (Lei & Govindaraju, 2005). Function-based approaches have traditionally yielded better results than feature-based ones. Additionally, fusion of the feature- and function-based approaches has been reported to provide a better performance than the individual systems (Fierrez-Aguilar, Nanni, Lopez-Penalba, Ortega-Garcia & Maltoni, 2005).

3. **Enrollment**. *Reference-based* systems store the extracted features of each signature of the training set as templates. In the similarity computation step, the input signature is compared with each reference signature. In *model-based* systems, a statistical user model is computed using a training set of genuine signatures which is used for future comparisons in the matching step (Richiardi & Drygajlo, 2003). The typical amount of signatures used to estimate a model in the literature is five, although more signatures usually lead to lower error rates during verification.

4. **Similarity Computation.** This step involves two sub steps: pre-alignment between the

Figure 1. Diagram of the typical architecture of a dynamic signature verification system

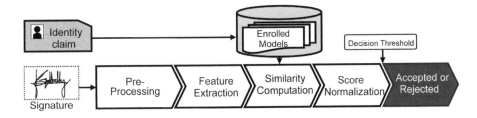

input signature and the enrolled model (if necessary) and a matching phase, which returns a *matching score*. In the case of feature-based systems, statistical or distance-based techniques such as the Mahalanobis distance, Parzen Windows or Neural Networks are used for matching (Nelson, Turin & Hastie, 1994). Function-based systems employ other techniques like Dynamic Time Warping (DTW) (Kholmatov & Yanikoglu, 2005) and Hidden Markov Models (HMM) (Dolfing, Aarts & van Oosterhout, 1998; Fierrez, Ramos-Castro, Ortega-Garcia & Gonzalez-Rodriguez, 2007) to measure the similarity between signatures and the enrolled signature models.

5. **Score Normalization.** The matching score can be normalized to a given range. This is a critical step when combining score outputs from multiple classifiers or in multibiometric systems (Ross, Nandakumar & Jain, 2006). More sophisticated techniques like target-dependent score normalization can lead to an improved system performance (Fierrez-Aguilar, Ortega-Garcia & Gonzalez-Rodriguez, 2005; Martinez-Diaz, Fierrez and Ortega-Garcia, 2007).

A signature will be considered from a claimed user if its corresponding matching score exceeds a given threshold. Once a threshold has been specified, a verification system can be described by its False Acceptance Rate (FAR) and its False Rejection Rate (FRR). The FAR describes the proportion of impostors that are accepted as genuine users, while the FRR represents the proportion of genuine users that are erroneously rejected by the system, as if they were impostors. The Equal Error Rate (EER) refers to the rate where FAR = FRR, and is commonly used to compare the verification performance among different systems.

Signature Verification on Handheld Devices

Applications

Dynamic signature verification on handheld devices improves usability in consumer applications and can reduce costs in corporate environments through workflow streamlining. For many years, electronic signature is a legally accepted method to bind parties to agreements and allow them to be enforceable. This was stated in the Electronic Signatures in Global and National Commerce (ESIGN) act, from the United States and the 1993/93/EU European directive. The term electronic signature often refers to keys or codes that identify an individual. In this context, captured dynamic signature data is also a valid electronic signature (Lenz, 2008).

Among the wide range of applications of signature verification on handheld devices, we cite the following:

*   **Payments in commercial environments.** The signature is used to validate a commercial transaction that is performed wirelessly by using technologies like WIFI, UMTS or CDMA. This enables ubiquitous access to commercial transactions wherever a wireless network is available. Moreover, signatures are not always visually verified by the cashier, so automatic verification could provide higher security levels.

*   **Legal transactions.** Legal documents or certificates are signed by the user adding additional security and fraud prevention, as the signature is verified. This can be a convenient user validation scheme for e-government and healthcare-related applications. Using on-line signature verification, the protection against repudiation (the case where an individual denies having signed) of signed documents is even increased over

traditional signature, as more information of the signature process is available.

- **Remote or local user login.** The signature is used to login into a local or remote system as an access control measure (e.g. bank account, personal records, etc.), instead of traditional methods such as PINs or passwords. This is a promising application in banking and healthcare scenarios that is already being tested in some banks and hospitals (Lenz, 2008).

- **Customer validation.** A customer is validated by its signature. A client that uses a service or receives a delivery (e.g. a parcel) signs in a portable device carried by the deliverer or service provider to certify his conformity.

- **Crypto-biometrics.** Signature is used as a cryptographic key (Freire-Santos, Fierrez-Aguilar & Ortega-Garcia, 2006) that identifies the user. A cryptographic key or hash can be generated by using data extracted from the signature. The signature data is secured as only the key or hash is stored, preventing the recovery of the original signature data by a malicious attacker.

- **Paperless office.** Documents can be electronically signed without printing them, allowing verification of the signatures and ubiquitous access to them. This allows business process and workflow automation where signatures are needed. The captured signature data can be embedded in a document, allowing future verification. This functionality has already been implemented by some vendors in PDF documents (http://www.cic.com).

In all these applications, the verification platform can be either remote or local. Local verification systems perform the matching process in the handheld device, while remote systems send the input signature over the network and the matching is performed on a remote server. A model of the two aforementioned architectures is presented in Figure 2. Security must be ensured in both architectures. While in local systems, the user template and matcher must be secured in the handheld device, in remote systems, the transmission channel and verification system on the server side, including the user model database, must be kept secure.

Signature verification does not always need to be instantly performed, for example, in the case of legal transactions or parcel delivery the captured signature may only need to be validated in case of any reclamation. In such cases, signature is used on demand as a mean against repudiation or against impostors. A key advantage for the deployment of such applications is that touch-screen handheld devices do not need any extra hardware for signature verification, as it is the case of sensors and cameras for fingerprint and face verification systems respectively. Consequently, no extra costs exist and the system complexity does not increase.

Handheld devices such as smartphones allow in most cases the easy combination of signature verification with other biometrics, leading to a multimodal biometric system. In general, multimodal systems are more robust than monomodal ones, reaching significantly lower error rates since they combine evidence from multiple biometrics (Ross, Nandakumar & Jain, 2006). Some approaches have already been proposed in the literature, such as voice and signature (Hennebert, Humm & Ingold, 2007). In that specific application, users must say what they are writing while they sign.

Some of the current devices incorporating touchscreens allow interfacing with them directly with the fingertips instead of using a stylus. Signature verification is also feasible in these devices by tracing the signature directly with the finger or by substituting signatures with graphical passwords. Graphical passwords have in general the disadvantage of tending to be less complex than signatures (and thus less discriminative) because

Figure 2. Examples of possible architectures of an automatic signature verification platform

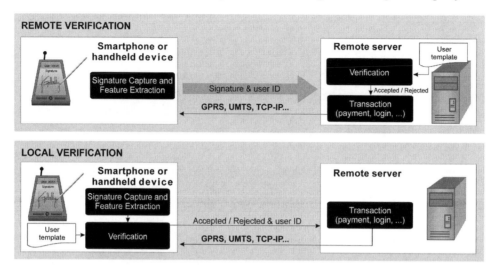

users tend to define them as simple pictograms. A practical example of graphical passwords is the draw-a-secret scheme, proposed by Jermyn, Mayer, Monrose, Reiter and Rubin (1999).

Market Overview

There is currently a significant amount of companies that offer products related to signature verification. The following is a list of vendors which offer specific products for signature verification:

- CIC provides secure signature verification applications and SDKs. A number of success stories and references from customers related to the financial sector among other industries stories are given in (http://www.cic.com).
- Crypto-Sign offers a PDA user authentication solution (http://www.crypto-sign.com).
- KeCrypt is aimed towards user authentication on workflows and payment systems. It provides both software and hardware solutions. Some examples of case studies and practical implementations are also given in (http://www.kecrypt.com).

- Nemophila provides a signature verification software named Cyber-SIGN, mainly aimed towards client-server architectures (http://www.nemophila.com).
- Romsey Associates have in their portfolio the PDALok software, which is a PDA authentication solution (http://www.romseyassoc.com).
- App Informatik Davos offers SignPlus, which is an online signature verification and management system, primarily oriented to bank transactions (http://www.app-davos.ch).
- SoftPro solutions are focused on fraud detection and process securing. Practical implementation examples by SoftPro can be found in (http://www.signplus.com).
- Topaz Systems provides software, SDKs and hardware for automatic signature verification (www.topazsystems.com).
- Xyzmo offers signature verification solutions to a wide range of industries (http://www.xyzmo.com).

As can be seen, only two from the listed companies have products directly focused on handheld devices. The rest are focused on digitizing tablets or tablet-PCs. Unfortunately, no vendor provides technical details about the signature matching process being thus not possible to systematically compare their verification performance with other systems proposed by research institutions.

Dynamic signature verification is seen by analysts as a market with a high growth potential. According to Frost & Sullivan (as cited in Miller, 2008), the market of dynamic signature verification earned in 2006 revenues of $14.4 million, and is predicted to grow until $85.7 million by 2013. Nevertheless, these predictions cut the expectative of growth of dynamic signature verification markets from previous estimations. In fact, Frost & Sullivan (as cited in Signature Verification Biometrics, 2004) had predicted back in 2003 revenues of $123.3 million in 2009.

The current interest in the dynamic signature verification field can be corroborated by the increased number of contract announcements, press releases and case studies published by the major dynamic signature software vendors (Lenz, 2008). Dynamic signature verification is seen as a mean to reduce costs, as it can accelerate workflows and bring the possibility of fully-paperless transactions. Consequently, the needs of printing, archiving, scanning or indexing are avoided. In the field of consumer applications, signature verification on handheld devices is a mean to improve usability and increase accessibility. Recently, KeCrypt Systems funded an independent survey among businesses, in which it was found that 83% of the respondents found desirable for their banks to offer dynamic signature as an alternative to their current verification method (Dale, 2008).

Challenges

Signature Verification system designers must face many challenges. As has been previously stated, inter- and intra-variability represent one of the main difficulties when trying to reach a good verification performance, especially in the case of skilled forgeries. A key challenge that has not already been successfully solved is the ability of a verification system to adapt to inconsistencies in signatures from an individual over a period of time.

Handheld devices such as smartphones or PDAs are affected by size and weight constraints due to their portable nature. While processing parts, memory chips and battery components are nowadays experimenting higher levels of miniaturization and integration, the input (e.g. keyboard, touchscreen) and output (e.g. display) parts must have reasonable dimensions in order to keep their usability, which is inevitably affected by these size constraints. Poor ergonomics and small input areas in mobile devices are two key factors that may increase the variability during the signing process. Moreover, the unfamiliar signing surface may affect the signing process of new users. The touchscreen digitizing quality must also be taken into account. Irregular sampling rates and sampling errors, which are common in low-end mobile devices, may worsen the verification performance and must be addressed during the preprocessing steps (Martinez-Diaz, Fierrez, Galbally & Ortega-Garcia, 2008). In these devices, only position signals are available in general. Pressure, pen-azimuth or other signals that may improve the verification performance (Muramatsu & Matsumoto, 2007), are not usually available in handheld devices.

The interest in security on smartphones has risen in the last few years (Khokhar, 2006). Security must be a critical concern while designing a signature verification platform as a breach could give an attacker access to personal data or bank accounts. Gaining access to the matcher could allow an attacker to perform software attacks such as brute force or hill-climbing attacks (Galbally, Fierrez & Ortega-Garcia, 2007). The user template must be appropriately secured (Maiorana, Martinez-Diaz, Campisi, Ortega-Garcia & Neri, 2008) and encrypted (Freire-Santos et al., 2006) as must be secured communication channels. This is a key issue in biometric systems, since end users are often concerned about the impact of having their biometric data compromised.

Resources, Initiatives and Standards

Recent initiatives related to signature verification specifically on handheld devices demonstrate that this field is in a growth stage. One of the main difficulties when comparing the performance of different biometric systems is the different experimental conditions under which each system is analyzed. This is primarily due to the lack of publicly available biometric databases, which has been a problematic issue in signature verification in the last decades. It is complex to make a signature database publicly available due to privacy and legal issues. To overcome these difficulties, in 2004, a database containing only invented signatures was published for the Signature Verification Competition (SVC) 2004 (Yeung et al., 2004). Nevertheless, invented signatures may be affected by the fact that they are not a natural and learned movement that has been performed by each individual for years. This database is still freely available on the SVC 2004 website (http://www.cse.ust.hk/svc2004). Other popular signature databases include the MCYT database (Ortega-Garcia et al., 2003) and the BIOMET database (Garcia-Salicetti et al., 2003) among others. Nevertheless, none of the currently available signature databases have been acquired using handheld devices.

In 2007, the BioSecure Multimodal Biometric Database was acquired in the context of the Bio-Secure Network of Excellence (Ortega-Garcia et al., 2009). The database has the novelty of including datasets of biometric traits captured using a handheld device, namely face, fingerprint, voice and signature. This database is one of the largest multimodal biometric databases ever captured and is currently being post-processed in order to make it available for research purposes. It consists of

Figure 3. Acquisition example for the PDA dataset of the BioSecure Multimodal Database

ca. 1000 users and was acquired at eleven different sites in Europe. In the case of signature, the database has the advantage of having two datasets with the same users, one captured with a digitizing tablet and the other one captured with a PDA under realistic conditions. The user was sitting while signing on the pen-tablet and standing and holding the PDA while signing on the PDA (see Figure 3). This allows a systematic comparison of the specificities of the signature verification process on both scenarios (e.g. useful features, preprocessing, etc.). Signatures captured with the PDA present sampling errors, produced by the touchscreen hardware. Due to the different nature of the acquisition devices, in the PDA scenario, only the x and y coordinates are captured, while for pen-tablet the x and y coordinates, pen-pressure, and pen inclination angles are available. Moreover, pen-tablets capture the whole pen trajectory during pen-ups (although the pen is not in contact with the tablet surface), which is not available on

Figure 4. Examples of PDA signatures from the BioSecure Multomodal Database. Missing points are due to sampling errors.

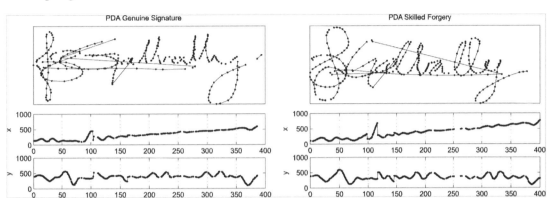

PDAs. Examples of signatures from the BioSecure database are presented in Figure 4.

Signatures were captured in two sessions separated by an average of two months. During each session, users were asked to perform 3 sets of 5 genuine signatures, and 5 forgeries between each set. Each user performed 5 forgeries for each the previous 4 users in the database. The users had visual access to the dynamics of the signing process of the signatures they had to forge. Thus, 30 genuine signatures and 20 forgeries are available for each user. On each dataset, forgeries are performed in a "worst case" scenario. In the pen tablet scenario, the forger had visual access to the dynamics of the signature in a screen. For the PDA dataset, forgers had access to the dynamics of the genuine signature in the PDA screen and were aided by a tracker tool to imitate the original strokes.

After the database was captured, an evaluation of state-of-the art verification systems was organized. Several European research institutions presented their systems and evaluated their verification performance on these new datasets. A subset of signatures corresponding to 50 users was previously released to participants, allowing them to adjust their systems. This was known as the BioSecure Multimodal Evaluation Campaign (BMEC). Eleven signature verification systems

from independent institutions were presented to the evaluation. The results of this evaluation and a description of each system that participated can be found in (Ortega-Garcia et al., 2009). The best EER in the evaluation was of 4.03% for random forgeries and of 13.43% for skilled forgeries. The relatively high EER for skilled forgeries reveals the high quality of the forgeries acquired in this database.

The BioSecure Signature Evaluation Campaign will be held in 2009. Signatures from the BioSecure database captured on pen-tablet and PDA will be used. This will make possible to measure the real impact of mobile conditions on the performance of the submitted algorithms.

Unlike other biometric modalities such as fingerprint, no reference system has been available for signature verification until very recently. A public reference system makes it possible to have a benchmark against which other systems can be compared. This issue has been recently overcome by the BioSecure Network of Excellence, with the development of an open-source signature verification framework (http://share. int-evry.fr/svnview-eph). The framework allows implementing feature- and function-based systems and will be modular, separating the different steps of the signature verification process such as preprocessing, feature extraction, etc. (Petrovska-Delacretaz, Chollet & Dorizzi, 2009).

Standardization of signature verification among other biometric modalities has been addressed by the ISO/IEC JTC1 SC37 subcommittee by developing the ISO/IEC 19794 standard. In Part 7 of this standard, the interchange and transmission formats for signatures or handwritten signs are addressed. A format for general use is defined as well a specific format for smart cards and other resource-limited tokens. A time series format that allows transmission and storage of raw captured data and of time-stamped standard features (e.g. x and y coordinates, speed, pressure, etc.) as well as property data is defined. A set of recommendations and best practices are also given in the standard, like the usage of cryptographic techniques to protect the authenticity and confidentiality of the biometric data. Part 11, currently under development, defines a set of typical statistical features extracted from the signature raw captured data, which can be further extended by another set of proprietary features. According to this standard, the whole feature set must allow interoperability at a feature level between systems, including samples collected on different types of devices.

CASE STUDY: A SIGNATURE VERIFICATION SYSTEM FOR HANDHELD DEVICES BASED ON FUSION OF LOCAL AND GLOBAL INFORMATION

A signature verification system based on fusion of two individual systems, one feature-based and another function-based is presented. In this system, feature selection is applied in two feature sets (local and global) that combine features proposed by many authors in the literature for systems based on digitizing tables. The proposed feature sets include a considerable amount of the features that have been presented in the literature in the last decades. Consequently, this case study gives the reader a practical overview of the state-of-the-art in signature verification, with application to the handheld scenario. The resulting system can be integrated in a mobile multimodal interface of user authentication system as a password replacement, or to integrate handwritten signatures in electronic documents. The signature dataset used for the experiments is captured with a PDA, so feature selection makes it possible to assess which features are the most discriminative for handheld devices. This is opposed to traditional research on signature verification, in which signatures captured with pen-tablets have been mostly used.

Global System

The feature-based signature verification system considers a set of 100 global features, extracted from each signature. The set of features is an extension of other sets presented in the literature and was first presented in (Fierrez et al., 2005). A complete description of the feature set is shown in Figure 5. These 100 features can be divided in four categories corresponding to the following magnitudes:

- Time (25 features), related to signature duration, or timing of events such as pen-ups or local maxima: 1, 13, 22, 32, 38, 40-42, 50, 52, 58-60, 62, 64, 68, 79, 81-82, 87-90, 94, 100.
- Speed and Acceleration (25 features), from the first and second order time derivatives of the position time functions, like average speed or maximum speed: 4-6, 9-11, 14, 23, 26, 29, 31, 33, 39, 44-45, 48, 69, 74, 76, 80, 83, 85, 91-92, 96.
- Direction (18 features), extracted from the path trajectory like the starting direction or mean direction between pen-ups: 34, 51, 56-57, 61, 63, 66, 71-73, 77-78, 84, 93, 95, 97-99.
- Geometry (32 features), associated to the strokes or signature aspect-ratio: 2, 3, 7-8,

Figure 5. Global feature set. Extracted from (Fierrez et al., 2005). T denotes time interval, N denotes number of events, and θ denotes angle. Note that some symbols are defined in different features of the table.

#	Time related feature	#	Direction related feature
#	Speed and Acceleration related feature	#	Geometry related feature

Ranking	Feature Description	Ranking	Feature Description
1	signature total duration T_s	2	$N(\text{pen-ups})$
3	N(sign changes of dx/dt and dy/dt)	4	average jerk $\bar{\jmath}$
5	standard deviation of a_y	6	standard deviation of v_y
7	(standard deviation of y)$/\Delta_y$	8	N(local maxima in x)
9	standard deviation of a_x	10	standard deviation of v_x
11	j_{rms}	12	N(local maxima in y)
13	$t(\text{2nd pen-down})/T_s$	14	(average velocity $\bar{v}$)$/v_{x,\text{max}}$
15	$A_{\min}=(y_{\max}-y_{\min})(x_{\max}-x_{\min})$ $(\Delta_x=\sum_{i=1}^{\text{pen-downs}}(x_{\max\,|i}-x_{\min\,|i}))\Delta_y$	16	$(x_{\text{last pen-up}}-x_{\max})/\Delta_x$
17	$(x_{\text{1st pen-down}}-x_{\min})/\Delta_x$	18	$(y_{\text{last pen-up}}-y_{\min})/\Delta_y$
19	$(y_{\text{1st pen-down}}-y_{\min})/\Delta_y$	20	$(T_w\bar{v})/(y_{\max}-y_{\min})$
21	$(T_w\bar{v})/(x_{\max}-x_{\min})$	22	(pen-down duration T_w)$/T_s$
23	$\bar{v}/v_{y,\text{max}}$	24	$(y_{\text{last pen-up}}-y_{\max})/\Delta_y$
25	$\frac{T((dy/dt)/(dx/dt)>0)}{T((dy/dt)/(dx/dt)<0)}$	26	$\bar{v}/v_{\max}$
27	$(y_{\text{1st pen-down}}-y_{\max})/\Delta_y$	28	$(x_{\text{last pen-up}}-x_{\min})/\Delta_x$
29	(velocity rms v)$/v_{\max}$	30	$\frac{(x_{\max}-x_{\min})\Delta_y}{(y_{\max}-y_{\min})\Delta_x}$
31	(velocity correlation $v_{x,y}$)$/v_{\max}^2$	32	$T(v_y>0|\text{pen-up})/T_w$
33	$N(v_x=0)$	34	direction histogram s_1
35	$(y_{\text{2nd local max}}-y_{\text{1st pen-down}})/\Delta_y$	36	$(x_{\max}-x_{\min})/x_{\text{acquisition range}}$
37	$(x_{\text{1st pen-down}}-x_{\max})/\Delta_x$	38	$T(\text{curvature}>\text{Threshold}_{\text{curv}})/T_w$
39	(integrated abs. centr. acc. a_{Ic})$/a_{\max}$	40	$T(v_x>0)/T_w$
41	$T(v_x<0|\text{pen-up})/T_w$	42	$T(v_x>0|\text{pen-up})/T_w$
43	$(x_{\text{3rd local max}}-x_{\text{1st pen-down}})/\Delta_x$	44	$N(v_y=0)$
45	(acceleration rms a)$/a_{\max}$	46	(standard deviation of x)$/\Delta_x$
47	$\frac{T((dx/dt)(dy/dt)>0)}{T((dx/dt)(dy/dt)<0)}$	48	(tangential acceleration rms a_t)$/a_{\max}$
49	$(x_{\text{2nd local max}}-x_{\text{1st pen-down}})/\Delta_x$	50	$T(v_y<0|\text{pen-up})/T_w$
51	direction histogram s_2	52	$t(\text{3rd pen-down})/T_s$
53	(max distance between points)$/A_{\min}$	54	$(y_{\text{3rd local max}}-y_{\text{1st pen-down}})/\Delta_y$
55	$(\bar{x}-x_{\min})/\bar{x}$	56	direction histogram s_5
57	direction histogram s_3	58	$T(v_x<0)/T_w$
59	$T(v_y>0)/T_w$	60	$T(v_y<0)/T_w$
61	direction histogram s_8	62	(1st $t(v_{x,\min})$)$/T_w$
63	direction histogram s_6	64	$T(\text{1st pen-up})/T_w$
65	spatial histogram t_4	66	direction histogram s_4
67	$(y_{\max}-y_{\min})/y_{\text{acquisition range}}$	68	(1st $t(v_{x,\max})$)$/T_w$
69	(centripetal acceleration rms a_c)$/a_{\max}$	70	spatial histogram t_1
71	$\theta(\text{1st to 2nd pen-down})$	72	$\theta(\text{1st pen-down to 2nd pen-up})$
73	direction histogram s_7	74	$t(j_{x,\max})/T_w$
75	spatial histogram t_2	76	$j_{x,\max}$
77	$\theta(\text{1st pen-down to last pen-up})$	78	$\theta(\text{1st pen-down to 1st pen-up})$
79	(1st $t(x_{\max})$)$/T_w$	80	$\bar{\jmath}_x$
81	$T(\text{2nd pen-up})/T_w$	82	(1st $t(v_{\max})$)$/T_w$
83	$j_{y,\max}$	84	$\theta(\text{2nd pen-down to 2nd pen-up})$
85	$j_{\max}$	86	spatial histogram t_3
87	(1st $t(v_{y,\min})$)$/T_w$	88	(2nd $t(x_{\max})$)$/T_w$
89	(3rd $t(x_{\max})$)$/T_w$	90	(1st $t(v_{y,\max})$)$/T_w$
91	$t(j_{\max})/T_w$	92	$t(j_{y,\max})/T_w$
93	direction change histogram c_2	94	(3rd $t(y_{\max})$)$/T_w$
95	direction change histogram c_4	96	$\bar{\jmath}_y$
97	direction change histogram c_3	98	$\theta(\text{initial direction})$
99	$\theta(\text{before last pen-up})$	100	(2nd $t(y_{\max})$)$/T_w$

12, 15-21, 24-25, 27-28, 30, 35-37, 43, 46-47, 49, 53-55, 65, 67, 70, 75, 86.

Features are normalized in the range of (0, 1) using tanh-estimators (Jain, Nandakumar & Ross, 2005). The Mahalanobis distance is used to compare a signature with a claimed user model during verification. Since the Mahalanobis distance is employed, user models are built by computing the mean and variance of the available training signatures during the enrollment phase. This distance measure has the advantage of being relatively simple to compute and thus appropriate for resource-limited devices.

Local System

This system is based on the one described by Fierrez et al., (2007), which participated in the Signature Verification Competition 2004 (Yeung et al., 2004), reaching the first and second positions against random and skilled forgeries respectively. The signals captured by the digitizer are used to extract a set of functions that model each signature. The original set of functions from (Fierrez et al., 2007) is extended here, adapting features from other contributions (Richiardi, Ketabdar & Drygajlo, 2005; Lei & Govindaraju, 2005; Van, Garcia-Salicetti, & Dorizzi, 2007). In Figure 6 we present the resulting set of 21 functions.

A preprocessing step consisting on position normalization is performed, by aligning the center of mass of each signature to a common coordinate. As can be observed in Figure 6, no functions related to pressure or pen inclination are used, since they are not available when signatures are captured using touchscreens. The extracted features are used to train a Hidden Markov Model (HMM) for each user. The HMMs used in this work have 2 states and 32 Gaussian Mixtures per state. Similarity scores are computed as the log-likelihood of the signature (using the Viterbi algorithm) divided by the total number of samples of the signature. Several implementations of Hidden Markov Models are freely

available from different research institutions. In particular, the PocketSphinx implementation (http://cmusphinx.sourceforge.net) is specifically designed for handheld devices.

Database and Experimental Protocol

A subset of the signature corpus of the BioSecure Multimodal Biometric Database is used. The set consists of 120 users, with 20 genuine signatures and 20 skilled forgeries per user. The set is divided in two subsets, one containing signatures from 50 users and the other one containing the remaining 70 users. The subset of 50 users is used to perform feature selection and will be referred to as the development set. The subset containing 70 users will be used to validate the results, and will be referred to as the validation set. Feature selection is performed with the Sequential Forward Floating Search (SFFS) algorithm (Pudil, Novovicova & Kittler, 1994). The feature selection algorithm is used to obtain the optimal feature vector that minimizes the system Equal Error Rate (EER) against skilled forgeries on the development set of signatures. Feature selection is performed separately on the global and the local system.

User models are trained with 5 genuine signatures of the first session (all from the same acquisition block), leaving the remaining 15 signatures from the second session to compute genuine user scores. This emulates the protocol of the BioSecure Multimodal Evaluation Campaign.

An example of the capture process is shown in Figure 3 and examples of a genuine captured signature and a skilled forgery are shown in Figure 4. Due to degraded capture conditions, a pre-processing step is first performed, where incorrectly detected samples (see Figure 4) are interpolated. As no pen pressure information is provided, pen-ups are assigned wherever a gap of 50 or more milliseconds between two consecutive samples exists.

Random forgery scores (the case where the forger uses his own signature but claims to be

Figure 6. Local feature set. The upper dot notation indicates time derivative.

#	Feature	Description
1	x-coordinate	x_n
2	y-coordinate	y_n
3	Path-tangent angle	$\theta_n = \arctan(\dot{y}_n/\dot{x}_n)$
4	Path velocity magnitude	$v_n = \sqrt{\dot{y}_n + \dot{x}_n}$
5	Log curvature radius	$\rho_n = \log(1/\kappa_n) = \log(v_n/\dot{\theta}_n)$, where κ_n is the curvature of the position trajectory
6	Total acceleration magnitude	$a_n = \sqrt{t_n^2 + c_n^2} = \sqrt{\dot{v}_n^2 + v_n^2\dot{\theta}_n^2}$, where t_n and c_n are respectively the tangential and centripetal acceleration components of the pen motion.
7-12	First-order derivative of features 1-6	$\dot{x}_n, \dot{y}_n, \dot{\theta}_n, \dot{v}_n, \dot{\rho}_n, \dot{a}_n$
13-14	Second-order derivative of features 1-2	$\ddot{x}_n, \ddot{y}_n$
15	Ratio of the minimum over the maximum speed over a window of 5 samples	$v_n^r = \min\{v_{n-4}, ..., v_n\}/\max\{v_{n-4}, ..., v_n\}$
16-17	Angle of consecutive samples and first order difference	$\alpha_n = \arctan(y_n - y_{n-1}/x_n - x_{n-1})$ $\dot{\alpha}_n$
18	Sine	$s_n = \sin(\alpha_n)$
19	Cosine	$c_n = \cos(\alpha_n)$
20	Stroke length to width ratio over a window of 5 samples	$r_n^5 = \dfrac{\sum\limits_{k=n-4}^{k=n} \sqrt{(x_k - x_{k-1})^2 + (y_k - y_{k-1})^2}}{\max\{x_{n-4}, ..., x_n\} - \min\{x_{n-4}, ..., x_n\}}$
21	Stroke length to width ratio over a window of 7 samples	$r_n^7 = \dfrac{\sum\limits_{k=n-6}^{k=n} \sqrt{(x_k - x_{k-1})^2 + (y_k - y_{k-1})^2}}{\max\{x_{n-6}, ..., x_n\} - \min\{x_{n-6}, ..., x_n\}}$

another user) are computed by comparing the user model to one signature of all the remaining users. Skilled forgery scores are computed by comparing all the available skilled forgeries per user with its own model. The system performance is evaluated using the Equal Error Rate (EER) separately for random and skilled forgeries.

Fusion of both systems is performed at the score level via weighted sum. The weighting coefficients are heuristically selected by optimizing the system performance on the development set against skilled forgeries.

Results and Discussion

The following features are selected by the feature selection algorithm on each system. Due to the nearly identical performance of several different global feature subsets, a feature vector of 40 features is selected among the best performing.

- Local. Features 1, 2, 4, 5, 8, 11, 19.
- Global.
  - Time features 1, 32, 52, 59, 62, 81, 88, 90, 94.
  - Speed and acceleration features 14, 29, 31, 39, 44, 48, 69, 85.

- ◦ Direction features 34, 51, 61, 66, 78, 95, 99.
- ◦ Geometry features 2, 3, 7, 8, 12, 15, 16, 18, 19, 24, 28, 30, 36, 46, 54, 86.

These results show that the optimal feature vector has a relatively small amount of features in the local system compared to previously published approaches in which feature vectors of 10-20 features are commonly considered. It must be taken into account that signatures captured with the PDA lack information during pen-ups, since resistive touch-screens (which are common in handheld devices) only acquire information when the stylus is in contact with them. Consequently, signatures captured with handheld devices may carry less information than signatures captured with pen-tablets, being thus possible to characterize with less features. In the case of global features, a prevalence of geometry features is observed, which can be due to the fact that the poor sampling quality is affecting features depending on speed or timing properties.

The verification performance of the original systems, which were designed considering

Table 1. Verification results on the validation set

Approach	System	Random forgeries EER (%)	Skilled forgeries EER (%)
Original systems	Global	7.81	15.22
	Local	7.86	20.94
	Fusion	5.99	15.35
PDA-optimized	Global	6.67	14.66
	Local	6.03	17.48
	Fusion	**4.70**	**12.29**

signatures captured with a pen-tablet, are also computed using the same signature subcorpus (i.e. captured with a PDA). Results are given in Table 1. As can be observed, a remarkable increase in the verification performance is obtained when the system is optimized for handheld devices with feature selection techniques. It is also observed that fusion increases notably the verification performance. In Figure 7, the DET plots of the original and the PDA-optimized systems using score fusion is presented. The results using feature selection and fusion are comparable to the ones

Figure 7. Fusion verification performance of the original and the PDA-optimized systems

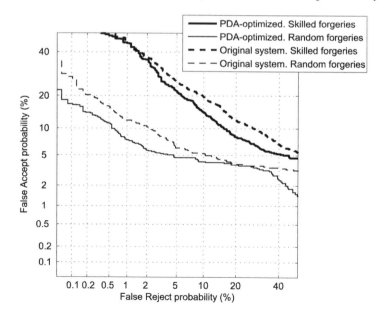

obtained by the best systems in the BioSecure Multimodal Evaluation Campaign. It must be taken into account that the experimental protocol is the same as in BMEC, although the signature database is not.

FUTURE RESEARCH DIRECTIONS

Dynamic signature verification is still an open issue, even more in handheld devices. The verification performance achieved by state-of-the-art systems in the recent BioSecure Multimodal Evaluation reveals that there is room for improvement, especially when skilled forgeries are considered. Another key issue that must be addressed is device interoperability. While in the last years signatures were mostly acquired using digitizing tables, the proliferation of touchscreen-enabled devices represents a completely new scenario. The effects of training a user model with signatures captured with a different device than the one used for verification must be analyzed. This is a well known issue by the speech recognition community, which is referred to as channel mismatch compensation.

The proliferation of touch- and multitouch-screens in handheld devices is becoming centered in applications where no PDA stylus is required and users interface with applications using their fingertips. In this context, new possibilities such as graphical passwords or even signatures traced with the fingertips must be explored. Little research has been carried out in this subject, representing an open field for new approaches.

Template protection is another incipient research field in signature verification (Maiorana et al. 2008). New methods must be proposed to allow the storage and transmission of signature data without allowing the reconstruction of the original traces if they are captured.

Template selection and update techniques have proven to be effective for other biometric traits, such as fingerprints (Uludag, Ross & Jain, 2004),

but have not been yet studied for signature verification. Such techniques may alleviate the problem of short- and long-term variability in signature verification and the scarcity of training samples, improving the system performance.

CONCLUSION

Automatic signature verification on handheld devices has been studied in this chapter. The typical architecture of a signature verification system has been described and references to popular approaches have been given. Current and promising applications have been listed, as well as some challenges that may slow the market adoption of products and solutions based on automatic signature verification. A case study has been analyzed, where a local and a global feature set that include a notable amount of features proposed in the signature verification literature have been presented. Feature selection has been applied to extract the most discriminative features for handheld devices. Results show that the performance of the proposed system is aligned with other state-of-the-art approaches.

As has already been stated, automatic signature verification is still an open problem. Despite its social acceptance, the error rates of biometric systems based on signature is still considerably higher than the one of other common biometrics such as fingerprint or iris. Researchers must deal with many issues that are specific to signature verification, such as a notable short- and long-term variability and the unpredictability of the accuracy of forgers.

Handheld devices represent a new scenario for signature verification that has not yet been tackled in depth. In the last few decades, research on this field has been focused on signatures captured with digitizing tables. Thus, more research is needed to find if the current assumptions about signature verification on pen-tablets are still valid for handheld devices.

REFERENCES

Ballagas, R., Borchers, J., Rohs, M., & Sheridan, J. (2006). The smart phone: a ubiquitous input device. *IEEE Pervasive Computing / IEEE Computer Society [and] IEEE Communications Society*, 5(1), 70–77. doi:10.1109/MPRV.2006.18

BioSecure multimodal evaluation campaign 2007 mobile scenario - experimental results. (2007). Retrieved February 1, 2008, from http://biometrics.it-sudparis.eu/BMEC2007/files/Results mobile.pdf

Dale, J. (2008). Signature Security. *Financial Services Technology.* (2008). Retrieved October 20, 2008, from http://www.fsteurope.com/currentissue/article.asp?art=270443&issue=207

Dolfing, J. G. A., Aarts, E. H. L., & van Oosterhout, J. J. G. M. (1998). On-line signature verification with hidden Markov models. In *Proceedings of the 14th International Conference on Pattern Recognition* (pp. 1309-1312). Washington, DC: IEEE CS Press.

Fierrez, J., & Ortega-Garcia, J. (2007). On-line signature verification. In A. K. Jain, A. Ross, & P. Flynn (Eds.), *Handbook of Biometrics* (pp. 189-209). Berlin, Germany: Springer.

Fierrez, J., Ramos-Castro, D., Ortega-Garcia, J., & Gonzalez-Rodriguez, J. (2007). HMM-based on-line signature verification: feature extraction and signature modelling. *Pattern Recognition Letters*, 28(16), 2325–2334. doi:10.1016/j.patrec.2007.07.012

Fierrez-Aguilar, J., Nanni, L., Lopez-Penalba, J., Ortega-Garcia, J., & Maltoni, D. (2005). An on-line signature verification system based on fusion of local and global information. In T. Kanade, A. Jain & N. K. Ratha (Eds.), *Audio- and Video-Based Biometric Person Authentication. Lecture Notes in Computer Science* (pp. 523-532). Berlin, Germany: Springer.

Fierrez-Aguilar, J., Ortega-Garcia, J., & Gonzalez-Rodriguez, J. (2005). Target dependent score normalization techniques and their application to signature verification. *IEEE Transactions on Systems, Man and Cybernetics, part C,* 35(3), 418-425.

Freire-Santos, M., Fierrez-Aguilar, J., & Ortega-Garcia, J. (2006). Cryptographic key generation using handwritten signature. In *Defense and Security Symposium, Biometric Technologies for Human Identification, BTHI* (pp. 225-231).

Galbally, J., Fierrez, J., & Ortega-Garcia, J. (2007). Bayesian hill-climbing attack and its application to signature verification. In S.-W. Lee & S.Z. Li (Eds.), *Advances in Biometrics. Lecture Notes in Computer Science* (pp. 386-395). Berlin, Germany: Springer.

Garcia-Salicetti, S., Beumier, C., Chollet, G., Dorizzi, B., Jardins, J. L.-L., Lanter, J., et al. (2003). BIOMET: A multimodal person authentication database. In J. Kittler & M. S. Nixon (Eds.), *Audio- and Video-Based Biometric Person Authentication. Lecture Notes in Computer Science* (pp. 845-853). Berlin, Germany: Springer.

Hennebert, J., Humm, A., & Ingold, R. (2007). Modelling spoken signatures with Gaussian mixture models adaptation. In *IEEE International Conference on Acoustics, Speech, and Signal Processing, ICASSP 2007*, Honolulu, USA (pp. 229-232). Washington, DC: IEEE Press.

Jain, A. K., Nandakumar, K., & Ross, A. (2005). Score normalization in multimodal biometric systems. *Pattern Recognition*, 38(12), 2270–2285. doi:10.1016/j.patcog.2005.01.012

Jain, A. K., Ross, A., & Pankanti, S. (2006). Biometrics: a tool for information security. *IEEE Transactions on Information Forensics and Security,* 1(2), 125–143. doi:10.1109/TIFS.2006.873653

Jermyn, I., Mayer, A., Monrose, F., Reiter, M. K., & Rubin, A. D. (1999). The design and analysis of graphical passwords. In *8th conference on USENIX Security Symposium, Vol. 8,* Washington, D.C.

Khokhar, R. (2006). Smartphones - a call for better safety on the move. *Network Security*, (4): 6–7. doi:10.1016/S1353-4858(06)70354-3

Kholmatov, A., & Yanikoglu, B. (2005). Identity authentication using improved online signature verification method. *Pattern Recognition Letters*, *26*(15), 2400–2408. doi:10.1016/j.patrec.2005.04.017

Lee, L. L., Berger, T., & Aviczer, E. (1996). Reliable on-line human signature verification systems. *IEEE Transactions on Pattern Analysis and Machine Intelligence*, *18*(6), 643–647. doi:10.1109/34.506415

Lei, H., & Govindaraju, V. (2005). A comparative study on the consistency of features in online signature verification. *Pattern Recognition Letters*, *26*(15), 2483–2489. doi:10.1016/j.patrec.2005.05.005

Lenz, J. M. (2008). Biometric signature verification – a sign of the times? *Biometric Technology Today*, *16*(4), 9–11. doi:10.1016/S0969-4765(08)70124-3

Maiorana, E., Martinez-Diaz, M., Campisi, P., Ortega-Garcia, J., & Neri, A. (2008). Template protection for HMM-based on-line signature authentication. In *IEEE Computer Society Conference on Computer Vision and Pattern Recognition Workshops, 2008. CVPR Workshops 2008* (pp. 1-6). Washington DC: IEEE Press.

Martinez-Diaz, M., Fierrez, J., & Galbally, J. Ortega-Garcia., J. (2008). Towards mobile authentication using dynamic signature verification: useful features and performance evaluation. In *Proceedings of 19th International Conference on Pattern Recognition 2008 (ICPR 2008),* Tampa, FL, (pp.1-5). Washington, DC: IEEE Press.

Martinez-Diaz, M., Fierrez, J., & Ortega-Garcia, J. (2007). Universal background models for dynamic signature verification. In *IEEE Conference on Biometrics: Theory, Applications and Systems, BTAS,* Washington DC (pp. 1-6). Washington DC: IEEE Press.

Miller, C. (2008, April 22). Dynamic signature verification market growth up. *SC Magazine*. Retrieved October 20, 2008, from http://www.scmagazineus.com/Dynamic-signature-verification-market-growth-up/article/109250/

Muramatsu, D., & Matsumoto, T. (2007). Effectiveness of pen pressure, azimuth, and altitude features for online signature verification. In S.-W. Lee & S.Z. Li (Eds.), *Advances in Biometrics. Lecture Notes in Computer Science* (pp. 503-512). Berlin, Germany: Springer.

Nelson, W., Turin, W., & Hastie, T. (1994). Statistical methods for on-line signature verification. *International Journal of Pattern Recognition and Artificial Intelligence*, *8*(3), 749–770. doi:10.1142/S0218001494000395

Ortega-Garcia, J., Fierrez, J., Alonso-Fernandez, F., Galbally, J., Freire, M. R., & Gonzalez-Rodriguez, J. (2009). (in press). The multi-scenario multi-environment biosecure multimodal database (BMDB). *IEEE Transactions on Pattern Analysis and Machine Intelligence.*

Ortega-Garcia, J., Fierrez-Aguilar, J., Simon, D., Gonzalez, J., Faundez-Zanuy, M., & Espinosa, V. (2003). MCYT baseline corpus: a bimodal biometric database. *IEE Proceedings. Vision Image and Signal Processing*, *150*(6), 391–401. doi:10.1049/ip-vis:20031078

Oviatt, S., & Lunsford, R. (2003). Multimodal interfaces for cell phones and mobile technology. [Berlin, Germany: Springer.]. *International Journal of Speech Technology*, *8*(2), 127–132. doi:10.1007/s10772-005-2164-8

Petrovska-Delacretaz, D., Chollet, G., & Dorizzi, B. (Eds.). (2009). *Guide to Biometric Reference Systems and Performance Evaluation*. Berlin, Germany: Springer.

Plamondon, R., & Lorette, G. (1989). Automatic signature verification and writer identification: the state of the art. *Pattern Recognition, 22*(2), 107–131. doi:10.1016/0031-3203(89)90059-9

Pudil, P., Novovicova, J., & Kittler, J. (1994). Floating search methods in feature selection. *Pattern Recognition Letters, 15*(11), 1119–1125. doi:10.1016/0167-8655(94)90127-9

Richiardi, J., & Drygajlo, A. (2003). Gaussian mixture models for on-line signature verification. In *ACM International Conference on Multimedia, Workshop on Biometric Methods and Applications, WBMA*, Berkeley, USA (pp. 115-122).

Richiardi, J., Ketabdar, H., & Drygajlo, A. (2005). Local and global feature selection for on-line signature verification. In *Proceedings of the Eighth International Conference on Document Analysis and Recognition 2005, Vol. 2*, Seoul, Korea (pp. 625-629). Washington, DC: IEEE Press.

Ross, A., Nandakumar, K., & Jain, A. K. (2006). *Handbook of Multibiometrics*. Berlin, Germany: Springer.

Signature verification biometrics market to overcome the effect of reduced IT Expenditures. (2008, Feb. 18). *BNET*. Retrieved October 20, 2008, from http://findarticles.com/p/articles/mi_m0EIN/is_/ai_113374152

Uludag, U., Ross, A., & Jain, A. K. (2004). Biometric template selection and update: a case study in fingerprints. *Pattern Recognition, 37*(7), 1533–1542. doi:10.1016/j.patcog.2003.11.012

Van, B. L., Garcia-Salicetti, S., & Dorizzi, B. (2007). On using the Viterbi path along with HMM likelihood information for online signature verification. *IEEE Transactions on Systems, Man, and Cybernetics . Part B, 37*(5), 1237–1247.

Yeung, D. Y., Chang, H., Xiong, Y., George, S., Kashi, R., Matsumoto, T., et al. (2004). SVC2004: First international signature verification competition. In *Biometric Authentication. Lecture Notes in Computer Science* (pp. 16-22). Berlin, Germany: Springer.

Compilation of References

411SONG. (2008). *ID songs anywhere.* Retrieved from http://www.411song.com/

Abdelmoty, A. I., Smart, P., & Jones, C. B. (2007). Building place ontologies for the semantic web: Issues and approaches. In *Proceedings of the 4th ACM workshop on Geographical information retrieval.*

Abowd, G. D., Mynatt, E. D., & Rodden, T. (2002). The human experience. *IEEE Pervasive Computing / IEEE Computer Society [and] IEEE Communications Society, 1*(1), 48–57. doi:10.1109/MPRV.2002.993144

Abowd, G. D., Wang, H. M., & Monk, A. F. (1995). A formal technique for automated dialogue development. *The first symposium of designing interactive systems - DIS'95,* (219-226). New York: ACM Press.

Abrial, J. R. (Ed.). (1996). *The B book: Assigning programs to meanings.* New York: Cambridge University Press.

Abrial, J.-R. (1996). Extending B without changing it for developing distributed systems. *First conference on the B-Method,* 169-190.

Adelstein, B. D., Begault, D. R., Anderson, M. R., & Wenzel, E. M. (2003). Sensitivity to haptic-audio asynchrony. In *Proceedings of 5th international Conference on Multimodal interfaces ICMI '03* (pp. 73-76). New York: ACM.

AIML. *(Artificial Intelligence Mark-up Language) Documentation.* Retrieved January 28, 2009, from http://www.alicebot.org/aiml.html.

Ait-Ameur, Y., & Ait-Sadoune, I., Baron, M. (2006). Étude et comparaison de scénarios de développements formels d'interfaces multimodales fondés sur la preuve et le raffinement. *MOSIM 2006 - 6ème Conférence Francophone de Modélisation et Simulation. Modélisation, Optimisation et Simulation des Systèmes: Défis et Opportunités.*

Ait-Ameur, Y., Girard, P., & Jambon, F. (1998). A uniform approach for the specification and design of interactive systems: the B method. In P. Markopoulos & P. Johnson (Eds.), *Eurographics Workshop on Design, Specification, and Verification of Interactive Systems (DSV-IS'98),* 333-352.

Aitenbichler, E., Kangasharju, J., & Mühlhäuser, M. (2007). MundoCore: A Light-weight Infrastructure for Pervasive Computing. *Pervasive and Mobile Computing.*

Aker, A., Fan, X., Sanderson, M., & Gaizauskas, R. (2008). *Automatic image captioning from web documents by image location.* Paper presented submitted to the 31st European Conference on Information Retrieval.

Alapetite, A. (2007). *On speech recognition during anaesthesia.* PhD thesis, Roskilde University.

Alexader, T. & Dixon, E., (2004). Usability Evaluation report for VPA 2. FASIL deliverable D.3.3.3.

Almeida, & P., Yokoi, S. (2003). Interactive character as a virtual tour guide to an online museum exhibition. In *Proceedings of Museum and the Web 2003.*

Amant, R., Horton, T., & Ritter, F. (2007). Model-based evaluation of expert cell phone menu interaction. *ACM Transactions on Computer-Human Interaction, 14*(1), 347–371. doi:10.1145/1229855.1229856

Ancona, M., Cappello, M., Casamassima, M., Cazzola, W., Conte, D., Pittore, M., et al. (2006). Mobile vision and cultural heritage: the AGAMEMNON project. In *Proceedings of the 1st International Workshop on Mobile Vision*, Austria.

André, E. (2000). The generation of multimedia presentations. In R. Dale, H. Moisl & H. Somers (Eds.), *A Handbook of Natural Language Processing* (pp. 305–327). New York: CRC.

André, E. (2003). Natural language in multimedia/multimodal systems. In R. Mitkov (Ed.), *Computational Linguistics* (pp. 650-669). New York: Oxford University Press.

André, E., Finkler, W., Graf, W., Rist, T., Schauder, A., & Wahlster, W. (1993). The automatic synthesis of multimodal presentations. In M. T. Maybury, (Ed.), *Intelligent Multimedia Interfaces* (pp. 75-93). Menlo Park, CA: AAAI Press.

Aoki, P. M., Grinter, R. E., Hurst, A., Szymanski, M. H., Thornton, J. D., & Woodruff, A. (2002). Sotto voce: Exploring the interplay of conversation and mobile audio spaces. In *Proceedings of ACM Conference on Human Factors in Computing Systems* (pp. 431-438). New York: ACM Press.

Apel, J., Neubarth, F., Pirker, H., & Trost, H. (2004). Have a break! Modelling pauses in German speech. In *Proceedings of KONVENS 2004, Österreichische Gesellschaft für Artificial Intelligence OEGAI* (pp. 5-12, 2004).

Arens, Y., & Hovy, E. H. (1995). The design of a model-based multimedia interaction manager. *Artificial Intelligence, 9*(3), 167–188. doi:10.1016/0954-1810(94)00014-V

Ask.com. (2008). *Ask mobile GPS: The next level of local.* Retrieved from http://www.ask.com

Augello, A., Santangelo, A., Sorce, S., Pilato, G., Gentile, A., Genco, A., & Gaglio, S. (2006). MAGA: A mobile archaeological guide at Agrigento. *Proceedings of the workshop Giornata nazionale su Guide Mobili Virtuali 2006, ACM-SIGCHI.* Retrieved from http://hcilab.uniud.it/sigchi/doc/Virtuality06/index.html.

Axelsson, J., Cross, C., Ferrans, J., McCobb, G., Raman, T. V., & Wilson, L. (2004). XHTML+Voice Profile 1.2. *W3C.* Retrieved from http://www.voicexml.org/specs/multimodal/x+v/12/

B'Far. R. (2005). *Mobile computing principles.* Cambridge, UK: Cambridge University Press.

Balke, W.-T., Kiebling, W., & Unbehend, C. (2003). *Personalized services for mobile route planning: a demonstration.* Paper presented at the Proceedings of 19th International Conference on Data Engineering.

Ballagas, R., Borchers, J., Rohs, M., & Sheridan, J. (2006). The smart phone: a ubiquitous input device. *IEEE Pervasive Computing / IEEE Computer Society [and] IEEE Communications Society, 5*(1), 70–77. doi:10.1109/MPRV.2006.18

Ballagas, R., Kuntze, A., & Walz, S. P. (2008). Gaming tourism: Lessons from evaluating REXplorer, a pervasive game for tourists. In *Pervasive '08: Proceedings of the 6th International Conference on Pervasive Computing.* Berlin, Germany: Springer-Verlag.

Ballagas, R., Memon, F., Reiners, R., & Borchers, J. (2006). iStuff Mobile: prototyping interactions for mobile phones in interactive spaces. In *PERMID 2006, Workshop at Pervasive 2006*, Dublin, Ireland.

Balme, L., Demeure, A., Barralon, N., Coutaz, J., & Calvary, G. (2004). CAMELEON-RT: A software architecture reference model for distributed, migratable, and plastic user interfaces. In *Proc. of EUSAI 2004, European symposium on ambient intelligence, No. 2, vol. 3295 of Lecture Notes in Computer Science* (pp. 291-302). Eindhoven, The Netherlands: Springer.

Banse, R., & Scherer, K. R. (1996). Acoustic profiles in vocal emotion expression. *Journal of Personality and Social Psychology, 70*, 614–636. doi:10.1037/0022-3514.70.3.614

Barnard, L., Yi, J., Jacko, J., & Sears, A. (2007). Capturing the effects of context on human performance in mobile computing systems. *Personal and Ubiquitous Computing, 11*(2), 81–96. doi:10.1007/s00779-006-0063-x

Basil, M. D. (1994). Multiple resource theory I: Application to television viewing. *Communication Research, 21*(2), 177–207. doi:10.1177/009365094021002003

Bass, L., Faneuf, R., Little, R., Mayer, N., Pellegrino, B., & Reed, S. (1992). A metamodel for the runtime architecture of an interactive system. *SIGCHI Bulletin, 24*(1), 32–37. doi:10.1145/142394.142401

Bastide, R., Palanque, P., Le, D., & Munoz, J. (1998). Integrating rendering specifications into a formalism for the design of interactive systems. In *Proceedings of DSV-IS '98*, Abington, UK. Berlin, Germany: Springer Verlag.

Baumberg, A. (2000, June). *Reliable feature matching across widely separated views.* Paper presented at the Proceedings of IEEE Conference on Computer Vision and Pattern Recognition, Hilton Head Island, USA.

Baumgarten, B. (1996). *Petri-Netze - Grundlagen und Anwendungen.* Berlin, Germany: Spektrum Verlag.

BCS Specialist Interest Group in Software Testing. (2001). Retrieved September 12, 2008, from http:/www.bcs.org.uk/siggroup/sigist/index.htm

Been-Lirn, D. H., & Tan, C. B. G., & Hsueh-hua Chen, V. (2006). Usability evaluation for mobile device: A comparison of laboratory and field tests. *8th Conference on Human-Computer Interaction with Mobile Device and Services, Mobile HCI'06*, 181-186. New York: ACM.

Bellik, Y. (1995), *Interfaces multimodales: Concepts, modèles et architectures.* Unpublished doctoral dissertation, Université Paris XI, Orsay, France.

Bellik, Y., & Teil, D. (1992). Multimodal dialog interface. In *Proceedings of WWDU'92*, Berlin.

Benford, S., & Fahlen, L. (1993). A spatial model of interaction in virtual environments. In *Proceedings of the Third European Conference on Computer Supported Cooperative Work (ECSCW'93).*

Benko, H., Ishak, E. W., & Feiner, S. (2004). Collaborative Mixed Reality Visualization of an Archaeological Excavation. *The International Symposium on Mixed and Augmented Reality (ISMAR 2004)*, November 2004.

Benoit, C., Martin, J. C., Pelachaud, C., Schomaker, L., & Suhm, B. (2000). Audio-visual and multimodal speech systems. In D. Gibbon (Ed.), *Handbook of Standards and Resources for Spoken Language Systems.* Amsterdam: Kluwer.

Beringer, N., Kartal, U., Louka, K., Schiel, F., & Turk, U. (2002). PROMISE: A procedure for multimodal interactive system evaluation. In *Proceedings of LREC Workshop on Multimodal Resources and Multimodal Systems Evaluation* (pp. 77–80).

Bernsen, N. O. (1994). Foundations of multimodal representations: a taxonomy of representational modalities. *Interacting with Computers, 6*(4). doi:10.1016/0953-5438(94)90008-6

Bernsen, N. O. (1995). A Toolbox of output modalities: representing output information in multimodal interfaces. *Esprit Basic Research Action 7040: The Amodeus Project*, document TM/WP21.

Berti, S., & Paternò, F. (2005). Migratory multimodal interfaces in multidevice environments. In *ICMI '05: Proceedings of the 7th International Conference on Multimodal Interfaces* (pp. 92–99). New York: ACM.

Beyer, H., & Holtzblatt, K. (1998). *Contextual design: A customer centered approach to systems design.* San Francisco, CA: Academic Press.

BioSecure multimodal evaluation campaign 2007 mobile scenario - experimental results. (2007). Retrieved February 1, 2008, from http://biometrics.it-sudparis.eu/BMEC2007/files/Results mobile.pdf

Blattner, M. M., Sumikawa, D. A., & Greenberg, R. M. (1989). Earcons and icons: Their structure and common design principles. *SIGCHI Bull, 21*(1), 123–124. doi:10.1145/67880.1046599

Block, H. U., Caspari, R., & Schachtl, S. (2004). Callable manuals - access to product documentation via voice. *Information Technology, 46*(6), 299–304.

Blumendorf, M., Feuerstack, S., & Albayrak, S. (2006). Event-based synchronization of model-based multimodal user interfaces. In A. Pleuss, J. V. den Bergh, H. Hussmann, S. Sauer, & A. Boedcher, (Eds.), *MDDAUI '06 - Model Driven Development of Advanced User Interfaces 2006, Proceedings of the 9th International Conference on Model-Driven Engineering Languages and Systems: Workshop on Model Driven Development of Advanced User Interfaces*, CEUR Workshop Proceedings.

Blumendorf, M., Feuerstack, S., & Albayrak, S. (2007). Multimodal user interaction in smart environments: Delivering distributed user interfaces. In M. Mühlhäuser, A. Ferscha, & E. Aitenbichler, (Eds.), *Constructing Ambient Intelligence: AmI 2007 Workshops Darmstadt* (pp. 113-120). Berlin, Germany: Springer.

Blumendorf, M., Lehmann, G., Feuerstack, S., & Albayrak, S. (2008). Executable models for human-computer interaction. In T. C. N. Graham & P. Palanque, (Eds.), *Interactive Systems - Design, Specification, and Verification* (pp. 238-251). Berlin, Germany: Springer.

Boersma, P., & Weenink, D. (2001). Praat, a system for doing phonetics by computer. *Glot International, 5*(9/10), 341–345.

Bohn, J., & Mattern, F. (2004), Super-distributed RFID tag infrastructures. In *Proceedings of the 2nd European Symposium on Ambient Intelligence (EUSAI 2004)* (pp. 1–12). Berlin, Germany: Springer-Verlag.

Bolognesi, T., & Brinksma, E. (1987). Introduction to the ISO specification language LOTOS. *Computer Networks and ISDN Systems, 14*(1), 25–59. doi:10.1016/0169-7552(87)90085-7

Bolt, R. A. (1980). "Put-that-there": Voice and gesture at the graphics interface. In *Proceedings of 7th Annual Conference on Computer Graphics and interactive Techniques SIGGRAPH '80* (pp. 262-270). New York: ACM

Bolt, R. A. (1980). Put-that-there: Voice and gesture at the graphics interface. *Computer Graphics, 14*(3), 262–270. doi:10.1145/965105.807503

Bordegoni, M., Faconti, G., Maybury, M. T., Rist, T., Ruggieri, S., Trahanias, P., & Wilson, M. (1997). A standard reference model for intelligent multimedia presentation systems. *Computer Standards & Interfaces, 18*(6-7), 477–496. doi:10.1016/S0920-5489(97)00013-5

Bouchet, J., & Nigay, L. (2004). ICARE: a component-based approach for the design and development of multimodal interfaces. In *CHI'2004 Conference on Human Factors in Computing Systems,* Vienna, Austria (pp. 1325-1328).

Bouchet, J., Madani, L., Nigay, L., Oriat, C., & Parissis, I. (2007). Formal testing of multimodal interactive systems. *DSV-IS2007, the XIII International Workshop on Design, Specification and Verification of interactive systems,* Lecture Notes in Computer Science. Berlin, Germany: Springer-Verlag.

Bouchet, J., Nigay, L., & Ganille, T. (2004). ICARE software components for rapidly developing multimodal interfaces. In *Proceedings of the 6th international conference on Multimodal interfaces* (251-258).

Bouchet, J., Nigay, L., & Ganille, T. (2005). The ICARE Component-based approach for multimodal input interaction: application to real-time military aircraft cockpits. *The 11th International Conference on Human-Computer Interaction.*

Bourguet, M. L. (2003). How finite state machines can be used to build error free multimodal interaction systems. In *Proceedings of the 17th British HCI Group Annual Conference* (pp. 81-84).

Bourguet, M. L. (2003). Designing and prototyping multimodal commands. In *Proceedings of Human-Computer Interaction INTERACT, 03,* 717–720.

Bourguet, M. L. (2004). Software design and development of multimodal interaction. In *Proceedings of IFIP 18th World Computer Congress Topical Days* (pp. 409-414).

Bourguet, M. L., & Chang, J. (2008). Design and usability evaluation of multimodal interaction with finite state machines: A conceptual framework. *Journal on Multimodal User Interfaces*, *2*(1), 53–60. doi:10.1007/s12193-008-0004-2

Braffort, A., Choisier, A., Collet, C., Dalle, P., Gianni, F., Lenseigne, B., & Segouat, J. (2004). Toward an annotation software for video of sign language, including image processing tools and signing space modeling. In *Proceedings of the Language Resources and Evaluation Conference (LREC'04)*.

Branco, G., Almeida, L., Beires, N., & Gomes, R. (2006). Evaluation of a multimodal virtual personal assistant. Proceedings from *20ᵗʰ International Symposium on Human Factors in Telecommunication*.

Braun, E., & Mühlhäuser, M. (2005). Interacting with Federated Devices. In A. Ferscha, R. Mayrhofer, T. Strang, C. Linnhoff-Popien, A. Dey, A. Butz, & A. Schmidt (Eds.), *Advances in pervasive computing, adjunct proceedings of the third international conference on pervasive computing* (pp. 153–160). Österreich, Austria: Austrian Computer Society.

Bregler, C., Manke, S., Hild, H., & Waibel, A. (1993). Bimodal sensor integration on the example of 'speechreading'. In *Proceedings of IEEE Int'l Conference on Neural Networks* (pp. 667-671).

Bregman, A. S. (1990). *Auditory scene analysis*. Cambridge, MA: MIT Press.

Bresciani, J.-P., Ernst, M. O., Drewing, K., Boyer, G., Maury, V., & Kheddar, A. (2005). Feeling what you hear: auditory signals can modulate tactile tap perception. *Experimental Brain Research*, *162*(2), 172–180. doi:10.1007/s00221-004-2128-2

Breton, E., & Bézivin, J. (2001). Towards an understanding of model executability. In N. Guarino, B. Smith & C. Welty (Eds.) *Proceedings of the international conference on Formal Ontology in Information Systems 2001* (pp. 70–80). New York: ACM Press.

Brewster, S. A. (1997). Navigating Telephone-Based Interfaces with Earcons. In *Proceedings of HCI on People and Computers XII H* (pp. 39-56). London: Springer-Verlag.

Brewster, S. A., & Brown, L. M. (2004). Non-visual information display using tactons. In *Proceedings of CHI '04 Extended Abstracts on Human Factors in Computing Systems* (pp. 787-788). New York: ACM.

Brewster, S., Faraz, C., & Brown, L. (2007). Tactile Feedback for Mobile Interactions. In [New York: ACM.]. *Proceedings of SIGCHI Conference on Human Factors in Computing Systems CHI*, *2007*, 159–162. doi:10.1145/1240624.1240649

Brewster, S., Wright, P., & Edwards, A. (1995). Experimentally derived guidelines for the creation of earcons. In *Proceedings of the British Computer Society Human-Computer Interaction Group Annual Conference* (pp. 155-159). Huddersfield, UK: Cambridge University Press.

Broadbent, D. E. (1958). *Perception and communication*. London: Pergamon.

Burges, C. J. C., Platt, J. C., & Jana, S. (2003). Distortion discriminant analysis for audio fingerprinting. *IEEE Transactions on Speech and Audio Processing*, *11*(3), 165–174. doi:10.1109/TSA.2003.811538

Burmeister, R., Pohl, C., Bublitz, S., & Hugues, P. (2006). SNOW - a multimodal approach for mobile maintenance applications. In *Proceedings of the 15th IEEE International Workshops on Enabling Technologies: Infrastructures for Collaborative Enterprises (WETICE 2006)* (pp. 131–136). Manchester, United Kingdom: IEEE Computer Society.

Calvary, G., Coutaz, J., Thevenin, D., Limbourg, Q., Bouillon, L., & Vanderdonckt, J. (2003). A unifying reference framework for multi-target user interfaces. *Interacting with Computers, 15*(3), 289–308. doi:10.1016/S0953-5438(03)00010-9

Calvary, G., Coutaz, J., Thevenin, D., Limbourg, Q., Souchon, N., Bouillon, L., et al. (2002). Plasticity of user interfaces: A revised reference framework. In C. Pribeanu & J. Vanderdonckt (Eds.), *Proceedings of the First International Workshop on Task Models and Diagrams for User Interface Design 2002* (pp. 127–134). Bucharest, Rumania: INFOREC Publishing House Bucharest.

Campione, E., & Véronis, J. (2002). A Large-Scale Multilingual Study of Silent Pause Duration. In *Proceedings of Speech Prosody 2002 Conference* (pp. 199-202).

Campos, J., & Harrisson, M. D. (2001). Model checking interactor specifications. *Automated Software Engineering, 8*, 275–310. doi:10.1023/A:1011265604021

Campos, J., Harrison, M. D., & Loer, K. (2004). Verifying user interface behaviour with model checking. *VVEIS*, (pp. 87-96).

Cañadas-Quesada, F. J., & Reyes-Lecuona, A. (2006). Improvement of perceived stiffness using auditory stimuli in haptic virtual reality. In [Piscataway, NJ: IEEE Mediterranean.]. *Proceedings of Electrotechnical Conference, MELECON, 2006*, 462–465. doi:10.1109/MELCON.2006.1653138

Cano, P., Batle, E., Kalker, T., & Haitsma, J. (2002). *A review of algorithms for audio fingerprinting*. Paper presented at the IEEE Workshop on Multimedia Signal Processing 2002.

Card, S. K., Moran, T. P., & Newell, A. (1983). *The psychology of human-computer interaction*. Hillsdale, NJ: Lawrence Erlbaum Associates.

Carriço, L., Duarte, C., Lopes, R., Rodrigues, M., & Guimarães, N. (2005). Building rich user interfaces for digital talking books. In Jacob, R.; Limbourg, Q. & Vanderdonckt, J., (Eds.), *Computer-Aided Design of User Interfaces IV* (pp. 335-348). Berlin, Germany: Springer-Verlag

Carroll, J. M., & Rosson, M. B. (1987). *The paradox of the active user. Interfacing Thought: Cognitive Aspects of Human-Computer Interaction*. Cambridge, MA: MIT Press.

Carter, S., Mankoff, J., & Heer, J. (2007). Momento: Support for situated ubicomp experimentation. In *Proceedings of the SIGCHI Conference on Human Factors in Computing Systems* (pp. 125-134). New York: ACM Press.

ChaCha. (2008). *ChaCha. Ask away*. Retrieved from http://www.chacha.com/

Chang, A., & O'Sullivan, C. (2005). Audio-haptic feedback in mobile phones. In *Proceedings of CHI '05 Extended Abstracts on Human Factors in Computing Systems* (pp. 1264-1267). New York: ACM.

Chee, Y. S., & Hooi, C. M. (2002). C-VISions: socialized learning through collaborative, virtual, interactive simulations. *Proceedings of CSCL 2002 Conference on Computer Support for Collaborative Learning*, Boulder, CO, USA, 2002, pp. 687-696

Chetali, B. (1998). Formal verification of concurrent programs using the Larch prover. *IEEE Transactions on Software Engineering, 24*(1), 46–62. doi:10.1109/32.663997

Chevrin, V., & Rouillard, J. (2008). Instrumentation and measurement of multi-channel services systems. *International Journal of Internet and Enterprise Management (IJIEM), Special Issue on . Quality in Multi-Channel Services Employing Virtual Channels, 5*(4), 333–352.

Chion, M. (1990). *Audio-vision: Sound on screen*. New York: Columbia University Press.

Choi, D.-Y. (2007). Personalized local internet in the location-based mobile web search. *Decision Support Systems, 43*(1), 31–45. doi:10.1016/j.dss.2005.05.005

Clark, A. (1997). *Being there: Putting brain, body and world together again*. Cambridge, MA: MIT Press.

Clarke, E. M., Emerson, E. A., & Sistla, A. P. (1986). Automatic verification of finite-state concurrent systems using temporal logic specifications. *ACM Transactions on Programming Languages and Systems, 2*(8), 244–263. doi:10.1145/5397.5399

Clerckx, T., Luyten, K., & Coninx, K. (2004). DynaMo-AID: A design process and a runtime architecture for dynamic model-based user interface development. In R. Bastide, P. Palanque & J. Roth (Eds.), *Engineering Human Computer Interaction and Interactive Systems: Joint Working Conferences EHCI-DSVIS 2004* (pp. 77–95). Berlin, Germany: Springer.

Clerckx, T., Vandervelpen, C., & Coninx, K. (2007). *Task-based design and runtime support for multimodal user interface distribution.* Paper presented at Engineering Interactive Systems 2007 (EHCI-HCSE-DSVIS'07), Salamanca, Spain.

Cohen, M. C., Giangola, J., & Balogh, J. (2004). *Voice User Interface Design.* Boston, MA: Addison-Wesley.

Cohen, P. R., & Dalrymple, M. Moran, D. B., Pereira, F. C. & Sullivan, J. W. (1989). Synergistic use of direct manipulation and natural language. In *Proceedings of the SIGCHI conference on Human factors in computing systems* (pp. 227–233). New York: ACM Press.

Cohen, P. R., Johnston, M., McGee, D., Oviatt, S., Pittman, J., Smith, I., et al. (1997). QuickSet: Multimodal interaction for simulation set-up and control. In *Proceedings of the Fifth Conference on Applied Natural Language Processing* (pp. 20-24).

Coles, A., Deliot, E., Melamed, T., & Lansard, K. (2003). A framework for coordinated multi-modal browsing with multiple clients. In *WWW '03: Proceedings of the 12th International Conference on World Wide Web* (pp. 718–726). New York: ACM Press.

Colle, H. A., & Welsh, A. (1976). Acoustic masking in primary memory. *Journal of Verbal Learning and Verbal Behavior, 15*, 17–31. doi:10.1016/S0022-5371(76)90003-7

Coninx, K., Luyten, K., Vandervelpen, C., den Bergh, J. V., & Creemers, B. (2003). Dygimes: Dynamically generating interfaces for mobile computing devices and embedded systems. In L. Chittaro (Ed.), *Human-Computer Interaction with Mobile Devices and Services 2003* (pp. 256–270). Berlin, Germany: Springer.

Consolvo, S., & Walker, M. (2003). Using the experience sampling method to evaluate ubicomp applications. *IEEE Pervasive Computing / IEEE Computer Society [and] IEEE Communications Society, 2*(2), 24–31. doi:10.1109/MPRV.2003.1203750

Coursey, K. Living in cyn: Mating aiml and cyc together with program n, 2004. R. Wallace. The anatomy of a.l.i.c.e. Resources avalaible at: http://alice.sunlitsurf.com/alice/about.html.

Coutaz, J. (1987). PAC, an object-oriented model for dialog design. In H.-J. Bullinger, B. Shackel (Eds.), *Proceedings of the 2nd IFIP International Conference on Human-Computer Interaction (INTERACT 87)* (pp. 431-436). Amsterdam: North-Holland.

Coutaz, J. (2006). Meta-User Interfaces for Ambient Spaces. In K. Coninx, K. Luyten & K.A. Schneider (Eds.), *Task Models and Diagrams for Users Interface Design 2006* (pp. 1-15). Berlin, Germany: Springer.

Coutaz, J., & Rey, G. (2002). Foundation for a theory of contextors. In [New York: ACM Press.]. *Proceedings of CADUI, 02*, 283–302.

Coutaz, J., Nigay, L., Salber, D., Blandford, A., May, J., & Young, M. R. (1995). Four easy pieces for assessing the usability of multimodal interaction: the CARE properties. In *INTERACT* (pp. 115-120). New York: Chapman & Hall.

Csikszentmihalyi, M. (1996). *Creativity: Flow and the psychology of discovery and invention.* New York: HarperCollins.

CYC ontology documentation. Retrieved January 28, 2009, from http://www.cyc.com/.

d'Ausbourg, B. (1998). Using model checking for the automatic validation of user interface systems. *DSV-IS* (pp. 242-260).

Dalal, M., Feiner, S., McKeown, K., Pan, S., Zhou, M., Höllerer, T., et al. (1996). Negotiation for automated generation of temporal multimedia presentations. In *Proceedings of ACM Multimedia'96* (pp. 55-64).

Dale, J. (2008). Signature Security. *Financial Services Technology*. (2008). Retrieved October 20, 2008, from http://www.fsteurope.com/currentissue/article.asp?art=270443&issue=207

Damala, A., Cubaud, P., Bationo, A., Houlier, P., & Marchal, I. (2008). Bridging the gap between the digital and the physical: design and evaluation of a mobile augmented reality guide for the museum visit, *Proceedings of the 3rd international conference on Digital Interactive Media in Entertainment and Arts* (DIMEA '08), Athens, Greece, Pages 120-127

Damiano, R., Galia, C., & Lombardo, V. (2006). Virtual tours across different media in DramaTour project, Workshop Intelligent Technologies for Cultural Heritage *Exploitation at the 17th European Conference on Artificial Intelligence* (ECAI 2006), Riva del Garda, 2006, pp. 21-25.

Davis, R. C., Saponas, T. S., Shilman, M., & Landay, J. A. (2007). SketchWizard: Wizard of Oz prototyping of pen-based user interfaces. In *Proceedings of the 20th Annual ACM Symposium on User interface Software and Technology* (pp. 119-128). New York: ACM Press.

De Boer, A., Dias, E., & Verbree, E. (2008). *Processing 3D geo-information for augmenting georeferenced and oriented photographs with text labels*. Paper presented at the GIM Spatial Data Handling Conference 2008.

De Bra, P., Brusilovsky, P., & Houben, G. J. (1999). Adaptive Hypermedia: From Systems to Framework. [CSUR]. *ACM Computing Surveys, 31*, 1–6. doi:10.1145/345966.345996

de Götzen, A., Mion, L., Avanzini, F., & Serafin, S. (2008). Multimodal design for enactive toys. In R. Kronland-Martinet, S. Ystad, & K. Jensen (Eds.), *CMMR 2007, LNCS 4969* (pp. 212-222). Berlin, Germany: Springer-Verlag.

de Sa, M., & Carrico, L. (2008). Defining scenarios for mobile design and evaluation. In *Proceedings of CHI'08, SIGCHI Conference on Human Factors in Computing Systems* (pp. 2847-2852).

de Saussure, F. (1983). *Course in general linguistics* (R. Harris, ed.). London: Duckworth. (Original work publishes in 1916).

Delisle, S., & Moulin, B. (2002). User interfaces and help systems: from helplessness to intelligent assistance. *Artificial Intelligence Review, 18*(2), 117–157. doi:10.1023/A:1015179704819

Demeure, A., & Calvary, G. (2003). Plasticity of user interfaces: towards an evolution model based on conceptual graphs. In *Proceedings of the 15th French-speaking Conference on Human-Computer Interaction*, Caen, France, (pp. 80-87).

Demeure, A., Calvary, G., Coutaz, J., & Vanderdonckt, J. (2006). The comets inspector: Towards run time plasticity control based on a sematic network. In K. Coninx, K. Luyten & K.A. Schneider (Eds.), *Task Models and Diagrams for Users Interface Design 2006* (pp. 324-338). Berlin, Germany: Springer.

Demeure, A., Sottet, J.-S., Calvary, G., Coutaz, J., Ganneau, V., & Vanderdonkt, J. (2008). The 4c reference model for distributed user interfaces. In D. Greenwood, M. Grottke, H. Lutfiyya & M. Popescu (Eds.), *The Fourth International Conference on Autonomic and Autonomous Systems 2008* (pp. 61-69). Gosier, Guadeloupe: IEEE Computer Society Press.

Demumieux, R., & Losquin, P. (2005). Gathering customers's real usage on mobile phones. *Proceedings of MobileHCI'05*. New York: ACM.

Deng, L., & Huang, X. (2004). Challenges in adopting speech recognition. *Communications of the ACM, 47*(1), 69–75. doi:10.1145/962081.962108

Dennerlein, J., Martin, D., & Hasser, C. (2000). Force-feedback improves performance for steering and combined steering-targeting tasks. *CHI Letters*, *2*(1), 423–429.

Deutsch, J. A., & Deutsch, D. (1963). Attention: Some theoretical considerations. *Psychological Review*, *70*(1), 80–90. doi:10.1037/h0039515

Deuz, D. (1982). Silent and non-silent pauses in three speech styles. *Language and Speech*, *25*(1), 11–28.

Dey, A. K. (2000). *Providing architectural support for building context-aware applications*. Unpublished doctorial dissertation, Georgia Institute of Technology, College of Computing.

Dolfing, J. G. A., Aarts, E. H. L., & van Oosterhout, J. J. G. M. (1998). On-line signature verification with hidden Markov models. In *Proceedings of the 14th International Conference on Pattern Recognition* (pp. 1309-1312). Washington, DC: IEEE CS Press.

Dong, H., Keates, S., & Clarkson, P. J. (2002). Accommodating older user's functional capabilities. In S. Brewster & M. Zajicek (Eds.), *HCI BCS*.

Dourish, P. (2001). *Where the action is: The foundations of embodied interaction*. Cambridge, MA: MIT Press.

Doyle, J., Bertolotto, M., & Wilson, D. (2007). A survey of multimodal interfaces for mobile mapping applications. In L. Meng, A. Zipf, & S. Winter (Eds.), *Map-based Mobile Services – Interactivity and Usability*. Berlin, Germany: Springer Verlag

Doyle, J., Bertolotto, M., & Wilson, D. (2008). Evaluating the benefits of multimodal interface design for CoMPASS – a mobile GIS. *GeoInformatica*.

Doyle, J., Bertolotto, M., & Wilson, D. (2008). Multimodal interaction – improving usability and efficiency in a mobile GIS context. In *Advances in Computer Human Interaction* (pp. 63-68). Washington, DC: IEEE Press.

du Bousquet, L., Ouabdesselam, F., Richier, J.-L., & Zuanon, N. (1999). Lutess: a specification driven testing environment for synchronous software. *International Conference of Software Engineering* (pp. 267-276). New York: ACM.

Duarte, C., & Carriço, L. (2005). Users and Usage Driven Adaptation of Digital Talking Books. In *Proceedings of the 11th International Conference on Human-Computer Interaction*. Las Vegas, NV: Lawrence Erlbaum Associates, Inc.

Duarte, C., & Carrico, L. (2006). A conceptual framework for developing adaptive multimodal applications. In *Proceedings of the 11th international conference on Intelligent User Interfaces*, (pp. 132-139).

Duarte, C., Carriço, L., & Morgado, F. (2007). Playback of rich digital books on mobile devices. In *Proceedings of the 12th International Conference on Human-Computer Interaction (HCI International 2007)*. Berlin, Germany: Springer.

Ducatel, K., Bogdanowicz, M., Scapolo, F., Leijten, J., & Burgelman, J.-C. (2001). *Scenarios for Ambient Intelligence in 2010, Final report*. Information Society Technologies Advisory Group (ISTAG), European Commission.

Duh, H. B., Tan, G. C., & Chen, V. H. (2006). Usability evaluation for mobile device: A comparison of laboratory and field tests. In *Proceedings of the 8th Conference on Human-Computer interaction with Mobile Devices and Services* (pp. 181-186). New York: ACM Press.

Duke, D., & Harrison, M. D. (1993). Abstract interaction objects. In *Eurographics conference and computer graphics forum*, *12* (3), 25-36.

Duke, D., & Harrison, M. D. (1995). Event model of human-system interaction. *IEEE Software Engineering Journal*, *12*(1), 3–10.

Dybkjær, L., Bernsen, N. O., & Minker, W. (2004). Evaluation and usability of multimodal spoken language dialogue. *Speech Communication*, *43*(2), 33–54. doi:10.1016/j.specom.2004.02.001

Edwards, A. D. N. (1992). Redundancy and adaptability. In A.D.N. Edwards & S. Holland (Eds.), *Multimedia interface design in education* (pp. 145-155). Berlin, Germany: Springer-Verlag.

Edworthy, J., & Hellier, E. (2006) Complex auditory signals and speech. In Wogalter, M. (Ed.), *Handbook of Warnings* (pp.199-220). Philadelphia, PA: Lawrence Erlbaum.

Elting, C., Rapp, S., Möhler, G., & Strube, M. (2003). Architecture and Implementation of Multimodal Plug and Play. In *ICMI-PUI `03 Fifth International Conference on Multimodal Interfaces*, Vancouver, Canada.

Elting, Ch., & Michelitsch, G. (2001). A multimodal presentation planner for a home entertainment environment. In *Proceedings of Perceptual User Interfaces (PUI) 2001*, Orlando, Florida.

Elting, Ch., Zwickel, J., & Malaka, R. (2002). Device-dependent modality selection for user-interfaces - an empirical study. In *International Conference on Intelligent User Interfaces IUI 2002*, San Francisco, CA.

Emerson, E. A., & Clarke, E. M. (1982). Using branching time temporal logic to synthesize synchronization skeleton. *Science of Computer Programming*.

European Telecommunications Standards Institute (ETSI). (2003). Distributed speech recognition; front-end feature extraction algorithm; compression algorithms. *ES 201 108*.

Fahlen, L., & Brown, C. (1992). The use of a 3D aura metaphor for computer based conferencing and tele-working. In *Proceedings of the 4th Multi-G workshop* (pp. 69-74).

Fan, X., Xie, X., Li, Z., Li, M., & Ma, W.-Y. (2005, Nov.). *Photo-to-Search: Using Multimodal Queries to Search the Web from Mobile Devices.* Paper presented at the 7th ACM SIGMM International Workshop on Multimedia Information Retrieval in conjunction with ACM Multimedia 2005, Singapore.

Farella, E., Brunelli, D., Benini, L., Ricco, B., & Bonfigli, M. E. (2005). Computing for interactive virtual heritage. *IEEE MultiMedia, 12*(3), 46–58. doi:10.1109/MMUL.2005.54

Fasciano, M., & Lapalme, G. (1996). PosGraphe: a system for the generation of statistical graphics and text. In *Proceedings of the 8th International Workshop on Natural Language Generation* (pp. 51-60).

Feiner, S. K., & McKeown, K. R. (1993). Automating the generation of coordinated multimedia explanations. In M. T. Maybury, (Ed.), *Intelligent Multimedia Interfaces* (pp. 117-139). Menlo Park, CA: AAAI Press.

Feki, M. A., Renouard, S., Abdulrazak, B., Chollet, G., & Mokhtari, M. (2004). Coupling context awareness and multimodality in smart homes concept. *Lecture Notes in Computer Science, 3118*, 906–913.

Ferrer, L., Shriberg, E., & Stolcke, A. (2002). Is the speaker done yet? Faster and more accurate end-of-utterance detection using prosody in human-computer dialog. In *Proceedings of LICSLP, 2002*, 2061–2064.

Feuerstack, S., Blumendorf, M., Schwartze, V., & Albayrak, S. (2008). Model-based layout generation. In P. Bottoni & S. Levialdi (Ed.), *Proceedings of the working conference on Advanced visual interfaces* (pp. 217-224). New York: ACM Press.

Fierrez, J., & Ortega-Garcia, J. (2007). On-line signature verification. In A. K. Jain, A. Ross, & P. Flynn (Eds.), *Handbook of Biometrics* (pp. 189-209). Berlin, Germany: Springer.

Fierrez, J., Ramos-Castro, D., Ortega-Garcia, J., & Gonzalez-Rodriguez, J. (2007). HMM-based on-line signature verification: feature extraction and signature modelling. *Pattern Recognition Letters, 28*(16), 2325–2334. doi:10.1016/j.patrec.2007.07.012

Fierrez-Aguilar, J., Nanni, L., Lopez-Penalba, J., Ortega-Garcia, J., & Maltoni, D. (2005). An on-line signature verification system based on fusion of local and global information. In T. Kanade, A. Jain & N. K. Ratha (Eds.), *Audio- and Video-Based Biometric Person Authentication. Lecture Notes in Computer Science* (pp. 523-532). Berlin, Germany: Springer.

Fierrez-Aguilar, J., Ortega-Garcia, J., & Gonzalez-Rodriguez, J. (2005). Target dependent score normalization techniques and their application to signature verification. *IEEE Transactions on Systems, Man and Cybernetics, part C,* 35(3), 418-425.

Flickner, M., Sawhney, H., Niblack, W., Ashley, J., & Huang, Q. (1995). Query by image and video content: the QBIC system. *IEEE Computer Special Issue on Content-Based Retrieval, 28*(9), 23–32.

Flippo, F., Krebs, A., & Marsic, I. (2003). A framework for rapid development of multimodal interfaces. In *Proceedings of the 5th International Conference on Multimodal Interfaces* (pp. 109-116).

Fodor, J. A. (1975). *The language of thought.* Cambridge, MA: Harvard University Press.

Fowler, M. (2003). *Patterns of enterprise application architecture.* Addison-Wesley: Reading, MA.

Freire-Santos, M., Fierrez-Aguilar, J., & Ortega-Garcia, J. (2006). Cryptographic key generation using handwritten signature. In *Defense and Security Symposium, Biometric Technologies for Human Identification, BTHI* (pp. 225-231).

Fritzke, B. (1994). Growing cell structures - a self-organizing network for unsupervised and supervised learning. *Neural Networks, 7*(9), 1441–1460. doi:10.1016/0893-6080(94)90091-4

Froehlich, J., Chen, M. Y., Consolvo, S., Harrison, B., & Landay, J. A. (2007). MyExperience: A system for in-situ tracing and capturing of user feedback on mobile phones. In *Proceedings of the 5th international Conference on Mobile Systems, Applications and Services* (pp. 57-70). New York: ACM Press.

Frohlich, D. M. (1991). The design space of interfaces. In L. Kjelldahl, (Ed.), *Multimedia Principles, Systems and Applications* (pp. 69-74). Berlin, Germany: Springer-Verlag.

Fu, G., Jones, C. B., & Abdelmoty, A. I. (2005). *Ontology-Based Spatial Query Expansion in Information Retrieval.* Paper presented at the On the Move to Meaningful Internet Systems 2005: CoopIS, DOA, and ODBASE, Agia Napa, Cyprus.

Fuggetta, A., Picco, G. P., & Vigna, G. (1998). Understanding code mobility. *IEEE Transactions on Software Engineering, 24*(5). doi:10.1109/32.685258

Fuhrmann, S., MacEachren, A., Dou, J., Wang, K., & Cox, A. (2005). Gesture and speech-based maps to support use of GIS for crisis management: A user study. In *Proceedings of AutoCarto.* Bethesda, MD: Cartography and Geographic Information Society.

Fujimura, N., & Doi, M. (2006). Collecting students' degree of comprehension with mobile phones. In *Proceedings of the 34th annual ACM SIGUCCS conference on User services table of contents,* Edmonton, Canada (pp. 123-127).

Galbally, J., Fierrez, J., & Ortega-Garcia, J. (2007). Bayesian hill-climbing attack and its application to signature verification. In S.-W. Lee & S.Z. Li (Eds.), *Advances in Biometrics. Lecture Notes in Computer Science* (pp. 386-395). Berlin, Germany: Springer.

Gallese, V. (2006). Embodied simulation: From mirror neuron systems to interpersonal relations. In G. Bock & J. Goode (Eds.), *Empathy and Fairness* (pp. 3-19). Chichester, UK: Wiley.

Gallese, V. (2008). Mirror neurons and the social nature of language: The neural exploitation hypothesis. *Social Neuroscience, 3*(3), 317–333. doi:10.1080/17470910701563608

Gallese, V., & Lakoff, G. (2005). The brain's concepts: The role of the sensory-motor system in reason and language. *Cognitive Neuropsychology, 22,* 455–479. doi:10.1080/02643290442000310

Ganneau, V., Calvary, G., & Demumieux, R. (2008). Learning key contexts of use in the wild for driving plastic user interfaces engineering. In *Engineering Interactive Systems 2008 (2nd Conference on Human-Centred Software Engineering (HCSE 2008) and 7th International workshop on TAsk MOdels and DIAgrams (TAMODIA 2008))*, Pisa, Italy.

Garavel, H., Lang, F., Mateescu, R., & Serwe, W. (2007). CADP: A toolbox for the construction and analysis of distributed processes. In *19th International Conference on Computer Aided Verification CAV 07* (pp. 158-163).

Garcia, D. (2000). Sound models, metaphor and mimesis in the composition of electroacoustic music. In *Proceedings of the 7th Brazilian Symposium on Computer Music*. Curitiba, Brasil: Universidade Federal do Paraná.

Garcia-Salicetti, S., Beumier, C., Chollet, G., Dorizzi, B., Jardins, J. L.-L., Lanter, J., et al. (2003). BIOMET: A multimodal person authentication database. In J. Kittler & M. S. Nixon (Eds.), *Audio- and Video-Based Biometric Person Authentication. Lecture Notes in Computer Science* (pp. 845-853). Berlin, Germany: Springer.

Gardner, D. K., & Helmes, E. (1999). Locus of control and self-directed learning as predictors of well-being in the elderly. *Australian Psychologist, 34*(2). doi:10.1080/00050069908257436

Gartner, Inc. (2008). *Press Release: Worldwide Smart-Phone Sales, 2008*. Retrieved from http://www.gartner.com.

Gaver, W. (1997). Auditory interfaces. In M. Helander. T Landauer, & P. Prabhu (Eds.), *Handbook of human-computer interaction (2nd ed.)* (pp. 1003-1042). Amsterdam: Elsevier.

Gaver, W. W. (1993). Synthesizing auditory icons. In [New York: ACM.]. *Proceedings of INTERCHI, 93*, 228–235.

Gelernter, B. (1998). Help design challenges in network computing. In *Proceedings of 16th Annual international Conference on Computer Documentation SIGDOC '98* (pp. 184-193). New York: ACM.

Gellersen, H., Kortuem, G., Schmidt, A., & Beigl, M. (2004). Physical prototyping with smart-its. *IEEE Pervasive Computing / IEEE Computer Society [and] IEEE Communications Society, 3*(3), 74–82. doi:10.1109/MPRV.2004.1321032

Geutner, P., Arevalo, L., & Breuninger, J. (2000). VODIS - voice-operated driver information systems: a usability study on advanced speech technologies for car environments. In *Proceedings of, ICSLP-2000*, 378–382.

Ghias, A., Logan, J., Chamberlin, D., & Smith, B. C. (1995). Query by humming - musical information retrieval in an audio database. *ACM Multimedia*, 231-236.

Gibson, J. J. (1979). *The ecological approach to visual perception*. Boston, MA: Houghton Mifflin.

Gilson, R. D., & Fenton, R. E. (1974). Kinesthetic-tactual information presentations - inflight studies. *IEEE Transactions on Systems, Man, and Cybernetics, 4*(6), 531–535. doi:10.1109/TSMC.1974.4309361

Göbel, S., Hartmann, F., Kadner, K., & Pohl, C. (2006). A Device-independent multimodal mark-up language. In *INFORMATIK 2006: Informatik für Menschen, Band 2* (pp. 170–177). Dresden, Germany.

Godøy, R. I. (2001). Imagined action, excitation, and resonance. In R. I Godøy, & H. Jørgensen (Eds.), *Musical imagery* (pp. 237-250). Lisse, The Netherlands: Swets and Zeitlinger.

Godøy, R. I. (2003). Motor-mimetic music cognition. *Leonardo, 36*(4), 317–319. doi:10.1162/002409403322258781

Godøy, R. I. (2004). Gestural imagery in the service of musical imagery. In A. Camurri & G. Volpe (Eds.), *Gesture-Based Communication in Human-Computer Interaction: 5th International Gesture Workshop, Volume LNAI 2915* (pp. 55-62). Berlin, Germany: Springer-Verlag.

Goldman-Eisler, F. (1972). Pauses, clauses, sentences. *Language and Speech, 15*, 103–113.

Goog411. (2008). *Find and connect with local businesses for free from your phone*. Retrieved from http://www.google.com/goog411/

Google. (2008). *Android.* Retrieved on November 12, 2008, from http://www.android.com/

GoogleMobile. (2008). *Google mobile search with my location.* Retrieved from http://m.google.com/

GPSWorld.com. (2007). *GPS Wins Out over Internet for Mobile Phone Users.* Retrieved from http://lbs.gpsworld.com/gpslbs/LBS+News/GPS-Wins-Out-over-Internet-for-Mobile-Phone-Users/ArticleStandard/Article/detail/480477

Grayling, T. (2002). If we build it, will they come? A usability test of two browser-based embedded help systems. *Technical Communication, 49*(2), 193–209.

Grolaux, D., Van Roy, P., & Vanderdonckt, J. (2002). FlexClock: A plastic clock written in Oz with the QTk toolkit. In *Proceedings of the Workshop on Task Models and Diagrams for User Interface Design (TAMODIA 2002).*

Große, D., Kuhne, U., & Drechsler, R. (2006). HW/SW co-verification of embedded systems using bounded model checking. In *The 16th ACM Great Lakes symposium on VLSI,* (pp 43-48). New York: ACM.

Guest, S., Catmur, C., Lloyd, D., & Spence, C. (2002). Audiotactile interactions in roughness perception. *Experimental Brain Research, 146*(2), 161–171. doi:10.1007/s00221-002-1164-z

Hackos, J. T., & Stevens, D. M. (1997). *Standards for online communication.* New York, NY: John Wiley & Sons, Inc.

Hagen, P., Robertson, T., Kan, M., & Sadler, K. (2005). Emerging research methods for understanding mobile technology use. In *Proceedings of the 17th Australia Conference on Computer-Human interaction: Citizens online: Considerations For Today and the Future* (pp. 1-10). New York: ACM Press.

Hampe, B. (2005). Image schemas in cognitive linguistics: Introduction. In B. Hampe (Ed.), *From Perception to Meaning: Image Schemas in Cognitive Linguistics* (pp. 1-14). Berlin, Germany: Mouton de Gruyter.

Hansson, R., Ljungstrand, P., & Redström, J. (2001). Subtle and public notification cues for mobile devices. In *Proceedings of 3rd international Conference on Ubiquitous Computing* (pp. 240-246). London: Springer-Verlag.

Harchani, M., Niagy, L., & Panaget, F. (2007). A platform for output dialogic strategies in natural multimodal dialogue systems. In *IUI'07,* (pp. 206-215). New York: ACM.

Hare, J. S., & Lewis, P. H. (2005, January). *Content-based image retrieval using a mobile device as a novel interface.* Paper presented at the Proceedings of SPIE Storage and Retrieval Methods and Applications for Multimedia 2005, San Jose, USA.

Harris, R. A., & Hosier, W. J. (1991). A taxonomy of online information. *Technical Communication, 38*(2), 197–210.

Hawthorn, D. (2000). Possible implications of aging for interface designers. *Interacting with Computers, 12,* 507–528. doi:10.1016/S0953-5438(99)00021-1

Healey, J., Hosn, R., & Maes, S. H. (2002). Adaptive content for device independent multi-modal browser applications. In *Lecture Notes In Computer Science; Vol. 2347, Proceedings of the Second International Conference on Adaptive Hypermedia and Adaptive Web-Based Systems* (pp. 401-405).

Heiser, M., Iacoboni, M., Maeda, F., Marcus, J., & Mazziotta, J. C. (2003). The essential role of Broca's area in imitation. *The European Journal of Neuroscience, 17,* 1123–1128. doi:10.1046/j.1460-9568.2003.02530.x

Hennebert, J., Humm, A., & Ingold, R. (2007). Modelling spoken signatures with Gaussian mixture models adaptation. In *IEEE International Conference on Acoustics, Speech, and Signal Processing, ICASSP 2007,* Honolulu, USA (pp. 229-232). Washington, DC: IEEE Press.

Ho, C., Tan, H. Z., & Spence, C. (2005). Using spatial vibrotactile cues to direct visual attention in driving scenes. *Transportation Research Part F: Traffic Psychology and Behaviour, 8,* 397–412. doi:10.1016/j.trf.2005.05.002

Hodjat, B., Hodjat, S., Treadgold, N., & Jonsson, I. (2006). CRUSE: a context reactive natural language mobile interface. In *Proceedings of the 2nd Annual international workshop on wireless internet,* Boston, MA (pp. 20). New York: ACM.

Hoggan, E., & Brewster, S. (2007). Designing audio and tactile crossmodal icons for mobile devices. In *Proceedings of the 9th International Conference on Multimodal Interfaces* (pp. 162-169). New York: ACM.

Holmquist, L. (2005). Prototyping: generating ideas or cargo cult designs? *Interaction, 12*(2), 48–54. doi:10.1145/1052438.1052465

Holmquist, L. E., Redström, J., & Ljungstrand, P. (1999). Token-based access to digital information. In *Proc. 1st International Symposium on Handheld and Ubiquitous Computing* (pp. 234-245). Berlin, Germany: Springer-Verlag.

Holzman, T. G. (1999). Computer-human interface solutions for emergency medical care. *Interaction, 6,* 13–24. doi:10.1145/301153.301160

Horton, W. (1994). *Designing and Writing Online Documentation.* Wiley: New York, NY.

Hübsch, G. (2008). *Systemunterstützung für Interaktorbasierte Multimodale Benutzungsschnittstellen.* Doctoral dissertation, Dresden University of Technology, Germany.

Hulkko, S., Mattelmäki, T., Virtanen, K., & Keinonen, T. (2004). Mobile Probes. In *Proceedings of the Third Nordic Conference on Human-Computer interaction* (pp. 43-51). New York: ACM Press.

Ibanez, J., Aylett, R., & Ruiz-Rodarte, R. (2003). Storytelling in virtual environments from a virtual guide perspective. *Virtual Reality (Waltham Cross), 7*(1). doi:10.1007/s10055-003-0112-y

International Telecommunication Union. (2005). *ITU Internet Reports 2005: The Internet of Things.* Geneva, Switzerland: ITU.

Ishii, H., & Ullmer, B. (1997). Tangible bits: towards seamless interfaces between people, bits and atoms. In *Proc. SIGCHI Conference on Human Factors in Computing Systems* (pp. 234-241). New York: ACM.

Jacobson, R. D. (2002). Representing spatial information through multimodal interfaces. In *Proceedings of the 6th International Conference on Information Visualisation* (pp. 730-734). Washington, DC: IEEE Press.

Jacquet, C. (2007). KUP: a model for the multimodal presentation of information in ambient intelligence. In *Proceedings of Intelligent Environments 2007 (IE 07)* (pp. 432-439). Herts, UK: The IET.

Jacquet, C., Bellik, Y., & Bourda, Y. (2005). An architecture for ambient computing. In H. Hagras, V. Callaghan (Eds.), *Proceedings of the IEE International Workshop on Intelligent Environments* (pp. 47-54).

Jacquet, C., Bellik, Y., & Bourda, Y. (2006), Dynamic Cooperative Information Display in Mobile Environments. In B. Gabrys, R. J. Howlett, L. C. Jain (Eds), *Proceedings of KES 2006, Knowledge-Based Intelligent Information and Engineering Systems,* (pp. 154-161). Springer.

Jain, A. K., Nandakumar, K., & Ross, A. (2005). Score normalization in multimodal biometric systems. *Pattern Recognition, 38*(12), 2270–2285. doi:10.1016/j.patcog.2005.01.012

Jain, A. K., Ross, A., & Pankanti, S. (2006). Biometrics: a tool for information security. *IEEE Transactions on Information Forensics and Security, 1*(2), 125–143. doi:10.1109/TIFS.2006.873653

Jensenius, A. (2007). *Action-Sound: Developing Methods and Tools to Study Music-Related Body Movement.* Ph.D. Thesis. Department of Musicology, University of Oslo.

Jermyn, I., Mayer, A., Monrose, F., Reiter, M. K., & Rubin, A. D. (1999). The design and analysis of graphical passwords. In *8th conference on USENIX Security Symposium, Vol. 8,* Washington, D.C.

Johnson, M. (1987). *The body in the mind: The bodily basis of meaning, imagination, and reason.* Chicago, IL: University of Chicago.

Johnson, M., & Rohrer, T. (2007). We are live creatures: Embodiment, American pragmatism and the cognitive organism. In: J. Zlatev, T. Ziemke, R. Frank, & R. Dirven (Eds.), *Body, language, and mind, vol. 1* (pp. 17-54). Berlin, Germany: Mouton de Gruyter.

Johnston, M. (1998). Unification-based multimodal parsing. In *COLINGACL* (pp. 624–630).

Johnston, M., & Bangalore, S. (2005). Finite-state multimodal integration and understanding. *Natural Language Engineering*, *11*(2), 159–187. doi:10.1017/S1351324904003572

Johnston, M., Cohen, P. R., McGee, D., Oviatt, S. L., Pittman, J. A., & Smith, I. (1997). Unification-based multimodal integration. In P. R. Cohen & W. Wahlster (Eds.), *Proceedings of the thirty-fifth annual meeting of the Association for Computational Linguistics and eighth conference of the European chapter of the Association for Computational Linguistics* (pp. 281–288). Somerset, NJ: Association for Computational Linguistics.

Jones, C. M., Lim, M. Y., & Aylett, R. (2005). Empathic interaction with a virtual guide. *AISB Symposium on Empathic Interaction*, University of Hertfordshire, UK.

Jones, D. M., Saint-Aubin, J., & Tremblay, S. (1999). Modulation of the irrelevant sound effect by organizational factors: Further evidence from streaming by location. *The Quarterly Journal of Experimental Psychology*, *52*(3), 545–554. doi:10.1080/027249899390954

Joo-Hwee Lim, Yiqun Li, Yilun You, & Chevallet, J.-P. (2007). Scene recognition with camera phones for tourist information access. In *2007 IEEE International Conference on Multimedia and Expo* (pp. 100-103).

Jorge, J. (2001). Adaptive tools for the elderly: New devices to cope with age-induced cognitive disabilities. In *Workshop on Universal Accessibility of Ubiquitous Computing,*(pp. 66-70). New York: ACM Press.

Jourde, F., Nigay, L., & Parissis, I. (2006). Test formel de systèmes interactifs multimodaux: couplage ICARE – Lutess. *ICSSEA2006, 19ème journées Internationales génie logiciel & Ingènierie de Systèmes et leurs Applications, Globalisation des services et des systèmes.*

Jung, J., & Choi, S. (2007). Perceived magnitude and power consumption of vibration feedback in mobile devices. In [Berlin, Germany: Springer-Verlag.]. *Proceedings of Human-Computer Interaction HCII, 2007*, 354–363.

Kadner, K. (2008). *Erweiterung einer Komponentenplattform zur Unterstützung multimodaler Anwendungen mit föderierten Endgeräten.* TUDpress Verlag der Wissenschaft.

Kadner, K., & Pohl, C. (2008). *Synchronization of distributed user interfaces.* European Patent no. 1892925, European Patent Bulletin 08/42, granted as of 2008-10-15.

Kapoor, A., & Picard, R. W. (2005). Multimodal affect recognition in learning environments. In *Proceedings of the ACM MM'05* (pp. 6–11).

Kawamura, T., Umezu, K., & Ohsuga, A. (2008). Mobile navigation system for the elderly – preliminary experiment and evaluation. In *Ubiquitous Intelligence and Computing* (pp. 579-590). Berlin, Germany: Springer LNCS.

Kearsley, G. (1988). *Online help systems: design and implementation*, Norwood, NJ: Ablex Publishing.

Kehoe, A., & Pitt, I. (2006). Designing help topics for use with text-to-speech. In *Proceedings of 24th Annual ACM international Conference on Design of Communication SIGDOC '06* (pp. 157-163). New York: ACM.

Kehoe, A., Neff, F., Pitt, I., & Russell, G. (2007). Improvements to a speech-enabled user assistance system based on pilot study results. In *Proceedings of 25th Annual ACM international Conference on Design of Communication SIGDOC '07* (pp. 42-47). New York: ACM.

Kernchen, K., Mossner, R., & Tafazolli, R. (2005). Adaptivity for multimodal user interfaces in mobile situations, autonomous decentralized systems. In *Autonomous Decentralized Systems, ISADS* (pp. 469–472). Washington, DC: IEEE.

Kerpedjiev, S., Carenini, G., Roth, S. F., & Moore, J. D. (1997). Integrating planning and task-based design for multimedia presentation. In *Proceedings of the International Conference on Intelligent User Interfaces* (pp. 145-152).

Keskin, C., Balci, K., Aran, O., Sankar, B., & Akarun, L. (2007). A multimodal 3D healthcare communication system. In *3DTV Conference.* Washington, DC: IEEE Press.

Khokhar, R. (2006). Smartphones - a call for better safety on the move. *Network Security,* (4): 6–7. doi:10.1016/S1353-4858(06)70354-3

Kholmatov, A., & Yanikoglu, B. (2005). Identity authentication using improved online signature verification method. *Pattern Recognition Letters, 26*(15), 2400–2408. doi:10.1016/j.patrec.2005.04.017

Kirste, T., Herfet, T., & Schnaider, M. (2001). EMBASSI: multimodal assistance for universal access to infotainment and service infrastructures. In *Proceedings of 2001 EC/NSF Workshop on Universal Accessibility of Ubiquitous Computing: Providing For the Elderly WUAUC'01* (pp. 41-50). New York: ACM.

Kjeldskov, J., & Graham, C. (2003). A review of mobile HCI research methods. In *Human-Computer Interaction with Mobile Devices and Services* (pp. 317-335). Berlin, Germany: Springer.

Kjeldskov, J., & Stage, J. (2004). New techniques for usability evaluation of mobile systems. *International Journal of Human-Computer Studies, 60*(5-6), 599–620. doi:10.1016/j.ijhcs.2003.11.001

Klemmer, S. R., Sinha, A. K., Chen, J., Landay, J. A., Aboobaker, N., & Wang, A. (2000). SUEDE: A Wizard of Oz prototyping tool for speech user interfaces. In *Proceedings of the 13th Annual ACM Symposium on User interface Software and Technology* (pp. 1-10). New York: ACM Press.

Klug, T., & Kangasharju, J. (2005). Executable task models. In A. Dix & A. Dittmar (Eds.), *Proceedings of the 4th international workshop on Task models and diagrams* (pp. 119–122). New York: ACM Press.

Koch, F., & Sonenberg, L. (2004). Using multimedia content in intelligent mobile services. In *Proceedings of the WebMedia & LA-Web Joint Conference 10th Brazilian Symposium on Multimedia and the Web 2nd Latin American Web Congress* (pp.1-43).

Krasner, G. E., & Pope, S. T. (1988). A cookbook for using the model-view controller user interface paradigm in Smalltalk-80. *Journal of Object Oriented Programming, 1*(3), 26–49.

Kratz, S., & Ballagas, R. (2007). Gesture recognition using motion estimation on mobile phones. In *Proceedings of 3rd Intl. Workshop on Pervasive Mobile Interaction Devices at Pervasive 2007.*

Kray, C., Wasinger, R., & Kortuem, G. (2004). Concepts and issues in interfaces for multiple users and multiple devices. In *Proceedings of MU3I workshop at IUI 2004* (pp. 7-11).

Kuhn, T. S. (1970). *The structure of scientific revolutions* (2nd ed). Chicago: University of Chicago Press.

Lafuente-Rojo, A., Abascal-González, J., & Cai, Y. (2007). Ambient intelligence: Chronicle of an announced technological revolution. *CEPIS Upgrade, 8*(4), 8–12.

Lakoff, G., & Johnson, M. (1980) *Metaphors we live by.* Chicago: University of Chicago Press.

Lakoff, G., & Johnson, M. (1999). *Philosophy in the flesh: The embodied mind and its challenge to Western thought.* New York: Basic Books.

Landay, J. (1996) SILK: Sketching interfaces like krazy. In *Conference Companion on Human Factors in Computing Systems: Common Ground* (pp.398-399). New York: ACM Press.

Larson, J. (2006). Standard languages for developing speech and multimodal applications. In *Proceedings of SpeechTEK West,* San Francisco.

Larson, J. A. (2002). *Voicexml: Introduction to developing speech applications.* Upper Saddle River, NJ: Prentice Hall.

Larson, J., Raman, T. V., & Raggett, D. (2003). Multimodal interaction framework. *W3C*. Retrieved from http://www.w3.org/TR/mmi-framework/

Latoschik, M. (2002). Designing transition networks for multimodal VR-interactions using a markup language. In *Proceedings of the Fourth IEEE International Conference on Multimodal Interfaces* (pp. 411-416).

Lazebnik, S., Schmid, C., & Ponce, J. (2004). *Semi-local affine parts for object recognition*. Paper presented at the Proceedings of British Machine Vision Conference 2004, Kingston, UK.

Lederman, S. J., Klatzky, R. L., Morgan, T., & Hamilton, C. (2002). Integrating multimodal information about surface texture via a probe: relative contributions of haptic and touch-produced sound sources. In *Proceedings of the 10th Symposium on Haptic Interfaces for Virtual Environments & Teleoperator Systems* (pp. 97-104). Piscataway, NJ: IEEE.

Lee, L. L., Berger, T., & Aviczer, E. (1996). Reliable on-line human signature verification systems. *IEEE Transactions on Pattern Analysis and Machine Intelligence*, *18*(6), 643–647. doi:10.1109/34.506415

Lei, H., & Govindaraju, V. (2005). A comparative study on the consistency of features in on-line signature verification. *Pattern Recognition Letters*, *26*(15), 2483–2489. doi:10.1016/j.patrec.2005.05.005

Leman, M. (2008). *Embodied music cognition and mediation technology*. Cambridge, MA: MIT Press.

Lenz, J. M. (2008). Biometric signature verification – a sign of the times? *Biometric Technology Today*, *16*(4), 9–11. doi:10.1016/S0969-4765(08)70124-3

Li, K. A., Sohn, T. Y., Huang, S., & Griswold, W. G. (2008). Peopletones: a system for the detection and notification of buddy proximity on mobile phones. In *Proceedings of 6th international Conference on Mobile Systems, Applications, and Services MobiSys '08* (pp. 160-173). New York: ACM.

Li, R., Taskiran, C., & Danielsen, M. (2007). Head pose tracking and gesture detection using block motion vectors on mobile devices. In *Proceedings of the 4th international conference on mobile technology, applications, and systems and the 1st international symposium on computer human interaction in mobile technology* (pp. 572–575). New York: ACM.

Li, Y., Hong, J., & Landay, J. (2007). Design challenges and principles for Wizard of Oz testing of location-enhanced applications. *IEEE Pervasive Computing / IEEE Computer Society [and] IEEE Communications Society*, *6*(2), 70–75. doi:10.1109/MPRV.2007.28

Liarokapis F., & Newman, Design, R. M. (2007). Experiences of multimodal mixed reality interfaces. In *Proceedings of the 25th annual ACM international conference on Design of communication* (pp. 34-41).

Liberman, A. M., & Mattingly, I. G. (1985). The motor theory of speech perception revised. *Cognition*, *21*, 136. doi:10.1016/0010-0277(85)90021-6

Limbourg, Q., Vanderdonckt, J., Michotte, B., Bouillon, L., & López-Jaquero, V. (2004). Usixml: A language supporting multi-path development of user interfaces. In R. Bastide, P. A. Palanque & J. Roth (Eds.), *Engineering Human Computer Interaction and Interactive Systems* (pp. 200–220). Berlin, Germany: Springer.

LiMo Foundation. (2008). *LiMo Foundation Platform*. Retrieved on November 12, 2208, from http://www.limofoundation.org/

Lin, J., Newman, M. W., Hong, J. I., & Landay, J. A. (2000). Denim: Finding a tighter fit between tools and practice for Web site design. In *Proceedings of the SIGCHI Conference on Human Factors in Computing Systems* (pp. 510-517). New York: ACM Press.

Live.com. (2008). *Windows Live Search for Mobile*. Retrieved from http://wls.live.com/

Loer, K., & Harrison, M. D. (2002). Towards usable and relevant model checking techniques for the analysis of dependable interactive systems. *ASE*, (pp. 223-226).

Lopez, J., & Szekely, P. (2001). Automatic web page adaptation. In *Proceedings of CHI2001-Workshop "Transforming the UI for anyone anywhere"*, Seattle.

Lorenz, A., Mielke, D., Oppermann, R., & Zahl, L. (2007). Personalized Mobile Health Monitoring for Elderly. In *Mobile* [New York: ACM Press.]. *Human-Computer Interaction, 07*, 297–304.

Lowe, D. G. (2004). Distinctive image features from scale-invariant keypoints. *International Journal of Computer Vision, 60*(2), 91–110. doi:10.1023/B:VISI.0000029664.99615.94

Lu, L., You, H., & Zhang, H.-J. (2001). *A new approach to query by humming in music retrieval*. Paper presented at the Proc. IEEE Int. Conf. Multimedia and Expo 2001.

Luk, J., Pasquero, J., Little, S., MacLean, K., Levesque, V., & Hayward, V. (2006). A role for haptics in mobile interaction: initial design using a handheld tactile display prototype. In [New York: ACM.]. *Proceedings of SIGCHI Conference on Human Factors in Computing Systems CHI, 06*, 171–180. doi:10.1145/1124772.1124800

Lynch, D., Bertolotto, M., & Wilson, D. (2005). Spatial annotations in a mapping environment. In *GIS Research UK* (pp. 524-528).

MacColl & Carrington, D. (1998). Testing MATIS: a case study on specification based testing of interactive systems. *FAHCI*.

Macken, W. J., Tremblay, S., Houghton, R. J., Nicholls, A. P., & Jones, D. M. (2003). Does auditory streaming require attention? Evidence from attentional selectivity in short-term memory. *Journal of Experimental Psychology. Human Perception and Performance, 29*(1), 43–51. doi:10.1037/0096-1523.29.1.43

Maiorana, E., Martinez-Diaz, M., Campisi, P., Ortega-Garcia, J., & Neri, A. (2008). Template protection for HMM-based on-line signature authentication. In *IEEE Computer Society Conference on Computer Vision and Pattern Recognition Workshops, 2008. CVPR Workshops 2008* (pp. 1-6). Washington DC: IEEE Press.

Malaka, M., & Zipf, A. (2000). Deep map – Challenging IT research in the framework of a tourist information system. In *Proceedings of the 7th International Congress on Tourism and Communication Technologies in Tourism (Enter 2000)* (pp. 15-27). Berlin, Germany: Springer LNCS.

Malerczyk, C. (2004). Interactive museum exhibit using pointing gesture recognition. In *WSCG'2004*, Plzen, Czech Republic.

Markopoulos, P. (1995). On the expression of interaction properties within an interactor model. In *DSV-IS'95: Design, Specification, Verification of Interactive Systems* (pp. 294-311). Berlin, Germany: Springer-Verlag.

Marshall, C. C., & Ruotolo, C. (2002). Reading-in-the-small: a study of reading on small form factor devices. In *Proceedings of 2nd ACM/IEEE-CS Joint Conference on Digital Libraries JCDL '02* (pp. 56-64). New York: ACM.

Martin, J. C. (1998). TYCOON: Theoretical framework and software tools for multimodal interfaces. In J. Lee, (Ed.), *Intelligence and Multimodality in Multimedia Interfaces*. Menlo Park, CA: AAAI Press.

Martinez-Diaz, M., Fierrez, J., & Galbally, J. Ortega-Garcia., J. (2008). Towards mobile authentication using dynamic signature verification: useful features and performance evaluation. In *Proceedings of 19th International Conference on Pattern Recognition 2008 (ICPR 2008)*, Tampa, FL, (pp.1-5). Washington, DC: IEEE Press.

Martinez-Diaz, M., Fierrez, J., & Ortega-Garcia, J. (2007). Universal background models for dynamic signature verification. In *IEEE Conference on Biometrics: Theory, Applications and Systems, BTAS,* Washington DC (pp. 1-6). Washington DC: IEEE Press.

May, A. J., Ross, T., Bayer, S. H., & Tarkiainen, M. J. (2003). Pedestrian navigation aids: Information requirements and design implications. *Personal and Ubiquitous Computing, 7*(6), 331–338. doi:10.1007/s00779-003-0248-5

Mayhew, D. J. (1999). *The usability engineering lifecycle*. San Francisco, CA: Morgan Kaufmann.

Mc Millan, K. (1992). *The SMV system*. Pittsburgh, PA: Carnegie Mellon University.

McCurdy, M., Connors, C., Pyrzak, G., Kanefsky, B., & Vera, A. (2006). Breaking the fidelity barrier: An examination of our current characterization of prototypes and an example of a mixed-fidelity success. In *Proceedings of the SIGCHI Conference on Human Factors in Computing Systems* (pp. 1233-1242). New York: ACM Press.

McGlashan, S., Burnett, D., Carter, J., Danielsen, P., Ferrans, F., Hunt, A., et al. (2004). Voice extensible markup language (VoiceXML) version 2.0. *W3C*. Retrieved from http://www.w3.org/TR/voicexml20/

McGurk, H., & MacDonald. (1976). Hearing lips and seeing voices. *Nature, 264*, 746–748. doi:10.1038/264746a0

McLuhan, M. (1966). *Understanding media: the extensions of man*. New York: New American Library.

McNeill, D. (2005). *Gesture and thought*. Chicago: University of Chicago Press.

McTear, M. F. (2002). Spoken dialogue technology: Enabling the conversational user interface. In *Proceedings of 1-85 ACM Computing Surveys (CSUR)*. New York: ACM Press.

Mellor, S. J. (2004). *Agile MDA*. Retrieved January 21, 2008, from http://www.omg.org/mda/mda_files/Agile_MDA.pdf

Mellor, S. J., Scott, K., Uhl, A., & Weise, D. (2004). *MDA Distilled: Principles of Model-Driven Architecture*. Boston: Addison-Wesley.

Mikolajczyk, K., & Schmid, C. (2004). Scale and affine invariant interest point detectors. *International Journal of Computer Vision, 60*(1), 63–86. doi:10.1023/B:VISI.0000027790.02288.f2

Mikolajczyk, K., & Schmid, C. (2005). A performance evaluation of local descriptors. *IEEE Transactions on Pattern Analysis and Machine Intelligence, 27*(10), 1615–1630. doi:10.1109/TPAMI.2005.188

Mikolajczyk, K., Tuytelaars, T., Schmid, C., Zisserman, A., Matas, J., & Schaffalitzky, F. (2005). A comparison of affine region detectors. *International Journal of Computer Vision, 65*(1-2), 43–72. doi:10.1007/s11263-005-3848-x

Miller, C. (2008, April 22). Dynamic signature verification market growth up. *SC Magazine*. Retrieved October 20, 2008, from http://www.scmagazineus.com/Dynamic-signature-verification-market-growth-up/article/109250/

MMIF. (2008). *W3C multimodal interaction framework*. Retrieved September 12, 2008, from http://www.w3.org/TR/2008/WD-mmi-arch-20080414/

Mobot. (2008). *Mobile visual search*. Retrieved from http://www.mobot.com/

Mori, G., Paterno, F., & Santoro, C. (2003). Tool support for designing nomadic applications. In *IUI '03: Proceedings of the 8th international conference on Intelligent user interfaces* (pp.141—148). New York: ACM.

Mori, G., Paternó, F., & Santoro, C. (2004). Design and Development of Multidevice User Interfaces through Multiple Logical Descriptions. *IEEE Transactions on Software Engineering, 30*(8), 507–520. doi:10.1109/TSE.2004.40

Moussa, F., Riahi, M., Kolski, C., & Moalla, M. (2002). Interpreted Petri Nets used for human-machine dialogue specification. *Integrated Computer-Aided Eng., 9*(1), 87–98.

Mueller, C., & Wasinger, R. (2002). Adapting multimodal dialog for the elderly. In *ABIS-Workshop on Personalization for the Mobile World* (pp. 31-34).

Mulholland, P., Collins, T., & Zdrahal, Z. (2004). Story fountain: intelligent support for story research and exploration. In *Proceedings of the 9th international conference on intelligent user interfaces*, Madeira, Portugal (pp. 62-69). New York: ACM.

Multimodal Interaction Framework. (2003). *MMI*. Retrieved February 02, 2009, from http://www.w3.org/TR/mmi-framework

Multimodal Interaction Working Group. (2002). *Multimodal Interaction Activity*. Retrieved on November 7, 2008, from http://www.w3.org/2002/mmi/

Multimodal Interaction Working Group. (2002). *Multimodal Interaction Framework*. Retrieved on November 7, 2008, from http://www.w3.org/TR/2003/NOTE-mmi-framework- 20030506/

Multimodal Interaction Working Group. (2003). *Multimodal Interaction Requirements*. Retrieved on November 7, 2008, from http://www.w3.org/TR/2003/NOTE-mmi-reqs-20030108/

Multimodal Interaction Working Group. (2008). *Multimodal Architecture and Interfaces*. Retrieved on November 7, 2008, from http://www.w3.org/TR/2008/WD-mmi-arch-20081016/

Muramatsu, D., & Matsumoto, T. (2007). Effectiveness of pen pressure, azimuth, and altitude features for online signature verification. In S.-W. Lee & S.Z. Li (Eds.), *Advances in Biometrics. Lecture Notes in Computer Science* (pp. 503-512). Berlin, Germany: Springer.

Nah, F. F.-H. (2004). A study on tolerable waiting time: how long are web users willing to wait? *Behaviour & Information Technology, 23*(3), 153–163. doi:10.1080/01449290410001669914

Nardelli, L., Orlandi, M., & Falavigna, D. (2004). A multi-modal architecture for cellular phones. In *ICMI '04: Proceedings of the 6th International Conference on Multimodal Interfaces* (pp. 323–324). New York: ACM.

Narzt, W., Pomberger, G., Ferscha, A., Kolb, D., Muller, R., Wieghardt, J., et al. (2003). Pervasive information acquisition for mobile AR-navigation systems. In *Proceedings of the Fifth IEEE Workshop on Mobile Computing Systems & Applications* (pp. 13-20).

Navarre, D., Palanque, P., Bastide, R., Schyn, A., Winckler, M. A., Nedel, L., & Freitas, C. (2005). A formal description of multimodal interaction techniques for immersive virtual reality applications. In *Proceedings of the IFIP Conference on Human-Computer Interaction (INTERACT'05)*.

Navarre, D., Palanque, P., Dragicevic, P., & Bastide, R. (2006). An approach integrating two complementary model-based environments for the construction of multimodal interactive applications. *Interacting with Computers, 18*(5), 910–941. doi:10.1016/j.intcom.2006.03.002

NayioMedia. (2008). Retrieved from http://www.nayio.com/

Neches, R. Foley, J. Szekely, P. Sukaviriya, P., Luo, P., Kovacevic, S. & Hudson, S. (1993). Knowledgeable development environments using shared design models. In *IUI '93: Proceedings of the 1st international conference on Intelligent user interfaces*. New York: ACM Press.

Neff, F., Kehoe, A., & Pitt, I. (2007). User modeling to support the development of an auditory help system. In *Lecture Notes in Computer Science. Text, Speech and Dialogue*. Berlin, Germany: Springer.

Nelson, W., Turin, W., & Hastie, T. (1994). Statistical methods for on-line signature verification. *International Journal of Pattern Recognition and Artificial Intelligence, 8*(3), 749–770. doi:10.1142/S0218001494000395

Nichols, J., & Myers, B. A. (2006). Controlling home and office appliances with smart phones. *IEEE Pervasive Computing / IEEE Computer Society [and] IEEE Communications Society, 5*(3), 60–67. doi:10.1109/MPRV.2006.48

Nickerson, M. (2005). All the world is a museum: Access to cultural heritage information anytime, anywhere. *Proceedings of International Cultural Heritage Informatics Meeting, ICHIM05*.

Nielsen, C. M., Overgaard, M., Pedersen, M. B., Stage, J., & Stenild, S. (2006). It's worth the hassle! The added value of evaluating the usability of mobile systems in the field. In *Proceedings of the 4th Nordic Conference on Human-Computer interaction: Changing Roles* (pp. 272-280). New York: ACM Press.

Nielsen, J. (1994). *Usability engineering (excerpt from chapter 5)*. San Francisco: Morgan Kaufmann. Retrieved September 7, 2008, from http://www.useit.com/papers/responsetime.html

Nigay, L. (1994). *Conception et modélisation logicielles des systèmes interactifs*. Thèse de l'Université Joseph Fourier, Grenoble.

Nigay, L., & Coutaz, J. (1993). A design space for multimodal systems: Concurrent processing and data fusion. In *Proceedings of ACM INTERCHI'93 Conference on Human Factors in Computing Systems, Voices and Faces* (pp. 172-178).

Nigay, L., & Coutaz, J. (1993). Espace problème, fusion et parallélisme dans les interfaces multimodales. In *Proc. of InforMatique'93*, Montpellier (pp.67-76).

Nigay, L., & Coutaz, J. (1995). A generic platform for addressing the multimodal challenge. In *Proceedings of the Conference on Human Factors in Computing Systems (CHI'95)* (pp. 98-105).

Nigay, L., & Coutaz, J. (1996). Espaces conceptuels pour l'interaction multimédia et multimodale. *TSI, spéciale multimédia et collecticiel*, 15(9), 1195-1225.

Niklfeld, G. Finan, R., & Pucher, M. (2001). Multimodal interface architecture for mobile data services. In *Proceedings of TCMC2001 Workshop on Wearable Computing*, Graz, Austria.

Nischelwitzer, A., Pintoffl, K., Loss, C., & Holzinger, A. (2007). Design and development of a mobile medical application for the management of chronic diseases: Methods of improved data input for older people. In *USAB 07 – Usability & HCI for Medicine and Health Care* (pp. 119-132). Berlin, Germany: Springer LNCS.

Noda, M., Sonobe, H., Takagi, S., & Yoshimoto, F. (2002, Sep.). *Cosmos: convenient image retrieval system of flowers for mobile computing situations*. Paper presented at the Proeedings of the IASTED Conf. on Information Systems and Databases 2002, Tokyo, Japan.

Noë, A. 2004. *Action in perception*. Cambridge, MA: MIT Press.

Norman, D. A. (1988). *The psychology of everyday things*. New York: Basic Books.

Norman, D. A. (1999). Affordance, conventions, and design. *Interactions (New York, N.Y.), 6*(3), 38–42. doi:10.1145/301153.301168

Nuance (2008). *Nuance voice control*. Retrieved from http://www.nuance.com/voicecontrol/

O'Hara, K., Kindberg, T., Glancy, M., Baptista, L., Sukumaran, B., Kahana, G., & Rowbotham, J. (2007). Collecting and sharing location-based content on mobile phones in a zoo visitor experience. [CSCW]. *Computer Supported Cooperative Work, 16*(1-2), 11–44. doi:10.1007/s10606-007-9039-2

ObjectConnections. (2008). *Common Knowledge Studio and engine provided by ObjectConnections*. Retrieved February 02, 2009, from http://www.objectconnections.com

oneSearch (2008). *Get answers with search designed for mobile*. Retrieved from http://mobile.yahoo.com/onesearch

Open Mobile Alliance. (2007). *OMA Mobile Profiles*. Retrieved from http://www.openmobilealliance.org/

OpenInterface European project. (n.d.). *IST Framework 6 STREP funded by the European, Commission (FP6-35182)*. Retrieved February 02, 2009, from http://www.oiproject.org and http://www.openinterface.org

Openmoko. (2008). *Openmoko development portal*. Retrieved November 28, 2008, from http://www.openmoko.org/

Oppermann, R., & Specht, M. (2001). Contextualised Information Systems for an Information Society for All. In *Proceedings of HCI International 2001. Universal Access in HCI: Towards an Information Society for All* (pp. 850 – 853).

Ortega-Garcia, J., Fierrez, J., Alonso-Fernandez, F., Galbally, J., Freire, M. R., & Gonzalez-Rodriguez, J. (2009). (in press). The multi-scenario multi-environment biosecure multimodal database (BMDB). *IEEE Transactions on Pattern Analysis and Machine Intelligence*.

Ortega-Garcia, J., Fierrez-Aguilar, J., Simon, D., Gonzalez, J., Faundez-Zanuy, M., & Espinosa, V. (2003). MCYT baseline corpus: a bimodal biometric database. *IEE Proceedings. Vision Image and Signal Processing, 150*(6), 391–401. doi:10.1049/ip-vis:20031078

Oviatt, A., DeAngeli, S., & Kuhn, K. (1997). Integration and synchronization of input modes during multimodal human-computer interaction. *CHI, 97*, 415–422.

Oviatt, S. (1996). Multimodal interfaces for dynamic interactive maps. In *SIGCHI Conference on Human Factors in Computing Systems* (pp. 95-102). New York: ACM Press.

Oviatt, S. (1997). Multimodal interactive maps: Designing for human performance. *Human-Computer Interaction, 12*, 93–129. doi:10.1207/s15327051hci1201&2_4

Oviatt, S. (1999). Mutual disambiguation of recognition errors in a multimodal architecture. *Proceedings of the SIGCHI Conference on Human Factors in Computing Systems: the CHI is the Limit* (pp. 576-583).

Oviatt, S. (1999). Ten myths of multimodal interaction. *Communications of the ACM, 42*(11), 74–81. doi:10.1145/319382.319398

Oviatt, S. (2000). Multimodal signal processing in naturalistic noisy environments. In Yuan, B., Huang, T., & Tang, X. (Eds.) *Proceedings of the International Conference on Spoken Language Processing* (pp. 696–699). Beijing, China: Chinese Friendship Publishers.

Oviatt, S. (2000). Multimodal system processing in mobile environments. In *UIST '00: Proceedings of the 13th annual ACM symposium on User Interface Software and Technology* (pp. 21–30). New York: ACM.

Oviatt, S. (2000). Taming recognition errors with a multimodal interface. *Communications of the ACM, 43*, 45–51.

Oviatt, S. (2003). Multimodal interfaces. In Jacko, J., & Sears, A. (Ed), *Human-Computer interaction Handbook: Fundamentals, Evolving Technologies and Emerging Applications* (pp. 286-304). Hillsdale, NJ: L. Erlbaum Associates.

Oviatt, S. L. (2002). Breaking the robustness barrier: recent progress on the design of robust multimodal systems. In M. Zelkowitz (Ed.), *Advances in Computers* (pp. 56). San Diego, CA: Academic Press.

Oviatt, S., & Cohen, P. (2000). Multimodal interfaces that process what comes naturally. *Communications of the ACM, 43*(3), 45–53. doi:10.1145/330534.330538

Oviatt, S., & Kuhn, K. (1998). Referential features and linguistic indirection in multimodal language. In *Proceedings of the International Conference on Spoken Language Processing* (pp. 2339–2342). Sydney, Australia: ASSTA, Inc.

Oviatt, S., & Lunsford, R. (2003). Multimodal interfaces for cell phones and mobile technology. [Berlin, Germany: Springer.]. *International Journal of Speech Technology, 8*(2), 127–132. doi:10.1007/s10772-005-2164-8

Oviatt, S., & VanGent, R. (1996). Error resolution during multimodal human computer interaction. In *Proceedings of the Fourth International Conference on Spoken Language* (pp. 204–207).

Oviatt, S., Cohen, P., Wu, L., Vergo, J., Duncan, L., & Suhm, B. (2000). Designing the user interface for multimodal speech and pen-based gesture applications: State-of-the-art systems and future research directions. *Human-Computer Interaction, 15*, 263–322. doi:10.1207/S15327051HCI1504_1

Oviatt, S., DeAngeli, A., & Kuhn, K. (1997). Integration and synchronization of input modes during multimodal human-computer interaction. In *Proceedings of the SIGCHI Conference on Human Factors in Computing Systems* (pp. 415-422). New York: ACM Press.

Palanque, P., & Schyn, A. (2003). A model-based for engineering multimodal interactive systems. *9th IFIP TC13 International Conference on Human Computer Interaction (Interact'2003)*.

Palanque, P., Bastide, R., & Sengès, V. (1995). Validating interactive system design through the verification of formal task and system models. In L.J. Bass & C. Unger (Eds.), *IFIP TC2/WG2.7 Working Conference on Engineering for Human-Computer Interaction (EHCI'95)* (pp. 189-212). New York: Chapman & Hall.

Pascoe, J., Ryan, N., & Morse, D. (2000). Using while moving: human-computer interaction issues in fieldwork environments. *Annual Conference Meeting Transactions on Computer-Human Interaction,* (pp. 417-437). New York: ACM.

Patermo, F., Santoro, C., Mäntyjärvi, J., Mori, G., & Sansone, S. (2008). Autoring pervasive multimodal user interfaces. *International Journal Web Engineering and Technology, 4*(2), 235–261. doi:10.1504/IJWET.2008.018099

Paternò, F. (2004). Multimodality and Multi-Platform Interactive Systems. In *Building the Information Society* (pp. 421-426). Berlin, Germany: Springer.

Paternò, F., & Giammarino, F. (2006). Authoring interfaces with combined use of graphics and voice for both stationary and mobile devices. In *AVI '06: Proceedings of the working conference on Advanced Visual Interfaces* (pp. 329–335). New York: ACM.

Paterno, F., & Mezzanotte, M. (1994). Analysing MATIS by interactors and actl. *Amodeus Esprit Basic Research Project 7040, System Modelling/WP36.*

Paterno, F., & Santoro, C. (2003). Support for reasoning about interactive systems through human-computer interaction designers' representations. *The Computer Journal, 46*(4), 340–357. doi:10.1093/comjnl/46.4.340

Pearce, D. (2000). Enabling new speech driven services for mobile devices: An overview of the ETSI standards activities for distributed speech recognition front-ends. *Technical Report AVIOS 2000, Aurora DSR Working Group.*

Peirce, C. S. ([1894] 1998). What is a sign? In Peirce Edition Project (Ed.), *The essential Peirce: selected philosophical writings vol. 2* (pp. 4-10). Bloomington: Indiana University Press.

Peltonen, P., Kurvinen, E., Salovaara, A., Jacucci, G., Ilmonen, T., Evans, J., et al. (2008). It's mine, don't touch: Interactions at a large multitouch display in a city center. In *Proc. of the SIGCHI conference on human factors in computing systems (CHI'08)* (pp. 1285-1294). New York: ACM.

Pemberton, S., Austin, D., Axelsson, J., Celik, T., Dominiak, D., Elenbaas, H., et al. (2000). XHTML 1.0 - The extensible hypertext markup language (2nd ed.) 1.0. *W3C.* Retrieved from http://www.w3.org/TR/2002/REC-xhtml1-20020801.

Pereira, A. C., Hartmann, F., & Kadner, K. (2007). A distributed staged architecture for multimodal applications. In F. Oquendo (Ed.), *European Conference on Software Architecture (ECSA) 2007* (pp. 195–206). Berlin, Germany: Springer Verlag.

Petrelli, D., Not, E., Zancanaro, M., Strapparava, C., & Stock, O. (2001). Modeling and adapting to context. *Personal and Ubiquitous Computing, 5*(1), 20–24. doi:10.1007/s007790170023

Petridis, P., Mania, K., Pletinckx, D., & White, M. (2006). Usability evaluation of the EPOCH multimodal user interface: designing 3D tangible interactions. In *VRST '06: Proceedings of the ACM symposium on Virtual reality software and technology,* (pp. 116-122). New York: ACM Press.

Petrovska-Delacretaz, D., Chollet, G., & Dorizzi, B. (Eds.). (2009). *Guide to Biometric Reference Systems and Performance Evaluation.* Berlin, Germany: Springer.

Pfaff, G. (1983). User interface management systems. In *Proceedings of the Workshop on User Interface Management Systems.*

Pham, T.-L., Schneider, G., & Goose, S. (2000). A situated computing framework for mobile and ubiquitous multimedia access using small screen and composite devices. In *Multimedia '00: Proceedings of the eighth acm international conference on multimedia* (pp. 323–331). New York: ACM Press.

Pirhonen, A. (2007). Semantics of sounds and images - can they be paralleled? In W. Martens (Eds.), *Proceedings of the 13th International Conference on Auditory Display* (pp. 319-325). Montreal: Schulich School of Music, McGill University.

Pirhonen, A., & Palomäki, H. (2008). Sonification of directional and emotional content: Description of design challenges. In P. Susini & O. Warusfel (Eds.), *Proceedings of the 14th International Conference on Auditory Display*. Paris: IRCAM (Institut de Recherche et Coordination Acoustique/Musique).

Pirhonen, A., Tuuri, K., Mustonen, M., & Murphy, E. (2007). Beyond clicks and beeps: In pursuit of an effective sound design methodology. In I. Oakley & S. Brewster (Eds.), *Haptic and Audio Interaction Design: Proceedings of Second International Workshop* (pp. 133-144). Berlin, Germany: Springer-Verlag.

Plamondon, R., & Lorette, G. (1989). Automatic signature verification and writer identification: the state of the art. *Pattern Recognition, 22*(2), 107–131. doi:10.1016/0031-3203(89)90059-9

Plotkin, G., (1981). A structural approach to operational semantics. Technical report, Departement of Computer Science, University of Arhus DAIMI FN 19.

Portillo, P. M., Garcia, G. P., & Carredano, G. A. (2006). Multimodal fusion: A new hybrid strategy for dialogue systems. In *ICMI'2006*, (pp. 357-363). New York: ACM.

Prompt, U. (2008). Retrieved from http://www.promptu.com/

Pudil, P., Novovicova, J., & Kittler, J. (1994). Floating search methods in feature selection. *Pattern Recognition Letters, 15*(11), 1119–1125. doi:10.1016/0167-8655(94)90127-9

Puerta, A. R., & Eisenstein, J. (1999). Towards a general computational framework for model-based interface development systems. In M. Maybury, P. Szekely & C. G. Thomas (Eds.), *Proceedings of the 4th international conference on Intelligent user interfaces 1999* (pp. 171–178). Los Angeles: ACM Press.

Purves, R. S., Edwardes, A., & Sanderson, M. (2008). *Describing the where - improving image annotation and search through geography*. Paper presented at the proceedings of the workshop on Metadata Mining for Image Understanding 2008.

QRCode.com. (2006). *A homepage about QR code.* Retrieved from http://www.qrcode.com/

Raffa, G., Mohr, P. H., Ryan, N., Manzaroli, D., Pettinari, M., Roffia, L., et al. (2007). Cimad - a framework for the development of context-aware and multi-channel cultural heritage services. In *Proc. International Cultural Heritage Informatics Meeting (ICHIM07)*.

Raman, T. V. (2003). User Interface Principles for Multimodal Interaction. In *Proceedings of MMI Workshop CHI 2003*.

Reeves, L. M., Lai, J., Larson, J. A., Oviatt, S., Balaji, T. S., & Buisine, S. (2004). Guidelines for multimodal user interface design. *Communications of the ACM, 47*(1), 57–59. doi:10.1145/962081.962106

Reich, S. S. (1980). Significance of pauses for speech perception. *Journal of Psycholinguistic Research, 9*(4), 379–389. doi:10.1007/BF01067450

Reichl, P., Froehlich, P., Baillie, L., Schatz, R., & Dantcheva, A. (2007). The LiLiPUT prototype: a wearable lab environment for user tests of mobile telecommunication applications. In *CHI '07 Extended Abstracts on Human Factors in Computing Systems* (pp. 1833-1838). New York: ACM Press.

Reis, T., Sá, M., & Carriço, L. (2008). Multimodal interaction: Real context studies on mobile digital artefacts. In *Proceedings of the 3rd international Workshop on Haptic and Audio interaction Design* (pp. 60-69). Jyväskylä, Finland: Springer-Verlag.

Reithinger, N., Alexandersson, J., Becker, T., Blocher, A., Engel, R., Löckelt, M., et al. (2003). Smartkom: adaptive and flexible multimodal access to multiple applications. In S. Oviatt, T. Darrell, M. Maybury & W. Wahlster, *Proceedings of the 5th international conference on Multimodal interfaces 2003* (pp. 101–108). New York: ACM Press.

Richiardi, J., & Drygajlo, A. (2003). Gaussian mixture models for on-line signature verification. In *ACM International Conference on Multimedia, Workshop on Biometric Methods and Applications, WBMA*, Berkeley, USA (pp. 115-122).

Richiardi, J., Ketabdar, H., & Drygajlo, A. (2005). Local and global feature selection for on-line signature verification. In *Proceedings of the Eighth International Conference on Document Analysis and Recognition 2005, Vol. 2*, Seoul, Korea (pp. 625-629). Washington, DC: IEEE Press.

Rist, T. (2005). Supporting mobile users through adaptive information presentation. In O. Stock and M. Zancanaro (Eds.), *Multimodal Intelligent Information Presentation* (pp. 113–141). Amsterdam: Kluwer Academic Publishers.

Rizzolatti, G., & Arbib, M. A. (1998). Language within our grasp. *Trends in Neurosciences, 21*, 188–194. doi:10.1016/S0166-2236(98)01260-0

Rocchi, C., Stock, O., & Zancanaro, M. (2004). The museum visit: Generating seamless personalized presentations on multiple devices. In *Proceedings of the International Conference on Intelligent User Interfaces*.

Rodrigues, M. A. F., Barbosa, R. G., & Mendonca, N. C. (2006). Interactive mobile 3D graphics for on-the-go visualization and walkthroughs. In *Proceedings of the 2006 ACM symposium on applied computing* (pp. 1002–1007). New York: ACM.

Rohr, M., Boskovic, M., Giesecke, S., & Hasselbring, W. (2006). *Model-driven development of self-managing software systems*. Paper presented at the "Models@run.time" Workshop at the 9th International Conference on Model Driven Engineering Languages and Systems 2006, Genoa, Italy.

Rosenbaum, S., & Kantner, L. (2005). Helping users to use help: Results from two international conference workshops. In *Proceedings from Professional Communication Conference IPCC 2005* (pp. 181–187).

Ross, A., Nandakumar, K., & Jain, A. K. (2006). *Handbook of Multibiometrics*. Berlin, Germany: Springer.

Rouillard, J. (2008). Contextual QR codes. In *Proceedings of the Third International Multi-Conference on Computing in the Global Information Technology (ICCGI 2008)*, Athens, Greece (pp. 50-55).

Rouillard, J., & Laroussi, M. (2008). PerZoovasive: contextual pervasive QR codes as tool to provide an adaptive learning support. In *International Workshop On Context-Aware Mobile Learning - CAML'08, IEEE/ACM SIGAPP*, Paris (pp. 542-548).

Rousseau, C. (2006). *Présentation multimodale et contextuelle de l'information*. Unpublished doctoral dissertation, Paris-Sud XI University, Orsay, France.

Ruf, B., Kokiopoulou, E., & Detyniecki, M. (2008). Mobile museum guide based on fast SIFT recognition. In *6th International Workshop on Adaptive Multimedia Retrieval*.

Ruf, J., & Kropf, T. (2003). Symbolic verification and analysis of discrete timed systems. *Formal Methods in System Design, 23*(1), 67–108. doi:10.1023/A:1024437214071

Rupnik, R., Krisper, M., & Bajec, M. (2004). A new application model for mobile technologies. *International Journal of Information Technology and Management, 3*(2), 282–291. doi:10.1504/IJITM.2004.005038

Ryan, N. S., Raffa, G., Mohr, P. H., Manzaroli, D., Roffia, L., Pettinari, M., et al. (2006). A smart museum installation in the Stadsmuseum in Stockholm - From visitor guides to museum management. In *EPOCH Workshop on the Integration of Location Based Systems in Tourism and Cultural Heritage*.

Sá, M., & Carriço, L. (2006) Low-fi prototyping for mobile devices. In *CHI '06 Extended Abstracts on Human Factors in Computing Systems* (pp. 694-699). New York: ACM Press.

Sá, M., & Carriço, L. (2008). Lessons from early stages design of mobile applications. In *Proceedings of the 10th international Conference on Human Computer interaction with Mobile Devices and Services* (pp. 127-136). New York: ACM Press.

Sá, M., Carriço, L., & Duarte, C. (2008). Mobile interaction design: Techniques for early stage In-Situ design. In Asai, K. (Ed.), *Human-Computer Interaction, New Developments* (pp. 191-216). Vienna, Austria: In-teh, I-Tech Education and Publishing KG.

Sá, M., Carriço, L., Duarte, L., & Reis, T. (2008). A mixed-fidelity prototyping tool for mobile devices. In *Proceedings of the Working Conference on Advanced Visual interfaces* (pp. 225-232). New York: ACM Press.

Salber, D. (2000). Context-awareness and multimodality. In *Proceedings of First Workshop on Multimodal User Interfaces*.

Salden, A., Poortinga, R., Bouzid, M., Picault, J., Droegehorn, O., Sutterer, M., et al. (2005). Contextual personalization of a mobile multimodal application. In *Proceedings of the International Conference on Internet Computing, ICOMP*.

Salmon Cinotti, T., Summa, S., Malavasi, M., Romagnoli, E., & Sforza, F. (2001). MUSE: An integrated system for mobile fruition and site management. In *Proceedings of International Cultural Heritage Informatics Meeting*. Toronto, Canada: A&MI.

SALT. (n.d.). *Speech Application Language Tags*. Retrieved February 02, 2009, from http://www.phon.ucl.ac.uk/home/mark/salt

Santoro, C., Paternò, F., Ricci, G., & Leporini, B. (2007). A multimodal mobile museum guide for all. In *Proceedings of the Mobile Interaction with the Real World* (MIRW 2007). New York: ACM.

Schaffalitzky, F., & Zisserman, A. (2002, May). *Multiview matching for unordered image sets*. Paper presented at the Proceedings of the 7th European Conference on Computer Vision, Copenhagen, Denmark.

Schmid, C., & Mohr, R. (1997). Local grayvalue invariants for image retrieval. *IEEE Transactions on Pattern Analysis and Machine Intelligence*, *19*(5), 530–535. doi:10.1109/34.589215

Schomaker, L., Nijtmans, J., Camurri, A., Lavagetto, F., Morasso, P., Benoit, C., et al. &. Blauert, J. (1995). A taxonomy of multimodal interaction in the human information processing system. *Technical Report 8579, Esprit Project*.

Searle, J. R. (1969). *Speech Acts: An Essay in the Philosophy of Language*. New York, NY: Cambridge University Press.

Searle, J. R. (1983). *Intentionality: An essay in the philosophy of mind*. New York: Cambridge University Press.

Searle, J. R. (2004). *Mind: A brief introduction*. New York: Oxford University Press.

Sebe, N., Tian, Q., Loupias, E., Lew, M., & Huang, T. (2003). Evaluation of salient point techniques. *Image and Vision Computing*, *17*(13-14), 1087–1095. doi:10.1016/j.imavis.2003.08.012

Serrano, M., Juras, D., Ortega, M., & Nigay, L. (2008). OIDE: un outil pour la conception et le développement d'interfaces multimodales. *Quatrièmes Journées Francophones: Mobilité et Ubiquité 2008*, Saint Malo, France (pp. 91-92). New York: ACM Press.

Serrano, M., Nigay, L., Demumieux, R., Descos, J., & Losquin, P. (2006). Multimodal interaction on mobile phones: Development an evaluation using ACICARE. In *8th Conference on Human-Computer Interaction with Mobile Device and Services, Mobile HCI'06* (pp. 129-136). New York: ACM.

Shannon, C. E., & Weaver, W. (1949). *The mathematical theory of communication*. Urbana, IL: University of Illinois Press.

Shriberg, E., & Stolcke, A. (2004). Direct Modeling of Prosody: An Overview of Applications in Automatic Speech Processing. *Speech Prosody, 2004*, 575–582.

Signature verification biometrics market to overcome the effect of reduced IT Expenditures. (2008, Feb. 18). *BNET*. Retrieved October 20, 2008, from http://findarticles.com/p/articles/mi_m0EIN/is_/ai_113374152

Simon, R., Wegscheider, F., & Tolar, K. (2005). Tool-supported single authoring for device independence and multimodality. In *MobileHCI '05: Proceedings of the 7th international conference on Human computer interaction with mobile devices & services* (pp. 91–98). New York: ACM.

Singh, P., Ha, H. N., Kuang, Z., Oliver, P., Kray, C., Blythe, P., & James, P. (2006). Immersive video as a rapid prototyping and evaluation tool for mobile and ambient applications. *Proceedings of the 8th conference on Human-computer interaction with mobile devices and services.* New York: ACM.

Sinha, A. K., & Landay, J. A. (2003). Capturing user tests in a multimodal, multidevice informal prototyping tool. In *5th International Conference on Multimodal Interfaces* (pp. 117-124).

Smartkom Project. (n.d.). *Dialog-based Human-Technology Interaction by Coordinated Analysis and Generation of Multiple Modalities.* Retrieved February 02, 2009, from http://www.smartkom.org

Smith, J. R., & Chang, S.-F. (1996, Nov.). *VisualSEEk: a fully automated content-based image query system.* Paper presented at the Proceedings of the 4th ACM International Conference on Multimedia, Boston, USA.

Smith, M., Duncan, D., & Howard, H. (2003). *AURA: A mobile platform for object and location annotation.* Paper presented at the 5th Int. Conf. Ubiquitous Computing.

Sohn, T., Li, K. A., Griswold, W. G., & Hollan, J. D. (2008). A diary study of mobile information needs. In *Proceeding of the Twenty-Sixth Annual SIGCHI Conference on Human Factors in Computing Systems* (pp. 433-442). New York: ACM Press.

Sonnenschein, D. (2001). *Sound design: The expressive power of music, voice and sound effects in cinema.* Saline, MI: Michael Wiese Productions.

Sonobe, H., Takagi, S., & Yoshimoto, F. (2004, Jan.). *Image retrieval system of fishes using a mobile device.* Paper presented at the Proceedings of International Workshop on Advanced Image Technology 2004, Singapore.

Sottet, J.-S., Calvary, G., & Favre, J.-M. (2006). Mapping model: A first step to ensure usability for sustaining user interface plasticity. In A. Pleuss, J. V. den Bergh, H. Hussmann, S. Sauer, & A. Boedcher, (Eds.), *MDDAUI '06 - Model Driven Development of Advanced User Interfaces 2006, Proceedings of the 9th International Conference on Model-Driven Engineering Languages and Systems: Workshop on Model Driven Development of Advanced User Interfaces* (pp 51–54), CEUR Workshop Proceedings.

Sottet, J.-S., Calvary, G., & Favre, J.-M. (2006). *Models at runtime for sustaining user interface plasticity.* Retrieved January 21, 2008, from http://www.comp.lancs.ac.uk/~bencomo/MRT06/

Sottet, J.-S., Ganneau, V., Calvary, G., Coutaz, J., Demeure, A., Favre, J.-M., & Demumieux, R. (2007). Model-driven adaptation for plastic user interfaces. In C. Baranauskas, P. Palanque, J. Abascal & S. D. J. Barbosa (Eds.), *Human-Computer Interaction – INTERACT 2007* (pp 397–410). Berlin, Germany: Springer.

Spivey, J. M. (1988). Understanding Z: A specification language and its formal semantics. *Cambridge Tracts in Theoretical Computer Science.* New York: Cambridge University Press.

Stanciulescu, A. (2008). *A methodology for developing multimodal user interfaces of information system.* Ph.D. thesis, Université catholique de Louvain, Louvain-la-Neuve, Belgium.

Stanciulescu, A., Limbourg, Q., Vanderdonckt, J., Michotte, B., & Montero, F. (2005). A transformational approach for multimodal web user interfaces based on UsiXML. In *Proceedings of 7th International Conference on Multimodal Interfaces ICMI'2005* (pp. 259–266). New York: ACM Press.

Stent, A., Walker, M., Whittaker, S., & Maloor, P. (2002). User-tailored generation for spoken dialogue: An experiment. In *ICSLP '02* (pp. 1281-84).

Stephanidis, C., & Savidis, A. (2001). Universal access in the information society: Methods, tools, and interaction technologies. *UAIS Journal, 1*(1), 40–55.

Stephanidis, C., Karagiannidis, C., & Koumpis, A. (1997). Decision making in intelligent user interfaces. In *Proceedings of Intelligent User Interfaces (IUI'97)* (pp. 195-202).

Stock, O., & the ALFRESCO Project Team. (1993). AL-FRESCO: Enjoying the combination of natural language processing and hypermedia for information exploration. In M. T. Maybury (Ed.), *Intelligent Multimedia Interfaces* (pp. 197-224). Menlo Park, CA: AAAI Press.

Sumikawa, D., Blattner, M., & Greenberg, R. (1986). *Earcons: Structured audio messages*. Unpublished paper.

Sun Microsystems. (2003). *JSR-184: Mobile 3D graphics API for J2ME*. Retrieved November 13, 2008, from http://www.jcp.org/en/jsr/detail?id=184

Svanaes, D., & Seland, G. (2004). Putting the users center stage: role playing and low-fi prototyping enable end users to design mobile systems. In *Proceedings of the SIGCHI Conference on Human Factors in Computing Systems* (pp. 479-486). New York: ACM Press.

SVG. Scalable Vector Graphics (n.d.). *W3C Recommendation*. Retrieved February 02, 2009, from http://www.w3.org/Graphics/SVG

Synchronized Multimedia Working Group. (2008). *Synchronized Multimedia Integration Language (SMIL 3.0)*. Retrieved November 12, 2008, from http://www.w3.org/TR/2008/PR-SMIL3-20081006/

Tagg, P. (1992). Towards a sign typology of music. In R. Dalmonte & M. Baroni (Eds.), *Secondo Convegno Europeo di Analisi Musicale* (pp. 369-378). Trento, Italy: Università Degli Studi di Trento.

Tan, H. Z., Gray, R., Young, J. J., & Irawan, P. (2001). Haptic cueing of a visual change-detection task: Implications for multimodal interfaces. In *Usability Evaluation and Interface Design: Cognitive Engineering, Intelligent Agents and Virtual Reality* (pp. 678-682). Philadelphia, PA: Lawrence Erlbaum Associates.

Tarasewich, P. (2003). Designing mobile commerce applications. *Communications of the ACM, 46*(12), 57–60. doi:10.1145/953460.953489

Teil, D., & Bellik, Y. (2000). Multimodal interaction interface using voice and gesture. In M. M. Taylor, F. Néel & D. G. Bouwhuis (Eds.), *The Structure of Multimodal Dialog II* (pp. 349-366).

Tellme (2008). *Tellme: Say it. Get it*. Retrieved from http://www.tellme.com/

Thevenin, D., & Coutaz, J. (1999). Plasticity of user interfaces: Framework and research agenda. In *Proceedings of the 7th IFIP Conference on Human-Computer Interaction, INTERACT'99*, Edinburgh, Scotland (pp.110-117).

TRIPOD. (2008). *Project Tripod - Automatically captioning photographs*. Retrieved from http://www.projecttripod.org/

Tuuri, K., & Eerola, T. (2008). Could function-specific prosodic cues be used as a basis for non-speech user interface sound design? In P. Susini & O. Warusfel (Eds.), *Proceedings of the 14th International Conference on Auditory Display*. Paris: IRCAM (Institut de Recherche et Coordination Acoustique/Musique).

Uludag, U., Ross, A., & Jain, A. K. (2004). Biometric template selection and update: a case study in fingerprints. *Pattern Recognition, 37*(7), 1533–1542. doi:10.1016/j.patcog.2003.11.012

User Interface Markup Language, O. A. S. I. S. (UIML) Technical Committee. (2008). *User Interface Markup Language (UIML)*. Retrieved from http://www.oasis-open.org/committees/uiml/

Välkkynen, P. (2007). *Physical selection in ubiquitous computing*. Helsinki, Edita Prima.

Van Gurp, J., & Bosch, J. (1999). On the implementation of finite state machines. In *Proceedings of the 3rd Annual IASTED International Conference on Software Engineering and Applications* (pp. 172-178).

Van, B. L., Garcia-Salicetti, S., & Dorizzi, B. (2007). On using the Viterbi path along with HMM likelihood information for online signature verification. *IEEE Transactions on Systems, Man, and Cybernetics . Part B, 37*(5), 1237–1247.

Vanderdonckt, J. (2008). *Model-driven engineering of user interfaces: Promises, successes, failures, and challenges.* Paper presented at 5th Romanian Conference on Human-Computer Interaction, Bucuresti, Romania.

Van-Eijk, P. (1991). The lotosphere integrated environment. In *4th international conference on formal description technique (FORTE91)*, (pp.473-476).

Varela, F., Thompson, E., & Rosch, E. (1991). *The embodied mind.* Cambridge, MA: MIT Press.

VibeTonz System. (2008). *Immersion Corporation.* Retrieved September 12, 2008, from http://www.immersion.com/mobility

Vilimek, R., & Hempel, T. (2005). Effects of speech and non-speech sounds on short-term memory and possible implications for in-vehicle use. In *Proceedings of ICAD 2005.*

Vlahakis, V., Ioannidis, M., Karigiannis, J., Tsotros, M., Gounaris, M., & Stricker, D. (2002). Archeoguide: an augmented reality guide for archaeological sites. *IEEE Computer Graphics and Applications, 22*(5), 52–60. doi:10.1109/MCG.2002.1028726

Vlahakis, V., Karigiannis, J., Tsotros, M., Gounaris, M., Almeida, L., Stricker, D., et al. (2001). Archeoguide: first results of an augmented reality, mobile computing system in cultural heritage sites. In *Proceedings of the 2001 conference on Virtual reality, archeology, and cultural heritage* (pp. 131-140). New York: ACM.

Vo, M. T., & Waibel, A. (1997). Modeling and interpreting multimodal inputs: A semantic integration approach. *Technical Report CMU-CS-97-192.* Pittsburgh: Carnegie Mellon University.

Voice, X. M. L. 2.0. (2004). *W3C Recommendation.* Retrieved February 02, 2009, from http://www.w3.org/TR/voicexml20

Voice, X. M. L. Forum. (2004). *XHTML + voice profile 1.2.* Retrieved on November 12, 2008, from http://www.voicexml.org/specs/multimodal/x+v/12/

Wahlster, W. (2003). Towards symmetric multimodality: Fusion and fission of speech, gesture and facial expression. In Günter, A., Kruse, R., Neumann, B. (eds.), *KI 2003: Advances in Artificial Intelligence, Proceedings of the 26th German Conference on Artificial Intelligence* (pp. 1-18). Hamburg, Germany: Springer.

Wahlster, W. (2006). SmartKom: Foundations of multimodal dialogue systems. *Series: Cognitive Technologies,* 644. X+V, XHTML + Voice Profile 1.2. (2004). *W3C Recommendation.* Retrieved February 02, 2009, from http://www.voicexml.org/specs/multimodal/x+v/12

Want, R., Fishkin, K. P., Gujar, A., & Harrison, B. L. (1999). Bridging physical and virtual worlds with electronic tags. In *Proc. SIGCHI Conference on Human factors in Computing Systems* (pp. 370-377), New York: ACM Press.

Ward, K., & Novick, D. G. (2003). Hands-free documentation. In *Proceedings of 21st Annual international Conference on Documentation (SIGDOC '03)* (pp. 147-154.). New York: ACM.

Wasserman, A. (1985). Extending state transition diagrams for the specification of human-computer interaction. *IEEE Transactions on Software Engineering, 11*(8), 699–713. doi:10.1109/TSE.1985.232519

Watanabe, Y., Sono, K., Yokoizo, K., & Okad, Y. (2003). *Translation camera on mobile phone.* Paper presented at the IEEE Int. Conf. Multimedia and Expo 2003.

Waters, K., Hosn, R., Raggett, D., Sathish, S., Womer, M., Froumentin, M., et al. (2007). Delivery Context: Client Interfaces (DCCI) 1.0. *W3C.* Retrieved from http://www.w3.org/TR/DPF

Waterworth, J. A., & Chignell, M. H. (1997). Multimedia interaction. In Helander, M. G., Landauer, T. K. & Prabhu, P. V. (Eds.): *Handbook of Human-Computer Integration* (pp. 915-946). Amsterdam: Elsevier.

Weimer, D., & Ganapathy, S. K. (1989). A synthetic visual environment with hand gesturing and voice input. *Proceedings of the SIGCHI conference on Human factors in computing systems* (pp. 235–240). New York: ACM Press.

Weiser, M. (1991). The computer for the 21st century. *Scientific American, 265*(3), 94–104.

Weiser, M. (1993). Some computer science issues in ubiquitous computing. *Communications of the ACM, 36*(7), 75–84. doi:10.1145/159544.159617

Weiser, M. (1999). The computer for the 21st century. *SIGMOBILE Mobile Computing and Communications Review, 3*(3), 3–11. doi:10.1145/329124.329126

West, D., Apted, T., & Quigley, A. (2004). A context inference and multi-modal approach to mobile information access. In *Artificial intelligence in mobile systems* (pp. 28–35). Nottingham, England.

Wickens, C. D. (1984). Processing resources in attention. In: R. Parasuraman & D.R. Davies (Eds.), *Varieties of attention* (pp. 63-102). Orlando, FL: Academic Press.

Wilson, M., & Knoblich, G. (2005). The case for motor involvement in perceiving conspecifics. *Psychological Bulletin, 1*(3), 460473.

Winkler, S., Rangaswamy, K., Tedjokusumo, J., & Zhou, Z. (2007). Intuitive application-specific user interfaces for mobile devices. In *Proceedings of the 4th International Conference on Mobile Technology, Applications, and Systems and the 1st International Symposium on Computer Human Interaction in Mobile Technology* (pp. 576–582). New York: ACM.

Wireless Application Forum. (2001). *User Agent Profile*. Retrieved from http://www.openmobilealliance.org/tech/affiliates/wap/wap-248-uaprof-20011020-a.pdf

Woodruff, A., et al. (2001). Electronic guidebooks and visitor attention. *Proceedings of the 6th Int'l Cultural Heritage Informatics Meeting* (pp. 623-637). Pittsburgh: A&MI.

Wrigley, S. N., & Brown, G. J. (2000). A model of auditory attention. *Technical Report CS-00-07, Speech and Hearing Research Group.* University of Sheffield, UK.

Xie, X., Lu, L., Jia, M., Li, H., Seide, F., & Ma, W.-Y. (2008). Mobile Search With Multimodal Queries. *Proceedings of the IEEE, 96*(4), 589–601. doi:10.1109/JPROC.2008.916351

Yamakami, T. (2007). Challenges in mobile multimodal application architecture. In *Workshop on W3C's Multimodal Architecture and Interfaces.*

Yang, J., Stiefelhagen, R., Meier, U., & Waibel, A. (1998). Visual tracking for multimodal human computer interaction. In *Proceedings of the Conference on Human Factors in Computing Systems (CHI-98): Making the Impossible Possible* (pp. 140–147). New York: ACM Press.

Yeh, T., Tollmar, K., & Darrell, T. (2004, June). *Searching the Web with mobile images for location recognition.* Paper presented at the Proceedings of IEEE Conference on Computer Vision and Pattern Recognition 2004 Washington D.C., USA.

Yeh, T., Tollmar, K., Grauman, K., & Darrell, T. (2005, April). *A picture is worth a thousand keywords: Image-based object search on a mobile platform.* Paper presented at the Proceedings of the 2005 Conference on Human Factors in Computing Systems, Portland, USA.

Yeung, D. Y., Chang, H., Xiong, Y., George, S., Kashi, R., Matsumoto, T., et al. (2004). SVC2004: First international signature verification competition. In *Biometric Authentication. Lecture Notes in Computer Science* (pp. 16-22). Berlin, Germany: Springer.

Yuan, X., & Chee, Y. S. (2003). Embodied tour guide in an interactive virtual art gallery. In *Proceedings of the 2003 international Conference on Cyberworlds* (pp. 432). Washington, DC: IEEE Computer Society

Zaykovskiy, D. (2006). Survey of the speech recognition techniques for mobile devices. In *Proceedings of the 11th International Conference on Speech and Computer (SPECOM)* (pp. 88–93). St. Petersburg, Russia

Zellner, B. (1994). Pauses and the temporal structure of speech. In E. Keller (Ed.), *Fundamentals of Speech Synthesis and Speech Recognition: Basic Concepts, State-of-the-Art and Future Challenges* (pp. 41-62). Chichester, UK: John Wiley and Sons Ltd.

Zheng, Y., Liu, L., Wang, L.-H., & Xie, X. (2008). *Learning transportation mode from raw GPS data for geographic applications on the Web.* Paper presented at the 17th International World Wide Web Conference Beijing, China.

Zhu, Y., & Shasha, D. (2003). *Warping indexes with envelope transforms for query by humming.* Paper presented at the Proceedings of the 2003 ACM SIGMOD, San Diego, California, USA

About the Contributors

Stan Kurkovsky is an associate professor at the Department of Computer Science at Central Connecticut State University. Stan earned his PhD from the Center for Advanced Computer Studies of the University of Louisiana in 1999. Results of his doctoral research have been applied to network planning and industrial simulation. Stan's current research interests are in mobile and pervasive computing, distributed systems and software engineering. He published over 40 papers in refereed proceedings of national and international conferences, scientific journals and books. Stan serves as a reviewer and a member of program committees on a number of national and international conferences. During his academic career, Stan received over a million dollars in funding from private and federal sources.

* * *

Sahin Albayrak is the chair of the professorship on Agent Technologies in Business Applications and Telecommunication (AOT). He is the founder and head of the DAI-Labor, currently employing about 100 researchers and support staff. He is member of "The Institute of Electrical and Electronics Engineers" (IEEE), "Association for Computing Machinery" (ACM), "Gesellschaft für Informatik" (German Computer Science Society, GI), and "American Association for Artificial Intelligence" (AAAI). Prof. Albayrak is one of the founding members of Deutsche Telekom Laboratories (T-Labs) and currently member of the steering board. He was the initiator of many reputable research projects, e.g.: E@MC2, Sun-Trec, in which he has been supervising research networks at national and international levels. He is also a member of various industrial and political advisory committees, e.g.: Impulskreis "Vernetzte Welten".

Agnese Augello received her "cum laude" Laurea degree in "Ingegneria Informatica" at the University of Palermo. In April 2008 she achieved the Ph.D. degree in Computer Science at DINFO (Dipartimento di Ingegneria Informatica) of Palermo University. Currently she is research fellow at DINFO and she is working on the processing of semantic spaces for natural language systems. Her research activity deals with knowledge representation for interactive, intelligent systems.

Yacine Bellik is an assistant professor at Paris-South University (IUT d'Orsay). He conducts his research activities at LIMSI-CNRS laboratory (Laboratoire d'Informatique pour la Mécanique et les Sciences de l'Ingénieur). He holds a PhD and HDR in Computer Science. He is the head of the « Multimodal Interaction » research topic at LIMSI. His research interests include the study of multimodal human-computer communication (both in input and output, taking into account the interaction context to allow intelligent multimodal interaction), aid for the blind and ambient intelligence. He is the coordinator of the IRoom project, a joint project between Supélec and LIMSI-CNRS whose aim is to conduct research about "smart room" environments. He is also involved in the ATRACO (Adaptive and TRusted Ambient eCOlogies) European project.

Michela Bertolotto received a PhD in Computer Science from the University of Genoa (Italy) in 1998. Subsequently she worked as a Postdoctoral Research Associate in the National Center for Geographic Information and Analysis (NCGIA) at the University of Maine. Since 2000 she has been a faculty member at the School of Computer Science and Informatics of University College Dublin. Her research interests include web-based and wireless GIS, spatio-temporal data modelling and 3D interfaces.

Marco Blumendorf is a researcher and PhD student at the DAI-Labor of the Technische Universität Berlin. He is leading the HCI working group of the DAI-Labor and participated in several research projects funded by the German government, the Deutsche Telekom, Sun Microsystems and others. His current research interests include multimodal human computer interaction, focussing on model-based development of user interfaces for smart home environments. He is a member of different organisations, including the "Association for Computing Machinery" (ACM) and "Gesellschaft für Informatik" (German Computer Science Society, GI).

Marie-Luce Bourguet is a lecturer in the School of Electronic Engineering and Computer Science, at Queen Mary, University of London, where she teaches multimedia and computer graphics. Her main research interests are multimodal interaction, e-learning, and multilingualism. She is currently working on developing tools for the design and implementation of multimodal applications for language learning. Prior to joining Queen Mary, Marie-Luce Bourguet has been a senior research scientist at Canon Research Center Europe (UK), NHK (Japanese National Broadcasting Corp, Japan), Secom Intelligent Systems Lab (Japan), and the Canadian Workplace Automation Center (Canada). She is currently based in Tokyo, Japan, and teaches on the Queen Mary / BUPT (Beijing University of Post and Telecommunication) joint programme in China.

Luís Carriço is an associate professor at the University of Lisbon's Department of Informatics. His main research interests are human computer interaction, user-centred design, mobile-interface design, pervasive computing, ubiquitous computing, groupware, accessibility and e-health. He received his PhD in Electrical and Computer Engineering from the Technical University of Lisbon. He has published 118 articles in journals, books and conferences, most of them international, and participated in more than 15 projects, 5 of which international. He is a member of more than 50 program committees and he is an invited reviewer and evaluator for the European Commission. He is the leader of the AbsInt team, of the HCIM Group, for which he is the executive coordinator, of the LaSIGE research unit. Contact him at LaSIGE/Departamento de Informática, Edifício C6, 1749 - 016 Campo Grande, Lisboa, Portugal.

Jaeseung Chang is an interaction architect at Handmade Mobile Entertainment Ltd in UK, where he takes charge of UI design and user testing for its mobile-based social networking service, Flirtomatic. Before Flirtomatic, he has been a mobile UI designer at Samsung Electronics in Korea. He got MSc in Computer Science by Research degree at Queen Mary, University of London and GDip in HCI degree at University College London. His research interests are multimodal interaction and mobile device interaction.

Julie Doyle received a PhD in Computer Science from University College Dublin, Ireland. Her thesis focused on improving the usability and interactivity of mobile geospatial applications. Subsequently she worked as a postdoctoral fellow at the School of IT and Engineering, University of Ottawa, Canada, where she developed a framework to ensure the long term preservation of 3D digital data. Currently she is a postdoctoral fellow in HCI and Systems Engineering in the TRIL Centre, UCD School of Computer Science and Informatics. Her main research interests include the design and evaluation of independent living technologies for the elderly, the design of interfaces for mobile geospatial applications, multimodal HCI and usability.

Carlos Duarte is a computer science researcher and an assistant professor at the Faculty of Sciences of the University of Lisbon, where he has taught since 2000. He got his PhD from the University of Lisbon, titled "Design and Evaluation of Adaptive Multimodal Systems", in 2008. His main research area is Human-Computer Interaction. On this subject his research interests include multimodal interaction, adaptive interfaces, touch and gesture based interaction, mobile computing, usability evaluation, accessibility, hypermedia systems and digital books. He has published over 30 papers, including journal articles, book chapters and conference proceedings. He has participated in research projects and is currently a team member in the EU funded ACCESSIBLE project.

Xin Fan is a Research Associate in the Department of Information Studies at the University of Sheffield. He received the B.E. and M.E. degrees in pattern recognition and intelligent system and the Ph.D. degree in Electronic Engineering from the University of Science and Technology of China, respectively in 2001, 2004 and 2007. He worked as a research intern in the Web Search and Mining Group at Microsoft Research Asia from 2002 to 2005 and visited the Interactive Communication Media and Contents Group at NICT, Japan as an embedded researcher from 2005 to 2006. His research interests include multimedia information retrieval, human computer interaction and natural language processing.

Julian Fierrez received the MSc and the PhD degrees in telecommunications engineering from Universidad Politecnica de Madrid, Spain, in 2001 and 2006, respectively. Since 2002 he has been affiliated with Universidad Autonoma de Madrid in Spain, where he currently holds a Marie Curie Postdoctoral Fellowship, part of which has been spent as a visiting researcher at Michigan State University in USA. His research interests include signal and image processing, pattern recognition, and biometrics, with emphasis on signature and fingerprint verification, multi-biometrics, performance evaluation, and system security. He has been actively involved in European projects focused on biometrics, and is the recipient of a number of distinctions, including: best poster paper at AVBPA 2003, Rosina Ribalta award to the best Spanish PhD proposal in ICT in 2005, best PhD in computer vision and pattern recognition in 2005-2007 by the IAPR Spanish liaison AERFAI, Motorola best student paper at ICB 2006, EBF European biometric industry award 2006, and one of the IBM best student paper awards at ICPR 2008.

Salvatore Gaglio is full professor of Computer Science and Artificial Intelligence at DINFO (Dipartimento di ingegneria INFOrmatica), University of Palermo, Italy. He is member of various committees for projects of national interest in Italy and he is referee of various scientific congresses and journals. His present research activities are in the area of artificial intelligence and robotics.

Alessandro Genco attended the University of Palermo, where he studied Mathematics and obtained his degree. Since 1986 he has been Associate Professor at the Computer Science and Engineering Department. His main research interests deal with: parallel and distributed computing; mobile networks and pervasive system middleware design for augmented reality.

Antonio Gentile is associate professor at DINFO – Dipartimento di Ingegneria Informatica of the Universita` degli Studi di Palermo, Italy. He received the Laurea degree "cum Laude" in electronic engineering and the doctoral degree in computer science from the Universita` di Palermo, in 1992 and 1996, respectively. He also received the PhD degree in electrical and computer engineering from the Georgia Institute of Technology in 2000. His research interests include high throughput portable processing systems, image and video processing architectures, embedded systems, speech processing, and human-computer interfaces. Antonio is a senior member of the IEEE, of the IEEE Computer Society, and a member of the ACM and of the AEIT. He is currently serving as associate editor for *Integration - the VLSI Journal*, and he is a member of the editorial board for the *Journal of Embedded Computing*.

Habib Hamam obtained the Diploma of Engineering in information processing from the Technical University of Munich, Germany, 1992, and the PhD degree in Physics and applications in telecommunications from Université de Rennes I conjointly with France Telecom Graduate School, France 1995. He also obtained a postdoctoral diploma, "Accreditation to Supervise Research in Signal Processing and Telecommunications", from Université de Rennes I in 2004. He is currently a full Professor in the Department of Electrical Engineering at the University of Moncton and a Canada Research Chair holder in "Optics in Information and Communication Technologies". He is an IEEE senior member and a registered professional engineer in New-Brunswick. He is among others associate editor of the IEEE Canadian Review, member of the editorial boards of Wireless Communications and Mobile Computing - John Wiley & Sons - and of Journal of Computer Systems, Networking, and Communications - Hindawi Publishing Corporation. He is also member of a national committee of the NSERC (Canada). His research interests are in optical and Wireless telecommunications, diffraction, fiber components, optics of the eye, RFID and e-Learning.

Gerald Hübsch is a post-doctoral researcher at the Chair of Computer Networks at Dresden University of Technology, where he finished his PhD thesis in the field of system support for multimodal user interfaces in 2008. His research interests comprise middleware technologies for multimodal and context-aware applications, model-driven software engineering and service-oriented architectures. He has been participating in several research projects in the area of mobile computing, model-driven development of adaptive multimodal applications and adaptive content delivery.

Christophe Jacquet is an Assistant Professor at Supélec, a "Grande École" in Information Sciences and Power Systems. He holds an Engineering degree from Supélec and a Master of Science. He received a PhD in Computer Science from the Paris-South University in 2006. His research interests include the modeling of heterogeneous systems, in particular complex systems and embedded systems. He studies how this applies to ambient intelligent settings, in particular within the IRoom project, a joint project between Supélec and LIMSI-CNRS whose aim is to conduct research about "smart room" environments. He teaches computer architecture, programming languages, information systems and human-computer interaction at Supélec, École polytechnique and EFREI.

Kay Kadner received his Diploma in Computer Sciene in 2004 and his PhD in 2008 from TU Dresden. He joined SAP Research CEC Dresden in 2004 as PhD candidate where he worked in public-funded projects targeting the use of multimodal interaction with mobile devices by maintenance workers and model-driven development of multimodal and context aware applications for mobile devices. Besides that, his research interests are distributed systems and computer networks. Consequently, his PhD thesis is about a platform for multimodal application access through multiple, networked and distributed devices. He is currently involved in applying SOA principles for developing a platform-based approach to disaster management for public security.

Nadjet Kamel received the MS degree in Computer Science from the University of Science and Technology Houari Boumediene (USTHB), Algiers, Algeria, in 1995, and the PhD degree in Computer Science from the university of Poitiers, France, in 2006. She is a Postdoctoral Researcher at the University of Moncton, in Canada. She has been a Lecturer at the USTHB for 12 years and she has been involved in many research projects. Her main interests are related to software engineering techniques for the development of multimodal user interfaces, especially for mobile and pervasive systems. She is a member of program and organizing committees of many international conferences and workshops (CCECE'2008, CCECE'2009, SEPS'08, SOMITAS'08, …).

Aidan Kehoe worked in industry for more than 20 years in a variety of software development roles, most recently at Logitech USA as Director of Software Engineering. He has worked on product and middleware development for peripherals on a variety of gaming platforms including PC, Sony and Nintendo. He now works for Logitech in Ireland on a part-time basis while pursuing a PhD in the Computer Science Department at University College Cork. His area of research is multimodal user assistance.

Martin Knechtel received his diploma in Computer Science Engineering from the Dresden University of Technology in 2007. Parallel to his studies, he worked as student research assistant at Fraunhofer IWS in Dresden for 6 years. Furthermore, he worked as student research assistant at the chair for computer networks and in projects at Siemens AG in Munich and Dresden. During his studies, he collected expertise in LAN Security, Mobile Devices and Distributed User Interfaces. After writing his diploma thesis at SAP Research, he joined as Research Associate. He is interested in Semantic Web technologies, especially in Description Logics in enterprise applications. His research focus is Access Control Mechanisms for Description Logic Web Ontologies.

Grzegorz Lehmann is a researcher and Ph.D. student at the DAI-Labor of the Technische Universität Berlin. His research focuses on the model based development and design of multimodal user interfaces. In his Ph.D thesis Grzegorz intends to thoroughly explore the benefits and disadvantages of the application of executable models to user interface development, which he started to investigate in his diploma thesis titled "Model Driven Runtime Architecture for Plastic User Interfaces". As the gap between the design and run time is being closed, Grzegorz is particularly interested in the impact of this process on topics such as end-user development, self-adaptive UIs and UI plasticity.

Marcos Martinez-Diaz received the MSc degree in telecommunications engineering in 2006 from Universidad Autonoma de Madrid, Spain, and was awarded for the highest academic ranks in his graduating year. He worked as an IT strategy consultant in Deloitte until 2007. Currently he is working in Business Intelligence IT project management in Vodafone. Since 2005 he is with the Biometric Recognition Group - ATVS at the Universidad Autonoma de Madrid, where he is collaborating as part-time student researcher pursuing the PhD degree. He has collaborated in national and European projects focused on biometrics (e.g., Biosecure NoE). His research interests include biometrics, pattern recognition and signal processing primarily focused on signature verification. He has been recipient of some distinctions, including the Honeywell Honorable Mention at the Best Student Paper Award in BTAS 2007 and the Special Mention at the Spanish National Awards for Telecommunications Engineering studies.

Flaithrí Neff has attained a BMus degree at University College Cork and a first class honours MSc degree at the University of Limerick specializing in Audio Technology. He worked for almost two years with Riverdance in the United States and Europe, and has been a Teaching Assistant at University College Cork for the last four years. He is currently pursuing a PhD at the Computer Science Department at UCC in the area of Sonification and Interface Design.

Javier Ortega-Garcia received the M.Sc. degree in electrical engineering (Ingeniero de Telecomunicacion), in 1989; and the Ph.D. degree "cum laude" also in electrical engineering (Doctor Ingeniero de Telecomunicacion), in 1996, both from Universidad Politecnica de Madrid, Spain. Dr. Ortega-Garcia is founder and co-director of the Biometric Recognition Group - ATVS. He is currently a Full Professor at the Escuela Politecnica Superior, Universidad Autnoma de Madrid, where he teaches Digital Signal Processing and Speech Processing courses. He also holds a Ph.D. degree course in Biometric Signal Processing. His research interests are focused on biometrics signal processing: speaker recognition, fingerprint recognition, on-line signature verification, data fusion and multibiometrics. He has published over 150 international contributions, including book chapters, refereed journal and conference papers. He chaired "Odyssey-04, The Speaker Recognition Workshop", co-sponsored by IEEE, and is currently co-chairing "ICB-09, the 3rd IAPR International Conference on Biometrics".

Giovanni Pilato received his "cum laude" degree in "Ingegneria Elettronica" and the Ph.D. degree in "Ingegneria Elettronica, Informatica e delle Telecomunicazioni" from the University of Palermo, Italy, in 1997 and 2001, respectively. Since 2001 he is a staff research scientist of the ICAR-CNR (Istituto di CAlcolo e Reti ad alte prestazioni, Italian National Research Council). He is also Lecturer at the DINFO (Dipartimento di ingegneria INFOrmatica) of the University of Palermo. His research interests include machine learnig and geometric techniques for knowledge representation.

Antti Pirhonen is a senior researcher and a PhD in educational sciences, currently preparing second thesis in computer science at the University of Jyväskylä. He has been a scientific leader of several user-interface sound related projects since 1999, funded by the Finnish Funding Agency for Technology and Innovation (TEKES) and Finnish ICT industry. He was a visiting researcher in the University of Glasgow (UK), Department of Computing, in 2000-2001 and 2005, and has worked as a lecturer in the University of Jyväskylä, Department of Computer Science and Information Systems, 2001-2004 & 2006. He is a regular reviewer of several HCI-related conferences, and he has been a chair of the 3rd International Workshop of Haptic and Audio Interaction Design, HAID2008 and an associate editor of International Journal of Mobile Human-Computer Interaction. His research interests include human conceptualization processes, multimodal interaction, design and learning.

Ian Pitt lectures in Usability Engineering and Interactive Media at University College, Cork, Ireland. He took his D.Phil at the University of York, UK, then worked as a research fellow at Otto-von-Guericke University, Magdeburg, Germany, before moving to Cork in 1997. He is the leader of the Interaction Design, E-learning and Speech (IDEAS) Research Group at UCC, which is currently working on a variety of projects relating to multi-modal human-computer interaction across various application domains. His own research interests centre around the use of speech and non-speech sound in computer interfaces, and the design of computer systems for use by blind and visually-impaired people.

Christoph Pohl is currently involved in shaping SAP's next generation modeling infrastructure. He was previously coordinating a cluster of projects on Model-Driven Business Application Engineering within the SAP Research Program Software Engineering and Architecture. In this role he was also leading a number of research activities in the area of multimodal interactions and mobile computing. Prior to joining SAP in 2004, he finished his PhD at TU Dresden in the field of component-based middleware. Besides his strong background in software engineering, he also has previous experiences in e-learning, data base management systems, security, and digital rights management, among others.

Tiago Reis is a PhD student at the Department of Informatics from the University of Lisbon and a researcher at the HCIM (Human Computer Interaction and Multimedia Group) at LaSIGE (Large Scale Informatics Systems) research lab. He graduated in Computer Engineering (2006), has an MSc (2008), from the same institution where he has been working with mobile devices towards the development of universal accessible solutions for different domains. He has published more than 10 refereed scientific papers and book chapters concerning multimodal and mobile interaction design as well as accessible user interfaces for mobile devices on different international conferences. His research interests include Accessibility, Multimodal Interfaces, Mobile Interaction Design, Usability, User Experience, Context Awareness and Ambient Intelligence.

Dirk Roscher is a researcher and Ph.D. student at the DAI-Labor of the Technische Universität Berlin. He is working in the HCI group of the DAI-Labor and participated in several research projects funded e.g. by the German government or the Deutsche Telekom. His research focuses on the model based specification of ubiquitous user interfaces and their runtime adaptation in smart environments. Dirk currently examines issues of runtime distribution of user interfaces between different interaction devices and modalities, the fusion of user input at runtime as well as topics regarding meta user interfaces.

José Rouillard is an Associate Professor in Computer Science at the University of Lille 1 in the LIFL laboratory. The LIFL (Laboratoire d'Informatique Fondamentale de Lille) is a Research Laboratory in the Computer Science field of the University of Sciences and Technologies of Lille (USTL) linked to the CNRS, and in partnership with the INRIA Lille - Nord Europe. José Rouillard obtained his PhD in 2000 from the University of Grenoble (France) in the field of Human-Computer Interfaces. He is interested in Human-Computer Interaction, plasticity of user interfaces, multi-modality, multi-channel interfaces and human-machine dialogue. He has written one book on VoiceXML in 2004, another one on Software

Oriented Architecture (SOA) in 2007 and more than 70 scientific articles. He is now engaged in research on mobility, pervasive/ubiquitous computing and natural human-machine dialogue.

Cyril Rousseau is a consulting engineer at ALTEN group. He holds a PhD in Computer Science from Paris-South University. His research interests concern the analysis and the design of multimodal systems for the intelligent presentation of information (using several communication modalities to adapt the information presentation to the interaction context).

Marco de Sá is a PhD candidate at the Department of Informatics from the University of Lisbon and a researcher at the HCIM (Human Computer Interaction and Multimedia Group) at LaSIGE (Large Scale Informatics Systems) research lab. He graduated in Computer Science (2003), has a post-graduation on Informatics/Information Systems (2004), and an MSc (2005), from the same institution. He has been working as a researcher at the LaSIGE Research Lab, within HCIM research lines, for more than five years. He has published over 40 refereed scientific papers and book chapters in international journals and conferences, most of which focusing Mobile Applications and Mobile Interaction Design. He is a reviewer for SIGCHI's major conferences (CHI, MobileHCI, CSCW) and other well know conferences within the area of HCI and Mobile Applications (Ubicomp, ICPCA). His research interests include Mobile Interaction Design, Design Methodologies, Prototyping, Evaluation, Usability, User Experience, M-Health, M-Learning and Ambient Intelligence.

Mark Sanderson is a Reader in information retrieval (IR) in the Information Studies Department at the University of Sheffield. He has worked in the field of IR for twenty years and has published extensively. His interests are in evaluation of searching systems, cross language IR and use of geographic information in search. He is on the editorial board of four of the main IR journals and has been a programme chair for a number of the key conferences in the field, including ACM SIGIR.

Antonella Santangelo received her "cum laude" Laurea degree in "Ingegneria Informatica" from the University of Palermo on July, 2005. She received the Prize "Federcomin-AICA 2005/2006" for her thesis degree in 2006. She is currently a Ph.D. candidate in computer science at DINFO Dipartimento di Ingegneria Informatica, University of Palermo, Italy. Her research interests are in the area of Multimodal Human-Computer Interfaces, Software Pattern Design for Speech Dialogue Management and Interaction in Augmented Reality Systems.

Sid Ahmed Selouani received his B. Eng. degree in 1987 and his M.S. degree in 1991 both in electronic engineering from the Algiers University of Technology (U.S.T.H.B). He joined the Communication Langagière et Interaction Personne-Système (CLIPS) Laboratory of Université Joseph Fourier of Grenoble taking part in the Algerian-French double degree program and then he got a Docteur d'État degree in the field of signal processing and artificial intelligence in 2000. From 2000 to 2002 he held a post-doctoral fellowship in the Multimedia Group at the Institut National de Recherche Scientifique (INRS-Télécommunications) in Montréal. He is currently full professor and responsible of the Human-Machine Interaction Laboratory at the Université de Moncton, Campus de Shippagan. He is also an invited Professor at INRS-Télécommunications. He is a Senior member of IEEE and member of signal processing and computer science societies. His main areas of research include new models in Human-

System interaction, E-services, mobile communications, ubiquitous systems and assistive technologies by speech-enabled solutions.

Salvatore Sorce studied Computer Science and Engineering at the University of Palermo, and obtained his M.S. (2001) and Ph.D. (2005) degrees. Currently he is post-doc fellow with the DINFO – Dipartimento di Ingegneria Informatica (Computer Science and Engineering Department) at the University of Palermo. His main research interests deal with: pervasive systems; wearable computers; positioning systems; personal mobile devices programming.

Thomas Springer received his PhD in Computer Science from TU Dresden in 2004. Since 1998 he is a researcher at the department of computer networks in Dresden. His research interests comprise mechanisms, architectures, and software engineering concepts for adaptive applications and services in heterogeneous and dynamic infrastructures, especially the relations and interactions between context-awareness, adaptation, and software engineering. He has been involved in research and coordination of numerous projects dealing with mobile and ubiquitous computing, context-awareness, sensor systems, software engineering for component-based and service-oriented systems, and mobile collaboration. Dr. Springer is author and co-author of numerous publications, including a book about distributed systems. He is reviewer for several workshops and conferences and a lecturer in the area of Application Development for Mobile and Ubiquitous Computing and Distributed Systems.

Kai Tuuri is a researcher and a PhD student at the University of Jyväskylä, from which he received his MA degree (1998) in music education. After the graduation he has worked as a lecturer for multimedia studies and project-based learning in Department of Computer Science and Information Systems (2000-2004) and also in Department of Arts and Culture Studies (2004-2006). During these periods he has been involved in numerous different projects of multimedia application development on the basis of assignments form industry and other types of organizations. From 2007 he has worked as a researcher in projects which study the user interface design from the perspectives of non-speech sounds and haptics. His current main research interest concerns communicational use of sound in the context of multimodal human-computer interaction.

Pasi Välkkynen is a research scientist at VTT Technical Research Centre of Finland. He received his PhD degree in computer science in 2007, the topic being *Physical Selection in Ubiquitous Computing*. In VTT (from 2000 to present) he has participated in several European and national research projects as software engineer and user interaction designer. During the past few years his research has concentrated on mobile phone and tag based interactions in ubiquitous computing environments. His current research interests include also multi-device user interfaces and embedding digital services and user interfaces into the physical world.

Keith Waters is currently a director of research at Orange Labs Boston. He received his Ph.D (1988) in Computer Graphics from Middlesex University London and prior to joining Orange, he was a senior member of the research staff at the Cambridge Research Laboratory of Digital Equipment and later Compaq, where he developed Computer Graphics, Computer Vision and HCI interaction techniques. He has built many multimodal systems including wallable macro-devices, smart kiosks, state-of-the-art commercial interactive face animation and more recently mobile multimodal systems for testing by Pages Jaunes in France. Keith has published his work extensively and is the co-author of Computer Facial Animation the definitive book on face synthesis. He is currently a France Telecom senior expert specializing in mobile services and has been involved with several standards activities of the World Wide Web Consortium.

David C. Wilson is an Assistant Professor in the College of Computing and Informatics at the University of North Carolina at Charlotte. Dr. Wilson is an expert in intelligent software systems and a leading name in the field of Case-Based Reasoning. Dr. Wilson's research centers on the development of intelligent software systems to bridge the gaps between human information needs and the computational resources available to meet them. It involves the coordination of intelligent systems techniques (artificial intelligence, machine learning, etc.) with geographic, multimedia, database, internet, and communications systems in order to elicit, enhance, apply, and present relevant task-based knowledge.

Xing Xie is a lead researcher in the Web Search and Mining Group, Microsoft Research Asia. He received the B.S. and Ph.D. degrees in Computer Science from University of Science and Technology of China in 1996 and 2001, respectively. He joined Microsoft Research Asia in July 2001, working on mobile web search, location based search and mobile multimedia applications. He has served on the organizing and program committees of many international conferences such as WWW, CIKM, MDM, PCM. He is the chair of WWW 2008 browser and UI track. He has published over 50 referred journal and conference papers, including SIGIR, ACM Multimedia, CHI, IEEE Trans. on Multimedia, IEEE Trans. on Mobile Computing, etc.

Index

A

abstract user interface 84
action model 151, 152, 153, 154, 157, 158
action-relevant 139, 140, 143, 144, 145, 148, 149, 151, 153, 154, 159
adaptation 24, 25, 26, 28, 29, 30, 32, 41, 42, 43, 44, 45, 46, 47, 48, 50, 51
ambient computing 54
amodal completion 139
App Informatik Davos 326
audio fingerprinting 244, 257
audio fingerprints 244
audio search 243, 246
auditory stream 282, 290
auditory streaming 282, 299
augmented environment 224
augmented reality (AR) 223, 225, 239, 240, 241
aura 183, 192
authentication 321, 322, 326, 330, 336, 337
automatic speech recognition (ASR) 219, 235, 236, 237

B

barcode 244
biometrics 322, 325, 335, 336, 338
Bolt, Richard A. 2, 3, 5, 54, 71, 167, 191
book player 108, 122, 124, 125, 126, 128, 129, 131, 133

C

ChaCha 244, 257
Channel-orientation 138
CIC 326

cognitivist conceptualisation 138
communicative function 150, 153, 154, 159
communicative purpose 154
complementarity, assignment, redundance, equivalence (CARE) 56, 60, 61, 67, 70, 71, 196, 197, 199, 200, 201, 202, 203, 205, 206, 207, 212, 213, 214
component platform 77
computer supported cooperative work (CSCW) 2, 22
concrete user interface 84, 85, 86, 101
content based image retrieval (CBIR) 244, 245
context 1, 2, 3, 4, 5, 6, 9, 10, 15, 18
context-aware 218, 224, 241
context-aware systems 197
context of use 2, 5
controlled update 89, 90, 91, 92
conversational agent 221, 233, 234
cooking assistant (CA) 43, 44, 45, 46, 50
cross-modal integration 139
cross-modal interaction 139
CRUMPET 309, 320
Crypto-Sign 326

D

Deep Map 309
delivery context: client interfaces (DCCI) 260, 261, 269, 270, 271, 272, 273, 275
design 196, 197, 198, 199, 200, 201, 202, 210, 211, 213, 214
design paradigm 137, 147, 148, 150, 160, 161
design process 107, 108, 110, 111, 112, 115, 124, 133
device 168, 170, 178, 180, 183, 184, 185, 186, 187

device independence 264, 274

device independence field 3

digital book player 108, 122, 124, 133

digital books 106, 134

digital literacy 301

distributed speech recognition (DSR) 266, 268, 274

distributed user interface 87, 89, 90, 95, 96

distribution strategy 88, 89

dual tone multi-frequency (DTMF) 265, 272, 275

dynamic models 27, 31, 33

dynamic properties 264, 272

E

elder population 301, 302, 303, 304, 305, 306, 307, 308, 309, 310, 311, 312, 314, 315, 316, 317, 318, 319, 320

elementary information unit 172, 180, 189

ELOQUENCE software platform 166, 170, 175

embodied action-oriented processes 147

embodied approach 139

embodied cognition 137, 139, 140, 144, 145, 146, 147

equal error rate (EER) 324, 329, 332, 333, 334

error-robustness evaluation 196, 201

e-signature 324

evaluation 54, 55, 56, 58, 60, 61, 69, 71, 72, 73, 106, 107, 108, 110, 111, 113, 114, 115, 117, 119, 120, 126, 130, 133, 134, 135, 196, 197, 199, 201, 203, 204, 208, 209, 213, 214

executable models 26, 31, 32, 33, 41, 42, 44, 45, 46, 48

extensible hypertext markup language (XHTML) 265, 268, 269, 271, 273, 274

F

false acceptance rate (FAR) 324

false rejection rate (FRR) 324

FAME architecture 197

features 322, 323, 328, 330, 331, 332, 333, 334, 335, 337

federated devices 77, 83, 84, 85, 87, 93, 95, 96, 99, 101

finite state machines (FSM) 196, 199, 200, 201, 202, 203, 204, 205, 206, 207, 208, 209, 210, 211, 212, 213, 214, 215

formal methods 53, 54, 55, 59, 61, 69, 70

formal techniques 59

fusion 26, 28, 29, 30, 40, 41, 48, 263

G

general packet data radio service (GPRS) 266, 267, 275

geographical information systems (GIS) 301, 302, 303, 304, 305, 309, 310, 311, 312, 314, 317, 318, 319, 320

GIS applications 301, 303, 309, 317

global positioning systems (GPS) 243, 245, 246, 247, 254, 255, 256, 257, 258, 259, 303, 309, 311, 312, 317

global system for mobile (GSM) 266, 267, 273, 275

Goog411 244, 257

H

haptic feedback 277, 278, 296, 297

haptics 277, 278, 279, 281, 284, 285, 286, 287, 288, 290, 292, 293, 294, 295, 296, 297

HCI-studies 139

heterogeneous devices 2, 6

hidden markov model (HMM) 324, 332, 336, 337, 338

human-computer interaction (HCI) 137, 138, 139, 150, 151, 154, 160

human-human communication 196

I

image matching 247

image schema 148

inferential capabilities 234

information retrieval 221, 233, 234, 239

input devices 138

input fusion 75, 78, 83

intelligent computing 54

intentionality 144, 145, 146, 148, 150, 151, 153, 154, 157, 158, 162

interaction CARE (ICARE) 197, 214

interaction context 167, 168, 169, 170, 171, 172, 173, 174, 175, 177, 188

interaction design 137, 138, 140, 143, 150, 155, 160, 161
interaction designers 141
interaction model 26, 29, 30, 35, 37, 39, 40, 44, 45, 48, 151, 152
Internet channel 9

K

KeCrypt 326, 327
knowledge, users, & presentation device (KUP) architectural model 166, 179, 180, 181, 183, 184, 186, 193

L

local search 243, 244, 246
location based search 242
location based services (LBS) 269
location-based system (LBS) 309, 317

M

mapping 27, 30, 32, 33, 34, 35, 38, 39, 44, 45, 48
matching score 324
medium 167, 168, 172, 173
metamodel 26, 28, 31, 32, 33, 34, 35, 37, 41, 42, 43, 48
Microsoft's Tellme 244
migration 28
MMI-framework 77, 78, 79, 80, 85, 87, 97, 99
mobile application 106, 111
mobile book player 125
mobile computing 1, 2, 3
mobile computing devices 277
mobile computing, multimodality in 1, 2
mobile device-based sensors 269
mobile devices 1, 4, 5, 6, 18, 107, 108, 110, 111, 112, 113, 114, 115, 119, 120, 126, 131, 133, 134, 135, 136, , 242, 243, 244, 245, 246, 247, 256
mobile geospatial applications 303, 310
mobile GIS 302, 303, 304, 309, 310, 311, 312, 314, 317, 318, 320
mobile multimodal applications, design of 108
mobile multimodal applications (MMA) 106, 107, 108, 109, 113, 133

mobile multimodal interfaces 301
mobile network 264, 271
mobile phones 3, 242, 243, 245, 259
mobile presence 261
mobile search 242, 243, 244, 245, 246, 252, 253, 254, 256, 257
mobile systems 53, 54, 55, 56, 57, 58, 69, 70
mobile technology 302, 305, 306
mobile users 242, 244, 245, 252, 256
Mobot 245, 258
MobPro 107, 108, 114, 115, 120, 122, 130, 131, 132
modalities 196, 197, 198, 199, 200, 201, 207, 208, 209, 210, 211, 213, 242, 243, 256
modality 53, 54, 55, 56, 57, 62, 64, 65, 66, 67, 68, 70, 167, 168, 170, 171, 172, 173, 174, 175, 181, 182, 185, 186, 189, 190, 191, 192, 194
mode 167, 168
model adaptation 41
model-based user interface development (MBUID) 25, 26, 27, 48, 49
model-checking 53, 59, 60, 61, 67, 70
model of interaction 150, 151, 160
models at runtime 24, 25, 26, 28, 31, 32, 33, 48
multi-access service platform (MASP) 26, 28, 29, 30, 31, 33, 35, 40, 41, 44, 45, 46, 47, 48
multimodal 301, 302, 303, 304, 305, 306, 307, 308, 309, 311, 312, 315, 316, 317, 318, 319, 320
multimodal architecture (MMA) 77, 79, 103
multimodal commands 196, 197, 200, 201, 202, 203, 204, 205, 206, 207, 208, 209, 210, 211, 212, 213, 214
multimodal cue 283, 284
multimodal design 148, 150, 153, 159, 160
multimodal HCI 302
multimodal input 232
multimodal interaction 217, 237
multimodal interaction activity 6
multimodal interaction (MMI) 76, 77, 78, 79, 80, 85, 87, 97, 99, 100
multimodal interaction platform 81, 95, 96, 99
multimodal interaction systems 196, 197, 199, 214

multimodal interfaces 4, 6, 107, 108, 109, 142, 162, 196, 197, 214, 215, 302, 307, 308, 319

multimodalities 106, 107, 110, 133

multimodality 1, 2, 3, 5, 9, 22, 137, 138, 139, 140, 141, 142, 143, 144, 147, 148, 159, 160, 260, 261, 262, 264, 265, 266, 267, 268, 271, 272, 273, 274

multimodal markup language (M3L) 10, 11

multimodal mobile devices 197

multimodal mobile GIS 303, 304, 312, 317, 318, 320

multimodal system 56, 58

multimodal systems 5, 13, 302, 317

multimodal user-interfaces 137, 160, 161

multimodal user interfaces (MUI) 24, 53, 54, 55, 56, 57, 60, 61, 62, 70, 73

multiple access channels 2

mutual disambiguation 263

N

NayioMedia 244, 258

nimbus 183

non-speech sounds 277, 279, 281, 285, 287, 288, 289, 295, 296, 300

O

online help 278

Open Mobile Alliance (OMA) 264, 274

output devices 138

output fission 75, 78

output generation 85

output synchronization 90, 99

P

pause intervals 277, 279, 280, 285, 287, 288, 289, 290, 293, 295, 296

perceptual space 181, 182, 183, 194

personal digital assistants (PDA) 1, 4, 5, 13, 242, 277

pervasive computing 54, 239

pervasive service 223

pervasive system 218, 219

platform architecture 83, 84

points of interest (POI) 303, 309

presentation of information in ambient environments (PRIAM) platform 167, 186, 187

proactivity 230, 231

PromptU 244

prototypes 106, 107, 108, 110, 113, 114, 115, 116, 117, 118, 119, 120, 122, 126, 127, 128, 129, 130, 131, 132, 133, 135

prototyping tool 108, 133, 135, 136,

"Put That There" paradigm 2, 3, 5, 7, 54, 55, 56, 167

Q

QR code 244, 258

query modalities 242, 256

R

radiance space 181, 182, 183, 185, 194

radio-frequency identification (RFID) 217, 219, 222, 227, 230, 231, 233, 234, 235, 236, 237

ramp up 286, 287

robustness 196, 200, 201, 204, 209, 213, 214

Romsey Associates 326

runtime models 35

S

scenario framework 106, 122, 133

scenario generation and context selection framework 107, 133

semantics 140, 142, 148, 150, 151, 152, 154, 158, 159, 160

semantic unit 180, 184, 185, 186, 194, 195

sensory-motor imagery 145

server-based interaction management 266

signatures 321, 322, 323, 324, 325, 326, 327, 328, 329, 330, 332, 333, 334, 335, 336, 337, 338

simulation 110, 118

smart environments 25, 31, 44, 47, 48

smart mobile devices 278, 295, 296

smart phones 242, 277

SoftPro 326

speech 277, 278, 280, 281, 284, 290, 296, 297, 298, 299, 300

static properties 264

synchresis 143, 150, 151
synchronized 285, 286

T

temporal logic 59, 61, 65, 66, 67, 70, 71, 72
tentative update 89, 91
text-to-speech (TTS) 219, 236, 237
Topaz Systems 326
touchscreen 321, 322, 323, 325, 327, 328, 335
traditional user interfaces 196

U

ubiquitous computing 1, 3, 54, 218, 224, 241, 277, 296
ubiquitous context 142
ubiquitous user interfaces (UUI) 24, 25, 26, 28, 44, 47
ubiquitous Web applications (UWA) 262
usability framework 196, 197, 199, 200, 201, 202, 203, 204, 205, 208, 210, 211, 212, 213
usability properties 53, 54, 55, 56, 59, 60, 61, 67, 70
user assistance 277, 278, 279, 280, 281, 282, 283, 284, 285, 287, 288, 292, 295, 296, 299
user interfaces (UI) 53, 54, 138, 140, 141, 143, 148
user, system, environment triplet 168
user-testing 196, 201, 205

V

validation 321, 322, 324, 325, 332, 334
V-ENABLE 244

verification 53, 59, 61, 67, 70, 71, 72, 73, 321, 322, 323, 324, 325, 326, 327, 328, 329, 330, 332, 334, 335, 336, 337, 338
virtual environment 238
virtual reality (VR) 219, 220, 221, 223, 225, 230, 240
virtual world 239
visual part 250, 252
visual queries 242, 247
visual search 242, 243, 245, 246, 248, 256, 258
voice 1, 3, 4, 5, 6, 7, 8, 9, 10, 11, 13, 14, 16, 20, 21, 23
voice recognition 222, 224, 237, 238
voice search 242, 256
voice synthesis 224
VoiceXML 2, 6, 7, 9, 10, 14, 265, 271, 273, 274

W

W3C Multimodal Interpretation Group (MMI) 261
what, which, how, then (WWHT) model 166, 170, 171, 172, 173, 175, 177, 178, 180
windows, icon, menu, pointing device (WIMP) interface 1
World Wide Web Consortium (W3C) 261, 269, 273, 274, 275

X

XHTML+VoiceML (X+V) 264, 265
Xyzmo 326

Y

Yahoo! oneSearch 244